DECOMPOSITION OF LARGE-SCALE PROBLEMS

DECOMPOSITION
OF
LARGE-SCALE PROBLEMS

NATO INSTITUTE ON DECOMPOSITION,
CAMBRIDGE, ENG., 1972

Editor

David M. HIMMELBLAU

Department of Chemical Engineering
The University of Texas
Austin, Texas

1973

NORTH-HOLLAND PUBLISHING COMPANY — AMSTERDAM • LONDON
AMERICAN ELSEVIER PUBLISHING CO., INC. — NEW YORK

Library of Congress Catalog Card Number: 72-90429
ISBN North-Holland 07204 2072 5
ISBN American Elsevier 0444 10427 5

PUBLISHERS:

NORTH-HOLLAND PUBLISHING COMPANY – AMSTERDAM
NORTH-HOLLAND PUBLISHING COMPANY, LTD. – LONDON

SOLE DISTRIBUTORS FOR THE U.S.A. AND CANADA:

AMERICAN ELSEVIER PUBLISHING COMPANY, INC.
52 VANDERBILT AVENUE
NEW YORK, N.Y. 10017

PRINTED IN THE NETHERLANDS

TABLE OF CONTENTS

PREFACE

The NATO Institute on Decomposition as a tool for solving large-scale problems was an interdisciplinary conference held in Cambridge, England, from July 16 to July 27, 1972. It was intended to familiarize individuals of varying backgrounds with decomposition techniques that have proved to be useful in the areas of mathematics, economics, structures, networks, and various engineering disciplines. The conference was designed to

(1) stimulate interest in new techniques of decomposition

(2) exchange information and concepts concerning decomposition from diverse fields and

(3) identify common features in the theory and practice of decomposition.

The participants were able to gain insight and perspective into potential new applications of decomposition by listening to the discussion, both formal and informal, from individuals practicing in areas outside of their immediate specialization.

The idea of decomposing a problem in order to simplify its solution is not new. For example, Ostrowski mentions that C. L. Gerling in 1843 used groups of variables rather than single variables in solving systems of algebraic equations with predominant principle diagonals. However, the use of large-scale computers in recent years has led to rapid expansion of decomposition techniques for optimization, for process control, for precedence ordering in systems, for solving electrical network problems, and in a wide variety of seemingly unconnected problems.

One of the main features of the conference was the diversity of opinion as to how decomposition should be defined. Some individuals were familiar with matrix decomposition, or factorization, as a basis for many direct methods for solving systems of linear algebraic equations. Others had worked in the decomposition of networks in solving large-scale electrical problems.

Probably the fundamental underlying idea that evolved at the conference was that a large complex system representing interacting elements could be broken up into subproblems of lower dimensionality. Presumably these subproblems can be treated independently for the purposes of optimization, control, design, and so forth, and the union of the solutions of the subproblems is the solution of the large original problem. The general principles involved in treating such large-scale problems can be termed decomposition principles.

The specific topics that were included in the conference can best be obtained from the table of contents, but a general classification of the areas covered were:

(1) Decomposition techniques in the analysis and design of electrical networks, namely how to split up large networks into smaller components and resolve the values of the variables in the equations representing the network.

(2) Multi-level approaches to the control of large-scale systems, including the basic concepts of hierarchical systems, stratification, multi-layer decision systems and multi-goal systems.

(3) Application of multi-level modeling of large systems to the analysis of environmental problems.

(4) Sparse matrices in decomposition with applications to the solution of algebraic and differential equations, including pivotal strategy.

(5) Decomposition in linear and non-linear programming, including relationships among the various forms of the linear programming problem, the problems involved in large models, data management problems, and the like.

Excluded from the conference were the following topics:

Finite automata switching systems

Linguistics

Boolean functions

Coding

Computer systems

Applications of decomposition in these areas were of a rather specialized nature, and were not deemed to be appropriate.

It became clear from the various papers presented that a dual classification of decomposition could be made, one by discipline and the other by technique, that gave some insight into the general scope of the field. Table 1 lists the areas of application of decomposition by academic discipline, while Table 2 lists the various tools and techniques that were discussed at the conference.

The organizing committee of the Institute on Decomposition wishes to express their grateful acknowledgment to the NATO Scientific Affairs Division, whose sponsorship made the Institute possible. We are also grateful to the many speakers and other participants who took the time to prepare and review the papers that were presented at the conference. Finally, Professor J. C. R. Turner did a splendid job as the arrangements chairman.

D. M. Himmelblau
Austin, Texas, U. S. A.

TABLE 1

DECOMPOSITION: APPLICATION BY ACADEMIC DISCIPLINES

Automatic Control
Data Management
Chemical Engineering Plant Design
Computer Science
Economics
Electrical Engineering Networks
Environmental Engineering Water Resources
Mechanical Engineering
Mathematics
Operations Research; Management Science
Sociology

TABLE 2

DECOMPOSITION: TECHNIQUES AND TOOLS

- Discrete Approximation
- Duality Theory
- Graph Theory
- Iterative Solution of Equations
- Mathematical Programming
  - Dynamic
  - Geometric
  - Integer
  - Linear
  - Nonlinear
- Matrix Partitioning
- Multilevel or Hierarchical Control
- Network Analysis
- Sparse Matrix Techniques

MORPHOLOGY OF DECOMPOSITION

DAVID M. HIMMELBLAU
Department of Chemical Engineering
The University of Texas, Austin, Texas, 78712, U.S.A.

Abstract: The influence of information flow on decomposition is examined in three areas: the solution of sets of equations, linear programming, and dynamic programming. Common structures for the occurrence matrices of large systems are observed indicating that the information flow takes on similar patterns.

INTRODUCTION

Decomposition can be applied to mathematical models that represent any large scale system whether a chemical plant, electrical network, river basin, bridge structure or the like. One of the distinct characteristics of a large system, as opposed to a subsystem, is that it represents a complex network of interacting elements. In treating such large systems in the physical and social sciences, as well as in engineering, one is concerned with methods of decomposing or breaking up a large problem into a set subproblem of lower dimensionality the union of which is equivalent to the original problem. Presumably these subproblems can be treated independently for the purposes of optimization, control, design, etc. The general principles involved in treating such large problems are known as decomposition principles.

As our ability to solve problems expands, the scale of the problems attacked themselves seem to expand at a similar rate. As a result there always exists over the horizon new categories of problems of greater size to tackle. By placing emphasis on the tools used to attack large scale problems, and by identifying common features of the tools used in different fields, it should be possible (1) to apply the tools employed in one field to large scale problems in many new areas of application as well as (2) to visualize how the tools can be applied to even bigger problems in areas of past success. The purpose of this paper is to describe certain applications of decomposition techniques in order to reveal how information flow is a common feature among them. In this way we can provide a broad framework in which to show the interrelationships among specialized techniques.

1. STRUCTURE AND DECOMPOSITION

Decomposition of a large system of relations rearranges the relations and/or variables based on the information flow among the relations via the variables. Clearly, when each relation of the system contains each variable, no decomposition can take place solely by partitioning, that is by rearrangement alone. Fortunately, in most mathematical representations of real problems or processes the relations contain only a few common variables, hence decomposition can take place, in some cases extensively. The most effective form of decomposition is to form <u>disjoint</u> subsystems, that is to form subsets of relations that do not contain any common variables so that each subset can be treated independently.

In order to clarify what a disjoint subsystem is, consider the following equations:

$$
\begin{aligned}
f_1\ (x_1, x_3) &= 0\\
f_2\ (x_2, x_4) &= 0\\
f_3\ (x_2, x_4) &= 0\\
f_4\ (x_1) &= 0
\end{aligned}
$$

Functions f_1 and f_4 contain only variables x_1 and x_3, variables which do not appear in the remaining two equations, and therefore, f_1 and f_4 constitute a disjoint subsystem. (Equation f_1 itself is also disjoint.)

The information flow among the system relations can be represented either by information flow diagrams such as organizational charts and signal flow diagrams, or by associated Boolean matrices that have a one to one correspondence with the structure of the graph, but can easily be loaded into a digital computer for analysis. Decomposition of a large system can take place using either the graph or the associated Boolean matrix. In the Boolean matrix called the occurrence matrix:

(1) Each row of the occurrence matrix corresponds to a system equation, and each column corresponds to a system variable.
(2) An element of the matrix, s_{ij}, is either a Boolean 1 or 0 according to the rule

$$
s_{ij} = \begin{cases} 1 \text{ if variable } j \text{ appears in equation } i \\ 0 \text{ otherwise} \end{cases}
$$

This matrix, then, indicates the occurrence of the dependent variables in each of the system relations. The occurrence matrix for the four functions and variables is shown in Figure 1. If a column permutation $\pi^C(23)$ is performed on the occurrence matrix (a) of Figure 1 followed by a row permutation $\pi^R(124)$, the occurrence matrix (b) is obtained which is in block diagonal form. The off diagonal blocks

$$
\begin{array}{c|cccc}
 & x_1 & x_2 & x_3 & x_4 \\
\hline
f_1 & 1 & 0 & 1 & 0 \\
f_2 & 0 & 1 & 0 & 1 \\
f_3 & 0 & 1 & 0 & 1 \\
f_4 & 1 & 0 & 0 & 0
\end{array}
\qquad
\begin{array}{c|cc:cc}
 & x_1 & x_2 & x_3 & x_4 \\
\hline
f_1 & 1 & 0 & 0 & 0 \\
f_2 & 1 & 1 & 0 & 0 \\
\hdashline
f_3 & 0 & 0 & 1 & 1 \\
f_4 & 0 & 0 & 1 & 1
\end{array}
$$

(a) (b)

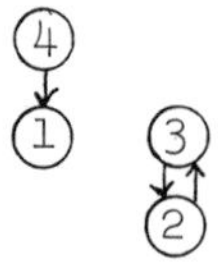

(c)

Figure 1 Occurrence Matrices

contain all zero elements and each block on the diagonal represents a disjoint subsystem. If a graph corresponding to the information flow were drawn, a separate graph would be obtained for each main disjoint subsystem with no edges connecting the two graphs; refer to Figure 1c.

Of course for large systems, it is impractical to ascertain by inspection which row and column permutations will yield a block diagonal form or to draw a graph to assist in the analysis of the system. Instead, partitioning methods executed on a digital computer achieve the same results.

A second effective type of decomposition can be obtained by partitioning a set of relations so that the information flow is serial, without recycle, within a block. For example, in matrix (b) of Figure 1, the 2 by 2 matrix in the upper left corner can be partitioned by solving f_4 first for x_1, and then solving f_1 for x_2. However, the matrix in the lower right corner is irreducible, that is it cannot be partitioned without tearing. Note in Figure 1c how the information flow is serial for the matrix that can be partitioned but is looped (recycled) for the irreducible matrix.

We will now examine various kinds of structures and the possibility of decomposing them in linear programming, dynamic programming, and in the solution of sets of equations.

2. STRUCTURE AND SPARCENESS IN THE SOLUTION OF LARGE SETS OF EQUATIONS

Sometimes the numerical solution of large sets of linear or nonlinear equations can be assisted by the systematic ordering of the equations and variables so they can be solved effectively. The chain of information flow among the system variables requires that (1) the solution of certain equations precede others, (2) certain of the equations be solved only for specified variables and not for others, (3) the computer storage and arithmetic cost be reduced as much as possible, and, finally, (4) if an iterative solution is required, the sensitivity of the system and numerical error propogation must be considered in the choice of iterates. All of these factors must be taken into account in setting up a precedence order to solve the system equations if the computer solution is to be efficient and economic.

We first consider solving a set of simultaneous linear equations

$$Ax = b$$

where A is a nonsingular sparce matrix of order n. If the matrix A can be factored as a product of lower and upper triangular matrices, A = LU, then a solution for x can be obtained by forward and backward substitution of the two triangular systems

$$Ly = b$$
$$Ux = y$$

In the triangularization of a sparce matrix some of the original zero elements become non-zero so that some preliminary partitioning (rearrangement) of the original system can take place that provides minimal fill-ins. Some trade-off must take place between the time required for the initial partitioning, the time saved by reducing the number of fill-ins, the storage, and the number of arithmetic operations.

During the forward course of the procedure new non-zero elements are created (the back substitution does not lead to new non-zero elements). Thus it should be possible to reduce the extent of the computer storage and reduce round off errors (because where there is a zero element no operation need be executed in the back substitution) by arranging the system equations in suitable form. Tewarson (1971) illustrated a number of structures (the occurrence matrices) for A, that might be obtained by partitioning or permutation, and that would generate a minimum number of new non-zero elements during the forward phase of the elimination. The possible non-zero elements fall in the shaded regions; the horizontal or vertical bordered regions (cross-hatched) can be superimposed as well. The information flow for

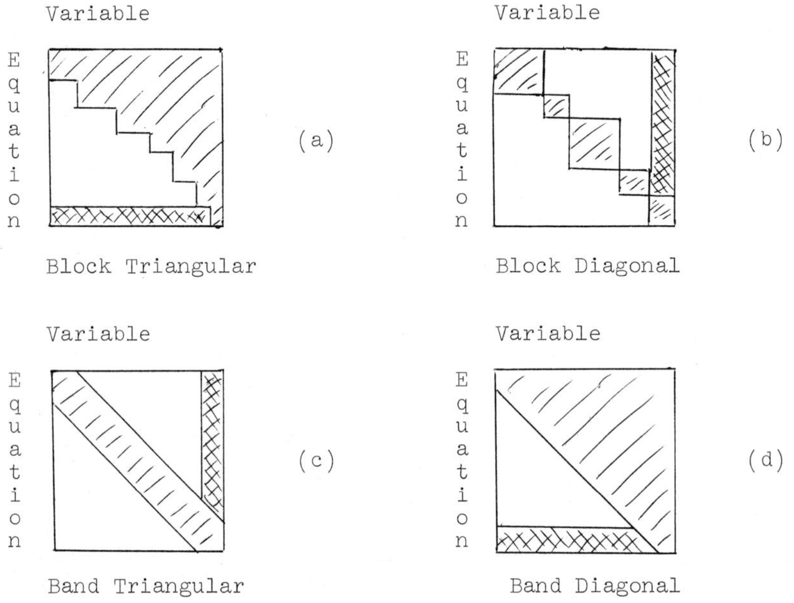

Figure 2 Desirable Structures to Avoid Creating Non-zero Elements

Figures 2a and 2d omitting the bordered regions involves much more recycle than does Figures 2b and 2c. Typical information flows of 2a and 2c are contrasted in Figure 3

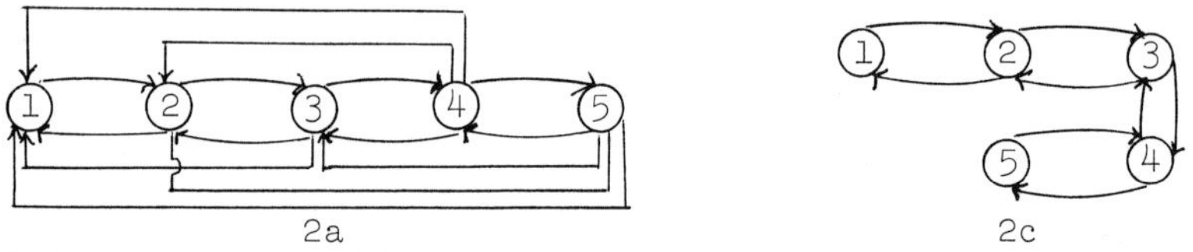

Figure 3 Information Flow Diagrams for Figures 2a and 2c

Band matrices such as are formed in solving differential equations by finite difference approximations have an especially simple structure if they can be decomposed in to L and U matrices. For example a tridiagonal matrix decomposes into

$$\begin{bmatrix} a_{11} & a_{12} & & \\ a_{21} & a_{22} & a_{23} & \\ & a_{32} & a_{33} & \\ & & & \ddots \end{bmatrix} = \begin{bmatrix} 1 & & & \\ \ell_{21} & 1 & & \\ & \ell_{32} & 1 & \\ & & & \ddots \end{bmatrix} \begin{bmatrix} u_{11} & u_{12} & & \\ & u_{22} & u_{23} & \\ & & u_{33} & \\ & & & \ddots \end{bmatrix}$$

Figure 4 Decomposition of a Tridiagonal Matrix A into A = LU

It also can be shown that pivoting is unnecessary for a diagonally dominant matrix, i.e., $|a_{ii}| \geq \sum_{j \neq i} |a_{ij}|$ for all i (and inequality for at least one i), that is irreducible.

We turn now to consideration of the solution of a set of n nonlinear equations. Usually iterative algorithms of various kinds are employed to execute a numerical solution. These algorithms generally start with an initial vector, x, and employ different strategies to modify x so as to converge to a solution by stages. Our purpose here is not to consider the specific technique of solution, whether it be Newton's method, a gradient method, a direct search method, or whatever. Instead we are concerned with how to decompose large sets of nonlinear equations based on the information flow among the equations so as to reduce the scale of the sets of equations that must be solved simultaneously. If the system is ordered into a sequence such that each equation or various groups of equations can be solved independently of the remaining equations in the sequence, the solution can proceed serially and the computer storage needed to effect a solution would correspond roughly to the size of the largest subsystem of equations instead of the entire system. The time to obtain a solution should also be substantially reduced. Finally, if the equations within each partitioned subsystem can be ordered so that the number of iterates which need be chosen to obtain a solution of the subsystem is smaller than the number of equations in the subsystem, it should be possible to reduce the computer time even further.

As an example, consider the set of equations in Figure 5a. Figure 5b is the corresponding occurrence matrix.

$$\begin{aligned} f_1\ (x_1, x_2) &= 0 \\ f_2\ (x_4) &= 0 \\ f_3\ (x_3, x_6) &= 0 \\ f_4\ (x_4, x_5) &= 0 \\ f_5\ (x_1, x_6) &= 0 \\ f_6\ (x_2, x_3, x_5) &= 0 \end{aligned}$$

Fig. 5(a)

Equation	Variable: x_1	x_2	x_3	x_4	x_5	x_6
f_1	(1)	1	0	0	0	0
f_2	0	0	0	(1)	0	0
f_3	0	0	(1)	0	0	1
f_4	0	0	0	1	(1)	0
f_5	1	0	0	0	0	(1)
f_6	0	(1)	1	0	1	0

Fig. 5(b)

In order to determine the direction of information flow in the system equations, one must first establish what information each equation is to supply, that is the identity of the variable whose value is to be obtained from the equation. The variable for which an equation is to be solved is called its <u>output variable</u> and the set of all of the variables assigned to the <u>equations as output</u> variables is called an output set. In terms of the occurrence matrix a (nonunique) output set is a set of non-zero elements in the occurrence matrix such that one and only one element appears in each row and simultaneously one and only one element appears in each column. In Figure 5b, such an output set has been encircled.

The information flow transmitted by the variable designated by the column number goes to all other equations which have non-zero elements in the column. Thus we can identify the non-zero elements in the adjacency matrix, i.e., the Boolean matrix that shows the flows of information between equations. The equations that are decendents are indicated by the non-zero entries in the columns and the equations that are predecessors by the non-zero entries in the rows. A quick way to form the adjacency matrix is to permute the rows of the occurrence matrix so that the output set falls on the main diagonal, and then replace the main diagonal with zeros; see Figure 6.

Variable

Equation	x_1	x_4	x_3	x_5	x_6	x_2
f_1	(1)	0	0	0	0	1
f_2	0	(1)	0	0	0	0
f_3	0	0	(1)	0	1	0
f_4	0	1	0	(1)	0	0
f_5	1	0	0	0	(1)	0
f_6	0	0	1	1	0	(1)

Occurrence Matrix
(a)

Equation

Equation	f_1	f_2	f_3	f_4	f_5	f_6
f_1	0	0	0	0	0	1
f_2	0	0	0	0	0	0
f_3	0	0	0	0	1	0
f_4	0	1	0	0	0	0
f_5	1	0	0	0	0	0
f_6	0	0	1	1	0	0

Adjacency Matrix
(b)

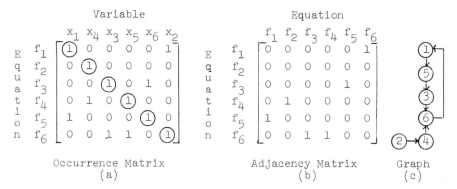

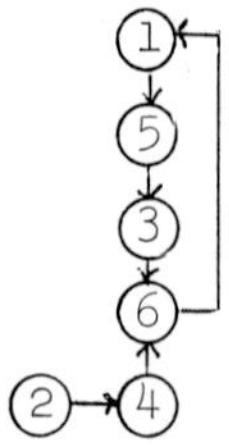

Graph
(c)

Figure 6 Occurrence Matrix and Associated Adjacency Matrix

Because recycle (feed back) of information creates loops, or cycles, the system of equations must next be partitioned into the smallest irreducible subsystems, that is the smallest groups of equations that must be solved simultaneously. Partitioning represents the first and easiest of the two phases of decomposition--tearing is more difficult. A number of methods of partitioning exist (Ledet and Himmelblau (1970)) of which that of Steward (1962) who traced the paths of information flow to find the maximal loops is probably the fastest. The result would be Figure 7. The graph corresponding to the adjacency matrix indicates the subsets of equations by inspection and illustrates the loop of information flow. Equation 2 is disjoint and can be solved first, then Equation 4, followed by the other four

	f_2	f_4	f_6	f_1	f_5	f_3
f_2	0	0	0	0	0	0
f_4	1	0	0	0	0	0
f_6	0	1	0	0	0	1
f_1	0	0	1	0	0	0
f_5	0	0	0	1	0	0
f_3	0	0	0	0	1	0

Figure 7 Partitioned Set of Equations

equations simultaneously that are tied in a loop.

Once the complete system of equations has been partitioned into the irreducible subsystems of simultaneous equations, it may be desirable to decompose further these irreducible blocks of equations so that their solution can be correspondingly simplified. The decomposition of the irreducible subsystems is called tearing. Specifically, the concept of tearing a variable consists of removing it from consideration as a variable in one or more of the equations in which it appears. More than one variable can be torn simultaneously. Tearing of a variable is effected by assuming its value, as determined, say, by physical principles or an educated guess, followed by reordering of the equations in the block. Because the value of the variable is assumed, it must be checked at some stage in the sequence of calculation for the block, and because the subsequent value rarely agrees with the assumed value, the torn variable becomes an iterated variable.

By inspection of the graph 6c the loop of recycled information can be cut by selecting the output variable of Equation (6), x_2, as the variable to tear. Then the block of four simultaneous equations is decomposed into a serial sequence for calculation. No general criteria exist as to what variables to tear but some trade off must be made in the times for tearing and iteration, numerical error, and convergence properties of the numerical solution technique all in terms of a nonunique selections of torn variables.

3. STRUCTURES THAT CAN BE DECOMPOSED IN LINEAR PROGRAMMING

Because large scale linear programming problems require excessive computer storage and execution time, a number of specialized techniques have been developed to overcome these difficulties. Most approaches to making the execution of the LP program more efficient involve the break up of the large problem into smaller subproblems and adjustment of the optimal solutions of the subproblems so as to produce an optimal solution for the large scale problem. We will examine a few of the structural features of large scale linear programming problems that prove to be decomposable. For a specific decomposition scheme to be computationally effective the subproblems must have considerably fewer variables and/or constraints than the original problem.

3.1 Decomposition Principle

Dantzig and Wolfe developed a decomposition principle that imposes on each subproblem additional constraints so that the optimal solution of one subproblem is independent of the solutions of the other subproblems. The additional constraints are selected so that the union of the solutions of the subproblems will be optimal for the large scale problem. For decomposition to be effective, the subproblems generated must be easy to solve. Applications of the decomposition principle have been reported for multicommodity networks, lot size scheduling, Markov chains, and linear control problems among others.

As a simple example consider the problem

$$\begin{array}{llcll} \text{Minimize:} & c^T x + d^T y & = & f & \\ \text{Subject to:} & Px & = & p, & x \geq 0 \\ & \qquad\quad Qy & = & q, & y \geq 0 \\ & Ax + By & = & r & \end{array} \qquad (1)$$

Problem (1) can be solved by solving the subproblems (2) and (3) if the vector r is partitioned in to two vectors: $r = a + b$.

$$\text{(2)} \quad \begin{array}{llcll} \text{Minimize:} & c^T x & = & f_1 & \\ \text{Subject to:} & Px & = & p, & x \geq 0 \\ & Ax & = & a & \end{array} \qquad\Bigg|\qquad \text{(3)} \quad \begin{array}{llcll} \text{Minimize:} & d^T y & = & f_2 & \\ \text{Subject to:} & Qy & = & q, & y \geq 0 \\ & By & = & b & \end{array}$$

What is needed is a suitable partition of r such that the union of the solution of problems (2) and (3) is a solution of (1).

In general if the structure for the problem

$$\begin{array}{ll} \text{Minimize:} & f = c^T x \\ \text{Subject to:} & \sum_{i=1}^{n} A_i x_i = b \\ & B_i x_i = b_i \\ & x \geq 0 \end{array} \qquad (4)$$

where A_i is an m by n_i matrix
b is an m by 1 vector
b_i is an m_i by 1 vector
B_i is an m_i by n_i matrix
c is an n_i by 1 vector

as is shown in Figure 8, then a number of algorithms have been proposed for the solution of the LP problem via decomposition. Note how

Variable

Equation

c_1	c_2	...	c_n
A_1	A_2	...	A_n
B_1	0	...	0
0	B_2	...	0
0	0	.....	B_n

(A)

Figure 8 Matrix for a Decomposable LP Problem

the information flow structure consists of a set of disjoint blocks B_i plus a set of blocks A_i that interconnect the B blocks.

Dantzig (1970) has summarized the essence of the decomposition principle for a nonlinear programming problem as follows. Let f(x) and $g(x)$ be convex functions of x. Given the problem

$$\begin{aligned} \text{Minimize:} \quad & f(x) \\ \text{Subject to:} \quad & g_1(x) \leq 0 \\ & g_2(x) \leq 0 \qquad i = 1,\ldots,m \qquad (5) \\ & \vdots \\ & g_m(x) \leq 0 \end{aligned}$$

by assigning Lagrange multipliers to any subset of the constraints, such as $i = 1,2,\ldots,n$, presumably to the connecting blocks, and forming a modified objective function adding these constraints, a subproblem can be formed with a set of the disjoint constraints

$$\begin{aligned} \text{Minimize:} \quad & \Phi(x) = f(x) + \sum_{i=1}^{n} \lambda_i g_i(x) \\ \text{Subject to:} \quad & g_i(x) \leq 0 \qquad , \; i = n+1,\ldots,m \end{aligned} \qquad (6)$$

If x^* is the solution to (6), if $g_i(x^*) \leq 0$ for $i = n+1,\ldots,m$, and if $\lambda_i g_i(x^*) = 0$ for $i = 1,\ldots,n$, then x^* is also the solution to (5). If the initial guesses for λ_i do not yield $x = x^*$ then λ_i can be improved by a suitable algorithm.

If decomposition is possible with the structure indicated, how can one determine if a general LP problem (in which the special structure above cannot be discerned by inspection) can be transformed into the desired structure? The question can be paraphrased as follows. Given an arbitrary matrix A, can the rows and/or columns be rearranged to provide a precedence order for information flow so that the desired (nonunique) structure (A) results?

Several algorithms based on graph theory exist [Weil and Kettler (1968), Weil and Steward (1967)] to find the precedence order for information flow given a set of equations that yields the structure (A). Each of the blocks B_i is termed an irreducible matrix and represents sets of equations (or inequalities) that must be treated simultaneously.

3.2 Parametric Variation

Parametric variation is closely related to the decomposition principle. By assigning a Lagrange multiplier λ to a weakly linked LP problem, a linking constraint can be removed and added to the objective function. The resulting system is then easier to solve. If the solution does not satisfy the constraint and the complementary slackness conditions, then λ is adjusted until these conditions are met. A number of methods of varying λ have been proposed including

(a) Rosen's partition programming (1960)
(b) Kron's diakoptics (1963)
(c) Beale's pseudo basic variables (1963)
(d) Gass'es dualplex method (1966)

In general the structure for parametric variation methods might be visualized as follows

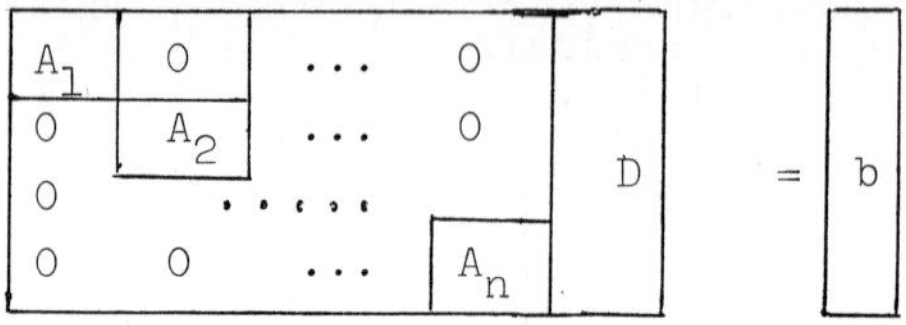

Pivoting anywhere but in the block D will not destroy the triangular
information flow scheme with disjoint blocks.

3.3 "Staircase" Blocks

In linear control problems and economic models a special "stair-
case" block structure exists. For example in control the input of a
given period is represented by

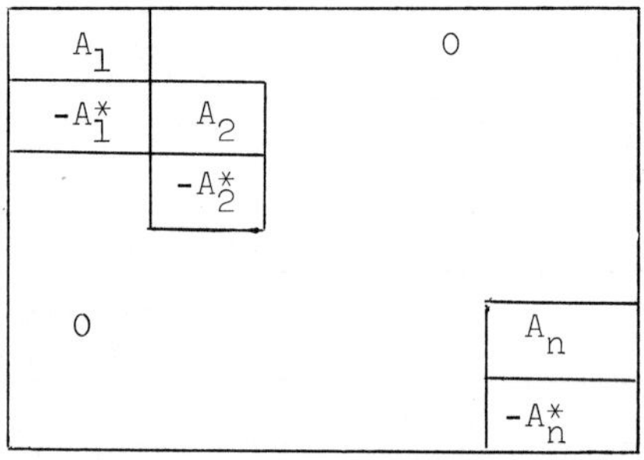

A_i and the output for the period by $-A_i^*$. Coupling exists because the
output of one period is directly related to the input of the next
period. Thus although it is not possible to obtain disjoint sub-
blocks, it is possible to represent the non-zero elements much more
compactly (see sparce matrices below), or the inverse of the full
system more compactly.

A simpler version of (C) is

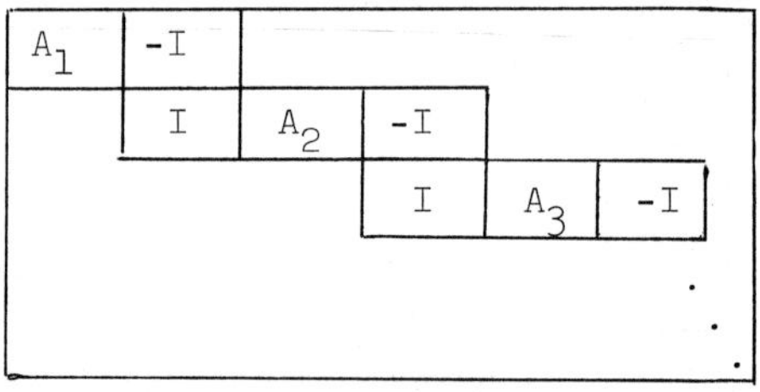

in which the matrices A_i may be identical.

In economics the matrices A_i might represent the input output
coefficients of various economies and the connection matrices I and
-I would represent import-export matrices as in the following structure for three systems

A_1			I	-I			I	-I
	A_2		-I	I	I	-I		
		A_3			-I	I	-I	I

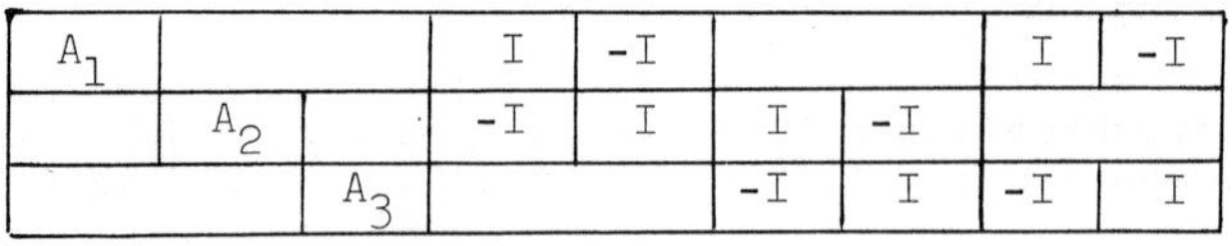

(D)

The general structure of these LP problems is that the blocks A_i are square, nonsingular irreducible matrices whose information flow diagrams might be (for a 3 by 3 matrix

$$A_i = \begin{bmatrix} a_{ii} & a_{ij} & a_{ik} \\ a_{ji} & a_{jj} & a_{jk} \\ a_{ki} & a_{kj} & a_{kk} \end{bmatrix}$$

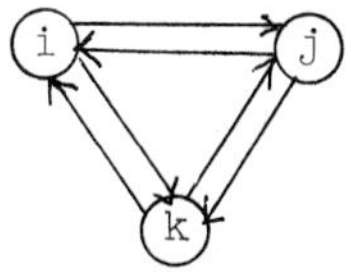

If the coupling matrices are eliminated the blocks A_i are disjoint.

3.4 Other Special Structures

Structure such as (E) consist of E sets of coupling linear equa-

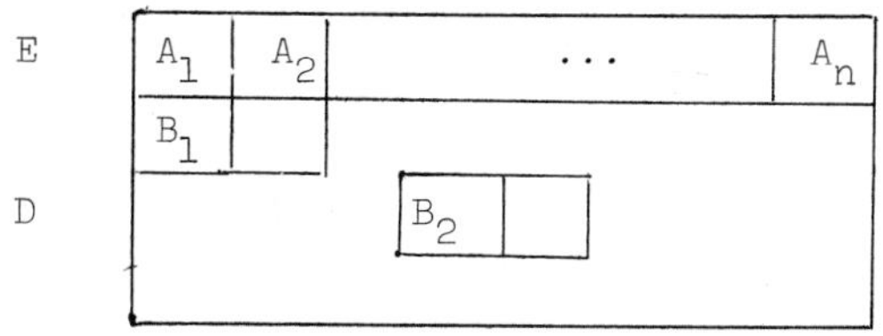

(E)

tions and D sets of disjoint equations that have no common variables. A combination of (A) and (B) consists of an L shaped or bordered array

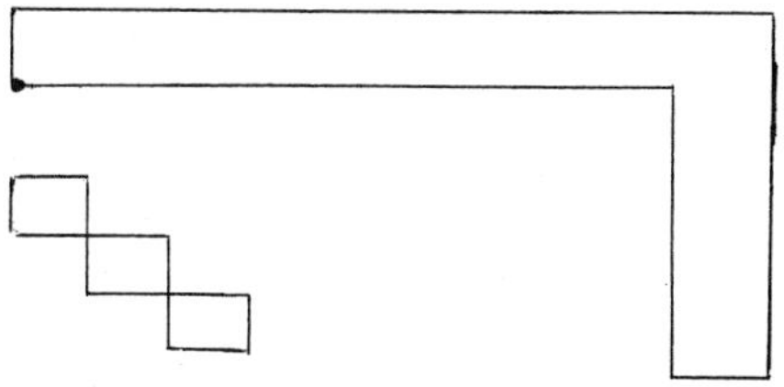

(F)

Structures such as (G), sometimes found interger programming,

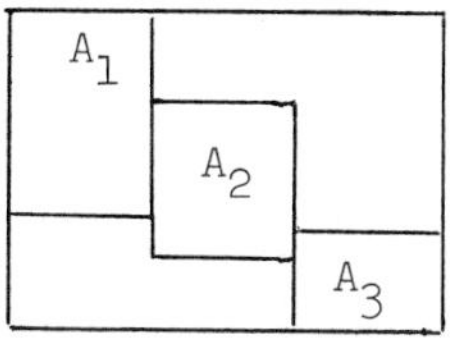

(G)

can be decoupled by specifying the variables in A_2, solving A_1 and A_3 separately. However, by inspection we can see that the connecting rows of (G) can be removed and put above the remaining rows to yield the form of (A). Which treatment proves best depends on the technique of solution used for the integer (or mixed integer) problem.

3.5 LU-Decomposition and Sparce Matrices

Sparceness in the matrices employed in linear programming implies not only less arithmetic but less computer storage if only the relavent non-zero elements and their row and column indices can be retained. In linear programming the computation time is greatly influenced by the method of finding an accurate and compact form of the inverse of a sparce matrix B. It has been recognized that if the product form of B is used, and in particular the LU form (B = LU), then by pivoting first on the diagonal elements of L in their natural order and then on the diagonal elements of U in the reverse order, a product form representation can be obtained that does not reduce the degree of sparceness. Tomlin (1971) indicates how the expansion of non-zero elements can be substantially reduced by use of the LU decomposition for large (such as 3500 rows, 9100 columns, and 75,000 elements) LP problems.

4. DECOMPOSITION AND STRUCTURE IN DYNAMIC PROGRAMMING

Dynamic programming is a synonym for a decomposition technique that utilizes the serial structure of information flow in optimization problems. If the information flow is serial (stagewise) as shown in Figure 8 (the physical flow may proceed in the reverse

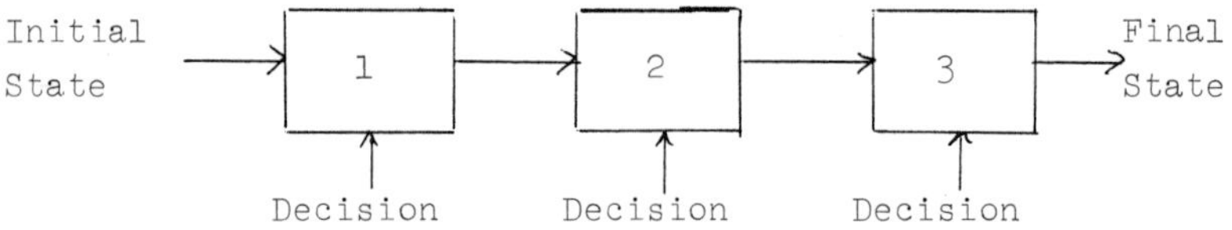

(H)

Figure 8 Serial Information Flow Used In Dynamic Programming

direction), then the principle of optimality indicates that decisions made at the end of the information flow do not affect upstream decisions. Consequently, one can break up an n-decision variable, one-state initial value problem, into a sequence of n one-decision, one-state subproblems. Stage 3 can be optimized, that is the optimal values of the final state can be determined as a function of the entering state. Then given the optimal decisions for stage 3, stage 2 can be optimized, and then stage 1, given the optimal decisions for 2 and 3 together.

Structure (H) can be viewed as a structure such as (C') except that the columns represent single variables x_0, x_1, x_2, and x_3, and the rows represent the models (stage transformations) for each stage.

Structure (H) can also be interpreted in terms of a tree of information flow in which the nodes correspond to the states and the arcs correspond to the decisions at any stage. Associated with each

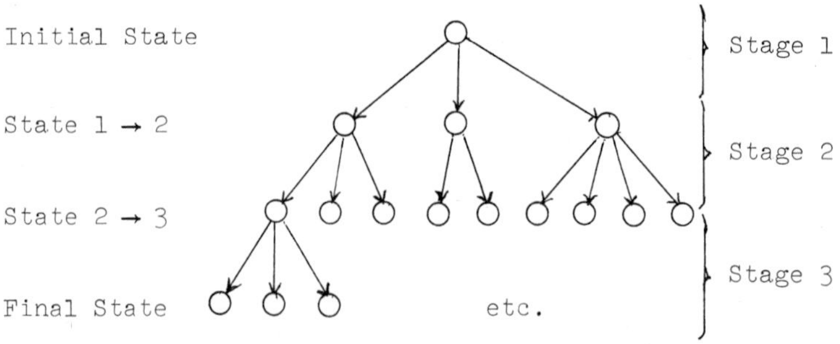

(I)

stage is an objective function that provides the return as a function of the entering and exit states and the decisions. Feasible solutions correspond to paths between the initial state and the final states while the return from the path is established from the sum of the sequence of returns from the objective functions. The serial flow of information makes the decomposition possible.

Recycle of information from down stream in the path to an upstream point causes a loop of information flow that corresponds to an irreducible occurrence matrix in graph theory. The loop of information flow prevents complete decomposition unless tearing takes place by cutting information flow stream at an appropriate state variable. Of course, the penalty for tearing is that an iterative scheme must be used to establish the value of the torn variable. Note how analogous tearing is in concept to the procedure of parametric variation.

REFERENCES

Beale, E.M.L., (1963), The Simplex method using pseudo-basic variables for structured linear programming problems, in Recent Advances in Math. Programming, ed. R. Graves and P. Wolfe, McGraw-Hill, p.133.

Beale, E.M.L., (1971), Sparceness in linear programming, in Large Sparce Sets of Linear Equations, ed. J.K. Reid, Academic Press London, p.1.

Dantzig, G. B., (1970), "Large Scale Systems and the Computer Revolution", in Proceed. Princeton Symp. Math. Programming 1967, ed. H. W. Kuhn, Princeton Univ. Press, p.51.

Gass, S. I., (1966), The Dualplex method for large scale linear programs, Ph.D. Thesis, University of Calif., Berkeley.

Kron, G., Diakoptics, McDonald Publishers, London (1963).

Ledet, W. P. and D. M. Himmelblau, (1970), Decomposition procedures for the solving of large scale systems, Advances in Chem. Engr., 8, p.186.

Rose, D. J., and J. R. Bunch, (1972), The role of partitioning in the numerical solution of sparce systems, in Sparce Matrices, ed. D. J. Rose and R. A. Willoughby, Plenum Press, N.Y., p.177.

Rosen, J. B., (1960), Partition programming, Notices Amer. Math. Soc., 7, p.718.

Steward, D. V., (1962), On an approach to techniques for the analysis of the structure of large systems of equations, Soc. Ind. Applied Math. Rev., 4, p.321.

Tewarson, R. P., (1971), Sorting and ordering sparce linear systems, in Large Sparce Sets of Linear Equations, ed. J. K. Reid, Academic Press, London, p.151.

Tomlin, J. A., (1972), Modifying triangular factors of the basis in the Simplex method, in Sparce Matrices, D. J. Rose and R. A. Willoughby, eds., Plenum Press, New York, p.77.

Weil, R. L. and P. C. Kettler, (1968), Transforming matrices to use the decomposition algorithm for linear programs, Report No. 6801, Graduate School of Business, University of Chicago.

Weil, R. L. and D. V. Steward, (1967), The question of determinancy in square systems of equations, Zeit fur Nationalokonomie, 27, p.261.

BRANCH AND BOUND METHODS AS DECOMPOSITION TOOLS

JEAN-PAUL BARTHES
Stanford University
Stanford, California 94305

ERNEST J. HENLEY
University of Houston
Houston, Texas 77004

Abstract: Branch and bound uses the concepts of trees, logic trees, and bounds to solve combinatorial problems. The method is a powerful alternative to exhaustive enumeration for decomposing large scale problems since the time and storage requirements for exhaustive enumeration increase exponentially with the number of variables.

Tree search methods go by a variety of labels including not only "branch and bound" but also "branch and exclude," "branch search," "bound and scan," "implicit enumeration," "branch and prune," etc. The particular choice of technique depends on the structure of the problem.

In this paper we apply the branch search method to a network decomposition problem, analyze a typical problem by a variety of search techniques, and offer some generalizations of these methods.

INTRODUCTION

Branch and bound methods are usually applied to systems synthesis rather than decomposition. Synthesis problems are basically assignment problems. One of the earliest chemical design problems solved by branch and bound was that of constructing an optimal heat exchange network (Lee, 1970). Here one has hot streams to be cooled and cold streams to be heated. Since the streams are at different temperatures, there are feasibility constraints imposed by virtue of thermodynamic limitations. Here, as elsewhere, the design problem has a unique structure which dictates the bounds and possible branches. Two limitations generally exist in applying branch and bound synthesis methods: Cyclic processes cannot be treated by conventional methods, and multiple assignments are deemed unfeasible.

Branch and bound methods are essentially enumeration techniques which use the concepts of logic trees to solve combinatorially explosive problems. In reality "branch and bound" techniques encompass a number of methods among which are 'branch and exclude', 'branch search', 'truncated enumeration', 'branch and prune', 'bound and scan','implicit enumeration', 'branch search', etc. In most situations, integer linear programming provides an alternative approach, the advantage of branch and bound being that only a very small subset of the 2^M (where the integer constraints have been reduced to M bivalent variables) need be examined.

There is relatively little mathematical formalism in the application of tree search methods to physical and engineering problems. Frequently algorithmic solutions are difficult to achieve and one resorts to heuristic reasoning; the branching and bounding depending on the structure of the problem. Cookbook algorithms are unavailable; logic, insight, clever programming, and creativity are required if one expects to apply branch and bound methods to new and meaningful problems.

In this paper we elaborate on this basic theme by:

1. Constructing a'branch and search' solution to a decomposition problem; that of finding the simple circuits in an oriented graph;

2. Presenting a formal mathematical definition of the branch and bound method which permits generalizations;

3. Solving the traveling salesman problem by four search algorithms, to show the simplicity and flexibility of this approach.

NETWORK DECOMPOSITION BY BRANCH SEARCH METHODS

Problem

Find all the simple circuits of an oriented graph.

Preliminary Remarks and Graph Reduction

A vertex that has a zero indegree or a zero outdegree cannot belong to a loop, it can therefore be erased from the graph, if the purpose is to find circuits.

Take for example the graph shown in Figure 1.

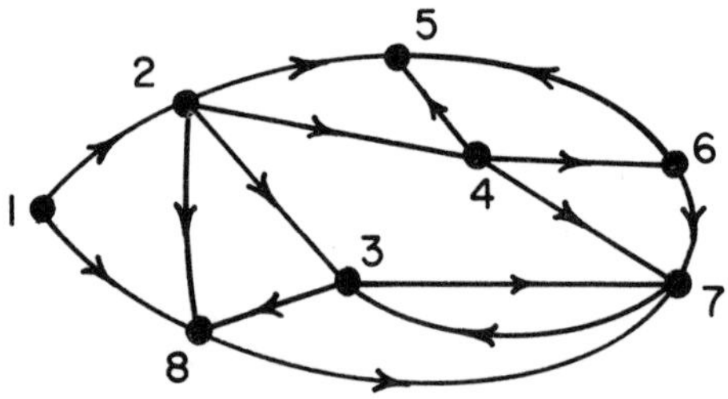

FIGURE 1. GRAPH WITH CIRCUITS

Reduction 1

The indegree of vertex 1 is 0, the outdegree of vertex 5 is zero, therefore they, and their adjacent arcs, can be erased (Figure 2).

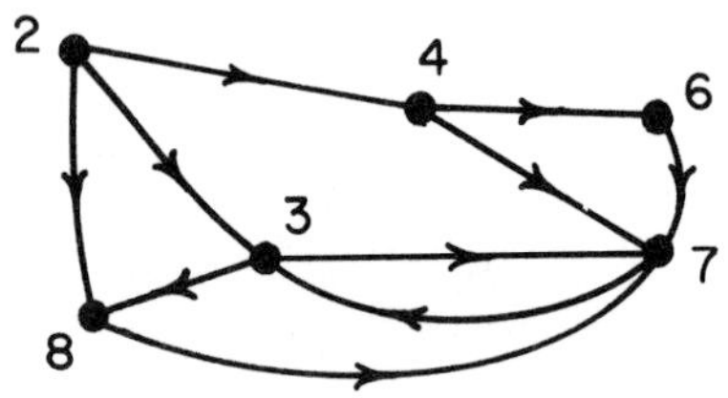

FIGURE 2. REDUCED GRAPH

Reduction 2

The indegree of vertex 2 is zero, so we can erase further (Figure 3).

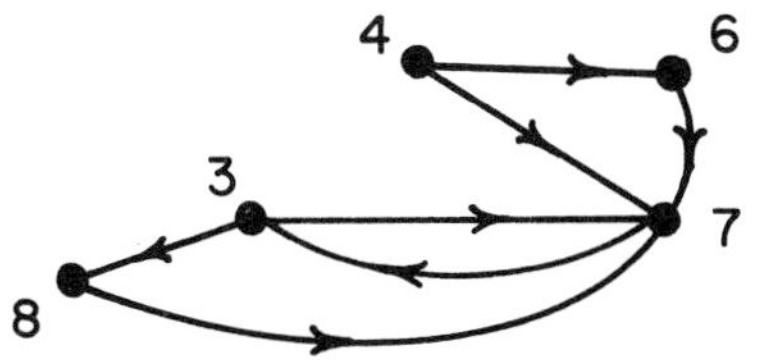

FIGURE 3. REDUCED GRAPH

Reduction 3

The indegree of vertex 4 is zero; we reduce further (Figure 4).

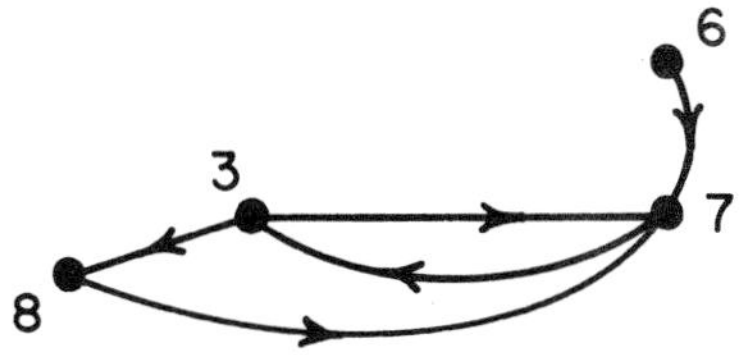

FIGURE 4. REDUCED GRAPH

Reduction 4

The final reduction consists of eliminating vertex 6 (Figure 5).

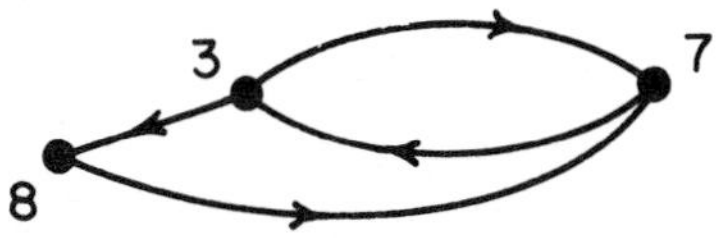

FIGURE 5. REDUCED GRAPH

Hence a problem with 8 vertices decomposes to a problem with only three. These reductions can be done easily using an edge incidence matrix, or any other suitable representation of the graph.

Branch Search Techniques to Enumerate Circuits (Labeling Procedure).

Step 0. All arcs and vertices are unused.

Step 1. Start with the vertex with the highest outdegree; in case of a tie, take any of the highest. For each vertex, arcs leading to other vertices are recorded in a given order. Take the first unused arc from the starting vertex. If it is a self loop, record it and take the next arc. Label the vertex extremity of the arc used, label the arc used, go to 2.

Step 2. From the previous vertex take the first unused arc, label it used and ascertain if its extremity is the starting vertex. If yes, record the cycle, and take the next unused arc. If no, check if the extremity is a used vertex; if yes, take the next unused arc: If no, label the extremity vertex as used and go on.

When no more arcs are available, unlabel the vertex under consideration, and all of its outgoing arcs, and go back to its predecessor, if one exists. If the vertex has no predecessor, go to 3.

Step 3. Delete the starting vertex from the graph, reduce the graph, and go to Step 1. If the reduced graph does not contain arcs, stop.

Example 1

Using the graph of Figure 6 we record the arcs as follows:

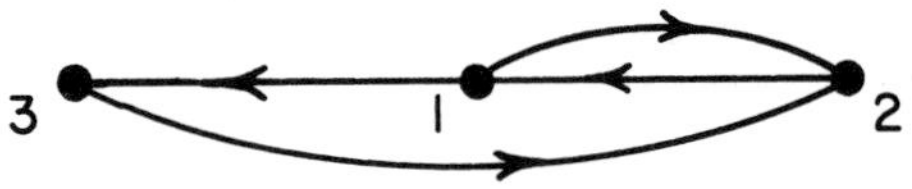

FIGURE 6. GRAPH WITH CIRCUITS

Starting Vertex	Extremity
1	3,2
2	1
3	2

The search, which is summarized on Figure 7, proceeds:

1. We pick vertex 1, arc a_{13}. Vertex 3 and a_{13} are labeled used.
2. From 3 we pick a_{32}. Vertex 2 and a_{32} are labeled used.
3. From 2 we pick a_{21} and record the circuit.
4. No more arc is available from 2; we unlabel 2 and backtrack to 3.
5. No more arc is available from 3, we unlabel 3 and a_{32} and go back to 1.
6. a_{13} is used, the next used arc is a_{12}. We label a_{12} and vertex 2 used.
7. From 2 we pick a_{21}, and record the circuit.
8. No more arc is available from 2, we backtrack to vertex 1.

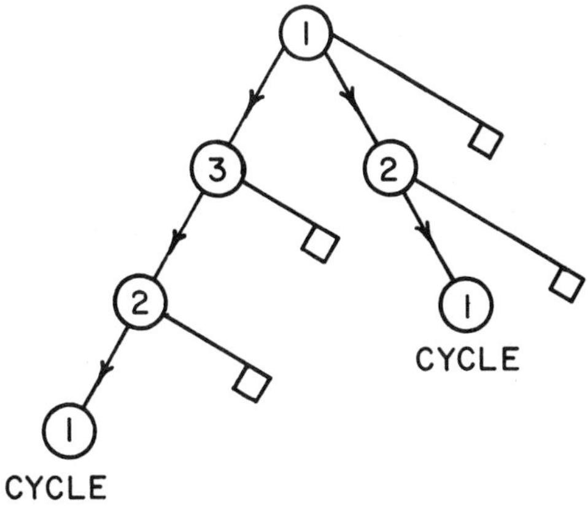

FIGURE 7. SEARCH TREE

9. No more arc is available from 1, we erase 1 from the graph and reduce it. No more arc remains in the reduced graph, so we stop.

Example 2

Figures 8 and 9 are offered, without explanation, to demonstrate a more complex case.

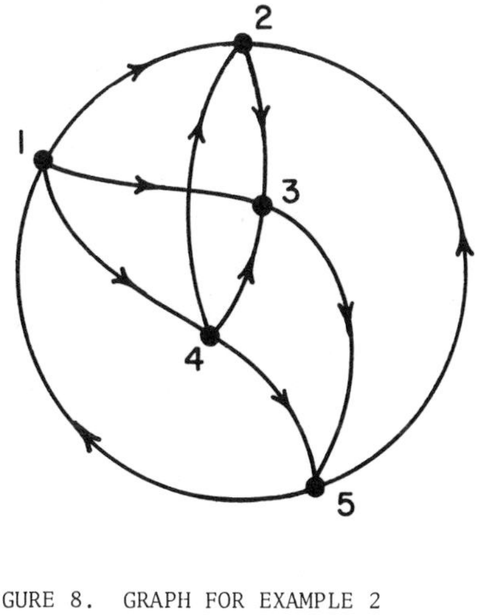

FIGURE 8. GRAPH FOR EXAMPLE 2

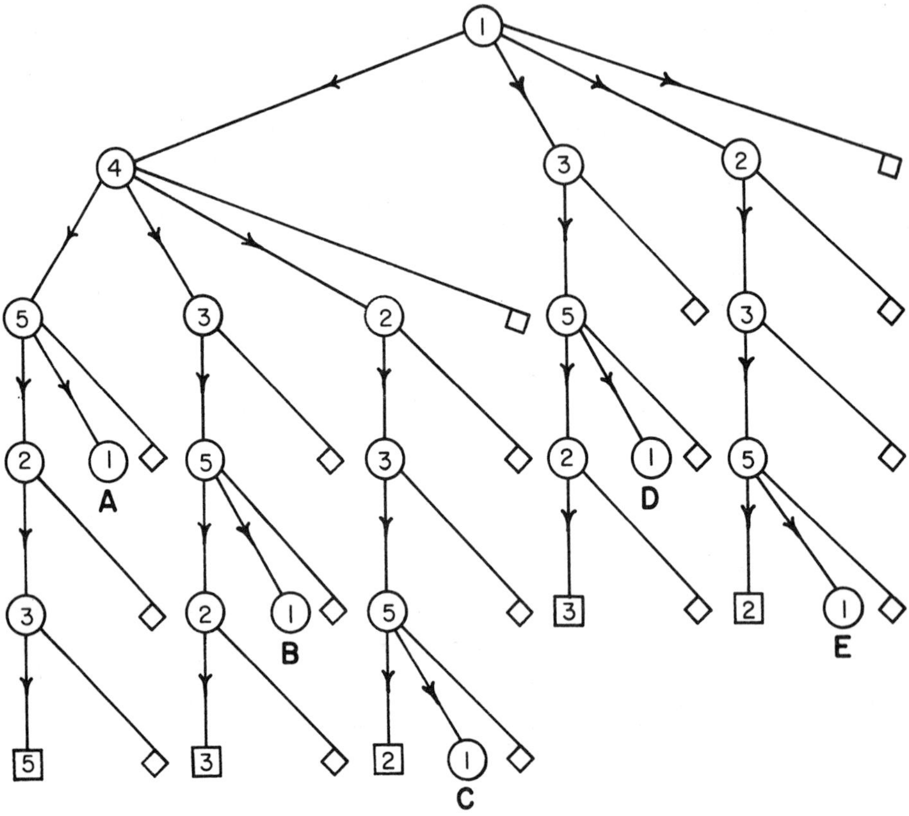

 means no more arcs going out of i are unused.

 means that vertex j is already used.

FIGURE 9. SEARCH TREE FOR FINDING LOOPS IN FIGURE 8

TREE PRUNING

The branch and search method generated thus far is in reality a direct combinatorial approach. We now endeavour to show how, by taking advantage of the structure of the problem, the search tree can be pruned, and the computational task simplified.

The first point to note is that certain patterns must, per force, repeat themselves during the search. (3-5-1 in loops B and C, of Figure 9, for example).

Once we have determined what loops are computed when we are at vertex k then the free (unlabeled) vertices constitute a set V_k, and in a later situation, whenever we encounter vertex k in the search process, if the remaining free vertices constitute a subset of V_k we can immediately deduce what new loops will be introduced.

Taking as an example the first branch of the tree in Figure 9 (1→4→5→2→3) we deduce that

1. When vertex 2 occurs in the search and $V_2=\{3\}$ meaning the only remaining vertex is 3, then no loop can be formed.

2. When vertex 5 is encountered in the search and $V_5=\{2,3\}$ meaning only 2 or 3 remain, the only loop that can be formed is with edge a_{51}.

It is possible to record loops in a list processing scheme such as shown below. Here we record loop A and take advantage of the fact that the origin for all loops is vertex 1.

TABLE 1

	ROW	END	POINTER
LOOP A	1	4	1
	2	5	1
	3	1	2

We also record the following information

TABLE 2

Vertex k	V_k Set of Remaining Edges	Sequences of Vertices for Possible New Loops
2	3	None
5	2,3	5→1

From this point on, every time we encounter 5 during the search, and the remaining set of free edges are 2,3 or 3,2, we can record the new loop without further exploration of the search tree. This can be done, for example, when we reach vertex 5 on the second branch from the left, as is shown in Figure 10.

Figure 11 shows the pruned tree one obtains when this concept is applied over all branches. Thirty-five percent fewer vertices need be recorded.

When all the loops containing vertex 1 are recorded, it is removed from the graph. The reduced graph, Figure 12, yields one additional circuit.

DEVELOPMENT OF COMPUTER ALGORITHMS

At this point, it is necessary to note a lack of consistency between Tables 1 and 2. Table 1 records loops using backward row pointers, Table 2 gives completions for new loops with a forward path.

To remain consistent, we can use forward row pointers for Table 1, as shown in Table 3. The zero indicates an end of list.

The complete tableau for the given example using backward row pointers is given in Table 4. The same tableau with forward row pointers is shown in Table 5.

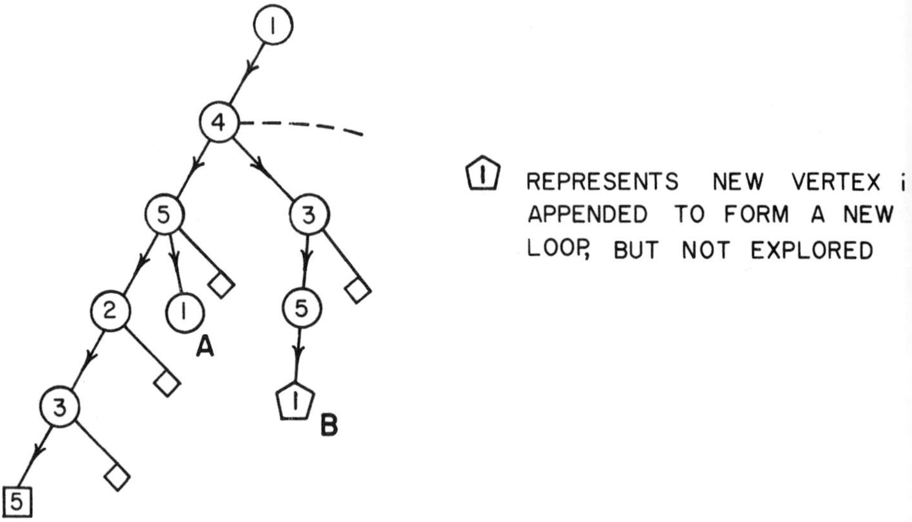

FIGURE 10. PARTIAL TREE FOR LOOP PROBLEM

TABLE 3

	Row (index)	Beginning	Forward Row Pointer
LOOP A	1	1	2
	2	4	3
	3	5	4
	4	1	0

TABLE 4

LOOP	ROW	END	ROW POINTER
A	1	4	1
	2	5	1
	3	1	2
B	4	3	1
	5	5	4
	6	1	5
C	7	2	1
	8	3	7
	9	5	8
	10	1	9

TABLE 4 (continued)

LOOP	ROW	END	ROW POINTER
D	11	3	11
	12	5	11
	13	1	12
E	14	2	14
	15	3	14
	16	5	15
	17	1	16

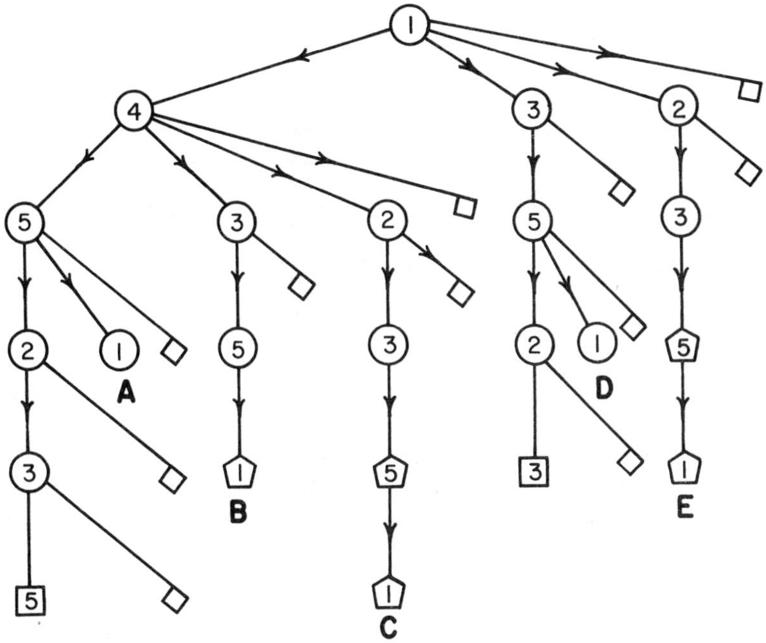

FIGURE 11. SEARCH TREE FOR EXAMPLE 2

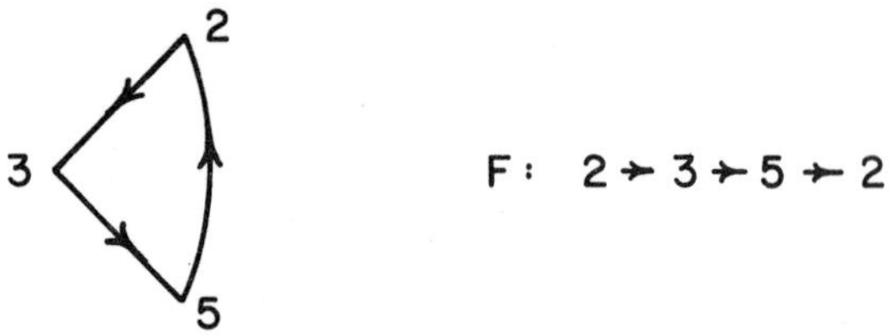

FIGURE 12. REDUCED GRAPH

TABLE 5

	Row Index	Starting Vertex	Forward Row Pointer
LOOP A	1	1	2
	2	4	3
	3	5	4
	4	1	0
LOOP B	5	1	6
	6	4	7
	7	3	3
LOOP C	8	1	9
	9	4	10
	10	2	7
LOOP D	11	1	7
LOOP E	12	1	10

The information corresponding to Table 2, but including the whole tree is:

TABLE 6

Vertex k	Set of Remaining Edges, V_k	Sequences of Vertices Leading To New Loops
2	3	None
5	2,3	5→1
3	2,5	3→5→1
2	3,5	2→3→5→1
4	2,3,5	4→5→1, 4→3→5→1 4→2→3→5→1
5	2,4	5→1
3	2,4,5	3→5→1
2	3,4,5	2→3→5→1

DOUBLE LINKED LISTS

Information storage can also be accomplished in a double linked tableau as in Table 7.

Retrievals from Table 7 are slightly more difficult in terms of bookeeping, but unambiguous. For example, Loop C is determined at Row 6. Its edges are

TABLE 7

Row Index	Backward Row Pointer	Starting Vertex	Forward Row Pointer
1	0	1	2
2	1	4	3
3	2	5	4
4	3	1	0
5	2	3	3
6	2	2	5
7	0	1	5
8	0	1	6

recovered by tracing two chains:

Edges from Backward Row Pointers	Row	Edges from Forward Row Pointers
1←4	2	→3→5→1

This yields

1→4→2→3→5→1

COMPUTER IMPLEMENTATION

Efficient implementation could be achieved in assembler language using bit patterns for the edges, the loops being stored as in a circuit incidence matrix such as

TABLE 8

Arcs / Circuits	1	1	1	2	3	4	4	4	5	5
	4	3	2	3	5	5	3	2	2	1
A	1	0	0	0	0	1	0	0	0	1
B	1	0	0	0	1	0	1	0	0	1
C	1	0	0	1	1	0	0	1	0	1
D	0	1	0	0	1	0	0	0	0	1
E	0	0	1	1	1	0	0	0	0	1

If there are N_c circuits, n edges and B bits/word the necessary storage is

$$\lceil n/B \rceil * N_c \quad \text{words}$$

where $\lceil x \rceil$ is the ceiling function of x

The information of Table 6 could be stored in linked form using bit patterns as in Table 9.

BRANCH AND BOUND

The name "Branch and Bound" comes from the particular approach used by Little et al. (1963) in their attempt to solve the famous operations research problem of the traveling salesman. Branch and bound methods use "search trees", each node of

TABLE 9

Vertices	Pointer or Row Index
1	0
2	1
3	3
4	5
5	2

Row	Pointer	Set of remaining edges V_k U{k}										Sequence of vertices bringing a new circuit									
(Edges→)		1-4	1-3	1-2	2-3	3-5	4-5	4-3	4-2	5-2	5-1	1-4	1-3	1-2	2-3	3-5	4-5	4-3	4-2	5-2	5-1
1	4	0	0	0	1	1	0	0	0	0	0	indicates none									
2	8	0	0	0	1	1	0	0	0	1	1	0	0	0	0	0	0	0	0	0	1
3	9	0	0	0	1	1	0	0	0	1	1	0	0	0	0	1	0	0	0	0	1
4	10	0	0	0	1	1	0	0	0	1	1	0	0	0	1	1	0	0	0	0	1
5	6	0	0	0	1	1	1	1	1	1	1	0	0	0	0	0	1	0	0	0	1
6	7	0	0	0	1	1	1	1	1	1	1	0	0	0	0	1	0	1	0	0	1
7	0	0	0	0	1	1	1	1	1	1	1	0	0	0	1	1	0	0	1	0	1
8	0	0	0	0	1	0	1	1	1	1	1	0	0	0	0	0	0	0	0	0	1
9	0	0	0	0	1	1	1	1	1	1	1	0	0	0	0	1	0	0	0	0	1
10	0	0	0	0	1	1	1	1	1	1	1	0	0	0	1	1	0	0	0	0	1

which represents a class of possible solutions to the problem. The union of all the pending nodes (hanging nodes not yet processed) represents the class of all possible solutions. The algorithm begins by assigning this last class to the root of the tree. Branching is used to replace one pending node by a set of pending nodes, the solution classes of which partition the solution class of the replaced node. A cost is computed for each pending node. This corresponds to a lower bound on the cost of any solution in the solution class of the node. If a node is known to contain no feasible solutions the cost bound is infinite. The cost bounds induce an ordering of desirability on the pending nodes, which determines the branching in subsequent steps. The algorithm stops when it is not possible to generate any new node, or when a feasible solution with associated cost less than the lower bounds of the pending nodes has been found.

FORMAL DEFINITION OF THE BRANCH AND BOUND TECHNIQUE

The procedure will be developed using the following nomenclature and symbols

f = function to be minimized (or maximized)
K_n = complete graph with n vertices for the problem
$\ell^{(i)}$ = lower bound of f at (i)
$L^{(i)}$ = upper bound of f at (i)
$S^{(i)}_{0\ell m\ldots pq}$ = node of search tree at the i-th level, and its class of solutions
$|S|$ = cardinality of S
σ^* = optimal solution
Closed node = node of a search tree which can no longer be processes
Current node (edge) = node (edge) under consideration
Forbidden edge = edge banned from all subsequent solutions
Free edge = edges which are not forbidden
Imposed edge = an edge which is part of all subsequent solutions
Pending node = not yet processed node of a search tree
Terminal node = a node representing one feasible solution

We would like to find the solution $\sigma^* \varepsilon S$ which minimizes the function f, and is feasible, i.e., satisfies a set of conditions {C}.

Suppose the problem has a property which allows us to make a partition Π of a subset $S^{(i-1)}_{0\ell m\ldots p}$ of S.

$$\Pi = \{S^{(i)}_{0\ell m\ldots p1}, S^{(i)}_{0\ell m\ldots p2}, \ldots, S^{(i)}_{0\ell m\ldots pq}\}, q>1$$

where the subsets are defined recursively by

$$S^{(i)}_{0\ell m\ldots pk} \neq \phi \quad , \quad k = 1,2\ldots, q$$

$$\bigcup_{k=1}^{q} S^{(i)}_{0\ell m\ldots pk} = S^{(i-1)}_{0\ell m\ldots p}$$

$$S^{(i)}_{0\ell m\ldots pk} \cap S^{(i)}_{0\ell m\ldots pj} = \phi, \quad k, j = 1,2,\ldots q, k \neq i$$

with the initial condition

$$S^{(1)}_0 = S$$

A graphical representation of this set of relationships is shown in Figure 13. Here we build a search-tree. Each node bears the name of a subset $S^{(i)}_{0\ell m\ldots pk}$ of the solutions. It is convenient to call the superscript i the level. The set of i indices $0\ell m\ldots pk$ indicates a path from the node to which they belong to the root of the tree.

For example, in Figure 13 node $S^{(r)}_{011r}$ is at level 4, and the path to the root of the tree is

$$S^{(4)}_{011r}, S^{(3)}_{011}, S^{(2)}_{01}, S^{(1)}_0 .$$

In Figure 13 $S^{(1)}_0$, $S^{(2)}_{01}$ are closed, and $S^{(4)}_{011r}$ is pending. Since each subset corresponding to each node is partitioned into two or more non-empty subsets, the cardinality of the subsets is monotonically decreasing along a branch of the search tree, i.e.,

$$\left|S^{(4)}_{011r}\right| < \left|S^{(3)}_{011}\right| < \ldots < \left|S^{(1)}_0\right| = \left|S\right|$$

$|S|$ is finite, so we will eventually reach a level at which one of the pending nodes $S^{(j)}_{0\ell m\ldots pq}$ contains only one element of S, i.e., contains a solution to the problem.

The goal of the method is to get the minimal (or one of the minimal) solution σ^* by enumerating as few nodes as possible. To do so, at each node we must be able to compute, over the subset assigned to the node, an upper and lower bound for the function f.

The strategy consists of branching from the pending node having the least lower bound. In other words, one uses a property inherent in the nature of the problem to make a partition of the most promising pending node. For a terminal node, the upper and lower bounds collapse to the value of f for the solution assigned to this node. The search is ended when a node contains a feasible solution, the value of which is less than the smallest value of the lower bounds of the pending vertices.

The next section develops an example of the method for the type of problem to which it was first applied.

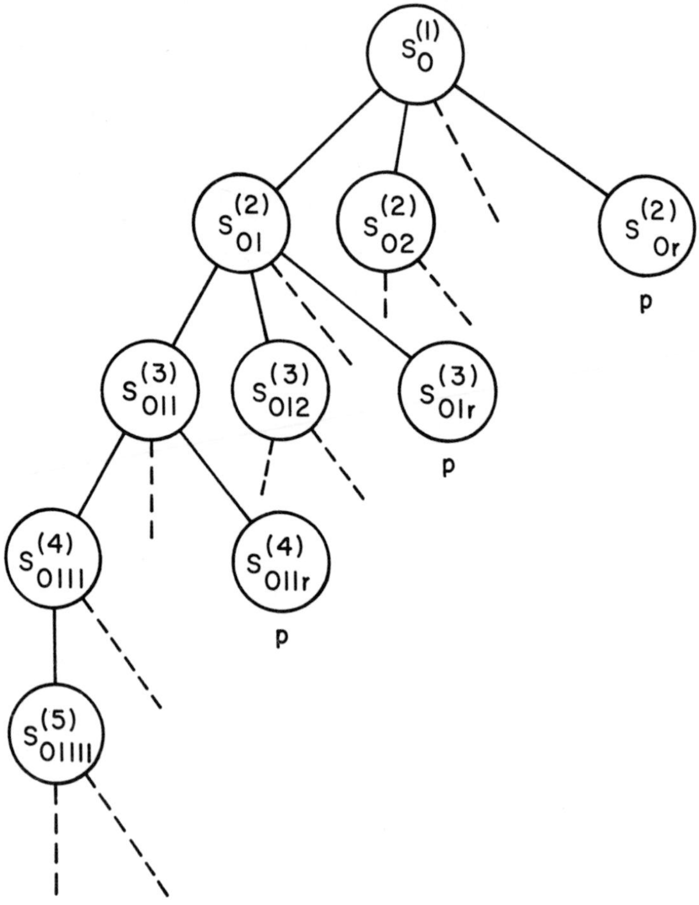

FIGURE 13. SEARCH TREE

The Traveling Salesman Problem

Several examples of branch and bound algorithms will be constructed for the traveling salesman problem to emphasize the fact that branch and bound is not one method, but a class of methods.

The statement of the problem is: "A traveling salesman, starting in one city, wishes to visit each of n-1 other cities once and only once and return to the start. In what order should he visit the cities to minimize the total distance traveled?"

A feasible solution to this problem is called a tour (Hamiltonian cycle). It is a cycle containing every vertex; a Hamiltonian circuit of the graph corresponding to the geographical map is as shown in Figure 14. The function f to be minimized might be the total distance or the total cost of traveling. For example, a mileage chart is shown in Table 10 for a symmetric, T.S. problem with n=5. In this case the set of feasible solutions contains all the 4! = 24 tours, the number of Hamiltonian circuits in the complete graph K5, (Berge, 1967, p. 8).

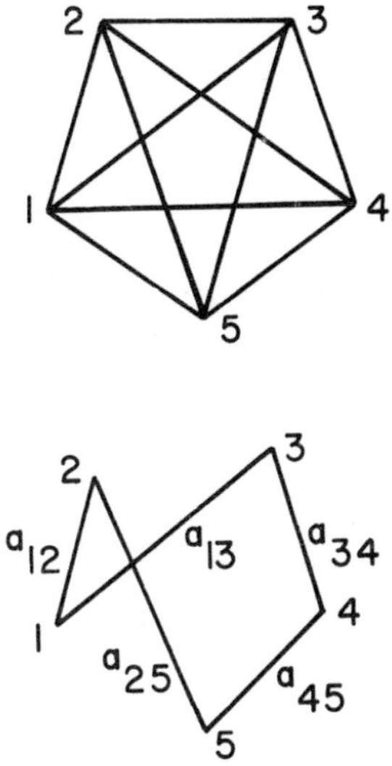

FIGURE 14. (a) GRAPH OF CONNECTIONS (b) A HAMILTONIAN CIRCUIT

Method I

To tour all the cities the salesman has to make n trips. Thus, we define any solution σ_j to be a set of n distinct edges. There are $\binom{10}{5}$ = 252 such sets for n=5. Let S be the set of solutions $\{\sigma_j\}$. We search for a solution which is a tour and minimizes the sum of the distances corresponding to its edges.

TABLE 10 (Distances Between the Cities)

	2	3	4	5
1	13	3	19	4
2		25	3	5
3			12	10
4				8

Algorithm

Step 1 - Initialization -

Pick any tour, call it $\sigma°$ and compute its total distance L°. This gives an initial upper bound for the problem.

The root of the search tree is assigned the set $S_0^{(1)} = S$ of all possible solutions. Set the current node to be $S_0^{(1)}$. All edges of Kn are free.

Step 2 - Computation of a New Lower Bound

Compute a lower bound ℓ for the current node by adding up the five smallest entries in the distance chart corresponding to imposed or free edges of Kn. Record the corresponding solution σ.

If α is not a tour, then, go to Step 3, otherwise go to 2.1.

Step 2.1

If ℓ is equal to L°, or ℓ is smaller than the smallest of the lower bounds corresponding to the pending nodes, stop; we have an optimal solution.

Step 2.2

If ℓ is smaller than L°, set L° to ℓ and $\sigma°$ to σ; go to Step 3.

Step 3 - Partitioning Criterion

Consider the pending node having the smallest lower bound ℓ. If the corresponding solution σ is a tour, stop; it is the optimal solution; otherwise go on.

Step 3.1

If σ contains a vertex of Kn of degree more than 2, set the current edge, e, to be one of the free edges incident to this vertex; go to Step 4.

Step 3.2

If σ does not contain a vertex of degree higher than two, but contains subtours, set e to be one edge of one of the subtours.

Step 4 - Partitioning

Partition $S_{0...k\ell}^{(i)}$ into two nodes, one $S_{0...k\ell 1}^{(i+1)}$ having σ as a solution and e as an imposed edge, the other $S_{0...k\ell 2}^{(i+1)}$ has e as a forbidden edge.

If in $S_{0...k\ell 1}^{(i+1)}$ the imposed edges form a subtour or are incident to a vertex of Kn of degree more than two, close $S_{0...k\ell 1}^{(i+1)}$, or equivalently set its lower bound to ∞, otherwise set the lower bound to ℓ. Close $S_{0...k\ell}^{(i)}$. Set the current node to be $S_{0...k\ell 2}^{(i)}$ and go to Step 2.

Summary of the Search for the Example with 5 Cities (Solution shown in Figure 20)

Iteration 1

Step 1

Take σ_o to be, for example, $(a_{12}, a_{23}, a_{34}, a_{45}, a_{51})$. Then L°=13+25+12+8+4=62.

$S = S_0^{(1)}$, all edges of K5 are free.

Step 2

Consider $S_0^{(1)}$.

$\sigma = (a_{13}, a_{24}, a_{25}, a_{15}, a_{45})$; $\ell = 23$ (Figure 15)

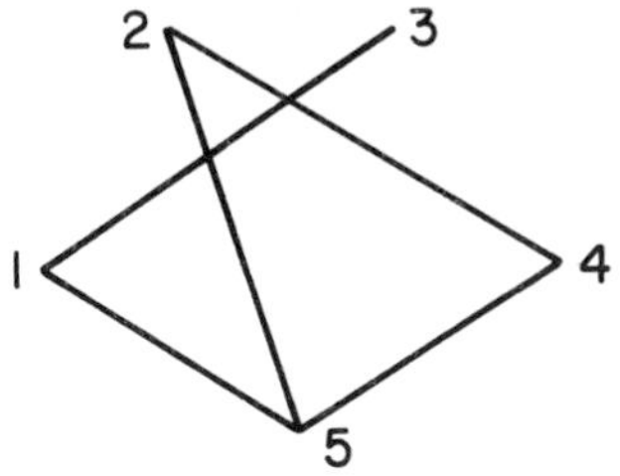

FIGURE 15. SOLUTION CORRESPONDING TO LOWER BOUND OF $S_0^{(1)}$

Step 3

Vertex 5 of the graph corresponding to σ (Figure 15) has degree 3; pick $e = a_{15}$ for example.

Step 4

Define $S_{01}^{(2)} = \{\sigma_j \varepsilon S_0^{(1)} \mid a_{15} \varepsilon \sigma_j\}$

$S_{02}^{(2)} = \{\sigma_j \varepsilon S_0^{(1)} \mid a_{15} \notin \sigma_j\}$

$S_0^{(1)}$ closed. Current node is $S_{02}^{(2)}$. The lower bound is 23.

Iteration 2

Step 2

Current node is $S_{02}^{(2)}$; a_{15} is forbidden.

$\sigma = (a_{13}, a_{24}, a_{25}, a_{35}, a_{45})$; $\ell = 29$ (Figure 16)

Step 3

Lowest lower bound on pending node is 23 for $S_{01}^{(2)}$.

Vertex 5 of the graph of the solution obtained in Step 2 of Iteration 1, Figure 15 has degree 3. Free edges incident to 5 are a_{25} and a_{45}, a_{15} being imposed. Pick $e=a_{25}$, for example.

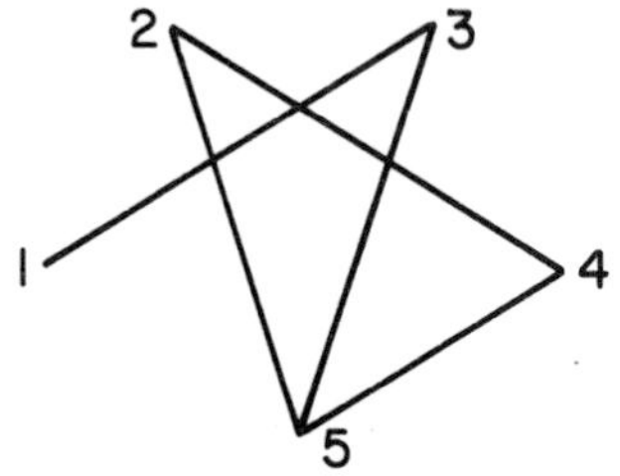

FIGURE 16. SOLUTION CORRESPONDING TO THE LOWER BOUND OF $S_{02}^{(2)}$

Step 4

Define $S_{011}^{(3)} = \{\sigma_j \varepsilon S_{01}^{(2)} \mid a_{25} \varepsilon \sigma_j\}$

$S_{012}^{(3)} = \{\sigma_j \varepsilon S_{01}^{(2)} \mid a_{25} \notin \sigma_j\}$

$S_{01}^{(2)}$ is closed. Current node is $S_{012}^{(3)}$. The lower bound is 23.

Iteration 3

Step 2

Current node is $S_{012}^{(3)}$. a_{15} is imposed, a_{25} is forbidden.

$\sigma = (a_{13}, a_{15}, a_{24}, a_{35}, a_{45})$; $\ell = 28$ (Figure 17)

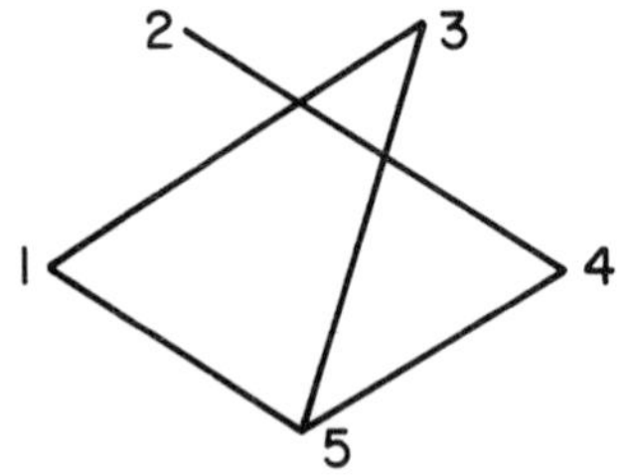

FIGURE 17. SOLUTION CORRESPONDING TO THE LOWER BOUND OF $S_{012}^{(3)}$

Step 3

Lowest lower bound is 23 on pending node $S_{011}^{(3)}$.

Vertex 5 of K5 of the solution obtained in Step 2 of Iteration 1 (Figure 15) has degree 3. The only free edge incident to 5 is a_{45}. Set $e = a_{45}$.

Step 4

Define $S^{(4)}_{0111} = \{\sigma_j \varepsilon S^{(3)}_{011} \mid a_{45} \varepsilon \sigma_j\}$

$S^{(4)}_{0112} = \{\sigma_j \varepsilon S^{(3)}_{011} \mid a_{45} \notin \sigma_j\}$

A lower bound for $S^{(4)}_{0111}$ is infinity since $a_{15}a_{25}a_{45}$ are imposed and cannot lead to a feasible solution. $S^{(4)}_{0111}$ is closed. $S^{(3)}_{011}$ is closed.

Current node is $S^{(4)}_{0112}$.

Iteration 4

Step 2

Current node is $S^{(4)}_{0112}$; a_{15}, a_{25} are imposed, a_{45} is forbidden.

$\sigma = (a_{13}, a_{15}, a_{24}, a_{25}, a_{35})$; $\ell = 25$ (Figure 18)

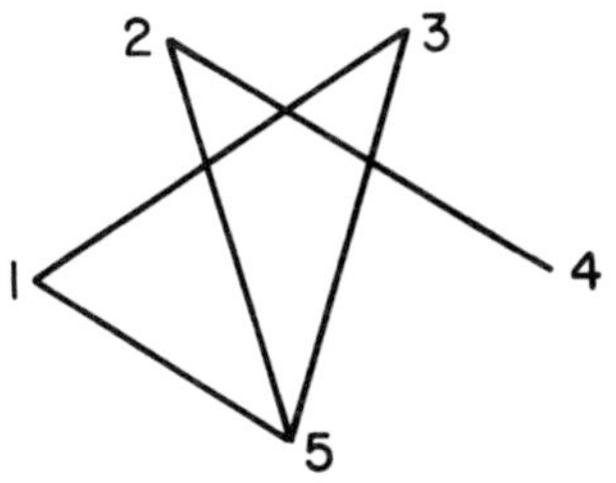

FIGURE 18. SOLUTION CORRESPONDING TO THE LOWER BOUND OF $S^{(4)}_{0112}$

Step 3

Lowest lower bound is 25 corresponding to $S^{(4)}_{0112}$.

Since vertex 5 is of degree 3, (Figure 18) and the only free edge incident to 5 is a_{35}, set $e = a_{35}$.

Step 4

Define $S^{(5)}_{01121} = \{\sigma_j \varepsilon S^{(4)}_{0112} \mid a_{35} \varepsilon \sigma_j\}$

$S^{(5)}_{01122} = \{\sigma_j \varepsilon S^{(4)}_{0112} \mid a_{35} \notin \sigma_j\}$

A lower bound for $S^{(5)}_{01121}$ is infinity, since a_{15}, a_{25}, a_{35} being imposed, cannot lead to a feasible solution.

Iteration 5

Step 2

Current node is $S^{(5)}_{01122}$; a_{15}, a_{25} are imposed a_{35}, a_{45} are forbidden.

$\sigma = (a_{13}, a_{15}, a_{24}, a_{25}, a_{34})$; $e = 27$ (Figure 19)

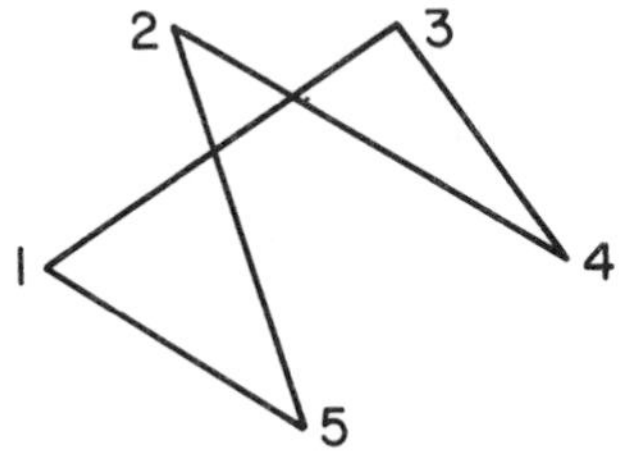

FIGURE 19. SOLUTION CORRESPONDING TO THE LOWER BOUND OF $S^{(5)}_{01122}$

σ is a tour; 27 is lower than the lower bounds 28 and 29, therefore, we have an optimal solution (Figure 20).

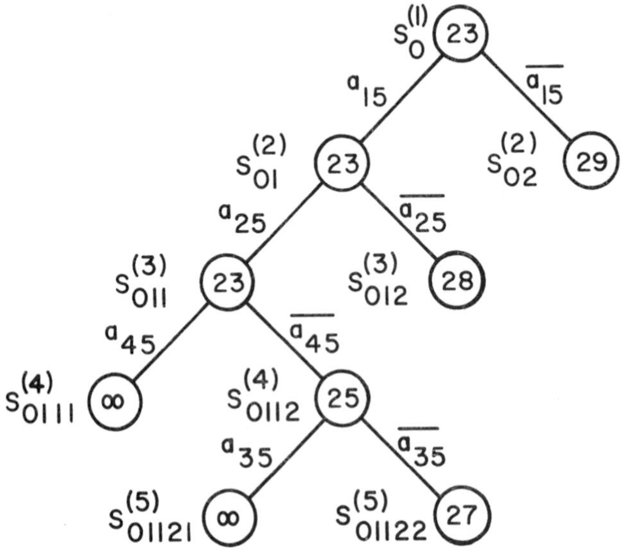

FIGURE 20. SEARCH TREE. (The encircled numbers are lower bounds. Newly imposed a_{ij} or forbidden $\overline{a}_{ij}$ edges are indicated)

Method 2

Another approach is to restrict the solution σ_i to partial tours, i.e. sets of edges of Kn which do not contain any vertex of degree more than two and any circuit of length less than n. Then one branches from each pending node by adding to the partial tour already obtained the shortest (or least expensive) compatible edge, i.e., an edge such that the resulting new set is a partial tour. To carry out this procedure we list the edges by increasing cost, as in Table 11.

TABLE 11

Ranking of the Edges

Rank	Edge	Cost
1	a_{13}	3
2	a_{24}	3
3	a_{15}	4
4	a_{25}	5
5	a_{45}	8
6	a_{35}	10
7	a_{34}	12
8	a_{12}	13
9	a_{14}	19
10	a_{23}	25

The root of the search tree is then partitioned into n(n-1)/2 subsets corresponding to each partial tour built by taking as the first edge one of the edges in the list, and then adding edges further down in the list. Thus, for our example (Figure 21) $S_{01}^{(2)}$ would contain solutions having a_{13} imposed, and edges below a_{13} in the list (Table 11). $S_{02}^{(2)}$ would contain solutions having a_{24} imposed and edges below a_{24} in the list, i.e. solutions of $S_{02}^{(2)}$ will not contain edge a_{13}, etc.

A lower bound can be computed for the solution class of node $S_{0p\ldots q}^{(k+1)}$ of the search tree by adding the costs of the first k imposed edges to the costs of the n-k edges following the last edge chosen from the list. For example, a lower bound for $S_{01}^{(2)}$ is (3) + (3+4+5+8) = 23; a lower bound for $S_{02}^{(2)}$ is (3) + (4+5+8+10) = 30, etc. If, at the bottom of the list, not enough edges remain to compute a lower bound, then the bound is set to infinity.

During the search we always branch from the smallest lower bound. At the beginning of the search an upper bound may be infinity, or else the cost of any tour. The upper bound is updated as in Method 1, when tours are found. The terminal criteria are the same as in Method 1.

The complete search tree is shown on Figure 21 for the example corresponding to the mileage chart of Table 10, (which gave rise to the list of Table 11).

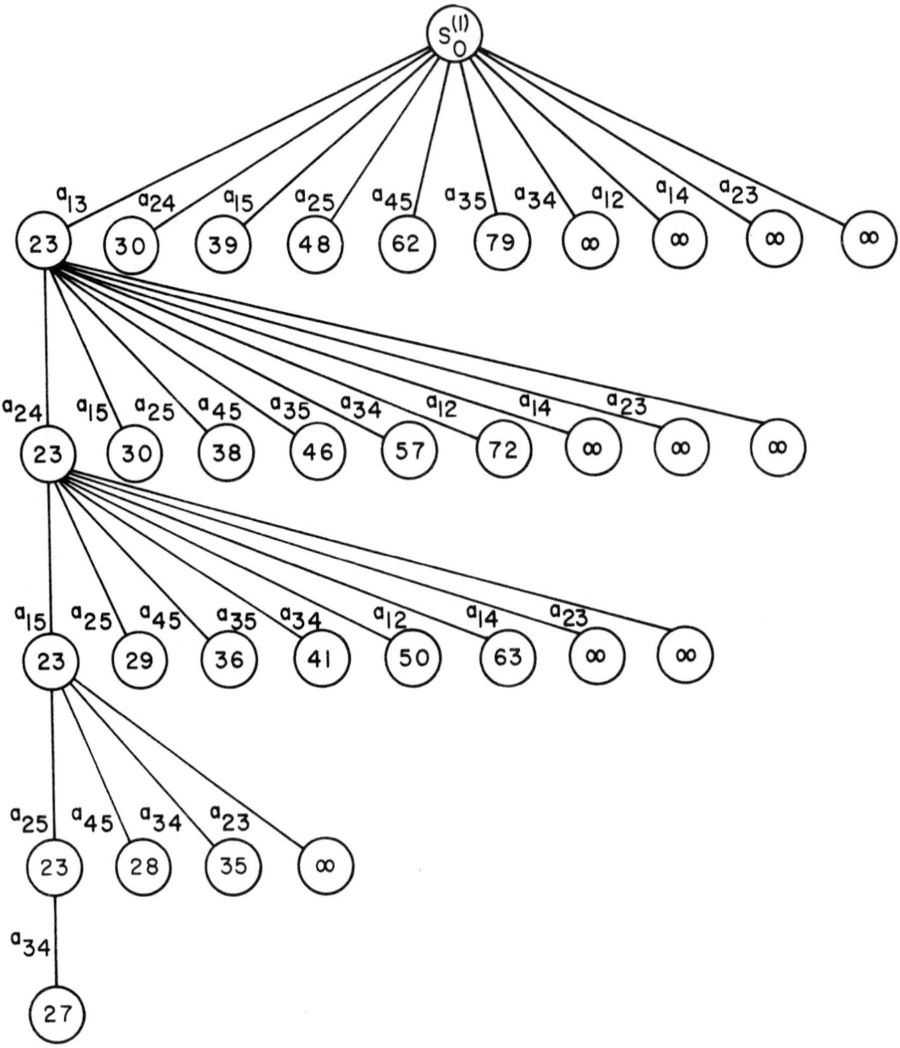

FIGURE 21. SEARCH TREE, METHOD 2

Notice that the number of nodes in the search tree can be prohibitively high if n is large. This can be reduced by partitioning each pending node into only two new nodes $S'^{(i)}_{0k\ldots m1}$ and $S'^{(i)}_{0k\ldots m2}$, with

$$\begin{cases} S'^{(i)}_{0k\ldots m1} = S^{(i)}_{q0k\ldots m1} \\ S'^{(i)}_{0k\ldots m2} = \bigcup_{p=2}^{q} S^{(i)}_{0k\ldots mq} \end{cases}$$

We assign to $S'^{(i)}_{0k\ldots m2}$ the lower bound corresponding to $S^{(i)}_{0k\ldots m2}$. The new search tree (Figure 22) is more compact and requires the computation of only two bounds at each level.

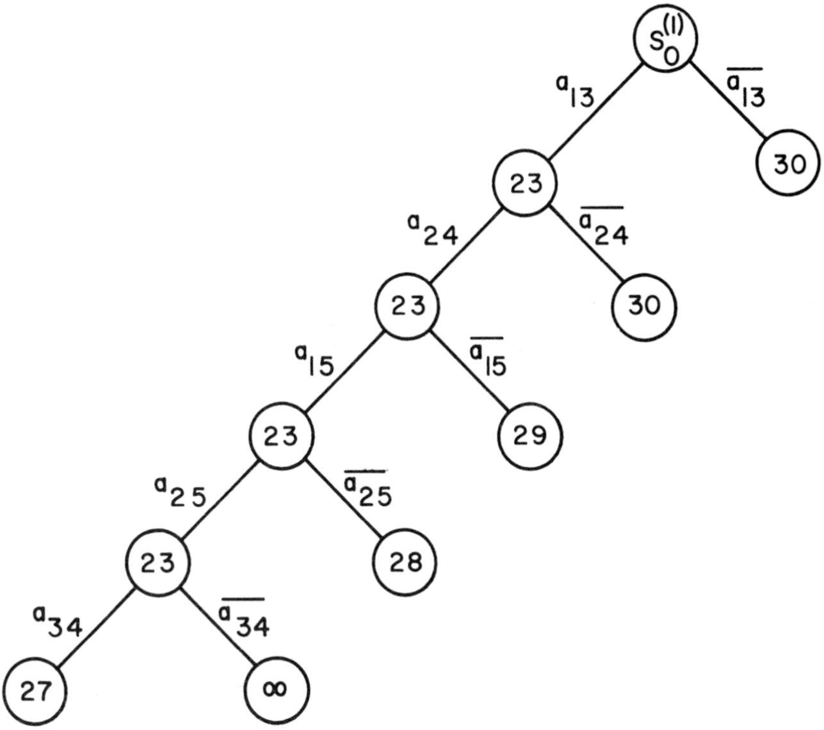

FIGURE 22. COMPACT SEARCH TREE, METHOD 2

Notice that the search tree in Figure 22 is very similar to the tree of Method one. The difference lies in that each solution in Method two has to be a partial tour.

Method 3

Method 3 which is a sophisticated version of Method 1, was proposed by Eastman (1958). Unlike in Method 1, where a solution σ_i may be any set of n edges of Kn, σ_i has to be a solution to the assignment problem (Hillier, 1967, p. 198) corresponding to the matrix of costs (or distances). σ_i does not contain any vertex of degree higher than 2, but may contain subtours, each edge of which will become a forbidden edge when branching. There are no imposed edges. The cost attached to the solution of the assignment problem yields a lower bound for each node. The corresponding search tree is shown in Figure 22. Since our example problem is symmetric, and the method was derived for unsymmetric cost matrices, one can remove the second half of the search tree ($S_{02}^{(2)}$ and followers) from further considerations. Although there are only four nodes in the search tree, this does not mean that Method 3 is best, since at each branching step, one has to solve the assignment problem. Method 1 and 3 go under the name of subtour elimination procedures.

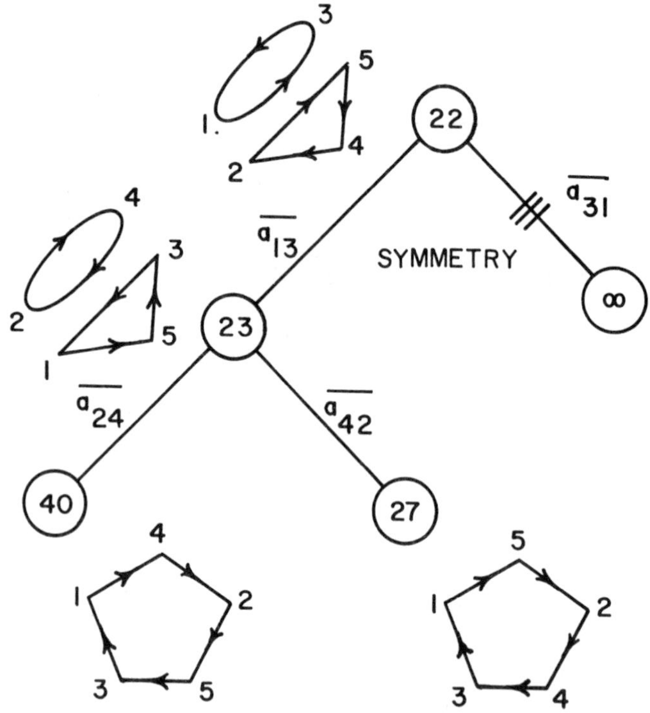

FIGURE 23. SEARCH TREE FOR METHOD 3

Method 4

Method 4, the approach of Little, et al. (1963) is essentially the same as Method 2. The computation of bounds is more complicated, but more efficient, and some heuristic rules determine the choice of the imposed edges. The search tree is shown in Figure 24. As in Method 3, one can close $S_{02}^{(3)}$ for symmetric problems. Method 2 and 4 belong to the class of "tour building" algorithms.

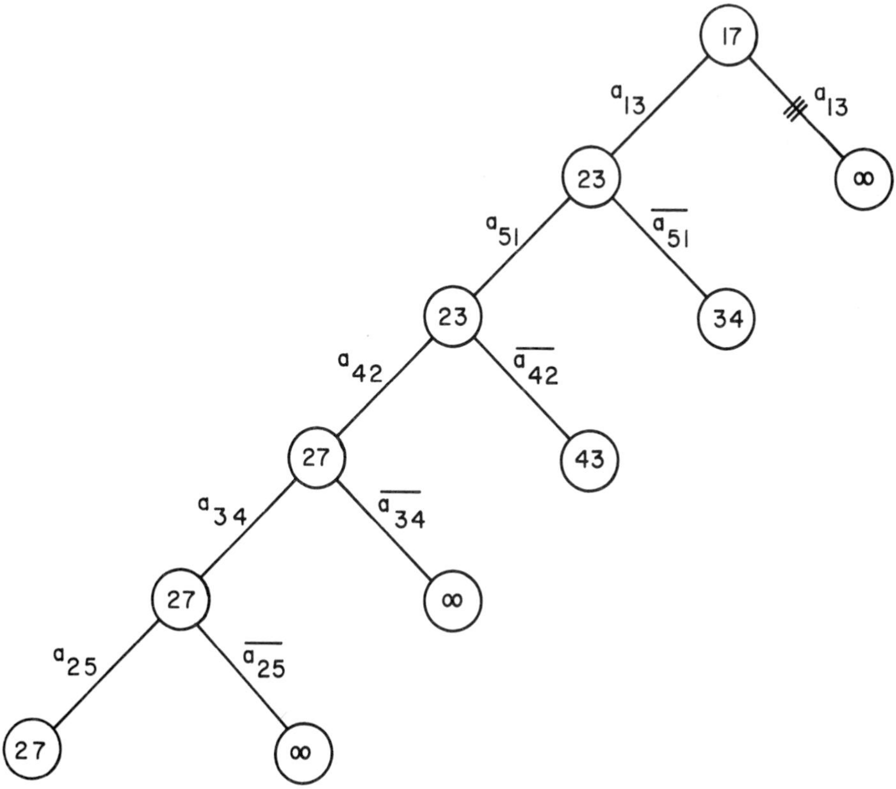

FIGURE 24. SEARCH TREE FOR METHOD 4

Generalization of the Branch and Bound Method

Branch and bound is a structural method containing the seeds of many possible generalizations. One of them is to base the choice of a node from which to branch not on the value of its lower bound, but rather on a special function, g, which provides a local evaluation of the situation. g has a wide use in artificial intelligence in game playing and theorem-proving programs. It is referred to as the scoring or evaluation function; a good example of this is the bidding contest problem formulated by Dr. Werner Burckhardt and solved by Jean-Paul Barthes (Henley and Williams, 1972).

Conclusion

The traveling salesman example illustrated the compromises between the length of the computations at each node (suboptimization) and the number of nodes of the search tree that have to be generated in order to obtain an optimal solution. It is also important to thoroughly understand the structure of a particular problem before designing a branch and bound scheme to solve it. If the structure is poorly understood, this may lead to a prohibitive amount of stored data. Some methods are becoming available for dealing with combinatorial explosive problems (McRoberts, 1971); namely, the use of decision analysis approaches.

Heuristic methods differ from branch and bound algorithms in the sense that one is not sure of having the optimal solution at the end of the search; only a "good" solution. Heuristic searches have been tried on mathematical models as well as on real-life problems. Sometimes this is the only way to obtain a result. Heuristic criteria are also often used to accelerate the branch and bound procedure. Again the structure of the problem must be relatively well understood before applying this kind of criteria.

In the last few years, many applications have been reported in the literature. These can be divided into two categories.

The first deals with theoretical combinatorial problems that have been around for some time but for which good solution methods are yet to be found. Among them are:

- the Traveling Salesman problem (Bellmore,1968);(Eastman, 1958);(Held, 1970); (Little, 1963).
- Integer Programming (Balinski, 1965); (Benayoun, 1970);(Bradley, 1971); (Geoffrion, 1969);(Hillier, 1969);(Land, 1960);(Shaftel, 1971).
- Machine Scheduling (Balas, 1969);(Charlton, 1969);(Florian, 1971)
- Pseudo-Boolean Programming (Balas, 1965);(Balas, 1967); (Hammer, 1969).
- Set Partitioning (Garfinkel, 1969).
- The Knapsack Problem (Greenberg, 1970).
- Non-convex (Florian, 1970); or non-linear Problems (Lawler, 1963).

The second category deals with more specific cases, where branch and bound techniques were applied successfully

- Synthesis of Integrated Process Design (Lee, 1970).
- Computer Aided Synthesis of Chemical Plants (Stewart, 1965).
- Ordering of Recycle Calculations (Lee, 1966).
- Modular Design (Shaftel, 1971).
- Generation of NAND Structures (Davidson, 1969).

REFERENCES

Ashour, S., (1970), "Computational Experience on 0-1 Programming Approach to Various Combinatorial Problems", Operations Research, Japan, 13, 78.

Balas, E., (1965), "An Additive Algorithm for Solving Linear Programs with 0-1 Variables", Operations Research, 13, 517.

———, (1967), "Discrete Programming by the Filter Method", Operations Research, 15, 915.

———, (1969), "Machine Sequencing Via Disjunctive Graphs: An Implicit Enumeration Algorithm", Operations Research, 17, 941.

Balinski, M. L., (1965), "Integer Programming, Methods, Uses, Computations", Management Science, 12, 253.

Bellmore, M., and G. L. Nemhauser, (1968), "The Traveling Salesman, A Review", Operations Research, 16, 538.

Benayoun, R., B. Roy, and J. Tergny, (1970), "De La Procedure S.E.P. au Programme Ophélie Mixte", Metra 9, 141.

Berge, C., (1967), "Théorie des Graphes et ses Applications", Dunod, Paris.
Bradley, G. H., and P. N. Washi, (1971), "An Algorithm for Integer Linear Programming, A Combined Algebraic and Enumeration Approach", Report #29, Administrative Sciences, Yale University, New Haven, Connecticut.
Charlton, J. M., and C. C. Death, (1969), "A Method of Solution for General Machine Scheduling Problems", Operations Research, 18, 689.
Davidson, E. S., (1969), "An Algorithm for NAND Decomposition Under Network Constraints", IEEE Trans. on Computers, 12, 1098.
Eastman, W. L., (1958), "A Solution to the Traveling Salesman Problem", Am. Summer Meeting of the Econometric Society.
Florian, M., and P. Robillard, (1970), "An Implicit Enumeration for the Concave Cost Network Flow Problem", Publication #12, Département d'Informatique, Montreal, Canada.
Florian, M., P. Trepant, and G. McMahon, (1971), "An Implicit Enumeration Algorithm for the Machine Sequencing Problem", Management Science, 17, B782.
Freeman, R. J., (1965), "Computational Experience with Balas Integer Programming Algorithm:, Rand Corporation Report, P-3241, Santa Monica, Calif.
Garfinkel, R. S. and G. L. Nemhauser, (1969), "The Set Partitioning Problem: Set Covering with Equality Constraints", Operations Research, 17, 848.
Geoffrion, A. M., (1969), "An Improved Implicit Enumeration Approach for Integer Programming", Operations Research, 17, 437.
Greenberg, H. and R. Hegerich,(1970), "A Branch Search Algorithm for the Knapsack Problem", Management Science, 16, 327.
Gutterman, M. M., (1969), "Efficient Implementation of a B and B Algorithm", Standard Oil Company, Whiting, Indiana, Operations Research Division.
Hammer, P. L., (1969), "A B-B-B Method for Linear and Nonlinear Bivalent Programming", Report #48, Operations Research, Statistics and Economics, Technion, Haifa, Israel.
Held, M. and R. M. Karp, (1970), "The Traveling Salesman and Minimum Spanning Tree", Operations Research, 18, 1138.
Henley, E. J., and R. F. Williams, (in press), "Graph Theory in Computer Aided Design, Control, and Optimization", Academic Press, N.Y., N.Y.
Hillier, F. S., (1969), "Efficient Heuristic Procedure for Integer Linear Programming with an Interior", Operations Research, 17, 600.
Hillier, F. S. and J. Lieberman, (1967), "Introduction to Operations Research", Holden-Day, Inc., New York, New York.
Kaufmann, A., (1968), "Introduction à la Combinatorique", Dunod, Paris.
Land, A. H. and A. Doig, (1960), "Automatic Method for Solving Discrete Programming Problems", Econometrica 28, 497.
Lawler, E. L., (1963), "The Quadratic Assignment Problem", Management Science, 9, 586.
Lawler, E. L. and D. E. Wood, (1966), "Branch and Bound Methods: A Survey", Operations Research, 14, 699.
Lee, K. F., A. H. Masso and D. F. Rudd, (1970), "Branch and Bound Synthesis of Integrated Process Designs", Ind. and Eng. Chem. Fundamentals, 9, 1.
Lee, W., J. Christensen, and D. Rudd, (1966), "Design Variable Selection to Simplify Process Calculations", Am. Inst. of Chem. Eng. J., 12, No. 6, 1104.
Little, J. D. C., K. G. Murty, D. W. Sweeney, and G. Karel, (1963), "An Algorithm for the Traveling Salesman Problem", Operations Research, 11, 972.
Mason, T. and C. L. Moodie, (1971), "A Branch and Bound Algorithm for Minimizing Cost in Project Scheduling", Management Science, 18, No. 4, B158.
McRoberts, K. L., (1971), "A Search Model for Evaluating Combinatorially Explosive Problems", Operations Research , 19, 1331.
Nilsson, N. J., (1971), "Problem Solving in Artificial Intelligence", McGraw-Hill Book Company, New York, New York.
Shaftel, T., (1971), "An Integer Approach to Modular Design", Operations Research, 19, 130.
Siirola, J. S. and D. F. Rudd, (1971), "Computer Aided Synthesis of Chemical Process Designs", Chemical Engineering Department, University of Wisconsin, Madison, Wisconsin.

Steward, D. V., (1965), "Partitioning and Tearing Systems of Equations", SIAM, Numerical Analysis, 2, 345.

DECOMPOSITION OF PROVISIONS OF DESIGN SPECIFICATIONS

STEVEN J. FENVES
Department of Civil Engineering
Carnegie-Mellon University
Pittsburgh, Pennsylvania 15213, USA

Abstract: Building codes and design specifications are written so as to be interpreted by experienced design engineers. Their conversion to computer programs, as well as their consistent textual representation, first requires a careful interpretation and decomposition of the pertinent provisions, and the creation of a network of interrelations and references. The development of such a network will be described, and it will be shown how this network can be used to generate both efficient processing strategies as well as more consistent textual organizations.

1. Background

In order to relate this paper to the topic of the Institute, a short historical sketch is in order. The author's professional activity has dealt primarily with computer-aided design in Civil Engineering, with heavy emphasis on structural analysis, a topic intimately related to the solution of large-scale systems of linear equations and linear programming tableaus, as well as to network and graph theory (Fenves and Branin (1963), Fenves (1966), Fenves and Gonzalez-Caro (1971)).

A few years ago, the author was approached by the American Institute of Steel Construction, a professional body responsible for the major US specifications dealing with steel construction, to develop a series of computer programs based on the new 1969 version of the Specification (AISC (1969)). Upon reflection and study, it was decided not to produce specific programs, but to develop a rigorous basis of representation of the provisions of the Specification, so as to minimize the subsequent work of programming. This study, and its later extensions, have revealed a completely new outlook on the nature and format of specifications, with far-reaching implications for both computer processing and textual organization.

The purpose of this paper is to summarize the results of the above studies. Although no formal decomposition methods have been applied in this study, an attempt will be made to indicate how these techniques could (or should have been) applied.

2. Purpose and Structure of Design Specifications

2.1 Purpose

The detailed design of many engineered facilities and systems is governed by specifications or codes. Initially conceived as compendia or guides of good practice, many of the specifications have acquired a legal character, through incorporation, verbatim or by reference, into the appropriate building or safety codes or laws.

The organization, format and contents of most specifications reflect their original intent, and are based on the implicit assumption that the specifications will be used by experienced engineers as a guide or "aide-memoire" in the design process. For this type of use, absolute precision in wording has not been necessary, as the engineer could be expected to supplement the information contained in the specifications from the background of his experience and judgement.

Aside from the potential legal problems arising from this paucity of precision, two major trends are causing a fundamental review of the organization and format of specifications. First, an increasingly large portion of the design operations covered by specifications is being performed by digital computer programs. Obviously, such a program cannot make any inferential judgements or follow implicit crossreferences. Thus, the programming of specification provisions requires a great deal of interpretation of the text proper. Unfortunately, such interpretation is repeated over and over by every design office, often by junior engineers whose prime qualification for the job is their knowledge of computer programming, and not experience in design. Yet the programs based on this interpretation are likely to process many more designs than an experienced engineer can accomplish in a lifetime.

A second factor has been the desire by most code-writing bodies to update the specifications at much more rapid intervals than in the past, in order to take advantage of new methods or products. Revisions of this type are usually inserted into the old text at some point judged most appropriate, without regard to their interaction with other provisions, old or new. As a consequence, even experienced engineers become reluctant to accept the revisions, which may ask them to forego familiar procedures for new ones, with little or no guidance. Thus, improvements and rationalizations embodied in the revisions are often negated by the increased difficulty of relating new provisions to old ones.

2.2 Structure

Basically, a design specification consists of a collection of design criteria to be satisfied. A criterion may be defined as "a functional relationship intended to provide an adequate margin of safety with respect to a particular mode of failure (Wright, Boyer and Melin (1971))." The terms "safety" and "mode of failure" are, of course, defined in different terms by different design professions: in building design, modes of failure range from collapse due to overloads to undesirable vibrations, each with its own margin of safety; in piping and pressure vessel design, modes of failure again range from explosion due to overpressure, through excessive thermal displacements, to wear and tear resulting in leakage, etc.

One class of specifications, called the performance type, limits itself to listing the pertinent criteria, leaving to the designer the task of determining the satisfactory behavior for each criterion. Performance-type specifications are gaining increased acceptance in several fields, because they encourage innovation in materials, processes and design techniques. Conversely, they make the official and legal acceptance process considerably more difficult.

The much larger class of specifications is of the prescription type, where the specification contains a hierarchial sequence of provisions, limits, computations, checks, etc. leading to the evaluation of each criterion. The discussion that follows deals with this class of specifications.

For both the textual organization and computer implementation, it is convenient to decompose the specification into three hierarchial levels:

a) the top level, dealing with the basic classification of subject matter, and acting as an index or directory to the pertinent criteria that must be checked and satisfied in each given case;

b) the intermediate level, dealing with the sequence of provisions leading to the evaluation of the criteria, i.e. the prescription part of each specific criterion; and

c) the detailed level, concerning the specific provisions for evaluating a particular data item, or datum.

The detailed level pertains essentially to individual paragraphs of the specification. It has been shown in previous papers (Goel and Fenves (1971), Goel, Fenves and Gaylord (1971)) that an ideal representation of the provisions at this level is in the form of decision tables (Montalbano (1962), McDaniel (1968)). Concepts applicable at this level will only be briefly mentioned.

3. Logic of the Specifications

The objective of this section is to develop an abstract representation of the logic of the specification, independent of both textual representation and computer program implementation. Such an abstract representation is afforded by a graph or network. The succeeding subsections deal with the development of the salient features of this network. The examples used are taken from the AISC Specification for steel buildings, but the reader should have no difficulty in substituting terms from other specifications.

3.1. Components

The specification deals with a large number of data. An individual item of data, or datum, may be one of the following:

a) the status of a design criterion, which has possible values of "satisfied" or "violated;" such as "compression member satisfactory";

b) a property which has numerical value, such as the allowable axial stress F_a;

c) a quantity which has as its value one of several alphabetic names, such as "section shape" which may be "rod," "angle," "channel," wide flange," etc.; or

d) a logical variable which has possible values "true" or "false," such as the variable "sidesway prevented".

There are two sets of relationships between the data items. The functional relationship deals with the source of data, i.e., the manner in which a value is assigned to the data item in question. A data source may be:

a) given by the design data, tables, or other source (e.g., the area of a standard section or the designer's knowledge whether sidesway is prevented or not);

b) generated by an explicit function, implemented in a program as a subroutine, such as $C_c = \sqrt{2\Pi^2 E/F_y}$; or

c) derived following a logical procedure which may be described by a decision table. For example, the criterion "compression member satisfactory" is given by the decision table T1 below, whereas the value of the allowable stress F_a is generated by the table T2.

T1. Compression Member Check

Section 1.9. satisfied	Y	E
$K\ell/r \leq 200$	Y	
$f_a/F_a \leq 1.0$	Y	
Criterion satisfied	Y	
Criterion violated		Y

T2. Allowable Compressive Stress

Main member?	Y	Y	N	N	N	N
$K\ell/r \leq C_c$?	Y	N	Y	Y	N	N
$\ell/r < 120$?	I	I	Y	N	Y	N
F_a = Formula 1.5-1	Y		Y			
F_a = Formula 1.5-2		Y			Y	
$F_a = \dfrac{\text{Formula 1.5-1}}{(1.6 - \ell/200r)}$				Y		
$F_a = \dfrac{\text{Formula 1.5-2}}{(1.6 - \ell/200r)}$						Y

These three classes of data sources appear adequate for representing specifications proper. A fourth class, that of data assigned by a subjective expert opinion, may be included in a more general logical network of design.

From the standpoint of the logic of a specification, a more important relation between items of data is their precedence in evaluation, irrespective of the algorithms or procedures of evaluation themselves. These logical relations between items of data may be concisely expressed by two related lists, the ingredience and dependence lists for each datum. The generation of each datum may be characterized as a transformation T_i, of the form

$$T_i : y_i = f_i \; I(y_i)$$

$$\text{where} \qquad I(y_i) = \{ x_{i1}, x_{i2}, \ldots x_{ik}, \ldots x_{in(i)} \}. \tag{1}$$

In Eq. (1) y_i is an output quantity or derived datum and the x_{ik} are input quantities. The collection of $I(y_i)$ of x_{ik} will be called the ingredience list of the output variable y_i.

The transformations given as examples above can be represented in the form of Eq. (1) as:

$$I \text{ (Compression criterion)} = \{\text{"Section 1.9 satisfied"}, K, \ell, r, f_a, F_a\} \tag{2}$$

$$I\,(F_a) = \{\text{"member type"}, K, \ell, r, C_c, ..\} \tag{3}$$

$$\text{and} \qquad I(C_c) = \{E, F_y\} \tag{4}$$

The interconnection of transformations arises from the fact that the data items used by one transformation may themselves be output variables of another transformation, or in general

$$x_{ik} = y_j \tag{5}$$

where y_j is the output of some other transformation T_j, as for example, F_a in Eqs. (2)-(3) and C_c in Eqs. (3)-(4).

Conversely, an output such as F_a of Eq. (3) becomes the input to another transformation, in this case to Eq. (2). In general, each output y_j may appear as the input or ingredient x_{ik} of several transformations, and thus have several dependents (in the AISC Specification, the yield stress F_y has 39 separate dependents). All dependents of a data element x_i can be collected into a dependence list:

$$D(x_i) = \{y_{i1}, y_{i2}, \ldots, y_{im(i)}\} \tag{6}$$

The use of dependence lists is discussed further in Section 4.1.

3.2. The network and its properties

Using the above definitions, the abstract logical network can be readily created by:

a) assigning a node (point) to each data item, regardless of source; and

b) drawing an oriented branch (arrow) from each datum to each of its dependents.

An abstract network corresponding to the criterion "Compression Member Check", Section 1.5.1.3 of the AISC Specification, is shown in Figure 1. It is to be noted that such a network can be automatically generated from the local ingredience lists of each datum. The complete global network of the AISC Specification is too large to be presented here; it consists of over 450 data items, of which over 120 are derived quantities generated by as many separate decision tables.

The global logical network permits one to trace out two sets of relations or influences:

a) the global dependence of a datum, consisting of all data items depending on it, is obtained by traversing the network from the given datum in the direction of the arrows until all data items which may be reached in this fashion are enumerated; and

b) the global ingredience of a datum, consisting of all data items influencing it, is obtained by traversing the network from the datum in the direction opposite to that of the arrows until all data items which may be reached are enumerated.

It is obvious that input items are identified in the global network as those which have no ingredients (all arrows point out); similarly, terminal or output items are those which have no dependents (all arrows point in). More generally, each datum in the network can be assigned two numbers:

a) the global level from input, that is, the number of branches on the longest path from any input item to the datum in question; and

b) the global level from output, that is, the number of branches on the shortest path from the datum in question to any terminal or output item.

These definitions explicitly refer to the longest and shortest paths, respectively, because a given datum may have several ingredients and dependents and the path lengths for each may be different.

In summary, the functional network derived in this section, defined by the data items and their lists of ingredience and dependence, is independent of the textual organization of the specification. The functional network displays all transformations from the input data to the outputs, the latter being defined as the criteria discussed in Section 3.1.

3.3. Organizational network

A designer, or any other user of a specification, must accomplish two things:

a) determine which criterion or criteria are applicable to the problem at hand; and

b) for each of the applicable criteria, perform the necessary steps to evaluate the criterion.

The functional network defined in the previous section is insufficient for the above two purposes, as it contains only the list of design criteria, without any relation between them.

The organization of the specification must therefore provide a grouping and ordering to the individual criteria so as to provide a logical and consistent entry to the various criteria which must be evaluated according to the Specification.

From an abstract standpoint, the grouping and ordering of the criteria corresponding to the specification may be represented as a second, or organizational network, which starts from a single entry node and proceeds through a series of branch points, corresponding to the various major categories of the specification, to the appropriate criteria which must be evaluated.

4. Application of Logic Networks

It is the thesis of this paper that the logical structure of a specification can be represented in an abstract form by means of the two interrelated networks (functional and organizational) discussed in the previous section.

The application of this representation to two fundamentally different problems will now be discussed.

4.1. Computer implementation

Let us assume that we have available a series of subroutines corresponding to all of the transformations T_i. We seek a sequencing of the transformations such that any given problem can be efficiently executed. By execution we mean to proceed from a given set of input data, containing both mandatory input data and optional data overriding derivable values, and execute all the necessary transformations encountered. It can be shown that a specific problem solution involved the traversing of a particular tree in the logic graph.

One possible sequencing can be obtained on the basis of partial ordering, i.e., using relations (5) to set:

$$j \prec i \text{ if } x_{ik} = y_j \text{ for any } i \qquad (7)$$

Thus, any transformation j which provides an ingredient for T_i must be executed before executing T_i. This sequence corresponds to a traversal of the graph according to increased levels from input, following the longest path. This sequence is essentially equivalent to the longest-path algorithm, as well as to endorder processing of trees (Knuth (1968)).

There are three reasons why such a sequencing scheme is undesirable:

a) it does not readily accommodate the input of optional data, which eliminate the need for executing certain lower-level transformations;

b) because of the presence of immaterial condition entries in transformations represented by decision tables, certain conditions may not have to be tested at all in order to locate the governing rule. The pre-computation of such conditions is definitely wasteful; and

c) it is possible that a condition which turns out to be immaterial according to b) is undefined on the basis of the data supplied, even though the conditions actually used are defined. It is thus extremely wasteful to require as input all possible data for all possible conditions. Under many circumstances, this is not even possible.

It follows from the preceding argument that for efficient execution, the sequencing of the tables must be determined at execution time on the basis of information at hand. Such a sequencing can be implemented by means of a recursive procedure, made possible by associating with each data item, y_i, a boolean flag or <u>status indicator</u>, s_i, which has possible values of <u>valid</u> or <u>void</u>. The input of every data element, as well as every evaluation (execution of T_i) causes the corresponding status indicator to be set to <u>valid</u>.

The recursive process of evaluating or SEEKing a needed data item y_i can be conveniently represented by the decision table shown below.

T3. SEEK for data items y_i

	(1)	(2)	(3)	(4)
y_i present (s_i = <u>valid</u>)?	Y	N	N	N
input parameter (I_i empty)?	I	Y	N	N
all ingredients present (all s_{ik} = <u>valid</u>)?	I	I	Y	N
Evaluate T_i, store y_i, mark s_i = <u>valid</u>			Y	
SEEK missing ingredient x_{ik}				Y
Error		Y		
Exit	Y	Y	Y	

Whenever rule (4) governs, processing of the current data item y_i is suspended, and the process reinitiated for its first missing ingredient, x_{ik}. Eventually (barring the case of missing input parameters), all ingredients x_{ik} of y_i are evaluated, so that rule (3) governs. Then y_i itself can be evaluated and its flag s_i set to <u>valid</u>.

It should be noted that the above process permits the designer to input any computable or derivable quantity directly; in such cases, rule (1) will govern and the evaluation of T_i will simply be bypassed.

The SEEK process described can be visualized as a top-to-bottom traversal of the logic graph, recursively descending to lower levels of hierarchy only when needed, and thus tracing out the shortest-path subtree from the node in question through its global ingredience. The process is thus equivalent to the shortest path algorithm and to preorder processing of trees.

In actual computer implementation, the two strategies are combined: the organizational graph is traversed in direct, or endorder, sequence until the applicable criteria are identified, and then the y_i pertaining to these criteria are recursively evaluated (Goel and Fenves (1968)).

In iterative, trial-and-error design, the case frequently arises where a data item is modified after its value has been used in evaluating other, higher-level data items which depend on it. Alternatively, one may wish to repeat chains of calculations with only a few input parameters modified. In either case, derived quantities based on the previous values are no longer applicable.

There are several strategies that may be invoked in such cases:

a) erase all derived quantities (i.e., set their status to void); this would be inefficient, as some of these quantities may be unaffected by the change;

b) recalculate automatically all derived quantities; this may be even more inefficient, as several cumulative changes may force repeated recalculations; or

c) selectively erase only those quantities which are directly affected, leaving all unaffected data unchanged.

The third strategy can be implemented by a second recursive process, called WARN, which uses the dependence lists defined by Eq. (6). The process can be represented by the decision table shown below.

T4. WARN for data item x_i

terminal item (D_i empty)?	Y	N	N
all dependents erased (all s_{ik} = void)?	I	Y	N
Mark s_i = void	Y	Y	
WARN dependent y_{ik}			Y
Exit	Y	Y	

WARN can thus be visualized as a bottom-to-top traversal of the logic graph, following the longest path until either the terminal node is reached or all dependent nodes are already set to void.

It is important to emphasize that the network represents only logical ingrediences and dependences, and that it does not have information on the numerical sensitivity of any dependent with respect to a change in an ingredient. It is for this reason that WARN simply sets the dependents to void; if and when these need to be re-evaluated, SEEK can be used to "backtrack" to the changes and re-establish new values for the dependents.

4.2. Textual organization

The network representation of the specifications provides an ideal basis for investigating alternate organizations and formats of the textual presentation of the specification.

The treatment of the top-level organization is too specific to be described here. Basically, it is assumed that all criteria comprising a specification can be labelled according to several independent bases. (In the AISC Specification, these bases are: component type, e.g. beams, columns, connections, fasteners, etc.; stress state, e.g. tension, compression, shear, etc.; and limit states or modes of failure, e.g. yielding, buckling, excessive deflection, rupture, etc.). Once the criteria are thus labelled, it is a simple matter to generate with a computer program outlines according to any of the possible permutations of independent bases.

At the intermediate level, dealing with the cascades of definition, provisions, formulae, etc. pertaining to a given criterion, the two execution strategies discussed in the previous section provide the basis for two alternative textual organization.

The textual format produced by the "bottom-to-top" endorder traversal of the longest path in the graph implies that all lower-level definitions or provisions are listed before any higher-level provisions referring to them. For example, the textual format resulting from endorder processing of the graph shown on Fig. 1 would appear as follows (indentations are shown to reflect the hierarchial order):

Level: 1 2 3 4 5 6 7 8

- Compute effective width
  - Compute effective area
  - Compute actual area
    - Compute Q_a
  - Check geometrical constraints
    - Compute Q_s
      - Compute C_c
      - Compute K
        - Compute allowable stress F_a
          - Compute modified stress F_{as}
        - Compute gross area A_g
          - Compute allowable stress f_a
            - Compute $R_a = f_a/F_{as}$
        - Check unstiffened elements
        - Check stiffened elements
            - Check section 1.9
              - Check compression member

It is to be noted that this format is consistent, in the sense that cross-references always "point to" previously defined data items or provisions; furthermore all such definitions will have been given at a lower or equal hierarchial level. This format, however, has the same weakness from a textual standpoint that the corresponding algorithmic sequence had from an execution standpoint, namely, that the most complex case must be covered first, before shortcuts (corresponding to "immaterial" entries in the decision tables) can be identified.

By contrast, the textual format corresponding to "top-to-bottom" preorder traversing would start with the highest-level definitions, namely the criteria to be satisfied, and then proceed to list intermediate and lowest-level provisions, according to the shortest reverse (dependent) paths in the graph. The textual format of presentation of the previous example would be:

```
Level:   1   2   3   4   5   6   7

         Check compression member
             Compute K
             Check Section 1.9
                 Check stiffened elements
                 Check unstiffened elements
                     Check geometrical constraints
             Compute R_a
                 Compute f_a
                     Compute gross area A_g
                 Compute modified stress
                     Compute allowable stress F_a
                         Compute Q_s
                         Compute Q_a
                             Compute actual area
                             Compute effective area
                                 Compute effective width
                         Compute C_c
```

This layout is to be read as follows:

In order to check a compression member, you must compute K, check Section 1.9, and compute R_a.

In order to check Section 1.9, you must check all unstiffened and all stiffened elements.

Etc.

By comparing the above outline to Fig. 1, it can be seen that cross references still "point to" previously defined items, but the latter now occur at higher levels, i.e. closer to the output datum. The primary advantages of this format are again precisely those of the corresponding algorithmic sequence:

a) a user need to read only as "deeply" as necessary (for example, all standard rolled beams in the US automatically satisfy Section 1.9 of the AISC Specification; therefore a user checking a standard beam need not even read the detailed provisions of that Section); and

b) since the format proceeds from the general to the specific, the simpler, usual cases will automatically be listed first, followed by the more complex unusual cases.

It is somewhat questionable whether systematic aids should also be developed for the detailed level of the specification text, dealing with single provisions or paragraphs. If desired, the parsing techniques associated with decision table processing could easily be applied to produce consistent textual formats of provisions represented by decision tables.

5. Summary and Conclusions

The extensive use of computers in many branches of engineering has introduced jarring dichotomies at three levels. First, whereas the engineers' ability to analyze mathematical models of systems has increased by many orders of magnitude, their ability to design has hardly improved at all, due to the constraining influence of the governing codes and specifications. Second, improvements in products and methodologies, often resulting from computer-aided research and analysis, cannot be incorporated conveniently into specifications designed and formatted for an earlier era. Finally, the high cost of investment in programs based on current specifications is becoming an actual deterrent to the improvement and updating of the specification.

In this paper, a conceptual methodology which has the potential of alleviating all three of these problems is developed and illustrated. The methodology is based on the concept of an abstract logical network, which formally represents the intent of the specification and the interconnections between its provisions. It is conceivable that future changes and reorganizations of specifications will be processed as follows:

a) the provisions, concepts, etc., will be submitted to the appropriate code-writing body for discussion and conceptual approval;

b) the approved provisions and concepts will be converted to decision tables, to screen out inconsistencies, omissions, etc.;

c) the individual decision tables will be combined into a logic network, to evaluate their overall implications and connections;

d) the logic network will be reorganized so as to suit the organizational scheme chosen by the body.

e1) the detailed text will be developed.

e2) computer programs will be generated directly from the logic network and the constituent decision tables.

No formal decomposition techniques have been used in the study reported herein. It is highly questionable whether such techniques could in fact be used at the early levels, i.e. up to step b) above, due to the large amount of interpretation required at those levels. On the other hand, it appears obvious that at the combinatorial or network level, i.e. steps c) and d), such techniques could be of major value and impact. The author is anxious to begin applying the techniques discussed in this Institute to this problem.

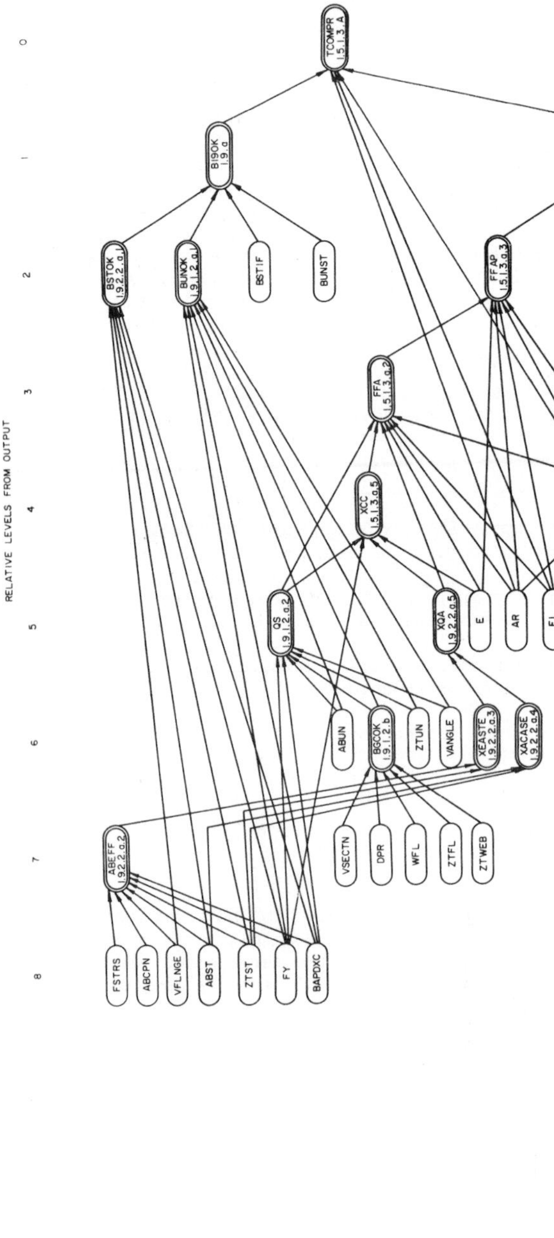

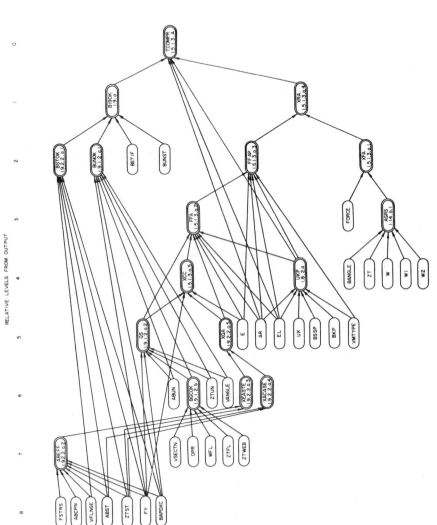

Figure 1. Ingredience Network for Compression Criterion of AISC Specification

References

American Institute of Steel Construction (1969), "Specification for the Design, Fabrication and Erection of Structural Steel for Buildings," New York, New York.

Fenves, S. J. (1966), "Structural Analysis by Networks, Matrices and Computers," J. of the Structural Division, ASCE, Vol. 92, No. ST1, pp. 199-221.

Fenves, S. J. and Branin, F. H. Jr. (1963), "Network-Topological Formulation of Structural Analysis," J. of Structural Division, ASCE, Vol. 89, No. ST4, pp. 483-514.

Fenves, S. J. and Gonzalez-Caro, A. (1971), "Network-Topological Formulation of Analysis and Design of Rigid-Plastic Framed Structures," International Journal for Numerical Methods in Engineering, Vol. 3, pp. 425-441.

Goel, S. K. and Fenves, S. J. (1968), "Computer-Aided Processing of Structural Design Specifications," SRS No. 348, Dept. of Civil Engineering, University of Illinois.

Goel, S. K. and Fenves, S. J. (1971), "Computer-Aided Processing of Design Specifications," J. of the Structural Division, ASCE, Vol. 97, No. ST1, pp. 439-461.

Goel, S. K., Fenves, S. J., and Gaylord, E. H. (1971), "Adapting the AISC Specification to Computer-Aided Design," AISC Engineering J., Vol. 8, No. 3, pp. 80-89.

Knuth, D. E. (1968), "The Art of Computer Programming," Vol. 1, Addison Wesley Co., Reading, Mass.

McDaniel, H. (1968), "An Introduction to Decision Logic Tables," John Wiley & Sons, Inc., New York.

Montalbano, M. (1961), "Tables, Flow Charts, and Program Logic," IBM Systems J., Vol. 1, No. 1.

Wright, R. N. Boyer, L. T. and Melin, J. W. (1971), "Constraint Processing in Design," J. of the Structural Division, ASCE, Vol. 97, No. ST1, pp. 481-494.

SPARSE MATRICES AND DECOMPOSITION WITH APPLICATIONS TO THE SOLUTION OF ALGEBRAIC AND DIFFERENTIAL EQUATIONS

J. K. REID
Theoretical Physics Division, A.E.R.E., Harwell, Didcot, Berks., England

Abstract: The meaning of "stability" in connection with Gaussian elimination is explained. Pivotal strategies in common use for the general case and for the symmetric and positive-definite case are described and compared on the grounds of stability, practicability and results on test problems. Similarly storage patterns are described and compared. Problems arising from elliptic partial differential equations are considered with a view to the use of both sparsity techniques and direct band-matrix techniques.

1. INTRODUCTION

We consider the solution of a general sparse set of n linear equations

$$Ax = b. \tag{1.1}$$

In section 5 we look at the form that these equations take when they have arisen from the discretization of an elliptic partial differential equation, but in the main body of the paper no reference is made to the origin of the equations.

To solve the system (1.1) we decompose the matrix A into a product

$$A = PLUQ \tag{1.2}$$

where P and Q are permutation matrices, L is a lower-triangular matrix with units on its main diagonal and U is an upper triangular matrix. Given this decomposition the system (1.1) may be written in the equivalent form

$$\left.\begin{aligned} Px^{(1)} &= b \\ Lx^{(2)} &= x^{(1)} \\ Ux^{(3)} &= x^{(2)} \\ Qx &= x^{(3)}, \end{aligned}\right\} \tag{1.3}$$

readily verified by eliminating the subsidiary vectors $x^{(1)}$, $x^{(2)}$ and $x^{(3)}$. Each of the subsystems of (1.3) is easy to solve.

Since A is sparse we aim to choose the permutations in such a way that L and U are sparse too, and to save storage we normally keep only the non-zeros and corresponding indexing integers. We consider suitable storage patterns in section 4, and pivotal strategies (i.e. choice of P and Q) in section 3. Once P and Q have been chosen the triangular factors L and U are normally unique, but there are situations in which they do not exist even though the system (1.1) is well-conditioned, or where they exist but the decomposition is unstable and in the presence of round-off errors the factorization differs substantially from the original matrix. This possibility of instability is sometimes overlooked in practical codes, perhaps because it cannot happen for certain classes of matrices. We examine this problem in the next section.

Choosing the pivotal sequence is, in general, a far more costly operation than decomposing A once it is known or solving the equations (1.3). On the other hand it often happens that several matrices with the same sparsity pattern will require factorization and also that several systems of equations (1.1) with the same matrix A will require solution. For these reasons it is usual, while choosing pivots, to store information to allow these tasks to be performed rapidly. This may be in the form of integer arrays indexing the non-zeros, stored in a convenient order, or it may be in the form of a long string of uninterrupted floating-point instructions referring to fixed addresses for the non-zeros (Gustavson et al., 1970). The latter procedure cannot be implemented entirely in a high-level language and the code is bulky to store, but the advantage in speed is substantial if the code can be held in main store, typical gains being by a factor of $2\frac{1}{2}$ to 3. As an example of the bulk of the code, we estimate that the last example of Table 2 would require about a million bytes of code to factorise a further matrix and about a third of a million bytes to solve a set of equations (given the factorization). In such a case the code would presumably be held on backing-store but the cost of bringing it down before execution would probably exceed the cost saving of the intrinsically higher execution speed. A variant of the Gustavson procedure, proposed by Erisman (1972), mitigates these disadvantages. Erisman replaces in-line floating-point code by the use of integer arrays to control jumps to subroutines which perform such jobs as calculating the inner product between two vectors of which one has elements stored consecutively. In this way he reduced the length of code by a factor of ten for his problems. A further advantage of Erisman's technique is that departure from a high-level language is no longer necessary.

A final remark we wish to make in this section is that given a decomposition (1.2) for A we can at once write down a similar decomposition

$$A^T = Q^T U^T L^T P^T \tag{1.4}$$

for A^T. We therefore see that solution of the equation

$$A^T y = c \tag{1.5}$$

is available to us, given the decomposition (1.2). Also at various points in this paper we will discuss storage schemes and algorithms which are definitely biased towards rows or columns. Since, however, we could equally have worked with A^T there is always a dual in which the roles of rows and columns are interchanged. For simplicity we describe just one of such a pair and make no reference to the other, although each are equally important.

2. GAUSSIAN ELIMINATION AND WILKINSON'S BACKWARD ERROR ANALYSIS

In this section we describe Gaussian elimination since this is the process upon which the whole of this paper is based and then say something of Wilkinson's backward error analysis. This is explained in more detail by Wilkinson (1965).

At the first stage of elimination we choose a "pivot", say a_{pq}, and interchange rows 1 and p columns 1 and q so that this element is brought to the (1,1) position. Next multiples ℓ_{i1}, i = 2,3,..n, of the first row are subtracted from subsequent rows, these multiples being chosen so that the resulting matrix has zeros in its first column, apart from the first row. At the second stage exactly the same operations are applied to the matrix of the last (n-1) rows and columns, with the exception that the row permutation is applied to the vector of multipliers ℓ_{i1}, i = 2, ...n, so that these are as if the second row permutation had been applied before these numbers were calculated. Similarly, at the mth stage multipliers ℓ_{im}, i = m + 1, ...,n are found and the row permutation used is applied to all the previous vectors of multipliers. It is straightforward to show that the upper triangular matrix eventually obtained is the required matrix U, that the multipliers ℓ_{ij} make up the subdiagonal elements of L and that product of the row and column permutations used make up the matrices P and Q.

Now suppose that P and Q are fixed, but the arithmetic is carried out in floating-point with a fixed word-length. Wilkinson's work consists of bounding not the errors in L and U but the difference

$$F = PLUQ - A \tag{2.1}$$

between the exact product of the factors actually obtained and the original matrix. A minor extension, Reid (1971b), of Wilkinson's work gives the bounds

$$|f_{ij}| \leqslant (3.01)\,\varepsilon\, n g_{ij} \tag{2.2}$$

where $\varepsilon\ (\leqslant 10^{-3})$ is the relative accuracy of the computation, $g_{ij} = \max_k |a_{ij}^{(k)}|$ and $a_{ij}^{(k)}$ is the element corresponding to a_{ij} in the kth reduced matrix. This is a powerful result provided the elements do not grow drastically in size during the reduction. That such growth is likely to be disastrous is easy to see for the elements change by having numbers subtracted from them and can grow only by such a computation as (in 4 decimal arithmetic)

$$1.234 - (-123.2) = 124.4 \tag{2.3}$$

and in such a case it is clear that information (.034) has been lost; such information may be vital to the solution. For these reasons we refer to an elimination in which severe growth in the size of matrix elements takes place as unstable.

Because it is not practicable to monitor the growth of all matrix elements it is usual to try to control $\max_{i,j}|a_{ij}^{(k)}|$. For example, it is easily seen that if all pivots are chosen to have maximum modulus in their row or column then the inequality

$$\max_{i,j} |a_{ij}^{(k+1)}| \leqslant 2 \max_{i,j} |a_{ij}^{(k)}| \tag{2.4}$$

holds. Such control does not constitute a sensible strategy, however, unless the matrix is well-scaled. Fortunately, it often happens that matrices are well-scaled naturally but particular care should be exercised where mixed units (e.g. neutron fluxes and feet) are in use. No really satisfactory scaling algorithm is known at present but studies made with A. R. Curtis (Curtis and Reid, 1972) indicate that straightforward equilibration of row and column norms (e.g. scale all rows to have maximal element unity, then do the same by columns) can give very bad results for sparse matrices. Our suggestion, following Hamming (1971), is to choose ρ_i, c_j to minimize

$$\Sigma(\log_\beta |a_{ij}| - \rho_i - c_j)^2 \tag{2.5}$$

where β is the base of the floating-point computation in use and the sum is over all i,j such that $a_{ij} \neq 0$; we then work with $\{\beta^{-R_i} a_{ij} \beta^{-C_j}\}$ where R_i and C_j are nearest integers to ρ_i and c_j respectively. On our test sparse matrices we were able to solve the minimization problem in the equivalent of 7-10 sweeps through the matrix and satisfactory scalings were obtained.

There are two important classes of matrices for which Gaussian elimination is bound to be stable for any choice of pivots from the leading diagonal ($P = Q^{-1}$).

The first class is of those matrices which are diagonally dominant either by rows, where the relation

$$|a_{ii}| \geqslant \sum_{j \neq i} |a_{ij}| , \quad i = 1,\dots,n \tag{2.6}$$

holds or by columns, when the relation

$$|a_{ii}| \geqslant \sum_{j \neq i} |a_{ji}| , \quad i = 1,\dots,n \tag{2.7}$$

holds. In this case the maximal element of any reduced matrix is no greater than twice the maximal element of A. This result is proved by Wilkinson (1961, §8) for diagonal dominance by columns (relation (2.7)) and the corresponding result by rows may be proved similarly.

The second, and more important, class of matrices for which diagonal pivoting is always stable is those that are symmetric and positive-definite ($z^T Az > o$ unless $z = o$). These matrices frequently arise after discretization of elliptic partial differential equations (see §5). Wilkinson (1961, §6) has shown that for such a matrix A no element of any of the reduced matrices is greater than the greatest element of A. Also it is readily verified that if pivots are chosen from the diagonal then symmetry is preserved in the remaining uneliminated matrix so that both the work and storage may be halved; if D is the diagonal matrix with diagonal elements equal to those of U then the relation

$$DL^T = U \tag{2.8}$$

is easily proved so that the decomposition may be written

$$A = PLDL^T P^T \tag{2.9}$$

and there is no need to store the off-diagonal elements of U.

3. PIVOTAL STRATEGIES

At each stage of the elimination process a choice of pivot has to be made and this is usually based on a compromise between a good choice on stability grounds and a good choice on sparsity grounds. A zero pivot would give infinite growth, the limiting case of instability, and codes have always avoided such a choice. Sometimes they have done no more than this, although small pivots are obviously likely to lead to large growth. It is usual either to restrict pivots to those greater than a certain tolerance or to those which result in multipliers (elements of L) no greater than a certain limit. We have used the latter criterion (Curtis and Reid, 1971) on the ground that if all the multipliers are less than w then the largest matrix element cannot grow by a factor greater than (1 + w) at each elimination, although of course we hope that overall growth at this rate will not be maintained for long. We found the choice of w not very critical but recommended w = 4 on the basis of a limited set of tests. Tomlin (1972) recommends the use of w = 100 in linear programming applications but includes the safeguard of checking the size of the residuals

$$b - A\bar{x} \tag{3.1}$$

where $\bar{x}$ is the computed solution and repeating the decomposition with w = 10 if they are unsatisfactorily large. Wilkinson (1965, for example) has also shown that the triangular sets of equations (1.3) are solved accurately in the sense that the solution obtained is the exact solution of a slightly perturbed system; it follows that large residuals can only have been caused by unstable decomposition. We note in passing that small residuals do not necessarily imply an accurate solution for if the matrix is ill-conditioned then quite large errors in $\bar{x}$ may make little difference to $A\bar{x}$ and so leave the residuals small.

Next we consider sparsity criteria for the choice of pivot. Perhaps the most commonly used criterion (Markowitz, 1957) is to minimize the multiplication count, that is the product of the number of elements in the pivotal row and the number of elements in the pivotal column, excluding the pivot itself in each case. This is an upper bound for the number of non-zeros that may be created during this elimination. Alternatively we may minimize the number of non-zeros actually created. This latter strategy obviously is more labourious and in practice usually gives similar fill-in (see for example Tables 1 and 3 of §5), although the following example of Rose (1970) illustrates that it can give very much better results.

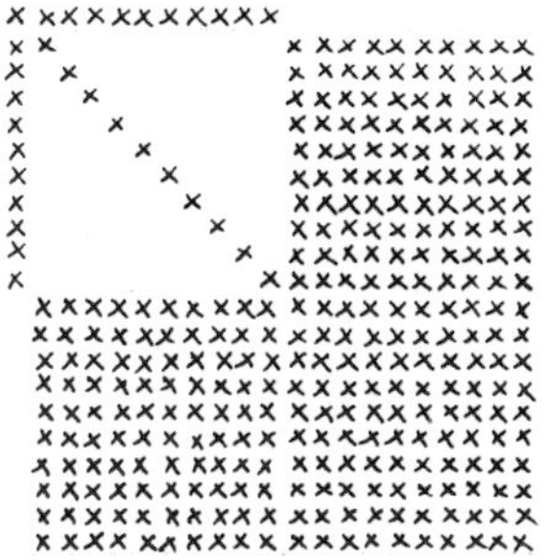

The conjecture that minimizing the fill-in at each stage of the elimination will minimize the overall fill-in has been proved false by an example of Duff (1972) which has order 13, and two large examples are contained in Tables 1 and 3 of §5.

An organisational disadvantage of these criteria is that they require the sparsity pattern of the matrix at each stage, whereas it is convenient to delay operations on each column until it is pivotal and then perform them all together. This is convenient if a two-level store is in use and the matrix is stored by columns for the column is only modified once and so needs writing to backing store only once. It is also convenient in a one-level-store program because new non-zeros are created only within a column and the need to store the whole matrix as a linked list is avoided (see also §4). For these reasons it is quite common to choose the pivotal column order a priori. A merit number is attached to each column and the columns are ordered in increasing merit number; two common choices for this merit number are the number of non-zeros in the column and the total number of non-zeros in those rows that have a non-zero in the column. Within the pivotal column the pivot is usually chosen as that whose row has minimal number of non-zeros in it, and subject to whatever stability criterion is in use. Duff (1972) has experimented with the rather more sophisticated criterion of choosing the element whose row originally contained the minimal number of non-zeros in those columns that have not yet been eliminated, but has found that this gives worse results, presumably because the original number of non-zeros in the eliminated part of the row is better than zero as an estimate of the number of zeros filled-in. Experiments carried out at Harwell (Curtis and Reid, 1971 and Duff, 1972) have indicated that significantly more fill-in occurs with a priori ordering.

We conclude this section by remarking that there is an important class of matrices for which a good pivotal strategy is to take pivots directly from the diagonal. This is for a matrix that is variable-band, symmetric and positive definite and is based on the observation that if $P = I$ in the decomposition (2.9) then all the non-zeros of each row of L are in columns not in front of that of the first non-zero of each corresponding row of A. Thus the following sparsity pattern, for example, is preserved without fill-in.

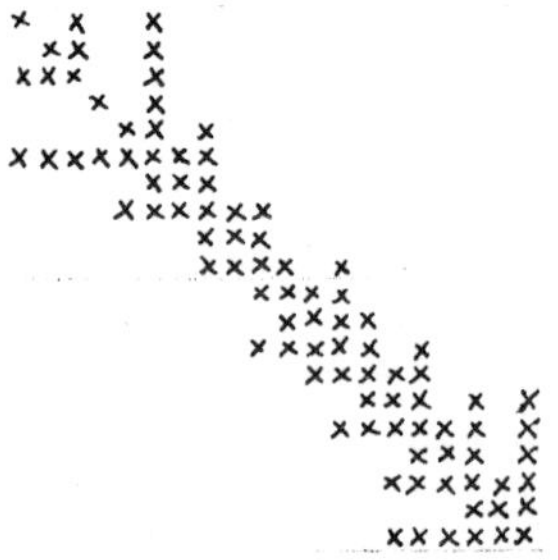

If it has maximal semi-band-width m, that is if $a_{ij} = o$ when $|i-j| \geqslant m$ ($m = 6$ in the above example) then to perform operations on a particular row involves that row and at most the previous $(m - 1)$ rows of L so that the earlier ones can conveniently be placed on backing store. A Harwell report (Reid, 1972) contains a Fortran subroutine implementing this algorithm. A more complicated procedure, allowing the algorithm to be implemented when very little main store is available or when m is very large, is described by Jennings and Tuff (1971).

4. STORAGE CONSIDERATIONS

There are two strong reasons for using sparse matrix techniques, namely, to save computational effort and to save storage. We save storage by holding only the non-zero elements but, of course, need additional integer indexing arrays so that we know where each non-zero belongs in the matrix. We discuss here some of the more commonly used storage patterns, viz:-

1. The matrix is stored by columns, e.g. a_{11}, a_{21}, a_{51}, a_{22}, a_{42}, a_{33}, a_{43}, a_{14}, a_{24}, a_{44}, with row numbers held in correspondence with each non-zero; also held are pointers to the beginnings of the columns (including an imaginary $(n + 1)$st column), e.g. 1, 4, 6, 8, 11 in the above example.

2. As 1., but instead of storing all row numbers, store just the first and last of each column and the bit patterns of the intermediate ones.

3. The matrix elements are not stored in any particular physical order but rather as a linked list. Associated with each non-zero is some combination of the following: pointers to the next and to the previous element in its row and in its column; row and column numbers. In addition there are pointers to the first elements of the rows and columns. The last forward pointer and the first backward pointer in a row or column is available for other purposes such as a pointer to the other end of the row or column or to hold the row or column number.

Structure 2 is very economical in storage if the non-zeros are close together within each column but the "bit-picking" necessary to work with this structure makes it rather clumsy, and it is generally rather out of favour. Structure 1 does not suffer from this disadvantage and yet is almost as economical in storage, requiring only one integer to be associated with each non-zero. Its principle disadvantage is that if it is used with a pivotal strategy that requires the sparsity pattern to be kept and updated then it is very clumsy because extensive shuffling is necessary after each creation of a new non-zero. Curtis and Reid, (1971, algorithm X) uses this structure with the a priori pivotal column ordering strategy described in §3, keeping the columns in (known) pivotal order; non-zeros are created within one column at a time so that shuffling is confined to this one column; essentially separate structures of type 1 are kept for the columns that have been pivotal and the rest, the first structure expanding as the second contracts and storage released by the second may later be used by the first.

The linked list structure 3 allows each elimination with associated creation of new non-zeros to be performed in a number of computer operations that is only a small multiple of the number of multiplications involved. The new non-zeros are placed physically at the end of the structure and the links are adjusted, thus avoiding any shuffling or reordering. Similarly, interchanges may be performed by adjusting links, avoiding any actual physical movement of the non-zeros. The reason for avoiding such shuffling is principally because the number of computer operations involved is likely to be a substantial fraction of the total number of non-zeros. It is, of course, equally important to avoid doing a similarly large number of operations in some other part of the calculation. Having all six integers (links and row/column numbers) associated with each non-zero is very convenient since it is then very straightforward to insert new non-zeros into the structure, to interchange a pair of rows or a pair of columns and to identify the row and column of a non-zero, but this is usually considered to be too wasteful of storage. The number of these integers may be reduced at a little expense in extra computation. For example, in our own work (Curtis and Reid, 1971) we have not used backward pointers but instead have scanned the row or column forwards until the element for which the backward pointer is needed is encountered and in the last of our algorithms (Curtis and Reid, 1971, algorithm Z) we have not used row and column numbers either but instead have scanned the row or column to its end and arranged that the last pointer contains the row or column number. Such scans are not unduly expensive since we do not expect the average row or column to contain more than 5 or 6 non-zeros, but of course it is important to arrange the calculation to do as few of them as possible.

If the pivotal strategy and the final sparsity pattern is known then the linked-list structure 3 shows no advantage over structure 1 which is the more economical of storage. We work with the final sparsity structure, which includes all the intermediate structures and so no fill-in takes place. We use this in our own algorithm (Curtis and Reid, 1971, algorithm Z). Unless the pivotal sequence is known to give a stable decomposition (e.g. in the positive-definite case) some check should be made on stability, such as recording the largest number encountered or checking the size of the residuals when using the decomposition to solve a system of equations.

If stability considerations can be ignored while choosing the pivotal sequence (e.g. in the positive-definite case, again) then the non-zeros themselves need not be stored at all during this phase of the computation, making substantial savings in storage for structure 3.

We conclude this section with some remarks on the use of backing store (e.g. disk). There is a good reason for avoiding backing store altgother for the general sparse matrix if at all possible. This is because during the elimination stage access is required to scattered portions of the matrix. It is usual to hold the matrix on backing store by columns and to use a priori column ordering so that all the operations on one column can be delayed until the column is pivotal, thus reducing the number of records needing to be written to backing store. However, several earlier columns, in general scattered, will be required in order to operate on that column. An exceptional case where a backing store may be used without excessive degradation of performance is when the Jennings variable band-width algorithm is in use (see end of §3).

5. ELLIPTIC PARTIAL DIFFERENTIAL EQUATIONS

We consider the solution by finite differences or by finite elements of elliptic partial differential equations in two and three dimensions. We will not draw any great distinction between finite differences and finite elements for our concern will be with the solution of the resulting system of algebraic equations rather than with their derivation. A mesh (e.g. triangular, rectangular) is placed over the region to give a number of nodes at which parameters determining the solution are required. We use as one of our test examples the problem of the determination of the solution of Laplace's equation

$$\frac{\partial^2 \phi}{\partial x^2} + \frac{\partial^2 \phi}{\partial y^2} = 0 \tag{5.1}$$

over a rectangle with known values for ϕ at its edge. Taking a regular rectangular mesh with ℓxm internal modes and using the standard 5-point finite difference approximation yields a matrix of the form

$$\begin{bmatrix} T & D & & & \\ D & T & D & & \\ & D & T & D & \\ & & \ddots & \ddots & D \\ & & & D & T \end{bmatrix} \tag{5.2}$$

where there are m blocks T, each of which is a tridiagonal matrix of order ℓ, and 2m-2 diagonal matrices D of order ℓ. We solve this to obtain function values at each node. In a more general case there are several parameters associated with each node, perhaps because we are solving a coupled system of elliptic equations or perhaps because more than one parameter is associated with each node (e.g. function values and derivatives) in a finite-element approximation. With care it is possible to arrange that the matrix is symmetric and positive-definite, a property that is obtained automatically if the variational formulation of the elliptic equation is approximated by finite elements. It is highly desirable to do this because (see §2) then the pivots may be chosen from the diagonal without fear of instability, so preserving symmetry and allowing both the storage and the computational work to be approximately halved; also it allows the possibility of the use of Jennings' variable band-width algorithm (end of §3).

To obtain the satisfactory band matrix form of (5.2) the points were numbered "pagewise". Such an ordering allows the use of the Jennings variable band-width algorithm (see §3) in an efficient way. A satisfactory way of numbering the variables in the general case is to number all the variables associated with each node consecutively, and to use the algorithm of Cuthill and McKee (1969) to order the nodes. We label a node 1 (taking a point on the edge of a thin side of the region is sensible), label its neighbours 2, 3, ..., then add to the end of the list all unlabelled neighbours of point 2 then of point 3 and so on. A simple example is the following:

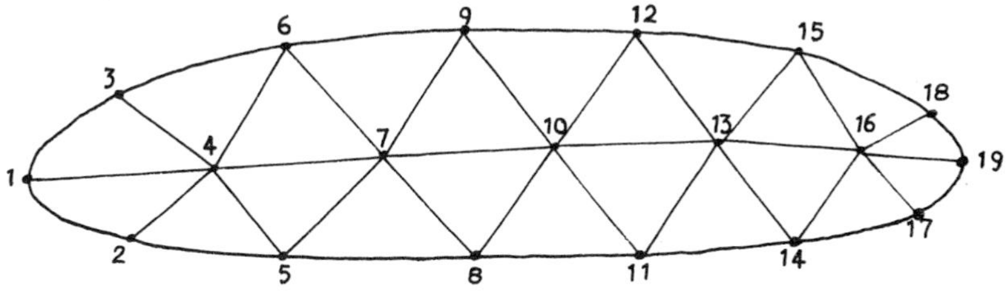

Two nodes are considered neighbours if they are both involved in a finite-difference expression or if they both belong to the same finite element. George (1971) reports particularly favourably of the use of the reverse ordering, calling this "reverse Cuthill-McKee".

The use of a sparse matrix code obviates the need for a good ordering of the nodes, but as we remarked at the end of the last section it is undesirable if it cannot be performed in the main store. The actual gains in storage are not as substantial as might perhaps have been hoped. Some examples are shown in Table 1; the pivotal strategies used for the sparse codes were to minimize the multiplication count (unbracketed figures) and the fill-in (bracketed figures)

m	3	4	5	6	7	8	9	10
Non-zeros (Jennings)	29	67	129	221	349	519	737	1009
Non-zeros (Sparse)	26 (26)	58 (58)	102 (102)	166	248 (248)	355	469 (472)	631

Table 1. Non-zeros stored for Laplacian in mxm square

at each elimination, choosing pivots from the diagonal. In Table 2 we show some results obtained on the IBM 370/165 when experimenting (Reid, 1972) with subroutines for Jennings' algorithm (MA15) and sparsity coding using a minimal multiplication count pivotal strategy (MA17). In interpreting the storage requirements allowance should be made for the fact that the compiled code for MA15 was about 3500 bytes shorter than that for MA17. It will be seen that the

ℓ	m	Sub-routine	No. of non-zeros D,L	Time (secs) in double length to			core storage (thousands of bytes)	
				Choose pivots	Factorize matrix	Solve equation	Single length	Double length
5	6	MA15 MA17	159 131	- .025	.0068 .0028	.0016 .0013	0.9 1.4	1.8 1.9
10	11	MA15 MA17	1119 728	- .20	.048 .018	.0091 .0066	5.4 7.1	10.8 10.1
15	16	MA15 MA17	3629 1987	- .80	.17 .063	.027 .019	16.5 18.8	33.0 26.7
30	31	MA15 MA17	27959 11328	- 12.8	2.0 .56	.20 .10	119.4 101.8	238.8 147.1

Table 2. Laplace's equation in a rectangle

two algorithms use comparable amounts of storage on small problems, but the Jennings' algorithm (MA15) uses more on large problems, although not all of this need be in core. In fact using disk backing store on the double-length version of the largest problem shown reduced the core storage required to 43.4 thousand bytes and increased the factorization time to 2.3 secs and the solution time to 0.35 secs.

In three dimensions the picture is far less encouraging for elimination because the matrices have far wider bands with far more zeros within their profile. In Table 3 we show figures for the Laplacian in a cube in the same format as those of Table 1.

m	3	4	5	6
Non-zeros (Jennings)	209	883	2729	6881
Non-zeros (Sparse)	147 (146)	531 (551)	1377 (1350)	3281

Table 3. Non-zeros stored for Laplacian in mxmxm cube

It will be seen that the storage required is substantial. The amount of compution is also large. In fact it is usually advantageous to use iterative methods for three-dimensional problems.

It is interesting to note also from Tables 1 and 3 that the two sparse matrix strategies give comparable results and in fact there is one example in each table of the minimum fill-in strategy giving an inferior result.

George (1972) makes some very useful comments on ordering for matrices arising from finite-element discretizations. An obvious and very worthwhile saving can be made by eliminating first all those variables corresponding to points internal to the elements (sometimes known as "static condensation"). Next, if variables corresponding to all the nodes on an edge between two elements are eliminated then the effects are confined to those two elements and it is as if they had been combined into one element, parameterized by the remaining variables. It may be convenient to stop this procedure early and go over to straightforward band-matrix elimination or we can continue it, obtaining successively larger "elements" until the whole region is one "element". George applies this procedure to the finite element solution of a problem over the unit square obtained by placing over it a square grid of side $1/q$, each square of which is divided into two triangular elements and taking α variables at each node and β variables on each edge between elements. In this way he reduces the number of multiplications from

$$\tfrac{1}{2}(\alpha+\beta)^2(\alpha+2\beta)q^4 + O(q^3) \tag{5.3}$$

to less than

$$22(\alpha+\beta)^3q^3 + 8(\alpha+\beta)^2q^2\,(\log_2 q - 4\alpha - 4\beta). \tag{5.4}$$

Qualitatively this appears to be a large gain, from $O(q^4)$ to $O(q^3)$, but the leading terms in (5.3) and (5.4) are equal at some value of q between 22 and 44, so the gains in practice are unlikely to be large. Actually George's procedure can be improved by taking lines of nodes to be eliminated right onto the boundary whenever the "elements" under combination are at the boundary but it remains true that for grids likely to be used in practice there will be a gain smaller than one might expect from the reduction of the number of operations from $O(q^4)$ to $O(q^3)$. The procedure generalizes to uneven grids and to higher dimensions without difficulty.

References

Curtis, A. R. and Reid, J. K. (1972). On the automatic scaling of matrices for Gaussian elimination. J. Inst. Maths. Applics., 10, 118.

Curtis, A. R. and Reid, J. K. (1971). The solution of large sparse unsymmetric systems of linear equations. J. Inst. Maths. Applics., 8, 344.

Cuthill, E. and McKee, J. (1969). Reducing the bandwidth of sparse symmetric matrices. Proc. ACM National Conference, 157.

Duff, I. S. (1972). Analysis of sparse systems. D. Phil. Thesis, Oxford. To appear.

Erisman, A. M. (1972). Sparse matrix approach to the frequency domain analysis of linear passive electrical networks. Rose and Willoughby (1972), 31.

George, J. A. (1972). Block elimination on finite element systems of equations. Rose and Willoughby (1972), 101.

George, J. A. (1971). Computer implementation of the finite element method. Ph. D. Thesis. Stanford University report STAN-CS-71-208.

Gustavson, F. G., Linger, W. M. and Willoughby, R. A. (1970). Symbolic generation of an optimal Crout algorithm for sparse systems of linear equations. J. Assoc. Comput. Mach., 17, 87.

Hamming, R. W. (1971). Introduction to applied numerical analysis. McGraw Hill.

Jennings, A. (1966). A compact storage scheme for the solution of symmetric linear simultaneous equations. Comp. J., 9, 281.

Jennings, A. and Tuff, A. B. (1971). A direct method for the solution of large sparse symmetric simultaneous equations. Reid (1971a), 97.

Markowitz, H. M. (1957). The elimination form of the inverse and its application to linear programming. Management Sci., 3, 255.

Read, R. (1971). Graph theory and computing. Academic Press.

Reid, J. K. (1971b). A note on the stability of Gaussian elimination. J. Inst. Maths. Applics., 8, 374.

Reid, J. K. (1972). Two Fortran subroutines for direct solution of linear equations whose matrix is sparse, symmetric and positive-definite. A.E.R.E. report R.7119. HMSO.

Reid, J. K. (1971a). Large sparse sets of linear equations. Academic Press.

Rose, D. J. and Willoughby, R. A. (1972). Sparse matrices and their applications. Plenum Press.

Rose, D. J. (1971). A graph-theoretic study of the numerical solution of sparse positive definite systems of linear equations. Read (1971).

Tomlin, J. A. (1972). Pivoting for size and sparsity in linear programming inversion routines. Submitted to J. Inst. Maths. Applics.

Wilkinson, J. H. (1965). The algebraic eigenvalue problem. Oxford University Press.

Wilkinson, J. H. (1961). Error analysis of direct methods of matrix inversion. J. ACM, 8, 281.

DECOMPOSITION METHODS USING SPARSE MATRIX TECHNIQUES WITH APPLICATION TO CERTAIN ELECTRICAL NETWORK PROBLEMS*

ALBERT M. ERISMAN

Boeing Computer Services, Inc.

Seattle, Washington 98124

Abstract: In this paper, decomposition methods are developed which are compatible with condensation and sparse matrix techniques. These methods are combined to provide an efficient approach to the solution of large systems of linear equations. The techniques are applied to frequency domain analysis of large (3000-4000 node) RLCM linear electrical networks.

1. INTRODUCTION

The solution of large systems of linear algebraic equations is an important tool in the analysis of many large scale physical systems. The use of special properties of the physical system can lead to significant savings in solving these equations. In this paper some rather general properties common to many physical systems are identified and their use in the solution of the associated algebraic equations is developed. Under consideration is the solution of

$$Yx = b \tag{1}$$

where Y is an $n \times n$ coefficient matrix, b is a given vector, and x is the vector of unknowns.

The properties to be used are:

(i) the large system is a composite of smaller subsystems;

(ii) in (1), only a portion of the solution vector x is required and most of b is zero;

(iii) Y is large and sparse (most of its entries are zero);

(iv) Y is a function of parameters; the matrix has the same zero, non-zero structure over the range of interest of the parameters;

(v) repeated solutions of the overall system are required where one or several of the subsystems are varying.

In the next section the numerical solution of (1) is discussed using the above properties. The third section contains an example of a large physical system, frequency domain analysis of large RLCM linear networks. It is shown that this problem satisfies the properties above, and some timing and size capabilities are given. The final section contains some comparisons with other methods in the literature.

---

* This work was performed under Air Force Contract F04701-69-C-0153 per C.O. P00123.

2. SOLUTION TECHNIQUES

The solution techniques of (1) discussed here are a modification of the basic LU factorization followed by forward and backward substitution.

Specifically, first a unit lower triangular matrix L and an upper triangular matirx U are determined satisfying

$$Y = LU \quad . \tag{2}$$

Then the solution of the equations

$$Lv = b \tag{3}$$

is obtained by forward substitution, followed by the solution of

$$Ux = v \tag{4}$$

by backward substitution.

The factorization (2), which is accomplished by Gaussian elimination or one of its variations, represents the most costly part of the computation. For a full matrix Y it requires on the order of $n^3/3$ multiplications, while (3) and (4) require on the order of $n^2/2$ multiplications each. For large n, a great deal of efficiency may be gained in the solution process by modifying the algorithm using properties (i) - (v).

Using property (i), a decomposition method is developed which makes it possible to analyze the subsystems and then interconnect these for the analysis of the overall system. The techniques are neither iterative nor approximate and are directed toward the LU factors (2) rather than the interconnection of solutions. This permits repeated solution of (1) with different b vectors where the interconnection problem is handled only once.

For this discussion a special form of matrix sum is used which does not require the individual matrices or their resulting sum to be of the same size. This is illustrated with the following example. For

$$A = \begin{bmatrix} b_{11} & b_{12} & 0 \\ b_{21} & b_{22}+c_{11} & c_{12} \\ 0 & c_{21} & c_{22} \end{bmatrix} ,$$

$$B = \begin{bmatrix} b_{11} & b_{12} \\ b_{21} & b_{22} \end{bmatrix} ,$$

$$C = \begin{bmatrix} c_{11} & c_{12} \\ c_{21} & c_{22} \end{bmatrix} ,$$

say

$$A = B \oplus C$$

where it is assumed the positioning of B and C within A is determined from context.

Suppose the matrix Y representing the overall system is a composite of k matrices Y_i representing subsystems such that

$$Y = Y_1 \oplus Y_2 \oplus \ldots Y_k \oplus E$$

where E is an additional contribution to the overall system not represented by summing the Y_i. Suppose further the Y_i matrices are partitioned

$$Y_i = \begin{bmatrix} A_i & B_i \\ C_i & D_i \end{bmatrix} \tag{5}$$

so that the portion of Y_i which connects to other subsystems is confined to D_i. Then by context the Y matrix is given by

$$Y = \begin{bmatrix} A_1 & & 0 & B_1 \\ & A_2 & & B_2 \\ & 0 & A_k & B_k \\ C_1 & C_2 & C_k & R \end{bmatrix} \tag{6}$$

where

$$R = D_1 \oplus D_2 \oplus \ldots D_k \oplus E \quad .$$

Now if the subsystems have been solved as in (1) - (4), the matrices L_i and U_i have been calculated where

$$Y_i = L_i U_i \quad .$$

Partitioning L_i and U_i in the same manner as Y_i, let

$$L_i = \begin{bmatrix} M_i & 0 \\ K_i & N_i \end{bmatrix} , \tag{7}$$

$$U_i = \begin{bmatrix} V_i & T_i \\ 0 & W_i \end{bmatrix} , \tag{8}$$

where M_i and N_i are unit lower triangular while V_i and W_i are upper triangular.

Now it follows from block multiplication that

$$Y = LU = \begin{bmatrix} M_1 & & 0 & \\ & M_2 & & \\ & 0 & M_k & \\ K_1 & K_2 \; . \; . \; . & K_k & M_{k+1} \end{bmatrix} \begin{bmatrix} V_1 & & 0 & T_1 \\ & V_2 & & T_2 \\ 0 & & V_k & T_k \\ & & & V_{k+1} \end{bmatrix} \tag{9}$$

where the only unknown blocks are M_{k+1} and V_{k+1}. Using block multiplication on (6) and (9),

$$M_{k+1}V_{k+1} = R - (T_1K_1 \oplus T_2K_2 \oplus \ldots \oplus T_kK_k) \quad . \tag{10}$$

But from (5), (7), and (8),

$$D_i = T_kK_k + N_iW_i \quad . \tag{11}$$

Thus combining (11) and (6), (10) becomes

$$M_{k+1}V_{k+1} = N_1W_1 \oplus N_2W_2 \oplus \ldots \oplus N_kW_k \oplus E \quad . \tag{12}$$

The important contrast between equations (10) and (12) should be emphasized. In general for a large problem, the portion involved in interconnection with other subsystems is small compared with the size of the subsystem. Thus the calculation of the product of two triangular matrices N_iW_i is small compared with the product of two larger rectangular matrices, T_iK_i.

Another importance of this calculation is that it yields the LU factors of Y. Thus solutions may be calculated for any number of b vectors.

The calculations required in (12) may be further reduced. In the course of calculating L_iU_i for subsystem Y_i, if Gaussian elimination is used the partially calculated factors look like

$$\begin{bmatrix} A_i & B_i \\ C_i & D_i \end{bmatrix} = \begin{bmatrix} M_i & 0 \\ K_i & I \end{bmatrix} \begin{bmatrix} V_i & T_i \\ 0 & \hat{D}_i \end{bmatrix} \tag{13}$$

where I is an identity matrix. It is easy to verify that $\hat{D}_i$ is the desired N_iW_i product. Thus by saving M_i, K_i, V_i, T_i, and $\hat{D}_i$ for subsequent computation, all that remains to be calculated for overall factors (9), is

$$M_{k+1}V_{k+1} = \hat{D}_1 \oplus \ldots \oplus \hat{D}_k \oplus E \quad .$$

Because of the ordering constraint imposed on the subsystems (5), it may be advantageous to combine some of the subsystems before calculating the overall factors if sparsity is to be exploited. The reason for this becomes clear in the discussion of the use of sparsity and in the use of property (ii).

Another savings which may be achieved in computing the overall response involves the use of property (ii), condensation. For this discussion, one subsystem is sufficient to identify the approach and so the subscripts will be dropped. Consider (1) where

$$Y = \begin{bmatrix} A & B \\ C & D \end{bmatrix} ;$$

again D contains the interconnect portion. Suppose, consistent with the assumption, that Y is further partitioned

$$Y = \begin{bmatrix} A_1 & A_2 & B_1 \\ A_3 & A_4 & B_2 \\ C_1 & C_2 & D \end{bmatrix} \tag{14}$$

corresponding to the partitions in x and b:

$$x = \begin{bmatrix} \text{Don't Care} \\ \hline \\ \hline \\ \end{bmatrix} \qquad b = \begin{bmatrix} 0 \\ \vdots \\ 0 \\ \hline \\ \hline \\ \end{bmatrix} .$$

To maintain original diagonal positions in the reordered matrix, Y was reordered so that the resulting "don't care" in x and 0 in b were made the same size. This is not a necessary restriction in theory but is important for some applications.

Let the factorization of (14) be denoted

$$Y = \begin{bmatrix} L_1 & & 0 \\ L_2 & L_4 & \\ L_3 & L_5 & L_6 \end{bmatrix} \begin{bmatrix} U_1 & U_2 & U_3 \\ & U_4 & U_5 \\ 0 & & U_6 \end{bmatrix} \tag{15}$$

Then it is clear that only

$$\hat{Y} = \begin{bmatrix} L_4 & 0 \\ L_5 & L_6 \end{bmatrix} \begin{bmatrix} U_4 & U_5 \\ 0 & U_6 \end{bmatrix}$$

is required to calculate the needed part of x. This produces a rather substantial savings in storage requirements for the required part of the factors of Y, and also leads to some reduction in computation in the forward and backward substitution parts, (3) and (4), of the solution process.

Note that condensation is compatible with decomposition as discussed here. For each Y_i, the corresponding L_4 replaces M_i, L_5 replaces K_i, L_6 replaces N_i; a similar correspondence exists for the partitions of U_i. This follows since $\hat{Y}$ (18) relates all positions corresponding to non-zero b_i and needed x_i. Furthermore it is already in usable factored form.

The key ingredients in the use of the sparsity of Y (property (iii)) are storage and data manipulation, ordering of computation, and preprocessing. This is a vast subject and cannot be discussed in detail here. [See Willoughby (1968), Reid (1970), Rose (1972b), for example.] Only enough of the subject is introduced to show its power and compatibility with decomposition and condensation.

By sparsity it is assumed the Y_i are sparse in addition to the zeroes in Y represented by (6). Then a great deal of storage may be saved by storing the non-zero values of Y in some compact form rather than in a full two-dimensional array. Two commonly used storage methods are given. In the first, three lists are stored which contain the non-zero value and its corresponding row and column index respectively. In the second, the data is ordered row-wise since this is a natural order for its use in Gaussian elimination [Gustavson (1972)]. Then two lists are required for the value and its column location; a third shorter list contains the starting position for each new row in the other two arrays.

Once the data is in this form, it is necessary to manipulate it to calculate the LU factors. Maintaining sparsity in the LU factorization usually requires reordering in the matrix to reduce fill-in ($y_{ij} = 0$, ℓ_{ij} or $u_{ij} \neq 0$) or multiplication count. Reordering is discussed in Hachtel (1972), Hachtel (1971), and Tinney (1967), for example. To make this reordering compatible with decomposition and condensation, it is necessary to apply constraints, as in Tinney (1967) and Erisman (1972b). That is, within the constraints of that portion of the subsystem connecting with others coming last, and that portion of the subsystem involving non-zero b_i or needed x_j coming next to last as indicated by decomposition and condensation, the parts of the matrix are reordered for reduced fill-in or reduced multiplication count.

Reordering is generally a part of a total preprocessing. This is especially important when property (iv) is valid. Then reordering and all zero, non-zero decisions may be made initially for the entire sequence of problems, and the cost may be distributed over the number of problems in the sequence.

Different schemes have been proposed for making these zero, non-zero decisions only once. These include:

(a) Generating a unique machine language program for each sparsity structure. This program is extremely fast and simple, has no loops or branches, and has absolute indexing. See Gustavson (1970) for details.

(b) Generating a table of indexing information in the ordering procedure. This table is then used interpretively by a recursive program. See Lee (1968).

The first procedure is basically faster, but the specialized program which is generated may be so large that it must be buffered in and out of core. In this case a shorter but not quite as fast program described by the second procedure may be faster [Erisman (1972a)].

If property (ii) does not hold, a program similar to the one which generates the program or indices of the above approaches may be used directly [Tinney (1967)].

In implementing this type of approach and making it compatible with decomposition and condensation, it is necessary to do the constrained reordering, be able to save portions of the decomposition, and to piece these blocks together for forward and backward substitution. Because the constraints on the reordering in general lead to additional fill-in, care must be taken in applying the decomposition and condensation. If two subsystems are not changing in subsequent analysis, for example, representing them as one is generally better. This follows because the interconnection between the two may now be absorbed leading to a smaller representation and a better ordering for sparsity.

Condensation and sparsity used together yield an additional benefit. Referring to (14) and (15), it is clear that L_1 and U_1 are no longer needed when the factorization process is calculating elements in L_4 and U_4. Because of the compact storage method discussed in connection with sparsity, these elements are all in a single rather than two-dimensional array. The area of storage in that array occupied by U_1 and L_1 is now free to be used by new values being calculated.

In all of this discussion, nothing has been said of reordering for numerical stability. Yet unless the coefficient matrix has very special properties (such as positive definiteness), this is a crucial requirement in achieving accuracy in the solution process.

Two approaches may be followed. One is to incorporate numerical considerations in the reordering along with the decomposition, condensation, and sparsity criteria. Incorporating numerical consideration in reordering is considered in Gustavson (1970). This may lead to larger D_i matrices (5) and less sparse factors, but may be the only way of achieving usable results.

Alternately, it may be determined from physical considerations that the arbitrary ordering from a numerical point of view is acceptable. Then to increase the confidence level in the solution process, a method of monitoring the numerical stability of the factorization process should be included in the computation, similar to Erisman (1972c).

The use of property (v) is perhaps best illustrated by the example of the next section.

III. AN EXAMPLE

For frequency domain analysis of linear RLCM networks, a mathematical model is

$$Y(\omega)x = b \tag{16}$$

where $Y(\omega)$ is the $n \times n$ nodal definite admittance matrix with elements which are a function of frequency ω, b is the vector of current inputs, and x is the vector of response node to datum voltages. Details on the formulation of this model may be found in Erisman (1972b). This matrix is complex-valued, symmetric, has the same zero, non-zero elements over a finite positive range of ω, and is sparse for large n. The input vector b usually has only a small percentage of non-zero entries. And only a small subset of the output voltages x are needed in the analysis.

The numerical solution of (16) involves the repeated solution over a discrete set of ω_i. Thus for each ω_i in the set, $Y(\omega_i)$ must be factored as in (2), followed by the forward and backward substitution to obtain the solution: (3) and (4). Typically n may be 3000 to 4000, between 50 and 250 ω_i values may be needed, and repeated analysis for each ω_i may be required due to changes in several subsystems.

The matrices $Y(\omega_i)$ are not positive definite and pathological examples can easily be constructed for which a diagonal reordering based on the constraints discussed in the previous section would be numerically disastrous. Under the assumption that these are pathological cases, however, numerical accuracy is not considered in the reordering process. Instead, the numerical stability of each factorization is bounded.

This is done by obtaining a bound on the matrix $\hat{E}$ satisfying

$$\hat{Y} + \hat{E} = \hat{L}\hat{U} \tag{17}$$

where $\hat{L}$ and $\hat{U}$ are calculated factors of $\hat{Y}$, a normalized form of Y. This calculation has verified numerical stability in the problems of this application. Details of the algorithm for doing this computation are in Erisman (1972c).

This solution process has been implemented in a program TRAFFIC. Timing and storage figures at the end of this section are based on a production run from this program.

To illustrate the use of the methods discussed in the previous section, a typical problem is defined and the steps of the analysis are considered. This problem is illustrated by the following figure.

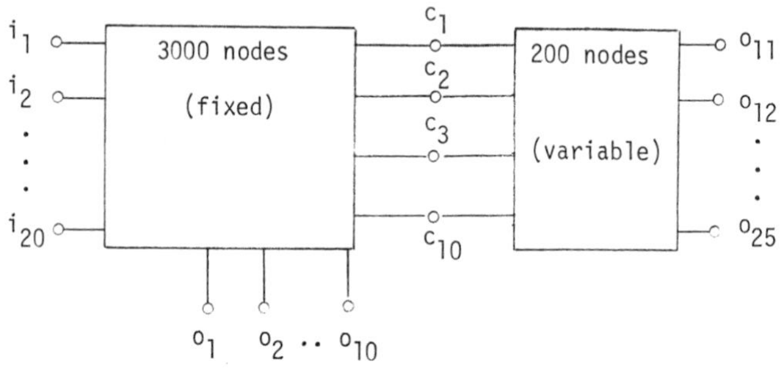

Referring to the figure, $i_1,\ldots,i_{20}$ are the 20 nodes at which non-zero input currents are defined, $o_1,\ldots,o_{25}$ are the 25 desired output voltages, and $c_1,\ldots c_{10}$ are the 10 common nodes between the two subsystems. The input and output nodes are referred to as IO nodes. In the analysis of the larger fixed network, denoted Y_1 and partitioned as (14), the factorization of

$$Y_1 = \begin{bmatrix} A_{11} & A_{21} & B_{11} \\ A_{21} & A_{41} & B_{21} \\ B_{11} & B_{21} & D_1 \end{bmatrix} = \begin{bmatrix} L_{11} & 0 & 0 \\ L_{21} & L_{41} & 0 \\ L_{31} & L_{51} & I \end{bmatrix} \begin{bmatrix} U_{11} & U_{21} & U_{31} \\ 0 & U_{41} & U_{51} \\ 0 & 0 & \hat{D}_1 \end{bmatrix} \tag{18}$$

is performed for each ω_i where A_{41} is 30×30 relating nodes $i_1,\ldots,i_{20}$, $o_1,\ldots,o_{10}$; D_1 relates nodes $c_1,\ldots c_{10}$ and is 10×10; A_{11} is 2960×2960 relating all other nodes in Y_1; the other blocks, using symmetry, are the inter-relations between A_{41}, A_{11}, and D_1. For each ω_i, the corners of the factors within the dotted lines relating the nodes of A_{41} and D_1 are saved on tape. If the outputs of the fixed subsystem are required without it being connected to the variable part, the factorization in (18) may be completed and the forward and backward substitution may be done.

Now for each variable 200 node subsystem denoted Y_2, output voltages $o_{11},\ldots,o_{25}$ are to be calculated, along with $o_1,\ldots o_{10}$ from Y_1 with the variable subsystem Y_2 attached. Let Y_2 be partitioned as in (14):

$$Y_2 = \begin{bmatrix} A_{12} & A_{22} & B_{12} \\ A_{32} & A_{42} & B_{22} \\ B_{12} & B_{22} & D_2 \end{bmatrix} \tag{19}$$

where as in the fixed network, block A_{42} relates the outputs, hence is order 15×15; D_2 relates connection nodes and is order 10×10; A_{12} relates other nodes and is 175×175; and the other blocks denote the interrelation between these.

The factorization portion for the overall solution requires factoring the 200 x 200 matrix Y_2 in (19) with D_2 replaced by $D_2 + \hat{D}_1$. (E is zero for this case.) Using condensation only the factors relating the corner 25 x 25 need to be preserved and these are denoted

$$\begin{bmatrix} L_{42} & 0 \\ L_{52} & L_{62} \end{bmatrix} \begin{bmatrix} U_{42} & U_{52} \\ 0 & U_{62} \end{bmatrix} .$$

Obtaining all of the desired output voltages using forward and backward substitution then requires solving

$$\begin{bmatrix} L_{41} & & \\ 0 & L_{42} & \\ L_{51} & L_{52} & L_{62} \end{bmatrix} \begin{bmatrix} \\ v \\ \\ \end{bmatrix} = \begin{bmatrix} b \\ \hline 0 \end{bmatrix} \tag{20}$$

$$\begin{bmatrix} U_{41} & 0 & U_{51} \\ & U_{42} & U_{52} \\ & & U_{62} \end{bmatrix} \begin{bmatrix} \\ x \\ \\ \end{bmatrix} = \begin{bmatrix} \\ v \\ \\ \end{bmatrix} . \tag{21}$$

It should be emphasized that the x produced in (21) for each ω_i is exactly, apart from round-off error, the segment of x which satisfies the 3200 x 3200 order problem (16).

A new x calculated for a new Y_2 would require repeating the factorization of the 200 x 200 matrix (22) with the same $\hat{D}_1$ matrix added to the new D_2, followed by the additional forward and backward substitution with the 55 x 55 LU matrices (20) and (21).

Observe also that if the input, b, is changed but Y_1 and Y_2 remain fixed, calculating the new x simply involves (20) and (21). No iteration or refactorization is involved.

Because of the shortened forward and backward substitution due to condensation, the dominant cost in the solution process is the factorization of Y_1. To get an idea of the computing cost, some sample timing and storage figures will be given for the solution of a sequence of problems (16). Those figures illustrate the use of condensation and sparsity, but do not reflect the repeated solution with the variable subsystem. It may be seen from development, however, that this is a minor part of the total computation.

The figures in the following table reflect the use of two other storage and time saving ideas which should be mentioned. Because in a physical system many data elements are the same, only unique values are stored for the matrix elements. Further because of symmetry, a symmetric version of the Crout formulas is used which requires calculating only the L rather than both L and U. The sparse matrix technique used is similar to the second approach described in the previous section. The generated index lists are used sequentially and hence may be conveniently stored on disk and buffered into core a piece at a time during the solution process. (These index lists are about one tenth as long as generated code using sparse matrix approach (a).) All other data is stored in core.

SAMPLE PROBLEM RUN ON CDC 6600	
Number of nodes	3,768
Number of non-zero values in matrix*	65,100
Number of non-zero values in LU*	105,600
Number of locations to store matrix	10,800
Number of locations to store needed LU	11,858
Preprocessing time (seconds)	353.
Solution time per frequency (seconds)	16.
Core region (thousand words)	300
Length of generated index lists (thousand words) =	450
Maximum $\lvert\hat{e}_{ij}\rvert$ from equation (17)	10^{-9}

* Includes both upper and lower half of matrix

4. CONCLUSIONS

Emphasis in this paper has been on efficient decomposition and interconnection methods compatible with sparsity and condensation in solving large linear systems of equations. The decomposition and interconnection has been directed toward the LU factors of the coefficient matrix rather than toward the solution or inverse. This is done for several reasons. Frequently repeated solutions with the same coefficient matrix but different right hand sides are required; some type of iteration or other interconnection of solution would have to be repeated for each case. Comparing the LU factors and the inverse, the factors are less expensive to obtain, usually more sparse, and just as efficient to use in completing the solution process.

In another application area, linear programming, modification of the LU factors is discussed by Bartels (1970) with an emphasis on numerical stability, and by Forrest (1971) with an emphasis on sparsity preservation. In these the main objective is to update the LU factors of the coefficient matrix based on the replacement of one column in the original matrix.

A more closely related method which can readily be applied to problems like those addressed in this paper is found in Iyer (1971) and Rose (1972a). It is based on an identity relating the inverse of a matrix and the inverse of one differing from it by a rank one matrix [Fadeeva (1959)]. Their application uses the LU factors rather than the inverse; this approach will be briefly reviewed.

Suppose

$$\hat{Y} = Y - \sigma uv^T \quad , \tag{22}$$

where σ is a scalar, u and v are vectors. Then using

$$(\hat{Y} + \sigma uv^T)^{-1} = \hat{Y}^{-1} - (\sigma^{-1} + v^T \hat{Y}^{-1} u)^{-1} \hat{Y}^{-1} uv^T \hat{Y}^{-1}$$

it is shown that for $\hat{Y} = \hat{L}\hat{U}$, the solution of

$$Yx = b$$

is given by

$$x = x_1 - \alpha x_2 \tag{23}$$

where

$$\hat{L}\hat{U}x_1 = b$$

$$\hat{L}\hat{U}x_2 = u$$

$$\alpha = (\sigma^{-1} + v^T x_2)^{-1} v^T x_1 \quad .$$

Thus if removing a rank one matrix from Y produces a matrix of simpler form (say block diagonal), the interconnection problem may be solved with two forward and backward substitution steps and a linear combination of the two solutions.

For example, with Y in the form

$$Y = \left[\begin{array}{c|c} A & * \\ \hline * & B \end{array}\right] \tag{24}$$

it is easy to remove the $*$'s by a rank one change in Y (this alters the corner positions of A and B). Then the solution of $Yx = b$ may easily be obtained by piecing together two solutions as in (23). Using the approach of this paper, the overall solution would be achieved using one forward and backward substitution on the modified LU matrix. The additional calculations required for this involve factoring an 2 x 2 matrix of the form

$$\begin{bmatrix} u_1 & * \\ * & u_2 \end{bmatrix}$$

where the $*$'s corresponds to (24) and play the role of the E matrix in (6), u_1 is the lower corner element in U in the factorization of A, and u_2 is the lower corner element in the U factorization of B.

This simplified form may be achieved provided the numerically stable reordering will permit the interconnection position to be last in A and B. This is not a requirement for the other method.

In the case where Y is denoted by

$$Y = \begin{bmatrix} a_{11} & a_{12} & & & a_{13} \\ a_{21} & a_{22} & & & a_{23} \\ & & b_{11} & b_{12} & b_{13} \\ & & b_{21} & b_{22} & b_{23} \\ a_{31} & a_{32} & b_{31} & b_{32} & a_{33}+b_{33} \end{bmatrix}$$

and the factors of A and B are known, this paper provides an easy way to interconnect them. Here it is not a matter of removing elements (like for example a_{31}, a_{32}, a_{13}, a_{23}) in the matirx to achieve a simpler form.

5. REFERENCES

Bartels, R. H., Golub, G. H., and Saunders, M. A., (1970), Numerical techniques in mathematical programming, SLAC-PUB-760, STAN-CS-70-162.

Erisman, A. M., (1972a), Sparse matrix approach to frequency domain analysis of linear passive electrical networks, in Sparse Matrices and Their Application, p. 31.

Erisman, A. M., and Spies, G. E., (1972b), Exploiting problem characteristics in the sparse matrix approach to frequency domain analysis, IEEE Trans. CT-19, p. 260.

Erisman, A. M., (1972c), Monitoring numerical stability in symmetric sparse matrix computation, Boeing Computer Services report.

Faddeeva, V. N., (1959), Computational Methods of Linear Algebra, Dover, New York, (Translated by C. D. Benster), p. 105.

Forrest, J. J. H., and Tomlin, J. A. (1971), Updating triangular factors of the basis to maintain sparsity in product form simplex method, SCICON report.

Gustavson, F. G., Liniger, W. M., and Willoughby, R. A., (1970), Symbolic generation of an optimal Crout algorithm for sparse systems of linear equations, J. Assoc. Comput. Mach. 17, p. 87.

Hachtel, G. D., (1972), Vector and matrix variability type in sparse matrix algorithms, in Sparse Matrices and their Application, p. 53.

Hachtel, G. D., Brayton, R. K., and Gustavson, F. G., (1971), The sparse tableau approach to network analysis and design, IEEE Trans. CT-18, p. 101.

Iyer, C., (1971), Computer analysis of large scale systems, Ph.d. Thesis, University of Hawaii.

Lee, H., (1968), An implementation for Gaussian elimination for sparse systems of linear equations, in Sparse Matrix Proceedings, p. 75.

Reid, J. K. (Editor), (1970), Large Sparse Sets of Linear Equations, Academic Press, London.

Rose, D. J., and Bunch, J. R., (1972a), The role of partitioning in the numerical solution of sparse systems, in Sparse Matrices and their Application, p. 177.

Rose, D. J., and Willoughby, R. A. (Editors), (1972b), Sparse Matrices and Their Applications, Plenum Press, New York.

Tinney, W. F. and Walker, J. W., (1967), Direct solutions of sparse network equations by optimally ordered triangular factorization, Proc. IEEE 55, p. 1801.

Willoughby, R. A. (Editor), (1968), Sparse Matrix Proceedings, IBM Research, RA1, Yorktown Heights, New York.

GENERAL MODIFICATIONS OF LARGE SETS OF SPARSE LINEAR EQUATIONS

JOHN R. ROY

Institut für Statik und Dynamik

Universität Stuttgart

7 Stuttgart-Vaihingen

Abstract:

Due to the high computation costs incurred in solving large sets of linear equations together with the normal practical requirement of arbitrarily modifying the problem during the design period, general modification procedures are necessary to circumvent a complete re-analysis. In order to preserve sparsity, such methods are based upon the existence of the triangularized factor of the original coefficient matrix, which is taken here to be positive-definite. Expressions are presented to directly give the change in the solution matrix caused by modifications including combinations of (i) removal and/or addition of complete rows (and the corresponding columns), (ii) modifying a group of terms in the coefficient matrix which can be reduced by an appropriate transformation to a symmetric sub-matrix along the diagonal (such as occurring upon the changing of properties of connecting elements or branches), (iii) the introduction of linear dependencies between unknowns and (iv) the disconnection of certain unknowns at nodes. Several common modification configurations are illustrated and shown to be combinations of the above classes. In the above procedures the ordering of the calculations is determined in order to minimize the amount of work. Back substitution is postponed as long as possible and positive-definiteness is maintained at all intermediate steps. Nevertheless, a break-even point is reached, depending on the ratio of the total number of modified rows to the average bandwidth of the coefficient matrix, when re-solution becomes cheaper than the modification approach.

1. INTRODUCTION

The high calculation speeds and large storage capacities of modern computers are permitting the solution of large sets of linear equations. Such equations often arise from mathematical models of complicated physical systems in which a set of trial parameters often needs to be initially estimated before the first solution under the imposed set of environmental conditions can be carried out. Especially for such large problems, this first trial solution may often indicate unsatisfactory behaviour in certain areas of the system. Thus it is often necessary, with the help of optimization techniques or trial-and-error methods, to repeat the analysis several times under a somewhat revised set of trial parameters and environmental conditions. In order to obviate a complete re-analysis in such situations, modification techniques can be often profitably used. This paper is related to the common case where the coefficient matrix to be modified is positive-definite.

The first decision is that of using an iterative or a direct procedure. Iterative methods are generally preferable when relatively minor modifications are to be introduced throughout a large part of the system. However, not only is the amount of calculation for such methods strongly dependent on the number of right-hand sides, but convergence difficulties can also at times arise, particularly for ill-conditioned problems. On the other hand, direct methods always give reliable results when sufficient decimal places are available in the computer for the representation and processing of the various matrices. Nevertheless, particularly for large sparse matrices, direct procedures possess a break-even point, depending on the amount of the system to be modified, where a re-analysis becomes cheaper than the modification approach. The present study is restricted to direct methods and is thus appropriate when the number of modifications is not too large.

Modification techniques have been in use for some years in the field of structural analysis. In the pioneering dual formulation of Argyris (2), (3), closed form expressions were developed for the change in the solution matrix resulting from arbitrarily changing the properties of the connecting elements. During the intervening period several rather similar procedures were proposed, as described in detail by Kavlie and Powell (7). In (4), Argyris, the writer and colleagues re-formulated the method in positive-definite form using the Sherman-Morrison formula, thus providing an algorithm especially adapted to the modification of large sparse systems. Particularly for coefficient matrices resulting from the discretization of a continuous system, desired modifications may include adding and/or subtracting complete rows (and the corresponding columns) to and from the coefficient matrix respectively,as well as the changing of the properties of the connecting elements or branches. To the author's knowledge, the first general algorithms for handling this situation were presented in (5) and (8). In addition, algorithms have been developed in (8) for treating as modifications the introduction of linear dependencies between unknowns and the de-coupling of branch unknowns at nodes. These procedures are re-formulated here, as fas as space allows.

Although the methods presented are capable of handling full matrices, special precautions have been taken to enable efficient treatment of sparse matrices. For instance, the original matrix is assumed to have been reduced to triangularized or decomposed form only. Although the availability of the original inverse may at times permit a faster run through the modification procedure, the initial calculation of the inverse is often prohibitively expensive and its storage on disk or drum may be impossible. Back-substitution, which also destroys sparsity, is postponed to the last possible moment, i.e. to the final calculation of the change of solution matrix. Wherever appropriate, symmetry is maintained in matrix products and intermediate matrices to be triangularized are derived in positive-definite form. In this way the number of calculations is minimized, particularly when the modifications occur at the higher numbered nodes.

As mentioned by Stewart and Baty (9) in their paper on the related topic of dissection of structures, simpler derivations result when the various matrices are treated in partitioned form. Although in the modification problem, as distinct from the dissection problem, the nodes to be modified can be distributed arbitrarily throughout the coefficient matrix, it is possible by introducing certain Boolean transformations to develop the method in terms of partitioned matrices. By appropriate sequencing of the matrix operations one can nevertheless avoid physical re-ordering on any of the large matrices and so reduce the data transfer costs for problems which cannot be handled completely within central memory. As the original or partly reduced sub-matrices along the diagonal of a positive-definite matrix are themselves positive-definite, the derivation in partitioned form guarantees the full exploitation of positive-definiteness.

2. PRELIMINARY RELATIONS

Consider the equation system of the original problem as

$$Kr = R \tag{1}$$

Although eq. 1 could represent any set of positive-definite linear equations, it is in fact the equation system for the Matrix Displacement Method (1) of structural analysis, solved for the unknown nodal point displacements r under the imposed nodal point loading system R for a structure with global stiffness matrix K. Matrix K has been assembled from the contributions of element or branch stiffnesses to the nodes, according to

$$K = a^t k a \tag{2}$$

where k is the diagonal super-matrix of element stiffnesses and a the Boolean connectivity matrix. Although the presented method does not require that the coefficient matrix be assembled from a relation analogous to eq. 2, some of the modification conditions can be interpreted as changes in the nodal point matrix K due to changes in properties of connecting elements or branches in k. During the original solution of eq. 1, it is assumed that matrix K has been already triangularized. As the Cholesky method has proved to be especially suitable for the handling of such large systems (6), K is regarded as having been triangularized into the upper triangular matrix U, i.e.

$$U^t U = K \tag{3}$$

Matrix K is considered as re-ordered by rows and columns into three separate families according to any proposed set of modifications: 1. family i for the n_i un-modified unknowns, 2. family m for the n_m unknowns occurring in a modified n_m by n_m symmetric block along the diagonal, such as arising from the change of certain element or branch properties through eq. 2, and 3. family r for the n_r freedoms which are either going to be removed or assigned given values, where $n = n_i + n_m + n_r$ is the total number of unknowns in K. If F is defined as the inverse of K and is taken as correspondingly re-ordered, the following can be written:

$$\begin{bmatrix} F_{rr} & F_{mr}^t & F_{ir}^t \\ F_{mr} & F_{mm} & F_{im}^t \\ F_{ir} & F_{im} & F_{ii} \end{bmatrix} \begin{bmatrix} K_{rr} & K_{mr}^t & K_{ir}^t \\ K_{mr} & K_{mm} & K_{im}^t \\ K_{ir} & K_{im} & K_{ii} \end{bmatrix} = \begin{bmatrix} I_r & 0 & 0 \\ 0 & I_m & 0 \\ 0 & 0 & I_i \end{bmatrix} \tag{4}$$

where the right-hand side represents the identity matrix. In order to solve for the submatrices K in terms of F, a forward pass of Gaussian elimination is performed on eq. 4. As we expect that $n_r << n_i$ and $n_m << n_i$, we avoid the expensive formation of F_{ii}^{-1} by placing family i last, as shown above. Thus eq. 4 can be reduced to

$$\begin{bmatrix} F_{rr} & F_{mr}^t & F_{ir}^t \\ 0 & V_{mm} & W_{im}^t \\ 0 & 0 & X_{ii} \end{bmatrix} \begin{bmatrix} K_{rr} & K_{mr}^t & K_{ir}^t \\ K_{mr} & K_{mm} & K_{im}^t \\ K_{ir} & K_{im} & K_{ii} \end{bmatrix} = \begin{bmatrix} I_r & 0 & 0 \\ -F_{mr}F_{rr}^{-1} & I_m & 0 \\ Y_{ir} & -W_{im}V_{mm}^{-1} & I_i \end{bmatrix} \tag{5}$$

where $V_{mm} = (F_{mm} - F_{mr}F_{rr}^{-1}F_{mr}^t)$; $\quad W_{im} = (F_{im} - F_{ir}F_{rr}^{-1}F_{mr}^t)$

$X_{ii} = (F_{ii} - F_{ir}F_{rr}^{-1}F_{ir}^t - W_{im}V_{mm}^{-1}W_{im}^t)$ and

$Y_{ir} = W_{im}V_{mm}^{-1}F_{mr}F_{rr}^{-1} - F_{ir}F_{rr}^{-1}$

in which V_{mm} can be readily shown to be positive-definite. By a back substitution process through the last two rows of eq. 5, the following expressions in the form required later are obtained:

$$K_{ii}^{-1} = X_{ii}\ ; \quad K_{ii}^{-1}K_{im} = -W_{im}V_{mm}^{-1}\ ; \quad K_{ii}^{-1}K_{ir} = Y_{ir} \tag{6}$$

$$(K_{mr} - K_{im}^tK_{ii}^{-1}K_{ir}) = -V_{mm}^{-1}F_{mr}F_{rr}^{-1}\ ; \quad (K_{mm} - K_{im}^tK_{ii}^{-1}K_{im}) = V_{mm}^{-1}$$

In general, matrices K and F are not originally ordered in the above partitioned form, but must be converted to such a form by means of Boolean transformations. Now in order to evaluate the terms of eq. 6, we need the submatrices of matrix F, which can be expressed by the following Boolean equations:

$$F_{ii} = \tilde{b}_i\, F\, \tilde{b}_i^t \qquad F_{im} = \tilde{b}_i\, F\, \tilde{b}_m^t \qquad F_{ir} = \tilde{b}_i\, F\, \tilde{b}_r^t$$

$$F_{mm} = \tilde{b}_m\, F\, \tilde{b}_m^t \qquad F_{mr} = \tilde{b}_m\, F\, \tilde{b}_r^t \qquad F_{rr} = \tilde{b}_r\, F\, \tilde{b}_r^t \tag{7}$$

where $\tilde{b}_i$ is a Boolean matrix (which can be stored as a one-dimensional array) containing n_i rows, each of which contains all zeros except for one unit value, located for each row at the column number, taken in increasing order, where that row of family i occurs in K or F. Boolean matrices $\tilde{b}_m$ and $\tilde{b}_r$ are the corresponding matrices for families m and r respectively. As the Cholesky reduction of eq. 3 has already performed part of the work towards forming matrix F, its result U can be used to carry out the following forward substitution operations on both $\tilde{b}_m^t$ and $\tilde{b}_r^t$, i.e.

$$Z_m = U^{-t}\,\tilde{b}_m^t \qquad\qquad Z_r = U^{-t}\,\tilde{b}_r^t \tag{8}$$

(where -t formally represents the transpose of the inverse). The results Z_m and Z_r are then used as follows to evaluate the small submatrices of F:

$$F_{mm} = Z_m^t Z_m \qquad\qquad F_{mr} = Z_m^t Z_r \qquad\qquad F_{rr} = Z_r^t Z_r \tag{9}$$

Note that as the large submatrices F_{ii}, F_{im} and F_{ir} are always associated in later expressions as pre-multipliers of matrices with l columns, where $l << n$, they should never be formed explicitly. The right-hand side matrix R and the solution matrix r can be correspondingly written in re-ordered form:

$$\begin{aligned} &R_i = \tilde{b}_i R\,; \quad R_m = \tilde{b}_m R\,; \quad R_r = \tilde{b}_r R\,; \quad R = \tilde{b}_i^t R_i + \tilde{b}_m^t R_m + \tilde{b}_r^t R_r \\ &r_i = \tilde{b}_i r\,; \quad r_m = \tilde{b}_m r\,; \quad r_r = \tilde{b}_r r\,; \quad r = \tilde{b}_i^t r_i + \tilde{b}_m^t r_m + \tilde{b}_r^t r_r \end{aligned} \tag{10}$$

3. VARIOUS MODIFICATION PROCEDURES

3.1 General Case

The partitioned form of eq. 1 can be expressed for the original system as

$$\begin{bmatrix} K_{ii} & K_{im} & K_{ir} \\ K_{im}^t & K_{mm} & K_{mr} \\ K_{ir}^t & K_{mr}^t & K_{rr} \end{bmatrix} \begin{bmatrix} r_i \\ r_m \\ r_r \end{bmatrix} = \begin{bmatrix} R_i \\ R_m \\ R_r \end{bmatrix} \tag{11}$$

and for the modified system as

$$\begin{bmatrix} K_{ii} & K_{im} & K_{ie} \\ K_{im}^t & \tilde{K}_{mm} & K_{me} \\ K_{ie}^t & K_{me}^t & K_{ee} \end{bmatrix} \begin{bmatrix} \tilde{r}_i \\ \tilde{r}_m \\ r_e \end{bmatrix} = \begin{bmatrix} R_i - K_{ir}\bar{r}_r \\ \tilde{R}_m - K_{mr}\bar{r}_r \\ R_e - K_{er}\bar{r}_r \end{bmatrix} \tag{12}$$

where family e represents the n_e rows (and columns) to be added to the system and $\bar{r}_r$ represents the matrix of prescribed solution values for family r. The change $\tilde{K}_\Delta$ imposed on the symmetric matrix K_{mm} and the change $\tilde{R}_\Delta$ specified for the right-hand side matrix R_m are seen to be respectively

$$\tilde{K}_\Delta = \tilde{K}_{mm} - K_{mm}\,; \qquad\qquad \tilde{R}_\Delta = \tilde{R}_m - R_m \tag{13}$$

where $\tilde{K}_\Delta$ can be singular even when the complete coefficient matrix of the modified system is non-singular. The present derivation handles this situation without any difficulty whatsoever. Note, that using the orthonormal properties of the $\tilde{b}$ matrices illustrated by eq. 10, we can express the modified coefficient matrix $\tilde{K}$ due to element modifications alone, as

$$\tilde{K} = K + \tilde{b}_m^t\,\tilde{K}_\Delta\,\tilde{b}_m \tag{14}$$

If the original matrix K has been assembled from a symmetric Boolean transformation as in eq.2, $\tilde{K}_\Delta$ could be interpreted as resulting from the following Boolean transformation on the change matrix k_Δ of element or branch properties, i.e.

$$\tilde{K}_\Delta = \tilde{a}_m^t\,k_\Delta\,\tilde{a}_m \qquad \text{with} \qquad \tilde{a}_m = a_m\,\tilde{b}_m^t \tag{15}$$

in which the Boolean matrix $\tilde{a}_m$ (which can be stored as a one-dimensional array) has the same number of rows as that part a_m of the Boolean matrix a encompassing the modified elements or branches. It has as many columns as $\widetilde{b}_m$ has rows and is built one row at a time from a_m, by searching $\widetilde{b}_m$ for a row r_1 identical to the current row in a_m, and inserting unity at column r_1 of the current row of $\widetilde{a}_m$ (see fig. 1). Note, that in order to build the submatrices of K associated with family e, it is necessary to update the parts of the a matrix, in an analogous way as for $\widetilde{a}_m$, for all elements containing e type nodes. The right-hand side results for the prescribed r family are expressed as

$$\widetilde{R}_r = K_{ir}^t \widetilde{r}_i + K_{mr}^t \widetilde{r}_m + K_{er}^t r_e + K_{rr} \overline{r}_r \tag{16}$$

However, if matrix K is formed similarly as in eq. 2, it is usually simpler to calculate $\widetilde{R}_r$ etc. by pre-multiplying the modified solution matrix $\widetilde{r}$, transformed into branch form, by the modified matrix $(k + k_\Delta)$ of branch properties. Defining the change of the solution matrices as $\widetilde{r}_{\Delta i}$, $\widetilde{r}_{\Delta m}$ and $\widetilde{r}_{\Delta r}$, i.e.

$$\widetilde{r}_{\Delta i} = \widetilde{r}_i - r_i ; \qquad \widetilde{r}_{\Delta m} = \widetilde{r}_m - r_m ; \qquad \widetilde{r}_{\Delta r} = \overline{r}_r - r_r \tag{17}$$

then eq. 11 can be subtracted from eq. 12 to yield

$$\begin{bmatrix} K_{ii} & K_{im} & K_{ie} \\ K_{im}^t & \widetilde{K}_{mm} & K_{me} \\ K_{ie}^t & K_{me}^t & K_{ee} \end{bmatrix} \begin{bmatrix} \widetilde{r}_{\Delta i} \\ \widetilde{r}_{\Delta m} \\ r_e \end{bmatrix} = \begin{bmatrix} - K_{ir} \widetilde{r}_{\Delta r} \\ \widetilde{R}_\Delta - \widetilde{K}_\Delta r_m - K_{mr} \widetilde{r}_{\Delta r} \\ R_e - K_{ie}^t r_i - K_{me}^t r_m - K_{er} \overline{r}_r \end{bmatrix} \tag{18}$$

Using the first of eq. 18 to express $\widetilde{r}_{\Delta i}$ as

$$\widetilde{r}_{\Delta i} = - K_{ii}^{-1} (K_{ir} \widetilde{r}_{\Delta r} + K_{im} \widetilde{r}_{\Delta m} + K_{ie} r_e) \tag{19}$$

then $\widetilde{r}_{\Delta i}$ can be eliminated from the last two equations of eq. 18, leading to

$$\begin{aligned} &(K_{mm} + \widetilde{K}_\Delta - K_{im}^t K_{ii}^{-1} K_{im}) \widetilde{r}_{\Delta m} + (K_{me} - K_{im}^t K_{ii}^{-1} K_{ie}) r_e = \\ &\qquad \widetilde{R}_\Delta - \widetilde{K}_\Delta r_m - (K_{mr} - K_{im}^t K_{ii}^{-1} K_{ir}) \widetilde{r}_{\Delta r} \\ &(K_{me} - K_{im}^t K_{ii}^{-1} K_{ie})^t \widetilde{r}_{\Delta m} + (K_{ee} - K_{ie}^t K_{ii}^{-1} K_{ie}) r_e = \\ &\qquad R_e - K_{me}^t r_m - K_{ie}^t (r_i - K_{ii}^{-1} K_{ir} \widetilde{r}_{\Delta r}) \end{aligned} \tag{20}$$

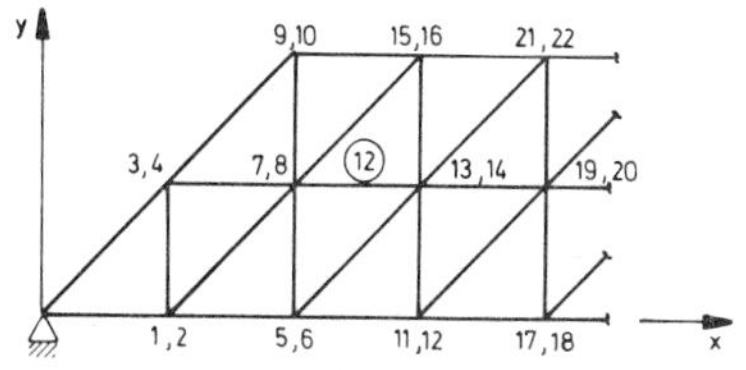

Typical Framework, in which the Stiffness of Bar 12 is to be modified by δk

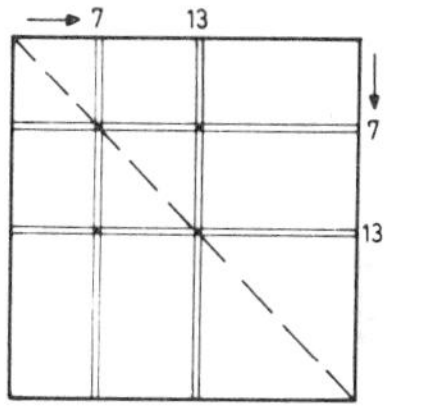

Positions of Changes within Matrix **K**

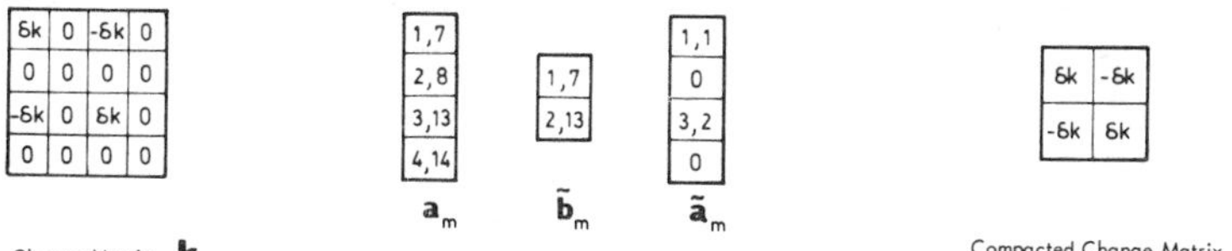

Change Matrix $\mathbf{k}_\Delta$ in Element Stiffness

Boolean Matrices stored as One-dimensional Arrays

Compacted Change Matrix $\widetilde{\mathbf{K}}_\Delta$

Fig. 1 MODIFICATION OF ELEMENTS

Now eq. 20 can be written in the following compact form

$$\begin{bmatrix} Y_{mm} & Y_{me} \\ Y_{me}^t & Y_{ee} \end{bmatrix} \begin{bmatrix} \tilde{r}_{\Delta m} \\ r_e \end{bmatrix} = \begin{bmatrix} P_m \\ P_e \end{bmatrix} \tag{21}$$

where the Y submatrices can be expressed using eqs. 5 to 10 as follows, with (22)

$$U_r^t U_r = F_{rr}\,;\quad X_{rm} = U_r^{-t} F_{mr}^t\,;\quad V_{mm} = (F_{mm} - X_{rm}^t X_{rm})\,;\quad Y_{mm} = (V_{mm}^{-1} + \widetilde{K}_\Delta)$$

for Y_{mm} (which can be shown to be positive-definite even when $\widetilde{K}_\Delta$ is singular), as well as

$$U_v^t U_v = V_{mm}\,;\quad K_e = \tilde{b}_i^t K_{ie}\,;\quad T_e = U^{-t} K_e\,;\quad X_{re} = U_r^{-t} (Z_r^t T_e) \tag{23}$$

$$X_{me} = U_v^{-t} (X_{rm}^t X_{re} - Z_m^t T_e)\,;\quad W_{me} = U_v^{-1} X_{me}\,;\quad Y_{me} = (K_{me} - W_{me})$$

for Y_{me}, and finally for Y_{ee}

$$Y_{ee} = (K_{ee} - T_e^t T_e + X_{re}^t X_{re} + X_{me}^t X_{me}) \tag{24}$$

The right-hand side submatrices P_m and T_e are similarly given as

$$W_r = U_r^{-t} \tilde{r}_{\Delta r}\,;\quad W_m = U_v^{-t} (X_{rm}^t W_r)\,;\quad P_m = \widetilde{R}_\Delta - \widetilde{K}_\Delta \tilde{b}_m r + U_v^{-1} W_m$$

$$P_e = R_e - K_e^t r - K_{er} \bar{r}_r - K_{me}^t \tilde{b}_m r - X_{re}^t W_r - X_{me}^t W_m \tag{25}$$

By block Gaussian or Cholesky reduction of eq. 21, r_e and then $\tilde{r}_{\Delta m}$ can be readily obtained (see (5) for details). From eq. 19 to 25 and by using eqs. 5 to 10, $\tilde{r}_{\Delta i}$ is given as

$$\tilde{r}_{\Delta i} = \tilde{b}_i U^{-1} \Big\{ Z_r U_r^{-1} (W_r + X_{re} r_e) - T_e r_e + (Z_r U_r^{-1} X_{rm} - Z_m) (W_{me} r_e - V_{mm}^{-1} \tilde{r}_{\Delta m} - U_v^{-1} W_m) \Big\} \tag{26}$$

in which the terms are evaluated from right to left.

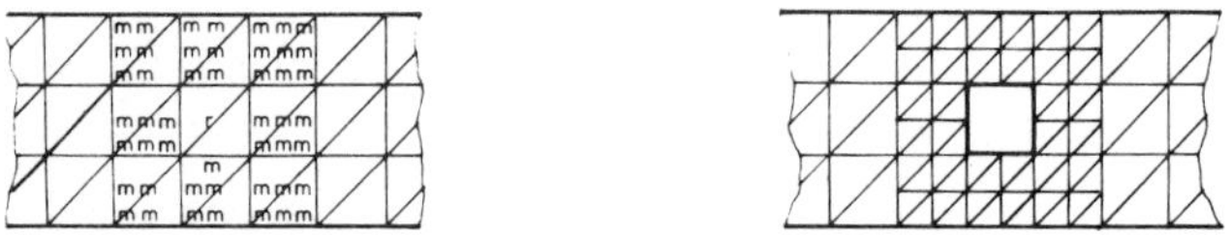

Original Continuum

Modified Continuum

Re-ordered Co-efficient Matrix of Original Continuum

Re-ordered Co-efficient Matrix of Modified Continuum

Fig. 2 INTRODUCTION OF A CUT-OUT INTO A CONTINUUM

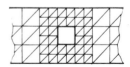

Before judging the above procedure as too complex for practical use, the reader should note that a large number of the matrix operations in eqs. 22 to 26 takes place on small matrices which will fit completely in the central memory of most computers. In making an approximate operation count for the procedure the assumption is made that the number of right-hand sides is considerably smaller than the average half-bandwidth b of matrix K. When an average distribution of corrections about the middle node numbers of K is assumed, we obtain the following number M_{mre} of multiply-accumulates:

$$M_{mre} = n\,(bn_c/2 + n_c^2/4) + n_c^3/6 + n_m n_e\,(2n_m + n_e)/2 + n_m^3/2 \qquad (27)$$

in which $n_c = n_m + n_r + n_e$ represents the total number of changed rows (or columns) in K. As the operations for re-triangularization number about $(n - n_r + n_e)\, b^2/2$, for large problems where $b << n$ the theoretical break-even point is about $n_c = 0.75\, b$. The actual break-even point could only really be determined by testing at a given computer installation.

Many practical modifications do not involve the full coupled situation derived above, but as shown in (8) allow the deletion of certain families and coupling terms. In fig. 2 above, the case of a cut-out including modification of the edge areas of the cut is illustrated. Here family e disappears and the coupling matrix K_{ir} is zero.

Upon observation of the principal terms within the operation count of eq. 27 in conjunction with eqs. 5 to 9 and eqs. 21 to 26, it can be appreciated that the major part of the calculation effort is expended in the formation of the Z and F submatrices in eqs. 8 and 9 respectively. As the content of these matrices depends on the location of the modifications and not on their magnitudes, repeated modifications of varying magnitudes on the same area of the coefficient matrix can be carried out very cheaply. Thus the method is most suitable for parameter studies over a limited part of the system. In addition, for the type of non-linear problems where the non-linear behaviour is confined to certain local areas and where a step-wise linearization solution approach is used, the method may be extended to a so-called "modification of modifications" procedure (8), in which merely those nodes entering the non-linear range at the previous step need be carried through the Z and F formation at the current step, before being finally coupled with the modified nodes already in the non-linear range.

3.2 Introduction of linear dependencies between certain unknowns. In structural analysis this situation usually occurs when rigid constraints are to be introduced between certain nodes, thus resulting in a reduction of the number of degrees of freedom within the system. The first task is to build the transformation matrix G fig. 3, which formally describes the linear dependencies between the unknowns. Here matrix K can be partitioned into three families:

1. family i for the n_i unmodified unknowns,
2. family m for the n_m unknowns that are chosen as independent within the transformation and
3. family r for the remaining n_r unknowns that are to be eliminated by the transformation.

It is practical to combine families m and r into the one family x. Thus the re-ordered equations of the original system are expressed as

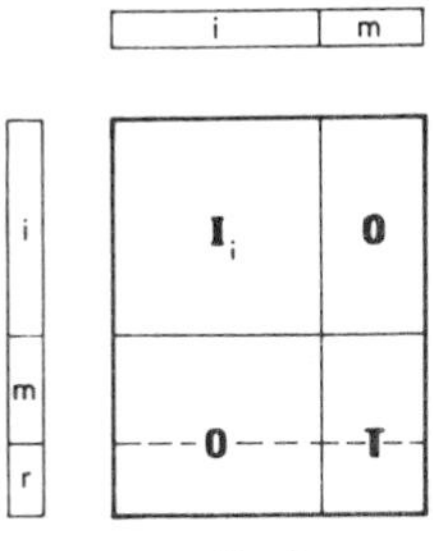

Fig. 3

TRANSFORMATION MATRIX **G** FOR THE INTRODUCTION OF LINEARLY DEPENDENT FREEDOMS IN THE RE-ORDERED COEFFICIENT MATRIX

$$\begin{bmatrix} K_{ii} & K_{ix} \\ K_{ix}^t & K_{xx} \end{bmatrix} \begin{bmatrix} r_i \\ r_x \end{bmatrix} = \begin{bmatrix} R_i \\ R_x \end{bmatrix} \tag{28}$$

The re-ordered equations of the modified system are written as:

$$\begin{bmatrix} K_{ii} & K_{ix} \\ K_{ix}^t & K_{xx} \end{bmatrix} \begin{bmatrix} \tilde{r}_i \\ \tilde{r}_x \end{bmatrix} = \begin{bmatrix} R_i \\ \tilde{R}_x \end{bmatrix} \tag{29}$$

in which we observe the part of right-hand side matrix $\tilde{R}$ at family x as modified to $\tilde{R}_x$ and define $\tilde{R}_{\Delta x} = \tilde{R}_x - R_x$, and where in the re-ordered modified solution matrix $\tilde{r}$ of eq. 29 the linear constraints according to

$$\tilde{r} = G\,\tilde{r}^* \tag{30}$$

are already taken into account. Expressing the reduced linear system for the $(n_i + n_m)$ unknowns as

$$\tilde{K}^* \; \tilde{r}^* = \tilde{R}^* \tag{31}$$

and noting that the corresponding pairs $\{\tilde{r}, \tilde{R}\}$ and $\{\tilde{r}^*, \tilde{R}^*\}$ must be equivalent with respect to their virtual work (or energy):

$$\tilde{r}^t \, \tilde{R} = \tilde{r}^{*t} \, \tilde{R}^* \tag{32}$$

then eqs. 29 to 32 yield

$$\tilde{R}^* = G^t \, \tilde{R}, \qquad \tilde{K}^* = G^t \tilde{K} G \tag{33}$$

where $\tilde{K}$ is the re-ordered co-efficient matrix of eq. 29. Thus eq. 31 may be expressed as:

$$\begin{bmatrix} K_{ii} & (K_{ix}\,T) \\ (K_{ix}\,T)^t & T^t K_{xx} T \end{bmatrix} \begin{bmatrix} \tilde{r}_i^* \\ \tilde{r}_m^* \end{bmatrix} = \begin{bmatrix} R_i \\ T^t \tilde{R}_x \end{bmatrix} \tag{34}$$

where as shown in fig. 3, matrix T is the $(n_m + n_r) \times n_r$ part of G containing the actual linear dependencies. Defining the change of solution matrix $(\tilde{r}_i^* - r_i)$ at family i as $\tilde{r}_{\Delta i}$ and reducing the number of rows in the second of eq. 28 by pre-multiplication with T^t, eq. 28 is subtracted from eq. 34, yielding

$$\begin{bmatrix} K_{ii} & (K_{ix}\,T) \\ (K_{ix}\,T)^t & T^t K_{xx} T \end{bmatrix} \begin{bmatrix} \tilde{r}_{\Delta i} \\ \tilde{r}_m^* \end{bmatrix} = \begin{bmatrix} K_{ix}\, r_x \\ T^t K_{xx}\, r_x + T^t \tilde{R}_{\Delta x} \end{bmatrix} \tag{35}$$

Matrix $\tilde{r}_{\Delta i}$ can be expressed from eq. 35 as:

$$\tilde{r}_{\Delta i} = - K_{ii}^{-1} \, K_{ix} \, (T \, \tilde{r}_m^* - r_x) \tag{36}$$

which is substituted into the second of eq. 35 to yield for $\tilde{r}_m^*$

$$\left[T^t (K_{xx} - K_{ix}^t K_{ii}^{-1} K_{ix}) \, T \right] \tilde{r}_m^* = T^t \left[\tilde{R}_{\Delta x} + (K_{xx} - K_{ix}^t K_{ii}^{-1} K_{ix}) \, r_x \right] \tag{37}$$

If $\tilde{b}_x$, Z_x and F_{xx} are defined for family x identically as $\tilde{b}_m$, Z_m and F_{mm} for family m, eqs. 5 to 9 in the absence of family r yield

$$(K_{xx} - K_{ix}^t K_{ii}^{-1} K_{ix}) = F_{xx}^{-1} \qquad K_{ii}^{-1} K_{ix} = -\tilde{b}_i \, U^{-1} Z_x \, F_{xx}^{-1} \tag{38}$$

If U_x is defined as the Cholesky factor of F_{xx} and V_x as the result of forward-substitution with U_x on T i.e. $V_x = U_x^{-t}\, T$, then eq. 37 can be reduced to the following form

$$(V_x{}^t V_x)\,\tilde{r}_m^* = V_x^t U_x^{-t} r_x + T^t \tilde{R}_{\Delta x} \tag{39}$$

which can be directly solved for $\tilde{r}_m^*$. Using eq. 38, then eq. 36 gives $\tilde{r}_{\Delta i}$ as

$$\tilde{r}_{\Delta i} = \tilde{b}_i U^{-1} Z_x U_x^{-1} U_x^{-t} (T\tilde{r}_m^* - r_x) \tag{40}$$

In order to avoid forming matrix $\tilde{b}_i$ and simultaneously obtain the total change of solution matrix $\tilde{r}_\Delta$ in the original order, the last of eq. 10 is used in incremental form to yield

$$\tilde{r}_\Delta = \tilde{b}_i^t \tilde{r}_{\Delta i} + \tilde{b}_x^t (T\tilde{r}_m^* - r_x) \tag{41}$$

By substituting eq. 40 into eq. 41, then using the identity $\tilde{b}_x U^{-1} Z_x F_{xx}^{-1} = I_x$ and the orthonormal properties of the Boolean matrices, $\tilde{b}_i^t \tilde{b}_i + \tilde{b}_x^t \tilde{b}_x = I$, eq. 41 may be put into the elegant form

$$\tilde{r}_\Delta = U^{-1} Z_x U_x^{-1} U_x^{-t} (T\tilde{r}_m^* - r_x) \tag{42}$$

which after solution for $\tilde{r}_m^*$ from eq. 39 and with the substitution $r_x = \tilde{b}_x r$ can be solved directly for $\tilde{r}_\Delta$. An operation count (8) leads to an average break-even point of $(n_m + n_r) = 0.75\,b$, which is seen to be related to that in Sec. 3.1.

3.3 Decoupling of unknowns at certain nodes. This situation arises in problems of crack propagation (fig. 4) or contact of elastic bodies, where certain element or branch unknowns, which were originally connected at their common nodes, become decoupled.

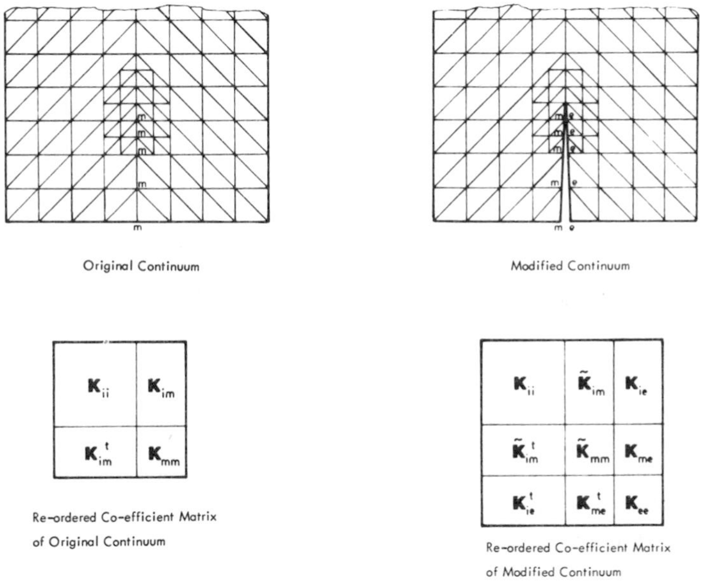

Fig. 4 DECOUPLING OF NODAL UNKNOWNS IN A CONTINUUM

The re-ordered equations of the original system can be written as

$$\begin{bmatrix} K_{ii} & K_{im} \\ K_{im}^t & K_{mm} \end{bmatrix} \begin{bmatrix} r_i \\ r_m \end{bmatrix} = \begin{bmatrix} R_i \\ R_m \end{bmatrix} \tag{43}$$

where family m represents the nodal unknowns where decouplings are to be introduced. As a result of the modification, the information in matrices K_{mm} and K_{im} (also K_{im}^t) is to be "stretched out" in the m direction into a new family e, thus leading for the modified system to

$$\begin{bmatrix} K_{ii} & \tilde{K}_{im} & K_{ie} \\ \tilde{K}_{im}^t & \tilde{K}_{mm} & K_{me} \\ K_{ie}^t & K_{me}^t & K_{ee} \end{bmatrix} \begin{bmatrix} \tilde{r}_i \\ \tilde{r}_m \\ r_e \end{bmatrix} = \begin{bmatrix} R_i \\ \tilde{R}_m \\ R_e \end{bmatrix} \quad (44)$$

In contrast to the case of element or branch modifications, we see that both K_{mm} and K_{im} are affected by the modification. In practice, matrices $\tilde{K}_{im}$, K_{ie}, $\tilde{K}_{mm}$, K_{me} and K_{ee} need to be assembled from the branch matrices k of branches incident at the nodes to be decoupled. For this purpose, the matrix $\tilde{a}_m$ of eq. 15 should be assembled for the three families i, m and e.

The first step in the derivation is to subtract the first of eq. 43 from that of eq. 44, i.e.

$$\begin{bmatrix} K_{ii} & \tilde{K}_{im} & K_{ie} \\ \tilde{K}_{im}^t & \tilde{K}_{mm} & K_{me} \\ K_{ie}^t & K_{me}^t & K_{ee} \end{bmatrix} \begin{bmatrix} \tilde{r}_{\Delta i} \\ \tilde{r}_m \\ r_e \end{bmatrix} = \begin{bmatrix} K_{im}\, r_m \\ \tilde{R}_m - \tilde{K}_{im}^t\, r_i \\ R_e - K_{ie}^t\, r_i \end{bmatrix} \quad (45)$$

The first equation of eq. 45 gives the following relation for $\tilde{r}_{\Delta i}$

$$\tilde{r}_{\Delta i} = K_{ii}^{-1} (K_{im}\, r_m - \tilde{K}_{im}\, \tilde{r}_m - K_{ie}\, r_e) \quad (46)$$

After eliminating $\tilde{r}_{\Delta i}$ from the rest of eq. 45, there results

$$\begin{bmatrix} (\tilde{K}_{mm} - \tilde{K}_{im}^t K_{ii}^{-1} \tilde{K}_{im}) & (K_{me} - \tilde{K}_{im}^t K_{ii}^{-1} K_{ie}) \\ (K_{me} - \tilde{K}_{im}^t K_{ii}^{-1} K_{ie})^t & (K_{ee} - K_{ie}^t K_{ii}^{-1} K_{ie}) \end{bmatrix} \begin{bmatrix} \tilde{r}_m \\ r_e \end{bmatrix} = \begin{bmatrix} \tilde{R}_m - \tilde{K}_{im}^t (K_{ii}^{-1} K_{im}\, r_m + r_i) \\ R_e - K_{ie}^t (K_{ii}^{-1} K_{im}\, r_m + r_i) \end{bmatrix} \quad (47)$$

Using eqs. 5 to 9 (in the absence of family r) there can be written

$$K_{ii}^{-1} = F_{ii} - F_{im} F_{mm}^{-1} F_{im}^t = \tilde{b}_i (K^{-1} - U^{-1} Z_m F_{mm}^{-1} Z_m^t U^{-t}) \tilde{b}_i^t$$
$$K_{ii}^{-1} K_{im} = - F_{im} F_{mm}^{-1} = - \tilde{b}_i U^{-1} Z_m F_{mm}^{-1} \quad (48)$$

As for eqs. 22 to 25, the following intermediate results can be obtained by using eqs. 5 to 10 and eq. 48, to yield in the solution of eq. 47

$$K_m = \tilde{b}_i^t \tilde{K}_{im} ; \quad T_m = U^{-t} K_m ; \quad K_e = \tilde{b}_i^t K_{ie} \qquad T_e = U^{-t} K_e \quad (49)$$
$$X_{me} = U_m^{-t} (Z_m^t T_e) ; \quad X_{mx} = U_m^{-t} (Z_m^t T_m) ; \quad Y_{mm} = (\tilde{K}_{mm} - T_m^t T_m + X_{mx}^t X_{mx})$$
$$Y_{me} = (K_{me} - T_m^t T_e + X_{mx}^t X_{me}) ; \quad Y_{ee} = (K_{ee} - T_e^t T_e + X_{me}^t X_{me})$$

and for the right-hand side

$$W_m = Z_m F_{mm}^{-1} r_m ; \quad P_m = (\tilde{R}_m - K_m^t r + T_m^t W_m) ; \quad P_e = (R_e - K_e^t r + T_e^t W_m) \quad (50)$$

Using the values in eqs. 49 and 50, eq. 21 can be solved directly for r_e and $\tilde{r}_m$ (replacing $\tilde{r}_{\Delta m}$) Using eq. 48, then eq. 46 can be solved finally for $\tilde{r}_{\Delta i}$ as

$$\tilde{r}_{\Delta i} = \tilde{b}_i U^{-1} \left\{ (Z_m F_{mm}^{-1} Z_m^t - I_m) (T_e\, r_e + T_m\, \tilde{r}_m) - W_m \right\} \quad (51)$$

This procedure is rather more complicated than that in Sec. 3.2. Nevertheless, as shown in (8), it possesses a break-even point of approximately $(2n_m + n_e) = 0.75$ b, and should be useful when the number of decoupled freedoms is not too large.

4. CONCLUSIONS

It is observed that quite a large variety of modifications can be handled by the procedures Secs. 3.1 to 3.3. No attempt has been made to couple the modification types of these separate sections. Although this would be theoretically possible, the resulting algorithms would be probably too cumbersome for practical use. The break-even point for the modification process is seen to depend directly on the average band-width of the coefficient matrix, and not on the total number of unknowns. The break-even points can be extended further if the modifications occur on the average at the higher numbered nodes. Note that if only the magnitude of a set of modifications is to change in future modification steps, the calculations for such steps become most rapid. This saving occurs because matrices such as Z_m, F_{mm}, Z_r etc., which are the most expensive to build, can be regarded as influence coefficients, dependent only on the location of the modifications. Lack of space has unfortunately precluded the provision of numerical examples for the procedures. The interested reader will find in (4), (5) and (8) a profusion of illustrative examples.

5. ACKNOWLEDGEMENTS

The author is above all heavily indebted to his thesis supervisor, J.H. Argyris, director of the Institut für Statik und Dynamik, University of Stuttgart. His continuing important contributions to the above field and his support in the present work were instrumental in bringing it to a successful conclusion.

6. REFERENCES

1. J.H. ARGYRIS (1954-55), Energy theorems and structural analysis, Aircraft Engineering, also (1960) p ublished in book form, Butterworths, London.

2. J.H. ARGYRIS (1956), The matrix analysis of structures with cut-outs and modifications, Communication to IX. IUTAM Conference, Brussels, p. 131.

3. J.H. ARGYRIS (1957), Die Matrizentheorie der Statik, Ingenieur-Archiv, 25, No. 3.

4. J.H. ARGYRIS, O.E. BRÖNLUND, J.R. ROY and D.W. SCHARPF (1971), A direct modification procedure for the displacement method, AIAA Journal, 9, p. 1861.

5. J.H. ARGYRIS and J.R. ROY (1972), General treatment of structural modifications, Journal of Structural Division ASCE, 98, ST2, p. 465.

6. G. von FUCHS, J.R. ROY and E. SCHREM (1972), Hypermatrix solution of large sets of symmetric positive-definite linear equations, to appear in second issue of Journal of computer methods in applied mechanics and engineering.

7. D. KAVLIE and G.H. POWELL (1972), Efficient reanalysis of modified structures, Journal of Structural Division ASCE, 97, ST1, p. 377.

8. J.R. ROY (1972) Allgemeine Modifikationsverfahren für die lineare und nichtlineare Berechnung von Tragwerken und Kontinua mit der Matrizenverschiebungsmethode, submitted to University of Stuttgart in partial fulfillment or requirements for degree of Doctor of Engineering.

9. K.L. STEWART and J.BATY (1967), Dissection of Structures, Journal of Structural Division ASCE, 93, ST5, p. 217.

A SURVEY OF DECOMPOSITION TECHNIQUES FOR ANALYSIS AND DESIGN OF ELECTRICAL NETWORKS

S. W. DIRECTOR

University of Florida
Gainesville, Florida 32601
U. S. A.

Abstract: A survey of decomposition methods which have proven useful for the analysis and design of electrical networks is presented. Special emphasis is given to sparse matrix methods and variable order, variable step-size stiff integration schemes.

1. INTRODUCTION

With the advent of integrated circuit technology, the size of the circuits which must be analyzed and designed has grown tremendously. For some classes of circuits, such as digital circuits, it is rather simple to decompose the entire circuit into smaller sections and analyze and design these individually. However, these smaller sections may still be very complex. For other circuits, such as most amplifier circuits, the circuit must be analyzed as a whole. Although some work is starting to be done on partitioning circuits into smaller blocks which can be analyzed individually, Wing (1972), most of the work in the area of analysis and design has been concerned with finding highly efficient numerical techniques.

This paper is a survey of some of the numerical techniques used for the computer analysis and design of electrical networks. Because of the limited space, not all the techniques that have been proposed are discussed, but rather only those methods which have proven to be the most useful are presented. The paper is aimed at persons who have some background in the field of decomposition methods. A strong background in circuit design is not assumed.

The paper is organized as follows. First, the types of equations which arise in the analysis and design of electrical networks are introduced. It is shown how these equations can be grouped into a set of simultaneous algebraic and differential equations. Then we discuss the three most prevalent methods of reducing these equations: nodal analysis, state-variable analysis, and the tableau approach. Next we introduce the variable-step size, variable-order numerical integration scheme which is used to discretize the derivative operator in the set of algebraic-differential equations. The Newton-Raphson algorithm is employed to ultimately transform the original problem to one of repeatedly solving sets of linear algebraic equations. We then consider the sparse matrix methods which make analysis of linear equations so efficient. Finally, we turn to the design of networks and show how this problem can also be decomposed into one of solving linear equations.

This research was supported in part by the National Science Foundation under Grant GK 27615.

2. TYPES OF EQUATIONS

Equations which arise in the analysis of electric networks fall into one of two classes. The first class contains <u>constraints of interconnection</u> and the second class the <u>branch relationships</u>. The constraints of interconnection are due to <u>Kirchhoff's laws</u>:

1) The algebraic sum of the instantaneous currents leaving any closed surface[1] must be zero for all time.
2) The algebraic sum of the instantaneous voltage drops around any loop must be zero for all time.

Consider an n_n node n_b branch network. If $\underset{\sim}{A}_a$ denotes the $(n_n \times n_b)$ <u>augmented nodal incidence</u> matrix with elements a_{ij} such that

$$a_{jk} = \begin{cases} +1 \text{ if branch k leaves node j} \\ -1 \text{ if branch k enters node j} \\ 0 \text{ if branch k doesn't touch node i} \end{cases}$$

and $\underset{\sim}{i}_b$ denotes the n_b-branch current vector, then the n_n Kirchhoff's current law (KCL) node equations can be expressed as

$$\underset{\sim}{A}_a \underset{\sim}{i}_b = \underset{\sim}{0}.$$

For example in the network graph[2] of Fig. 1, the KCL node equations are

$$\underbrace{\begin{bmatrix} -1 & 1 & 1 & 0 & 0 & 0 \\ 0 & 0 & -1 & 1 & 0 & 1 \\ 0 & -1 & 0 & 0 & -1 & -1 \\ 1 & 0 & 0 & -1 & 1 & 0 \end{bmatrix}}_{\underset{\sim}{A}_a} \underbrace{\begin{bmatrix} i_1 \\ i_2 \\ i_3 \\ i_4 \\ i_5 \\ i_6 \end{bmatrix}}_{\underset{\sim}{i}_b} = \underset{\sim}{0}$$

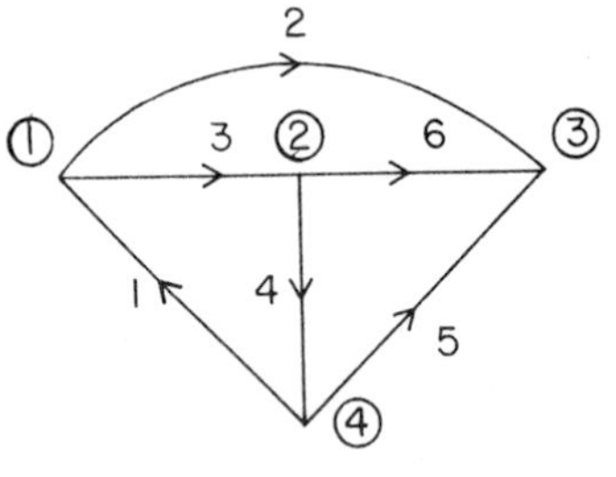

Fig. 1

[1] A node, or the point of connection of two or more branches is a special case of a closed surface.

[2] An interconnection of elements of a network is conveniently represented by a graph. The graph is obtained from the network by representing each element by a line segment or branch. Arrows are used to denote assumed directions of positive current flow. Voltage reference direction is specified by assuming current flows from + to -.

Since each column of $\tilde{A}_a$ contains one +1 and one -1, the sum of all rows equals zero and therefore $\tilde{A}_a$ is of a rank less than n_n. In fact, it can be shown, Desoer (1969), that $\tilde{A}_a$ is of rank n_n-1. The n_n-1 x n_b nodal incidence matrix $\tilde{A}$ is obtained from $\tilde{A}_a$ by designating one of the n_n nodes as the datum or reference node and eliminating the corresponding row from $\tilde{A}_a$. The KCL becomes

$$\tilde{A}\,\tilde{i}_b = \tilde{0}. \tag{1}$$

For the example above, if node ④ is chosen as the datum, then the nodal incidence matrix is

$$\tilde{A} = \begin{bmatrix} -1 & 1 & 1 & 0 & 0 & 0 \\ 0 & 0 & -1 & 1 & 0 & 1 \\ 0 & -1 & 0 & 0 & -1 & -1 \end{bmatrix}. \tag{2}$$

It is possible to express the Kirchhoff's voltage law (KVL) equations in a manner similar to that used for the KCL equations. In other words, we can define an $(n_b-n_n+1 \times n_b)$[3] incidence matrix $\tilde{B}$ whose elements b_{jk} are given by

$$b_{jk} = \begin{cases} +1 \text{ if branch } k \text{ is in loop } j \text{ and has the same orientation;} \\ -1 \text{ if branch } k \text{ is in loop } j \text{ and has the opposite orientation;} \\ 0 \text{ if branch } k \text{ is not in loop } j. \end{cases}$$

The KVL can be written as

$$\tilde{B}\,\tilde{v}_b = \tilde{0}, \tag{3}$$

where $\tilde{v}_b$ represents the n_b branch voltage vector.

Since we have n_b-n_n+1 linearly independent KVL equations in terms of n_b branch voltages, there are n_n-1 independent or basis voltages, in terms of which all other voltages can be expressed. The n_n-1 node to datum voltages, denoted by $v_{①}$, $v_{②}$, ..., or by the vector $\tilde{v}_n$ form a suitable basis. It is convenient to express KVL in terms of $\tilde{v}_n$. In particular, it is easy to show that the set of equations represented by

$$\tilde{v}_b = \tilde{A}^T \tilde{v}_n \tag{4}$$

implies (3) where T denotes transpose. Observe that for the network of Fig. 1

$$\begin{bmatrix} v_1 \\ v_2 \\ v_3 \\ v_4 \\ v_5 \\ v_6 \end{bmatrix} = \begin{bmatrix} -1 & 0 & 0 \\ 1 & 0 & -1 \\ 1 & -1 & 0 \\ 0 & 1 & 0 \\ 0 & 0 & -1 \\ 0 & 1 & -1 \end{bmatrix} \begin{bmatrix} v_{①} \\ v_{②} \\ v_{③} \end{bmatrix}$$

and the matrix premultiplying $\tilde{v}_n$ is indeed $\tilde{A}^T$ [see (2)]. In what follows, expression (4) will be used as the statement of KVL.

---

[3] It can be shown, Desoer (1969), that there are n_b-n_n+1 linearly independent KVL loop equations.

At this point we have a total of n_b+n_n-1 equations (n_n-1 KCL equations and n_b KVL equations) in terms of $2n_b+n_n+1$ unknowns (n_b branch currents, n_b branch voltages, n_n-1 node voltages). The additional n_b equations needed for solution of the unknowns $\tilde{v}_b$, $\tilde{i}_b$ and $\tilde{v}_n$ are provided by the elements in each branch. The most common types of circuit elements are described below.

The general resistive branch relationship is an algebraic expression of the form

$$f_R(v_R,i_R,p_R,t) = 0 \tag{5}$$

where the subscript R denotes the element type and p_R is an independent parameter. The two most usual types of resistive elements are the linear, time-invariant resistor, described by

$$v_R(t) - Ri_R(t) = 0,$$

and the nonlinear diode, described by

$$i_R(t) - I_S(e^{qv_R(t)/kT} -1) = 0.$$

The general capacitive branch relationships are a coupled pair of algebraic and differential equations of the form

$$f_C(q_C,v_C,p_C,t) = 0 \tag{6a}$$

and

$$i_C(t) - \frac{d}{dt} q_C(t) = 0 \tag{6b}$$

where q_C is capacitive charge. For the linear, time invariant capacitor, these equations become

$$q_C - Cv_C = 0$$

and

$$i_C - \frac{d}{dt} q_C = 0,$$

or

$$i_C - C\frac{d}{dt} v_C = 0.$$

The general inductive branch relationships are a coupled pair of algebraic and differential equations of the form

$$f_L(q_L,i_L,p_L,t) = 0 \tag{7a}$$

and

$$v_L(t) - \frac{d}{dt} q_L(t) = 0 \tag{7b}$$

where q_L is inductive flux. For the linear, time-invariant inductor, these equations become

$$q_L - Li_L = 0$$

and

$$v_L = \frac{d}{dt} q_L = 0,$$

or

$$v_L - L\frac{d}{dt} i_L = 0.$$

The independent voltage source is characterized by the algebraic equation

$$v_V(t) - \hat{v}_V(t) = 0 \tag{8}$$

where $\hat{v}_V(t)$ is some specified time function.

The independent current source is characterized by the algebraic equation of the form

$$i_I(t) - \hat{i}_I(t) = 0 \tag{9}$$

where $\hat{i}_I(t)$ is some specified time function.

All of the elements described thus far are two-terminal elements, i.e., they constitute a single branch of a network. A two-port, or four-terminal element, can be thought of as constituting two-branches of a network and is described by a coupled pair of equations which involve the two branch currents and voltages. Two-port elements can be used to characterize a wide variety of physical elements such as transistors and transformers. A general resistive two-port element is described by the equations

$$f_{T1}\ (v_{T1},\ v_{T2},\ i_{T1},\ i_{T2},\ p_{T1},\ p_{T2},\ t) = 0 \tag{10a}$$

and

$$f_{T2}\ (v_{T1},\ v_{T2},\ i_{T1},\ i_{T2},\ p_{T1},\ p_{T2},\ t) = 0 \tag{10b}$$

where the subscripts T1 and T2 are used to designate the primary and secondary branches, respectively. Some common two-ports include the voltage controlled current source, described by

$$i_{T1} = 0$$

and

$$i_{T2} - g_m\ v_{T1} = 0;$$

and the transformer, described by

$$v_{T1} - nv_{T2} = 0$$

and

$$ni_{T1} + i_{T2} = 0.$$

It is convenient to express all the branch relations by the algebraic-differential vector equations

$$\underset{\sim}{g}(\underset{\sim}{v}_b,\ \underset{\sim}{i}_b,\ \underset{\sim}{q},\ \underset{\sim}{p},\ t) = 0 \tag{11a}$$

and

$$\underset{\sim}{C}\underset{\sim}{v}_b + \underset{\sim}{D}\underset{\sim}{i}_b - \frac{d\underset{\sim}{q}}{dt} = 0 \tag{11b}$$

where $\underset{\sim}{C}$ and $\underset{\sim}{D}$ are matrices made up of +1's, and 0's and

$$\underset{\sim}{p}^T \equiv (\underset{\sim}{p}_R^T, \underset{\sim}{p}_C^T, \underset{\sim}{p}_L^T, \underset{\sim}{p}_{T1}^T, \underset{\sim}{p}_{T2}^T)$$

$$\underset{\sim}{v}_b^T \equiv (\underset{\sim}{v}_R^T, \underset{\sim}{v}_C^T, \underset{\sim}{v}_L^T, \underset{\sim}{v}_{T1}^T, \underset{\sim}{v}_{T2}^T, \underset{\sim}{v}_V^T, \underset{\sim}{v}_I^T)$$

$$\underset{\sim}{i}_b^T \equiv (\underset{\sim}{i}_R^T, \underset{\sim}{i}_C^T, \underset{\sim}{i}_L^T, \underset{\sim}{i}_{T1}^T, \underset{\sim}{i}_{T2}^T, \underset{\sim}{i}_V^T, \underset{\sim}{i}_I^T)$$

and

$$\underset{\sim}{q}^T \equiv (\underset{\sim}{q}_C^T, \underset{\sim}{q}_L^T).$$

To recapitulate, for an n_n node n_b branch network we have $2n_b + n_q + n_n - 1$ equations (n_n-1 KCL equations, n_b KVL equations and $n_b + n_q$, branch relationships, where n_q is the number of capacitors and inductors) in terms of $2n_b+n_q+n_n-1$ unknowns (n_b branch currents, n_b branch voltages, n_n-1 node voltages, and n_q capacitance charges and inductance fluxes.) Symbolically we will represent all these equations by the vector equation

$$\underset{\sim}{f}(\underset{\sim}{v}_b, \underset{\sim}{i}_b, \underset{\sim}{v}_n, \underset{\sim}{q}, \underset{\sim}{p}, t) = \underset{\sim}{0} \tag{12}$$

i.e.

$$\underset{\sim}{f}(\underset{\sim}{v}_b, \underset{\sim}{i}_b, \underset{\sim}{v}_n, \underset{\sim}{q}, \underset{\sim}{p}, t) \equiv \begin{bmatrix} \underset{\sim}{A}\underset{\sim}{i}_b \\ \underset{\sim}{v}_b - \underset{\sim}{A}^T \underset{\sim}{v}_n \\ \underset{\sim}{g}(\underset{\sim}{v}_b, \underset{\sim}{i}_b, \underset{\sim}{q}, \underset{\sim}{p}, t) \\ \underset{\sim}{C}\underset{\sim}{v}_b + \underset{\sim}{D}\underset{\sim}{i}_b - \frac{d\underset{\sim}{q}}{dt} \end{bmatrix} = \underset{\sim}{0} \tag{13}$$

This set of equations is referred to as the <u>tableau</u>.

It will be convenient in what follows to define the vector $\underset{\sim}{w}$,

$$\underset{\sim}{w}^T \equiv (\underset{\sim}{v}_b^T, \underset{\sim}{i}_b^T, \underset{\sim}{v}_n^T),$$

and the matrix $\underset{\sim}{E}$

$$\underset{\sim}{E} \equiv [\underset{\sim}{C} \,\vdots\, \underset{\sim}{D} \,\vdots\, \underset{\sim}{0}].$$

We also define the vector of algebraic and differential equations

$$\underset{\sim}{h}(\underset{\sim}{v}_b, \underset{\sim}{i}_b, \underset{\sim}{v}_n, \underset{\sim}{q}, \underset{\sim}{p}, t) \equiv \begin{bmatrix} \underset{\sim}{A}\underset{\sim}{i}_b \\ \underset{\sim}{v}_b - \underset{\sim}{A}^T \underset{\sim}{v}_n \\ \underset{\sim}{g}(\underset{\sim}{v}_b, \underset{\sim}{i}_b, \underset{\sim}{q}, \underset{\sim}{p}, t) \end{bmatrix}.$$

Then (13) may be rewritten as

$$\underset{\sim}{f}(\underset{\sim}{w}, \underset{\sim}{q}, \underset{\sim}{p}, t) = \begin{bmatrix} \underset{\sim}{h}(\underset{\sim}{w}, \underset{\sim}{q}, \underset{\sim}{p}, t) \\ \underset{\sim}{E}\,\underset{\sim}{w} - \dot{\underset{\sim}{q}} \end{bmatrix} = \underset{\sim}{0}. \tag{14}$$

3. REDUCTION OF NETWORK EQUATIONS

The term network analysis is used to designate the procedure which is used to solve the network equations represented by (14) given that the parameter vector $\underset{\sim}{p}$ is known. Later we will discuss <u>design</u> where the vector $\underset{\sim}{p}$ is allowed to vary so as to minimize some objective function. The three basic analysis procedures are described below.

The first type of analysis is based upon the reduction of the network equations (14) to a set of equations written solely in terms of an independent set of voltages and/or currents. It can be shown that for an n_n node n_b branch network there are n_n-1 independent, or basis voltages and $n_b - n_n + 1$ independent or basis currents. The n_n-1 non-datum node voltages are the most frequently used basis. Another basis set of voltages is the tree branch voltages.[4] Basis sets of currents include the mesh currents and the link currents.[5]

A distinct advantage of nodal analysis (analysis using a node voltage basis) over analysis using other basis voltages or currents is that the reduced set of equations can be written directly by inspection. Thus there is no need to perform any manipulation to arrive at the final set of equations to be solved. One disadvantage of any of the methods in this class is that not all variables of interest are computed. For example, if nodal analysis is used, and some branch current is desired two steps are required. First the branch voltage must be found by subtracting two node voltages and then the branch voltage must be substituted into the branch relationship to yield the branch current. Analysis schemes of this type are discussed further in Branin (1967), Jenkins (1971), Jensen (1968) and Nagel (1971).

The second type of analysis is based upon the reduction of the network equations to a set of simultaneous first-order differential or state equations of the form

$$\dot{\underset{\sim}{x}}(t) = \underset{\sim}{f}(\underset{\sim}{x}(t), \underset{\sim}{u}(t), t)$$

and a set of simultaneous algebraic equations of the form

$$\underset{\sim}{y}(t) = \underset{\sim}{g}(\underset{\sim}{x}(t), \underset{\sim}{u}(t), t)$$

where $\underset{\sim}{x}(t)$ is the state vector, usually capacitor charges and inductor fluxes, $\underset{\sim}{u}(t)$ the input vector and $\underset{\sim}{y}(t)$ the output vector. This form of analysis, although once quite popular, is beginning to lose interest because of the excessive amount of manipulation required to generate the state equations. Pottle (1966) gives a detailed account of this form of analysis.

Finally the last type of analysis requires no reduction at all but works directly with the set of equations (14). This last method, which turns out to be the most efficient, was proposed by Hachtel (1971a). Throughout the remaining part of the paper we assume our network equations are in the form of (14). However, the techniques to be discussed are applicable to the other analysis schemes mentioned above.

4. DISCRETIZATION OF THE DERIVATIVE OPERATOR

In Section 2 we have shown that all the equations in a network can be put in the following form

$$\underset{\sim}{f}(\underset{\sim}{w},\underset{\sim}{q},\underset{\sim}{p},t) = \begin{bmatrix} \underset{\sim}{h}(\underset{\sim}{w}, \underset{\sim}{q}, \underset{\sim}{p}, t) \\ \underset{\sim}{E}\,\underset{\sim}{w} - \dot{\underset{\sim}{q}} \end{bmatrix} = 0. \tag{15}$$

For the present discussion it is convenient to define the vector $\underset{\sim}{y}$

[4] Roughly speaking a tree is any connected set of branches of a network such that it includes all nodes but no loops.

[5] The links are the branches which are not chosen as tree branches.

$$\underset{\sim}{y} = (\underset{\sim}{w}, \underset{\sim}{q}, \underset{\sim}{p})^T$$

and allow the abuse of notation so that (15) may be written as

$$\underset{\sim}{f}(\underset{\sim}{y}, \dot{\underset{\sim}{y}}, t) = \underset{\sim}{0}. \tag{16}$$

In general (16) is a stiff set of algebraic and differential equations. That is, the equations have widely separated time constants.

Gear (1971) has proposed a very powerful algorithm for solving stiff systems of algebraic and differential equations such as (16). Included in this algorithm are the features of (1) varying the step size to preserve a prescribed error per unit time; and (2) varying the order to maximize the step size for a prescribed error. Gear's algorithm is based upon the multistep integration formula

$$\alpha_o \underset{\sim}{y}^{n+1} = h\beta_o \dot{y}^{n+1} - \sum_{j=1}^{k} (\alpha_j \underset{\sim}{y}^{n+1-j} + h\beta_j \dot{\underset{\sim}{y}}^{n+1-j}) \tag{17}$$

where $\alpha_o = 1$. If $\beta_o \neq 0$ then (17) cannot be solved explicitly for y^{n+1} because $\dot{y}^{n+1}$ appears on the right hand side. This equation would be solved iteratively for y^{n+1}, and is called a <u>corrector</u>. If the corrector iteration requires the solution of a set of simultaneous equations we say it is an <u>implicit</u> iteration. Otherwise we call it an <u>explicit</u> iteration. When $\beta_o = 0$, (17) can be solved explicitly for y^{n+1} and is called a predictor.

Consider solving (17) using an explicit iteration:

$$\alpha_o \underset{\sim}{y}^{n+1,m} = -h\beta_o \dot{\underset{\sim}{y}}^{n+1,\ m-1} - \sum_{j=1}^{k} (\alpha_j \underset{\sim}{y}^{n+1-j} + h\beta_j \dot{\underset{\sim}{y}}^{n+1-j}).$$

At the $(m+1)^{st}$ iteration this equation becomes

$$\alpha_o \underset{\sim}{y}^{n+1,\ m+1} = -h\beta_o \dot{\underset{\sim}{y}}^{n+1,\ m} - \sum_{j=1}^{k} (\alpha_j \underset{\sim}{y}^{n+1-j} + h\beta_j \dot{\underset{\sim}{y}}^{n+1-j}).$$

Subtracting these equations yields

$$\alpha_o (\underset{\sim}{y}^{n+1,\ m+1} - \underset{\sim}{y}^{n+1,\ m}) = -h\beta_o (\dot{\underset{\sim}{y}}^{n+1,m} - \dot{\underset{\sim}{y}}^{n+1,\ m-1}) \tag{18}$$

For a set of linear equations, $\dot{\underset{\sim}{y}} = \underset{\sim}{A}\, \underset{\sim}{y}$, this iteration would converge only if

$$|h\lambda_\ell| < 1, \text{ for all } \ell,$$

where the λ_ℓ are the eigenvalues of $\underset{\sim}{A}$. This implies that the step size must be chosen smaller than smallest time constant of the network. This situation is unacceptable for stiff equations as seen by the following example.

Consider the simple network of Fig. 2.

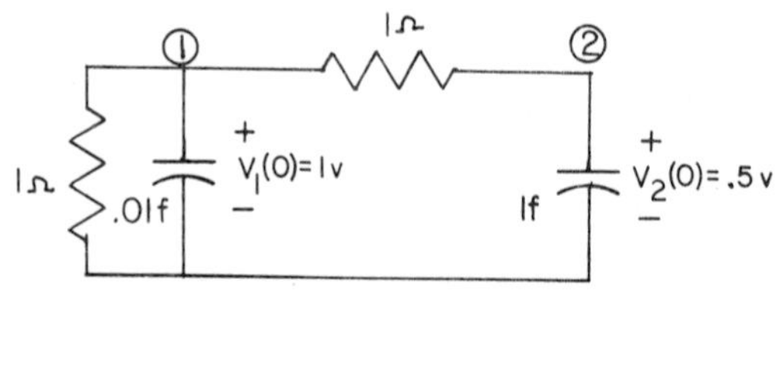

Fig. 2

The network equations in terms of the node voltage basis are

$$\begin{bmatrix} \dot{v}_{①} \\ \dot{v}_{②} \end{bmatrix} = \begin{bmatrix} -200 & 100 \\ 1 & -1 \end{bmatrix} \begin{bmatrix} v_{①} \\ v_{②} \end{bmatrix}, \quad \begin{bmatrix} v_{①}(0) \\ v_{②}(0) \end{bmatrix} = \begin{bmatrix} 1 \\ .5 \end{bmatrix}.$$

The analytic solution for $v_{②}(t)$ is

$$v_{②}(t) = .503e^{-.5t} - .00375e^{-200.5t}.$$

Suppose we wish to solve these equations numerically using the backward-Euler method and an explicit iteration.[6] Expression (18) becomes

$$\begin{bmatrix} \Delta v_{①}^{n+1,m+1} \\ \Delta v_{②}^{n+1,m+1} \end{bmatrix} = h \begin{bmatrix} -200 & 100 \\ 1 & -1 \end{bmatrix} \begin{bmatrix} \Delta v_{①}^{n+1,m} \\ \Delta v_{②}^{n+1,m} \end{bmatrix}$$

where $\Delta v^{n+1,m+1} \equiv (v^{n+1,m+1} - v^{n+1,\,m})$. This difference equation can be solved to yield

$$\begin{bmatrix} \Delta v_{①}^{n+1,m+1} \\ \Delta v_{②}^{n+1,m+1} \end{bmatrix} = \left[h \begin{bmatrix} -200 & 100 \\ 1 & -1 \end{bmatrix} \right]^{m} \begin{bmatrix} \Delta v_{①}^{n+1,0} \\ \Delta v_{②}^{n+1,\,0} \end{bmatrix}.$$

For this iteration to converge, the matrix

$$\left[h \begin{bmatrix} -200 & 100 \\ 1 & -1 \end{bmatrix} \right]^{m} \equiv (h\underset{\sim}{J})^{m}$$

must approach zero as $m \to \infty$. This will occur if the eigenvalues of the matrix $(h\underset{\sim}{J})$ lie within the unit circle. The characteristic equation of $\underset{\sim}{J}$ is

$$\lambda^2 + 201\lambda + 100 = 0$$

so that the eigenvalues are $\lambda_1 = -0.5$ and $\lambda_2 = -200.5$. For convergence of the corrector we must have

$$|-200.5h| < 1$$

or

$$h < \frac{1}{200.5}$$

which is less than the smallest time constant of the circuit. Even though the term $.0037e^{-200.5t}$ in the solution damps out after a few time steps, we are required to maintain a very small step size so that this term is constructed accurately. Therefore a considerable amount of computer time is required to determine the dominant response associated with the large time constant.

We conclude from this discussion that (17) must be solved implicitly for stiff equations. We return to this point later.

---

[6] The backward Euler method is a special case of (17) with $\beta_o = \alpha_1 = -1$ and $\beta_j = \alpha_{j+1} = 0$ for all $j \geq 1$.

There is no need to consider other methods of numerical integration such as Range-Kutta because they also require the use of pathologically small step sizes in the same way as with the use of an explicit iteration. (See Gear (1969) for a further discussion of this point.)

Stability

Let us consider the question of stability of (17). A method is termed A-stable if the difference equation which results from application of the method to a stable differential equation is stable, i.e., the solution of the difference equation doesn't depart radically from the solution of the original differential equation. It is relatively easy to show that the corrector (17) is A-stable for a linear differential equation of the form $\dot{y} = \lambda y$ if and only if the roots of the polynomial

$$\sum_{j=0}^{k} (\alpha_j + h\lambda\ \beta_j)\ \zeta^{k-j} = 0$$

all lie within the unit circle, or are of simple order if they lie on the unit circle. The order p of the multistep method (17) is defined as the highest degree of polynomial y(t) for which (17) yields the exact step by step solution. Dahlquist (1963) has shown the highest order A-stable method is of order 2 and that trapezoidal rule is the second-order method with the least error.

Since we would like to use a method with higher than second order accuracy, we must relax the necessity of A-stability. Gear (1968) defined a method to be stiffly stable if it is stable in the regions of the $h\lambda$ plane shown in Fig. 3, and demonstrates the superiority of such methods for the integration of stiff equations.

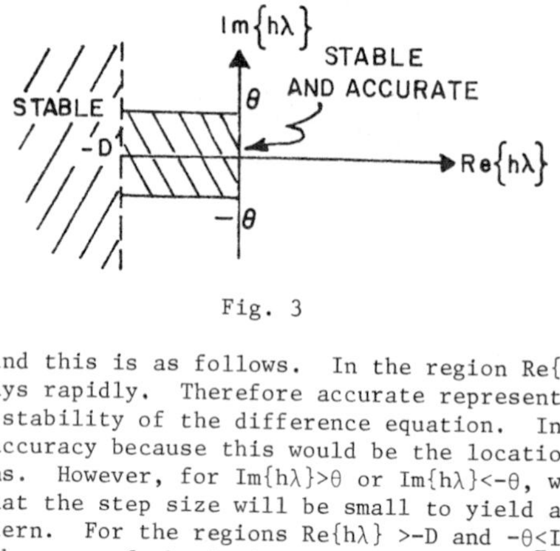

Fig. 3

The reasoning behind this is as follows. In the region $Re\{h\lambda\} < -D < 0$, the response term $ce^{\lambda t}$ decays rapidly. Therefore accurate representation of this term is not necessary, only stability of the difference equation. In the region $Re\{h\lambda\} > -D$ we require accuracy because this would be the location of the dominant long time constant terms. However, for $Im\{h\lambda\} > \theta$ or $Im\{h\lambda\} < -\theta$, we have rapidly oscillating terms so that the step size will be small to yield accuracy, and stability is of no concern. For the regions $Re\{h\lambda\} > -D$ and $-\theta \leq Im\{h\lambda\} \leq \theta$, we require both accuracy, because of the dominant behavior, and stability because of the large time steps possible. Gear then showed that multistep methods (17) up to and including order 6 are stiffly stable.

Implementation of the Corrector

We now turn to implementation of (17) in the solution of our original set of algebraic and differential equations (15). We first evaluate (15) at time t_{n+1}:

$$\underset{\sim}{f}(\underset{\sim}{w}^{n+1}, \underset{\sim}{q}^{n+1}, \underset{\sim}{p}, t_{n+1}) = \begin{bmatrix} \underset{\sim}{h}(\underset{\sim}{w}^{n+1}, \underset{\sim}{q}^{n+1}, \underset{\sim}{p}, t_{n+1}) \\ \underset{\sim}{E}\underset{\sim}{w}^{n+1} - \dot{\underset{\sim}{q}}^{n+1} \end{bmatrix} = \underset{\sim}{0}. \tag{19}$$

Now the multistep corrector (17) can be rewritten as

$$\dot{\underset{\sim}{y}}^{n+1} = -\frac{\beta_o}{h}\underset{\sim}{y}^{n+1} - \sum_{j=1}^{k} \alpha_j \underset{\sim}{y}^{n+1-j} \tag{20}$$

where we have set all $\beta_j = 0$, $j \neq 0$ as indicated by Gear (1971). (Note β_o here is the reciprocal of the β_o of (17)). The only differentiated quantities appearing in (19) are the charge variables $\underset{\sim}{q}$. From (20) we have

$$\dot{\underset{\sim}{q}}^{n+1} = \frac{\beta_o}{h}\underset{\sim}{q}^{n+1} - \sum_{j=1}^{k} \alpha_j y^{n+1-j} \tag{21}$$

Substitution of (21) into (19) yields

$$\begin{bmatrix} \underset{\sim}{h}(\underset{\sim}{w}^{n+1}, \underset{\sim}{q}^{n+1}, \underset{\sim}{p}, t_{n+1}) \\ \underset{\sim}{E}\underset{\sim}{w}^{n+1} - \frac{\beta_o}{h}\underset{\sim}{q}^{n+1} + \sum_{j=1}^{k} \alpha_j \underset{\sim}{q}^{n+1-j} \end{bmatrix} = \underset{\sim}{0} \tag{22}$$

which is a set of nonlinear algebraic equations which can be solved for w^{n+1} and q^{n+1}.

Linearizing (22) and solving by a Newton iteration yields:

$$\begin{bmatrix} \frac{\partial \underset{\sim}{h}}{\partial \underset{\sim}{w}}^{n+1,\,m} & \frac{\partial \underset{\sim}{h}}{\partial \underset{\sim}{q}}^{n+1,\,m} \\ \underset{\sim}{E} & \frac{\beta_o}{h}\underset{\sim}{1} \end{bmatrix} \begin{bmatrix} \Delta \underset{\sim}{w}^{n+1,\,m} \\ \Delta \underset{\sim}{q}^{n+1,\,m} \end{bmatrix} = - \begin{bmatrix} \underset{\sim}{h}(\underset{\sim}{w}^{n+1,\,m}, \underset{\sim}{q}^{n+1,\,m}, \underset{\sim}{p}, \underset{\sim}{t}_{n+1}) \\ \underset{\sim}{E}\underset{\sim}{w}^{n+1,\,m} - \dot{\underset{\sim}{q}}^{n+1,\,m}; \end{bmatrix} \tag{23}$$

where $\Delta w^{n+1,\,m} \equiv (w^{n+1,\,m+1} - w^{n+1,\,m})$ and $\Delta q^{n+1,\,m} \equiv (q^{n+1,\,m+1} - q^{n+1,\,m})$. Expression (23) represents the corrector iteration. Observe that for the network analysis problem we are considering, the Jacobian

$$\underset{\sim}{J}^{n+1,\,m} \equiv \begin{bmatrix} \frac{\partial \underset{\sim}{h}}{\partial \underset{\sim}{w}}^{n+1,\,m} & \frac{\partial \underset{\sim}{h}}{\partial \underset{\sim}{q}}^{n+1,\,m} \\ \underset{\sim}{E} & \frac{\beta_o}{h}\underset{\sim}{1} \end{bmatrix}$$

can be generated quite simply from the network equations -- no numerical differencing or other technique is required.

To this juncture we have not discussed the values of the coefficients α_j and β_o in (21). Gear (1967) computes the coefficients once assuming a fixed step size and requiring (17) to annihilate polynomials of degree $n \leq k$. In other words, by requiring (21) to be satisfied for each of the k+1 substitutions $q_\ell = (\ell h)^j$, j = 0, 1, . . ., k, we have a set of (k+1) equations in terms of the k+1 unknowns $\beta_o, \alpha_1, \alpha_2, \ldots, \alpha_k$. These equations can be solved to yield the α's and β_o for each order $1 \leq k \leq 6$. In a more recent implementation, Brayton (1972), recognizes that the coefficients β_o and α_ℓ of (21) are really dependent upon the time step and therefore must be recomputed whenever the time step and order changes. In particular, we require that (21) be satisfied for each of the k+1 substitutions $q_\ell = (t_{n+1} - t_{n+1-\ell})^j$, $j = 0, 1, \ldots, k$ which yields (k+1) equations in terms of the (k+1) unknowns $\beta_o, \alpha_1, \alpha_2 \ldots \alpha_k$. Brayton has devised an efficient algorithm for computation of these coefficients. We do not have sufficient space for a more detailed discussion of the computation of the coefficients of (21). The interested reader should consult the references.

The Predictor

We have described the corrector iteration and now turn to the predictor. There are two approaches:

1) At each time point store estimates of $\underset{\sim}{q}$ and its derivatives so that a Taylor series can be used as the predictor

$$\underset{\sim}{q}^{n+1,0} = \underset{\sim}{q}^n + \sum_{j=1}^{\hat{k}} \frac{h^j (\underset{\sim}{q}^{(j)})^n}{j!}$$

2) At each time point store $\underset{\sim}{q}$ and then use a form of (17) to predict

$$\gamma_o \underset{\sim}{q}^{n+1,0} = - \sum_{j=1}^{k} \gamma_j \underset{\sim}{q}^{n+1-j}$$

Gear (1971) uses the first formulation and has devised an efficient method for evaluating and updating the Nordseick vector (Nordseick (1962))

$$(h\dot{q}, \frac{h^2}{2} \ddot{q} \ldots \frac{h^{\hat{k}}}{\hat{k}!} q^{(\hat{k})})$$

This formulation is quite useful for interpolating between mesh points. Brayton (1972) uses the second approach and claims two advantages over Gear's method. First it is more stable than Gear's technique for rapidly varying time steps; and second, fewer operations are required for storing backward differences than in storing the Nordsieck vector.

Error Control

One of the most significant contributions made by Gear (1968b) is his error control strategy. Given an allowable per step truncation error, Gear's scheme finds that order which allows the maximum step size to be taken. We indicate briefly how the strategy works.

Let E_ℓ denote the total allowable error for the ℓth controlled variable and T the total time interval of integration. Therefore the allowable per step error is $E_\ell h_\ell / T$.

The approximate truncation error for a pth order method is of the form

$$e_\ell^p = h^{p+1} C_{p+1} q_\ell^{(p+1)}$$

where C_{p+1} is a constant. What we desire is

$$e_\ell^p = h^{p+1} C_{p+1} q_\ell^{(p+1)} = E_\ell h/T \tag{24}$$

Now assume we are currently using a method of order p with a step size of $h = t_n - t_{n-1}$. To determine if a larger step is possible if the order is increased or decreased by one, we must compute $(q_\ell^{(k)})^n$, k = p-1, p, p+1. In Gear's formulation we store the Nordsieck vector and therefore have an estimate of $(q^{(p-1)})^n$. We can estimate $(q^{(p)})^n$ and $(q^{(p+1)})^n$ using differencing. Then from (24), for orders p-1, p, and p+1 we find the smallest time step required by the variables being integrated which is consistent with the allowable truncation error. We define $h_{p-1} \equiv \min_\ell h_{\ell, p-1}$, $h_p \equiv \min_\ell h_{\ell, p}$, and $h_{p+1} \equiv \min_\ell h_{\ell, p+1}$. Then the new order chosen is the one which yields the largest step size. In practice, when a new order p is chosen, then p time steps are taken before a change in order is attempted again. A change in time step though, is possible at every time increment.

In Brayton's (1972) implementation, the Nordsieck vector is not available. However, he has shown that the local truncation error at the nth step is

$$e_\ell^p = \frac{h}{t_n - t_{n-k}} \quad (q_\ell^n - q_\ell^{n,o}).$$

The truncation error for orders p-1 and p+1 are computed by using predictors of order p-1 and p+1 and then the same algorithm as Gear proposes is used.

In summary, discretization of the derivative operator and use of the Newton-Raphson algorithm has reduced the problem of solving the set of nonlinear algebraic and differential equations

$$\underset{\sim}{f}(\underset{\sim}{y}, \dot{\underset{\sim}{y}}, t) = 0$$

to repeated solutions of a set of linear algebraic equations of the form

$$\underset{\sim}{A}\,\underset{\sim}{y} = \underset{\sim}{b}.$$

For a description of how this integration scheme can be efficiently implemented in a nodal analysis program see Calahan (1971).

5. SPARSE MATRIX METHODS

We have converted the analysis problem to one of repeatedly solving sets of linear equations of the form

$$\underset{\sim}{A}\,\underset{\sim}{y} = \underset{\sim}{b}.$$

In general if A is n x n, this system of equations can be solved in approximately mately $\frac{n^3}{3} + n^2$ operations using LU decomposition. More specifically, in $\frac{n^3}{3}$ operations $\underset{\sim}{A}$ can be factored into a product of two matrices

$$\underset{\sim}{A} = \underset{\sim}{L}\underset{\sim}{U}$$

where L is lower triangular and U is upper triangular with ones on the main diagonal. The elements of $\underset{\sim}{L}$ and $\underset{\sim}{U}$ are obtained from

$$\ell_{jk} = a_{jk} - \sum_{m=1}^{k-1} \ell_{jm}\, u_{mk}, \; j = k, k+1, \ldots, n$$

and

$$u_{kj} = \ell_{kk}^{-1} \; (a_{kj} - \sum_{m=1}^{k-1} \ell_{km}\, u_{mj}), \; j = k+1, k+2, \ldots, n.$$

An additional n^2 operations are required to perform the forward substitution

$$\underset{\sim}{L}\,\hat{\underset{\sim}{b}} = \underset{\sim}{b},$$

or

$$\hat{b}_k = \ell_{kk}^{-1} (b_k - \sum_{m=1}^{k-1} \ell_{km} \hat{b}_m), \; k = 1, 2, \ldots, n;$$

and backward substitution

$$\underset{\sim}{U}\,\underset{\sim}{y} = \hat{\underset{\sim}{b}}$$

or

$$y_k = \hat{b}_k - \sum_{m=k+1}^{m} u_{km} y_m, \; k = n, n-1, n-2, \ldots, 1.$$

It is important to recognize three facts:

1) A is sparse, i.e., a large number of the elements of A are zero;
2) The pattern of nonzero elements of $\underset{\sim}{A}$ remains unchanged throughout the entire analysis;
3) The system of equations $\underset{\sim}{A}\,\underset{\sim}{y} = \underset{\sim}{b}$ has to be solved many times.

We are therefore motivated to consider efficient implementation of the LU decomposition method outlined above.

Storage considerations

First consider storage of the matrix A. Since many of the elements A are zero a considerable saving in storage is possible if we store the nonzero elements of A as a single subscripted array. Two additional arrays are used to store the row and column indices of the corresponding nonzero elements of $\underset{\sim}{A}$.

For example, suppose

$$\underset{\sim}{A} = \begin{bmatrix} a_{11} & a_{12} & 0 \\ 0 & a_{22} & a_{23} \\ a_{31} & a_{32} & a_{33} \end{bmatrix}$$

Then we would store $\underset{\sim}{A}$ and the row and column pointers as

$A(1) = a_{11}$	PR(1) = 1	PC(1) = 1
$A(2) = a_{31}$	PR(2) = 3	PC(2) = 1
$A(3) = a_{12}$	PR(3) = 1	PC(3) = 2
$A(4) = a_{22}$	PR(4) = 2	PC(4) = 2
$A(5) = a_{32}$	PR(5) = 3	PC(5) = 2
$A(6) = a_{23}$	PR(6) = 2	PC(6) = 3
$A(7) = a_{33}$	PR(7) = 3	PC(7) = 3

Other techniques are also possible but this one is quite suitable for our purposes (see Reid (1972) or Calahan (1972)).

Generation of explicit code

Given that $\underset{\sim}{A}$ is stored in compacted form we now recognize that a vast saving in computation time can be realized if the operations involving multiplication and addition of zero elements were avoided. Two different approaches toward implementing a scheme which avoids manipulation of zero-valued elements have been proposed. Gustavson (1968) suggests generating a nonlooping, non-branching program which explicitly performs all the nontrivial arithmetic operations involved in solving

$$\underset{\sim}{A}\,\underset{\sim}{x} = \underset{\sim}{LUx} = \underset{\sim}{Ly} = \underset{\sim}{b}.$$

For example, given the $\underset{\sim}{A}$ matrix above, the nontrivial steps involved in the LU decomposition are:

$$\ell_{11} = a_{11}$$

$$\ell_{31} = a_{31}$$

$$u_{12} = a_{12}/\ell_{11}$$

$$\ell_{22} = a_{22}$$

$$\ell_{32} = a_{32} - \ell_{31}\, u_{12}$$

$$u_{23} = a_{23}/\ell_{22}$$

$$\ell_{33} = a_{33} - \ell_{32}\, u_{23}$$

These steps can be implemented by the following Fortran code

```
C(1) = A(1)
C(2) = A(2)
C(3) = A(3)/C(1)
C(4) = A(4)
C(5) = A(5) - C(2)*C(3)
C(6) = A(6)/C(4)
C(7) = A(7) - C(5)*C(6)
```

Since no fill was created the C array could just overlay the A array. If, however, fill was created the one to one correspondence between the C and A arrays would no longer exist. This approach can be extended one step further so that the program generated is in terms of compiled code which the computer can execute directly. Typically this program is initially written out on disk and then called back into core, overlaying the generating program, when completed. The major disadvantage to this approach is that for some problems excessively long programs are generated. Moreover, since the generated code is so efficient, the wait time in transfering code from the disk to core can become significant when compared with execution time of the code. This consideration though is not so important if the code will be used many times over, but only has to be transferred from disk to core once.

An alternative to generating explicit code is to generate lists of nontrivial operations and element address which are interpreted by another program (H. B. Lee (1969), Erisman (1972)). One possible implementation of this approach is as follows. An operation code is assigned to each of the operations or combinations of operations that could take place during LU decomposition. An example of such a code is

OP Code	Operation
1	$\alpha \to \delta$
2	$\alpha/\beta \to \delta$
3	$\alpha*\beta \to \delta$
4	$\alpha-\beta*\gamma \to \delta$
5	STOP

where α denotes an element of coefficient array A and β, γ and δ denote elements of the LU factors array C. A table with five columns, represented by a two dimensional array is then created. The first column denotes the operation count, the second thru fifth columns indicate the addresses of the elements α, β, γ and δ, respectively. The Fortran program given above could then be represented by Table 1.

OP Code	Addresses			
	α	β	γ	δ
1	1	0	0	1
1	2	0	0	2
2	3	1	0	3
1	4	0	0	4
4	5	2	3	5
2	6	4	0	6
4	7	5	6	7
5	0	0	0	0

Table 1

A program which could be used to interpret this table is shown in Fig. 4.

```
      SUBROUTINE INTERP (TABLE, A,C)
      INTEGER *2 TABLE, OPCODE, I, J, K, L
      DIMENSION TABLE (•, NZ), A(•), C(•)
      II=1
    7 OPCODE=TABLE(II,1)
      I=TABLE(II,2)
      J=TABLE (II,3)
      K=TABLE(II,4)
      L=TABLE(II,5)
      GΦ TΦ (1,2,3,4,5), OPCODE
    1 C(L)=A(I)
      GΦTΦ10
    2 C(L)=A(I)/C(J)
      GΦTΦ10
    3 C(L)=A(I)*C(J)
      GΦTΦ10
    4 C(L)=A(I)-C(J)*C(K)
   10 II=II+1
      GΦTΦ7
    5 RETURN
      END
```

Figure 4

A similar table can be constructed for the operations and addresses involved in the forward and backward substitutions.

Separation of code according to type

Additional savings in compute time can be realized by recognizing that the compiled or interpretable code can be decomposed into smaller programs or lists and that only part of the code need be executed during each Newton iteration. To see this, recognize that the elements in A and b, and therefore $\underset{\sim}{L}$, $\underset{\sim}{U}$, $\underset{\sim}{\hat{b}}$ and $\underset{\sim}{x}$, can be classified according to the following types:

Type 1: elements which are constants
Type 2: elements which depend upon circuit element parameters

Type 3: elements which depend upon time
Type 4: elements which depend upon the dependent variable x.

We then classify the type of operation performed in the LU factorization as the maximum of the types of the elements involved. For example, in a multiply and add operation:

$$\alpha - \beta \cdot \gamma ,$$

if α is type 1, β type 4 and γ type 2, then this operation is classified as type 4. Similarly, in a divide operation

$$\gamma / \beta ,$$

if γ is type 2 and β type 3, then this operation is a type 3.

The code or lists generated to perform the LU factorization and forward and backward substitution is now separated into the four programs P1, P2, P3, P4, according to the type of operation involved. All type 1 operations are put into program P1, all type 2 operations are put into program P2, and so on. Observe that program P1 need only be executed once during the analysis since all the operations in this program involve elements which are constants. Program P2 must be reexecuted whenever a parameter value changes. Program P3 must be reexecuted whenever a parameter changes and time increases. Finally, program P4 must be reexecuted whenever a parameter changes, time increases, or the unknown vector changes, i.e., every Newton iteration. Fig. 5 symbolically shows the order in which programs P1, P2, P3, P4 are executed.

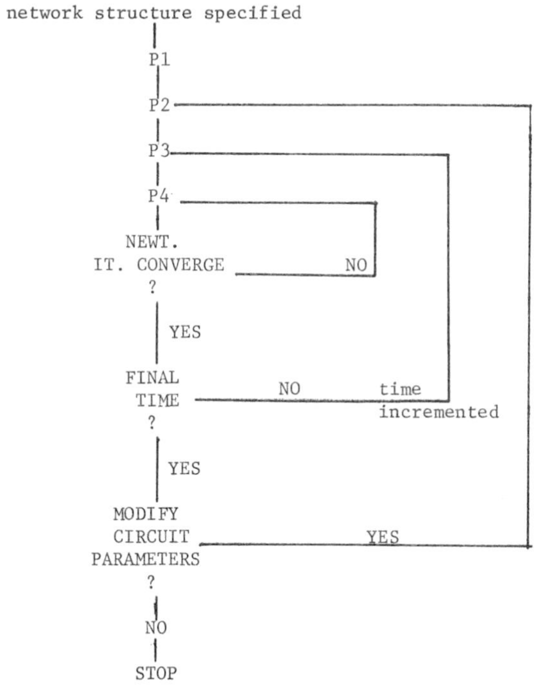

Figure 5

Choosing pivots

We now turn to the question of choosing pivots, viz, the ordering of the equations. There are basically three considerations in choosing a pivot: minimization of the creation of new nonzero elements, or fill; minimization of roundoff error, and minimization of the operation count.

First consider the question of fill. Given a set of linear algebraic equations

$$\underset{\sim}{A}\ \underset{\sim}{x} = \underset{\sim}{b}$$

with some pattern of nonzero elements of $\underset{\sim}{A}$ it is possible that during the LU decomposition new nonzero elements would be created. Since the amount of code generated to perform LU decomposition is a function of the number of nonzero elements, it is desirable to keep the amount of fill small. Tinney (1967) has proposed three schemes for ordering the equations so as to minimize fill. These schemes are:

1) Order the rows according to the number of nonzero off-diagonal terms before elimination.
2) Order the rows so that at the kth step in the factorization the kth pivot is in the row with fewest number of nonzero terms.
3) Order the rows so that at the kth step in the factorization the kth pivot is in the row which introduces the fewest number of nonzero terms.

The first scheme is the easiest to use and can be performed prior to entering the program which generates code. The second scheme is somewhat more difficult to program since it must be part of the program which generates code. However, it considerably reduces the amount of fill. The third scheme is the most difficult to program and takes the longest to use. Moreover, there is not a significant decrease in the fill when compared with the second method. Tinney (1967) concludes that the second method is the best compromise between time spent on ordering and the amount of fill.

The importance of ordering to minimize roundoff error can most easily be demonstrated with the following example, Forsythe (1967). Suppose we wish to solve the set of equations

$$\begin{bmatrix} .000100 & 1.00 \\ 1.00 & 1.00 \end{bmatrix} \begin{bmatrix} X_1 \\ X_2 \end{bmatrix} = \begin{bmatrix} 1.00 \\ 2.00 \end{bmatrix}$$

with three-decimal floating point arithmetic. The $\underset{\sim}{L}$ and $\underset{\sim}{U}$ matrices are

$$L = \begin{bmatrix} .0001000 & 0.0 \\ 1.00 & -10000. \end{bmatrix} \qquad U = \begin{bmatrix} 1.00 & 10000. \\ 0.0 & 1.00 \end{bmatrix}$$

because 10000. - 1.00 becomes 10000. The forward substitution step

$$\begin{bmatrix} .000100 & 0.0 \\ 1.00 & -10000. \end{bmatrix} \begin{bmatrix} \hat{b}_1 \\ \hat{b}_2 \end{bmatrix} = \begin{bmatrix} 1.00 \\ 2.00 \end{bmatrix}$$

yields b_1 = 10000. and b_2 = (2 - 10000.)/10000. = 1.00. The backward substitution step

$$\begin{bmatrix} 1.000 & 10000. \\ 0.0 & 1.0 \end{bmatrix} \begin{bmatrix} x_1 \\ x_2 \end{bmatrix} = \begin{bmatrix} 10000. \\ 1.00 \end{bmatrix}$$

yields $x_2 = 1.00$ and $x_1 = 0.0$. Clearly this is a wrong answer. If the order of the equations were changed, i.e.,

$$\begin{bmatrix} 1.00 & 1.00 \\ .000100 & 1.00 \end{bmatrix} \begin{bmatrix} x_1 \\ x_2 \end{bmatrix} = \begin{bmatrix} 2.00 \\ 1.00 \end{bmatrix}$$

then the correct answer of $x_1 = 1.00$ and $x_2 = 1.00$ would have been found.

The problem indicated above results because only a finite word length is used to store a number in the computer. Errors due to this mechanism are called roundoff errors. Roundoff errors can be minimized by choosing the pivots, i.e., the ℓ_{kk} terms, to be the element in the row or column with the largest magnitude. Of course, choosing pivots based solely on minimization of roundoff errors may cause considerable fill in.

Now consider minimization of operation count. In the discussion of generation of code to perform LU factorization we indicated the desirability of breaking the code into four programs P1, P2, P3, P4 depending upon the type of operation performed. Program P1 is executed once per analysis, P2 is executed only if parameter values changed, P3 is executed when time is incremented, and P4 is executed each Newton iteration. The more code that must be put in P4 the more operations must be performed each Newton iteration and the longer the analysis takes. Thus it is desirable to choose pivots so that operations involving type 4 elements be kept to a minimum. Once this is done we would like to minimize operations involving type 3 elements, and so on.

Hachtel (1971) has proposed an algorithm for choosing pivots which takes into consideration fill, roundoff error and operation count. Briefly this algorithm proceeds as follows:

1. For each row determine the subset of nonzero columns for which the elements have magnitudes sufficiently large to avoid excessive roundoff. These elements are candidates for pivots.

2. Determine the relative cost of using each of these candidates as the pivot. The cost is determined by counting the number of multiplications involved in eliminating the row and column of the pivot candidate. Associated with each multiplication is a variability or operation type. The higher the variability type the more times the operation must be performed during the analysis run and the higher the cost.

3. Choose as the pivot the candidate with the least cost.

4. Execute the next step in LU factorization.

5. Update the variability types of each nonzero element and return to 1.

A more detailed description of this procedure is given in Hachtel (1971). In a more recent paper, Hachtel (1972) has extended this technique to account for the variability type and sparsity of the unknown vector and right hand side. Reid (1972) presents additional material on pivoting.

6. NETWORK OPTIMIZATION

Although there have been many approaches to automating network design, (see Herskowitz (1968), Kuo (1966), Kuo (1969), Proceedings of the IEEE (1967) and (1972) and IEEE Transactions on Circuit Theory (1971)). the approach which has produced the best results is the one in which the design problem is recast into an optimization problem. For purposes of generality we consider the time domain design problem. (For frequency domain design see Director (1969a) and for dc design see Director (1969b) and Dowell (1971).) The basic steps involved are

1. Formulate a performance function which embodies the design criterion. The most general form of such a performance function is

$$\varepsilon = \int_0^T e(\underset{\sim}{w}, \underset{\sim}{q}, \underset{\sim}{p}, t)dt$$

(Some specialized forms of (25) are presented in Director (1971a)).

2. Construct an initial network as a starting point from which the design is to proceed. (This first feasible point problem has not really been attacked in the electrical engineering literature although some attempts have been made (Director (1969 a & b)). (For more information see Westerberg (1972)).

3. Use an optimization method such as Fletcher-Powell (Fletcher (1963)) or Fletcher-Reeves (Fletcher (1964)) to adjust the parameter vector $\underset{\sim}{p}$ so as to minimize ε.

The optimization schemes to be employed for minimization of ε require computation of the gradient. We will show that the gradient of ε with respect to all parameters can be obtained after only two network analyses. (Actually for linear, time-invariant design problems only a single network analysis is required, (Director (1971b)). Three different techniques for deriving this result have been proposed. One, Director (1969c), is based upon Tellegens Theorem, Desoer (1969); the second approach, Ho (1971), uses variational calculus; and the third method, Hachtel (1971a and b) uses direct differentiation. We present the approach used by Hachtel because of its notational simplicity.

Because the solution $\underset{\sim}{w}(t)$ must satisfy the original network's equations

$$\begin{bmatrix} \underset{\sim}{h}(\underset{\sim}{w}, \underset{\sim}{q}, \underset{\sim}{p}, t) \\ \underset{\sim}{E}\ \underset{\sim}{w} - \dot{\underset{\sim}{q}} \end{bmatrix} = \underset{\sim}{0} \tag{26}$$

the performance function (25) may be rewritten with the Lagrange multipliers $\hat{\underset{\sim}{w}}$ and $\hat{\underset{\sim}{q}}$ as

$$\varepsilon = \int_0^T [e(\underset{\sim}{w},\underset{\sim}{q},\underset{\sim}{p},t) + \hat{\underset{\sim}{w}}^T \underset{\sim}{h}(\underset{\sim}{w},\underset{\sim}{q},\underset{\sim}{p},t) + \hat{\underset{\sim}{q}}^T(\underset{\sim}{E}\underset{\sim}{w}-\dot{\underset{\sim}{q}})]dt.$$

Assume that the network described by (26) is in equilibrium, i.e., $\dot{\underset{\sim}{q}}(0)$, and that the final time T is independent of the designable parameters, i.e., $\frac{\partial T}{\partial \underset{\sim}{p}} = \underset{\sim}{0}$. By direct differentiation and integration by parts, we have

$$\begin{aligned} \frac{d\varepsilon}{d\underset{\sim}{p}} = \int_0^T \{ &[\frac{\partial e}{\partial \underset{\sim}{w}} + \hat{\underset{\sim}{w}}^T \frac{\partial \underset{\sim}{h}}{\partial \underset{\sim}{w}} + \hat{\underset{\sim}{q}}^T \underset{\sim}{E}] \frac{\partial \underset{\sim}{w}}{\partial \underset{\sim}{p}} + [\frac{\partial e}{\partial \underset{\sim}{q}} + \hat{\underset{\sim}{w}}^T \frac{\partial \underset{\sim}{h}}{\partial \underset{\sim}{q}} + \frac{d\hat{\underset{\sim}{q}}^T}{dt}] \frac{\partial \underset{\sim}{q}}{\partial \underset{\sim}{p}} \\ &+ [\frac{\partial e}{\partial \underset{\sim}{p}} + \hat{\underset{\sim}{w}}^T \frac{\partial \underset{\sim}{h}}{\partial \underset{\sim}{p}}]\} dt - \hat{\underset{\sim}{q}}^T(T) \frac{\partial \underset{\sim}{q}(T)}{\partial \underset{\sim}{p}} + \hat{\underset{\sim}{q}}^T(0) \frac{\partial \underset{\sim}{q}(0)}{\partial \underset{\sim}{p}} . \end{aligned} \tag{27}$$

We can choose the Lagrange multipliers so that they satisfy the equations

$$\frac{\partial e}{\partial \underset{\sim}{w}} + \hat{\underset{\sim}{w}}^T \frac{\partial \underset{\sim}{h}}{\partial \underset{\sim}{w}} + \hat{\underset{\sim}{q}}^T \underset{\sim}{E} = \underset{\sim}{0}, \tag{28a}$$

$$\frac{\partial e}{\partial \underset{\sim}{q}} + \hat{\underset{\sim}{w}}^T \frac{\partial \underset{\sim}{h}}{\partial \underset{\sim}{q}} + \frac{d\hat{\underset{\sim}{q}}^T}{\partial t} = 0 \tag{28b}$$

and $\hat{q}^T(T) = 0$ so that expression (27) can be written as

$$\frac{d\varepsilon}{d\underset{\sim}{p}} = \hat{\underset{\sim}{p}}^T(0) + \bar{\underset{\sim}{p}}^T(0) \tag{29}$$

where

$$\hat{\underset{\sim}{p}}^T(0) = \int_0^T \left(\frac{\partial e}{\partial \underset{\sim}{p}} + \hat{\underset{\sim}{w}}^T \frac{\partial \underset{\sim}{h}}{\partial \underset{\sim}{p}}\right)dt \tag{30}$$

and

$$\bar{\underset{\sim}{p}}^T = \hat{\underset{\sim}{q}}^T(0) \frac{\partial \underset{\sim}{q}(0)}{\partial \underset{\sim}{p}} . \tag{31}$$

Recognize that (30) can also be expressed in terms of the differential equation:

$$-\frac{d\hat{\underset{\sim}{p}}^T(t)}{dt} + \frac{\partial \underset{\sim}{e}}{\partial \underset{\sim}{p}} + \hat{\underset{\sim}{w}}^T \frac{\partial \underset{\sim}{h}}{\partial \underset{\sim}{p}} = 0 \tag{32}$$

with $\hat{\underset{\sim}{p}}^T(T) = 0$.

For purposes which will become clear later, we define one additional variable $\hat{\varepsilon}(t)$ such that $\hat{\varepsilon}(T) = -1$ and

$$-\frac{d\hat{\varepsilon}}{dt} = 0. \tag{33}$$

We now assemble expressions (28a), (28b), (33) and (32) into one matrix equation:

$$(\hat{\underset{\sim}{w}}^T, \hat{\underset{\sim}{q}}^T, \hat{\varepsilon}, \hat{\underset{\sim}{p}}^T) \begin{bmatrix} \frac{\partial \underset{\sim}{h}(t)}{\partial \underset{\sim}{w}} & \frac{\partial \underset{\sim}{h}(t)}{\partial \underset{\sim}{q}} & 0 & \frac{\partial \underset{\sim}{h}(t)}{\partial \underset{\sim}{p}} \\ \underset{\sim}{E} & \frac{d}{dt} & 0 & 0 \\ \frac{\partial e(t)}{\partial \underset{\sim}{w}} & \frac{\partial e(t)}{\partial \underset{\sim}{q}} & \frac{-d}{dt} & \frac{\partial e(t)}{\partial \underset{\sim}{p}} \\ 0 & 0 & 0 & \frac{-d}{dt} \end{bmatrix} = [\underset{\sim}{0}, \underset{\sim}{0}, 0, \underset{\sim}{0}] \tag{34}$$

with the "final" conditions $\hat{\underset{\sim}{w}}(T) = 0$, $\hat{\underset{\sim}{p}}(T) = 0$, $\hat{\underset{\sim}{q}}(T) = 0$ and $\hat{\varepsilon}(T) = -1$. Thus we may view $\hat{\underset{\sim}{p}}^T(0)$ as the solution of the set of algebraic-differential equations (34).

The implication of arranging the equations in this form will become clear shortly. Before proceeding though we wish to show that $\bar{p}^T(0) = \hat{q}^T(0) \frac{\partial \underset{\sim}{q}^T(0)}{\partial p}$ can be viewed as resulting from the solution of the algebraic set of equations

$$(\bar{\underset{\sim}{w}}^T, \bar{\underset{\sim}{q}}^T, \bar{\varepsilon}, \bar{\underset{\sim}{p}}^T) \begin{bmatrix} \frac{\partial \underset{\sim}{h}(0)}{\partial \underset{\sim}{w}} & \frac{\partial \underset{\sim}{h}(0)}{\partial \underset{\sim}{q}} & 0 & \frac{\partial \underset{\sim}{h}(0)}{\partial \underset{\sim}{p}} \\ \underset{\sim}{E} & 0 & 0 & 0 \\ \frac{\partial e(0)}{\partial \underset{\sim}{w}} & \frac{\partial e(0)}{\partial \underset{\sim}{q}} & 1 & \frac{\partial e(0)}{\partial \underset{\sim}{p}} \\ 0 & 0 & 0 & 1 \end{bmatrix} = [\underset{\sim}{0}, \hat{\underset{\sim}{q}}^T(0), 0, \underset{\sim}{0}] \tag{35}$$

To see this, observe that at time t = 0 the network constraints (26) become

$$\begin{bmatrix} \underset{\sim}{h}(\underset{\sim}{w}(0), \underset{\sim}{q}(0), p, 0) \\ \underset{\sim}{E}\,\underset{\sim}{w} \end{bmatrix} = \underset{\sim}{0}$$

because $\dot{\underset{\sim}{q}}(o) = 0$. Differentiating with respect to p yields

$$\begin{bmatrix} \dfrac{\partial \underset{\sim}{h}(0)}{\partial \underset{\sim}{w}} \dfrac{\partial \underset{\sim}{w}(0)}{\partial \underset{\sim}{p}} + \dfrac{\partial \underset{\sim}{h}(0)}{\partial \underset{\sim}{q}} \dfrac{\partial \underset{\sim}{q}(0)}{\partial \underset{\sim}{p}} + \dfrac{\partial \underset{\sim}{h}(0)}{\partial \underset{\sim}{p}} \\ \underset{\sim}{E} \dfrac{\partial \underset{\sim}{w}(0)}{\partial \underset{\sim}{p}} \end{bmatrix} = 0. \tag{36}$$

From (35) we have

$$\varepsilon = 0,$$

$$\bar{\underset{\sim}{p}}^T = -\bar{\underset{\sim}{w}}^T \frac{\partial \underset{\sim}{h}(0)}{\partial \underset{\sim}{p}}, \tag{37}$$

$$\bar{\underset{\sim}{w}}^T \frac{\partial \underset{\sim}{h}(0)}{\partial \underset{\sim}{q}} = \hat{\underset{\sim}{q}}^T(0) \tag{38}$$

and

$$\underset{\sim}{w}^T \frac{\partial \underset{\sim}{h}(0)}{\partial \underset{\sim}{w}} + \bar{\underset{\sim}{q}}^T \underset{\sim}{E} = 0. \tag{39}$$

From (37) and (36) we have

$$\bar{\underset{\sim}{p}}^T = -\bar{\underset{\sim}{w}}^T \left(- \frac{\partial \underset{\sim}{h}(0)}{\partial \underset{\sim}{w}} \frac{\partial \underset{\sim}{w}(0)}{\partial p} - \frac{\partial \underset{\sim}{h}(0)}{\partial \underset{\sim}{q}} \frac{\partial \underset{\sim}{q}(0)}{\partial \underset{\sim}{p}} \right). \tag{40}$$

But from (39)

$$\bar{\underset{\sim}{w}}^T = - (\bar{\underset{\sim}{q}}^T \underset{\sim}{E}) \left(\frac{\partial \underset{\sim}{h}(0)}{\partial \underset{\sim}{w}} \right)^{-1}$$

so that (4) becomes (using that $\underset{\sim}{E} \frac{\partial \underset{\sim}{w}(0)}{\partial p} = 0$ from (36))

$$\bar{\underset{\sim}{p}}^T = \bar{\underset{\sim}{w}}^T \frac{\partial \underset{\sim}{h}(0)}{\partial \underset{\sim}{q}} \frac{\partial \underset{\sim}{q}(0)}{\partial \underset{\sim}{p}}$$

or, after using (38)

$$\bar{\underset{\sim}{p}}^T = \hat{\underset{\sim}{q}}^T(0) \frac{\partial \underset{\sim}{q}(0)}{\partial \underset{\sim}{p}},$$

which demonstrates the conjecture stated at the beginning of this paragraph.

The flow of computation is easily established. First the original network is analyzed and the performance function (25) evaluated. Since (25) can be expressed as

$$e(\underset{\sim}{w}, \underset{\sim}{q}, \underset{\sim}{p}, t) - \dot{\varepsilon} = 0$$

and because the following equation must hold

$$-\dot{p} = 0$$

this step requires solution of the system of equations

$$\begin{bmatrix} h(\underset{\sim}{w}, \underset{\sim}{q}, \underset{\sim}{p}, t) \\ \underset{\sim}{E}\ \underset{\sim}{w} - \dot{\underset{\sim}{q}} \\ e(\underset{\sim}{w}, \underset{\sim}{q}, \underset{\sim}{p}, t) - \dot{\varepsilon} \\ -\dot{p} \end{bmatrix} = \underset{\sim}{0} \tag{41}$$

at time points t_n in the interval [0, T]. As discussed earlier, solution of (41) is carried out using a multi-step corrector and a Newton iteration which requires repeated solutions of the linear system

$$\begin{bmatrix} \frac{\partial \underset{\sim}{h}}{\partial \underset{\sim}{w}} & \frac{\partial \underset{\sim}{h}}{\partial \underset{\sim}{q}} & 0 & \frac{\partial \underset{\sim}{h}}{\partial \underset{\sim}{p}} \\ \underset{\sim}{E} & \beta_o/h & 0 & 0 \\ \frac{\partial e}{\partial \underset{\sim}{w}} & \frac{\partial e}{\partial \underset{\sim}{q}} & \beta_o/h & \frac{\partial e}{\partial \underset{\sim}{p}} \\ 0 & 0 & 0 & \beta_o/h \end{bmatrix} \begin{bmatrix} \underset{\sim}{w}^n \\ \underset{\sim}{q}^n \\ \underset{\sim}{\varepsilon}^n \\ \underset{\sim}{p}^n \end{bmatrix} = \underset{\sim}{b}, \tag{42}$$

where $\underset{\sim}{b}$ is a vector of constants. During the original analysis the values of $\underset{\sim}{w}^n$ and $\underset{\sim}{q}^n$ which are used to evaluate the partial derivatives are stored on disk. These values are recalled during the solution of (34) for $\hat{p}^T(0)$. (Note that (34) is solved in reverse time.) Since the time points t_m used for solution of (34) will not in general correspond to the time points t_n, a kth order interpolation must be used. At the "final time," t = 0, the value $\hat{q}$, i.e., $\hat{q}(0)$ is then used on the right hand side of (35), which is now solved to yield $\bar{p}^T$. Finally $\bar{p}^T$ and $\hat{p}^T(0)$ are added to yield $\partial\varepsilon/\partial p$. At this juncture both $\underset{\sim}{\varepsilon}$ and $\frac{\partial \varepsilon}{\partial \underset{\sim}{p}}$ are present and an optimization step may be taken.

It is important to observe that only a single set of code for sparse LU factorization need be generated for the solution of the original network equations (42), the adjoint equations (34) and the equations which yield the transversality conditions (35) because the same matrix structure is present in all three sets of equations. We must, of course, generate different sets of code for the forward and backward substitutions.

It is possible to make a meaningful physical interpretation of the adjoint equations (34) and the interested reader should consult the references (Director (1969c) and Hachtel (1971b)). This physical interpretation leads to other interesting results (Director (1972), Rohrer (1971)) which have application in electrical circuit analysis.

References

Branin, F. H., Jr. (1967), "Computer Methods of Network Analysis," Proc. IEEE, vol. 55, p. 1787.

Brayton, R. K., F. G. Gustavson, and G. D. Hachtel (1972), "A New Efficient Algorithm for Solving Differential-Algebraic Systems Using Implicit Backward Differentiation Formulas," Proc. IEEE, vol. 60, p. 98.

Calahan, D. A. (1971), "Numerical Considerations for Implementation of a Nonlinear Transient Circuit Analysis Program," IEEE Trans. Ckt. Th., vol. CT-18, p. 66.

Calahan, D. A. (1972), Computer Aided Network Design, McGraw-Hill, Chapt. 10.

Dahlquist, G. G. (1963), "A Special Stability Problem for Linear Multistep Methods," BIT, vol. 3, p. 27.

Desoer, C. A. and E. S. Kugh (1969), Basic Circuit Theory, McGraw-Hill.

Director, S. W. and R. A. Rohrer (1969a), "Automated Network Design - the Frequency Domain Case," IEEE Trans. Ckt. Th., vol. CT-16, p. 330.

Director, S. W. and R. A. Rohrer (1969b), "On the Design of Resistance N-port Networks by Digital Computer," IEEE Trans. Ckt. Th., vol. CT-16, p. 337.

Director, S. W. and R. A. Rohrer (1969c), "The Generalized Adjoint Network and Network Sensitivities," IEEE Trans. Ckt. Th., vol. CT-16, p. 318.

Director, S. W. (1971), "Survey of Circuit Oriented Optimization Techniques," IEEE Trans. Ckt. Th., vol. CT-18, p. 3.

Director, S. W. (1971b), "LU Factorization in Network Sensitivity Computations," IEEE Trans. Ckt. Th., vol. CT-18, p. 184.

Director, S. W. and D. A. Wayne (1972), "Computational Efficiency in the Determination of Thevenin and Norton Equivalents," IEEE Trans. Ckt. Th., vol. CT-19, p. 96.

Dowell, R. I. and R. A. Rohrer (1971), "Automated Design of Biasing Circuits," IEEE Trans. Ckt. Th., vol. CT-18, p. 85.

Erisman, A. M. and G. E. Spies (1972), "Exploiting Problem Characteristics in the Sparse Matrix Approach to Frequency Domain Analysis," IEEE Trans. Ckt. Th., vol. CT-19, p. 260.

Fletcher, R. and M. J. D. Powell (1963), "A Rapidly Convergent Descent Method for Minimization," Comp. J., vol. 6, p. 163.

Fletcher, R. and C. M. Reeves (1964), "Function Minimization by Conjugate Gradients," Comp. J., vol. 7, p. 149.

Forsythe, G. and C. B. Moler (1967), Computer Solution of Linear Algebraic Systems, Prentice Hall, p. 34.

Gear, C. W. (1966), "The Numerical Integration of Ordinary Differential Equations of Various Orders," Argonne National Lab. Report, Report #ANL 7126.

Gear, C. W. (1968a), "The Automatic Integration of Stiff Ordinary Differential Equations," Proc. IFIPS Conf., p. A81.

Gear, C. W. (1968b), "The Control of Parameters in Automatic Integration of Ordinary Differential Equations, Dept. of Comp. Sci. File #757, Univ. of Ill., Urbana, Illinois.

Gear, C. W. (1969), "The Automatic Integration of Large Systems of Ordinary Differential Equations," Dig. Record of Joint Conf. on Math and Comp. Aids to Design, p. 27.

Gear, C. W. (1971), "Simultaneous Numerical Solution of Differential-Algebraic Equations," IEEE Trans. Ckt. Th., vol. CT-18, p. 89.

Gustavson, F. G., W. M. Liniger, and R. A. Willoughby (1968), "Symbolic Generation of an Optimal Crout Algorithm for Sparse Systems of Linear Equations," Proc. of Sparse Matrix Symp., IBM Watson Research Center, p. 1.

Hachtel, G. D., R. K. Brayton and F. G. Gustavson (1971a), "The Sparse Tableau Approach to Network Analysis and Design," IEEE Trans. Ckt. Th., vol. CT-18, p. 101.

Hachtel, G. D., R. K. Brayton and F. G. Gustavson (1971b), "The Sparse Tableau Approach to Nonlinear Adjoint Sensitivity Computations," Proc. Int. Mexican Symp. on Systems, Networks and Comp., p. 903.

Hachtel, G. D. (1972), "Vector and Matrix Variability Type in Sparse Matrix Algorithms," Proc. Sparse Matrix Symp., Plenum Press.

Herskowitz, G. J. (1968), Computer Aided Integrated Circuit Design, McGraw-Hill, Chapt. 5, 7.

Ho, C. W. (1971), "Time Domain Sensitivity Computation for Networks Containing Transmission Lines," IEEE Trans. Ckt. Th., vol. CT-18, p. 114.

Jenkins, F. S. and S.-P. Fan (1971), "TIME - A Nonlinear dc and Time Domain Circuit Simulation Program," IEEE J. Solid State Ckts., vol. SC-6, p. 182.

Jensen, R. W. and Lieberman (1968), IBM Electronics Circuit Analysis Program, Prentice Hall.

Kuo, F. F. and J. F. Kaiser (1966), System Analysis by Digital Computer, Wiley, Chapt. 6.

Kuo, F. F. and W. G. Magnuson (1969), Computer Oriented Circuit Design, Prentice Hall.

Lee, H. B. (1969), An Implementation of Gaussian Elimination of Sparse Systems of Linear Equations, Proc. Sparse Matrix Symp., IBM Watson Research Center, p. 75.

Nagel, L. and R. Rohrer (1971), "Computer Analysis of Nonlinear Circuits, Excluding Radiation," IEEE J. Solid State Ckts., vol. SC-6, p. 166.

Nordseick, A. (1962), "On Numerical Integration of Ordinary Differential Equations," Math Comp., vol. 16, p. 22.

Pottle, C. (1966), "State-space Techniques for General Active Network Analysis," Systems Analysis by Digital Computer, F. F. Kuo and J. F. Kaiser, Eds., Wiley, Chapt. 3.

Reid, J. K. (1972), "Sparse Matrices and Decomposition with Applications to the Solution of Algebraic and Differential Equations," Proceedings of this meeting.

Rohrer, R., L. Nagel, R. Meyer and L. Weber (1971), "Computationally Efficient Electronic-Circuit Noise Calculation," IEEE J. Solid State Ckts., vol. SC-6, p. 204.

Tinney, W. F. and J. W. Walker (1967), "Direct Solutions of Sparse Network Equations by Optimally Ordered Triangular Factorization," Proc. IEEE, vol. 55, p. 1801.

Wing, O., and J. Gielchinsky (1972), "Computation of Time Response of Large Networks by Partitioning," Proc. Int. Symp. Ckt. Th., p. 125.

IEEE Transactions on Circuit Theory, (1971), Special Issue on Computer-Aided Circuit Design.

Proceedings of the IEEE, (1967), Special issue on Computer-Aided Design, vol. 55.

Proceedings of the IEEE, (1972), Special issue on Computers in Design, vol. 60.

Westerberg (1972), "Decomposition Methods for Solving Steady State Process Design Problems," Proceedings of this meeting.

SOLUTION OF LARGE ELECTRIC NETWORKS BY TEARING IN THE TIME DOMAIN*

OMAR WING

Department of Electrical Engineering and Computer Science

Columbia University

New York, New York

Abstract: A computational method is presented which obtains the time response of a network of larger order by tearing. The method is based on the decomposition of the network into subnetworks, each of which is characterized by a set of differential equations of low order. The dynamic variables of one subnetwork are coupled to those of another by a set of algebraic equations which match the voltages and currents at the common terminals of the subnetworks. Significant savings in computation time and storage are realized as a result of decomposition. The method is particularly suitable for analysis of electric filters and discretized network models of diffusion equations for heat flow and wave equations for longitudinal waves in an elastic medium.

BACKGROUND

The equations that govern the behavior of a linear electric network are a set of maximally linearly independent Kirchhoff voltage equations, a set of maximally linearly independent Kirchhoff current equations, and a set of element equations which relate the voltages and currents of the elements that constitute the network, the last set being composed of linear algebraic equations for the resistors and linear differential equations for the capacitors and inductors of the network. The three sets of equations can be combined, by eliminating all resistor voltages and currents, into a set of maximally linearly independent differential equations of the form (Bashkow, 1957):

$$\frac{d\underline{x}}{dt} = A\underline{x} + Bu \tag{1}$$

where $\underline{x}$ is a column vector of state variables, $x_1, x_2, \ldots, x_N$, A is an N×N matrix, and u is a scalar if there is but one excitation in the network. Equation (1) is known as the state equation of the network. Usually the state variables are identified with the linearly independent capacitor voltages and inductor currents. The order (degree of freedom) of the system of differential equations N, also called the complexity of the network, is equal to the total number of capacitors and inductors less the number of linear relations among the capacitor voltages and inductor currents (Bryant, 1959).

Given a network, its state equations can be formulated explicitly in terms of matrices that describe the topology and element values of the network (Bryant, 1962; Brown, 1963; Dervisoglu, 1964), and computer programs exist that set up the state equations from the network description (Pottle, 1966; Branin, 1967).

The solution of (1) is usually obtained by numerical integration. If an explicit integration scheme such as the fourth order Runge-Kutta is used, then the number of multiplications per integration step is about $4N^2$. To store the

* Research supported by National Science Foundation grant 31462x.

matrices A and B, a memory of capacity of N(N+1) words is required. Clearly if the network under consideration is of order greater than 1000, practical implementation of the solution scheme is in doubt.

If the network is topologically sparse, matrix A may or may not be sparse. Figure 1 shows two networks with identical topology, but in one, A is sparse and in the other A is almost full.

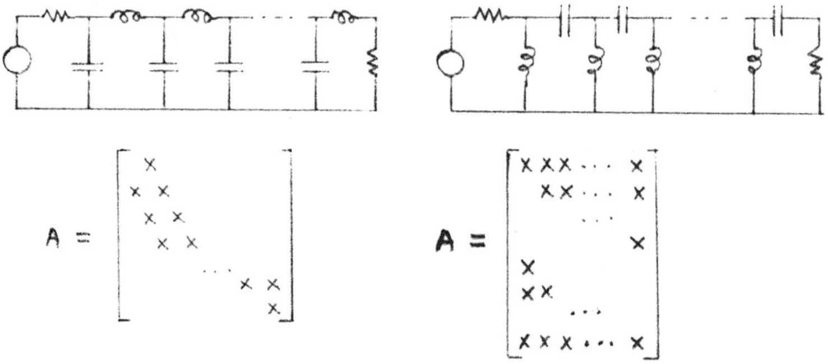

Fig.1 Two networks with identical topology but different A matrices, one being sparse and the other almost full.

SCOPE

In this paper, a computation method is described which obtains the time response of a large network by tearing. The network is decomposed into subnetworks, each characterized by its own state equations of low order. The state variables of one subnetwork are coupled to those of another by a set of algebraic equations which match the voltages and currents at the common terminals of the subnetworks. In effect, the large system of differential equations is replaced by a number of decoupled, low order systems and a system of spare algebraic equations. If the network is decomposed into M subnetworks of order $N_1, N_2, \ldots, N_M$, then the number of multiplications is about $4(N_1^2 + N_2^2 + \ldots + N_M^2)$ plus that required to solve the algebraic system, the total being usually much less than $4(N_1 + N_2 + \ldots + N_M)^2$, which is what is required if the network had not been decomposed. Similar results apply to the storage requirements.

As will be apparent later, the method is particularly suitable for analysis of electric filters, and discretized electric network models of heat flow in a conducting slab and longitudinal waves in an elastic medium.

It must be noted that the proposed method is fundamentally different from Kron's method of tearing (Kron, 1953; Branin, 1959; Pinel and Blotstein, 1967). The latter is based on the algebraic properties of networks and makes use of certain matrix inversion techniques (Householder, 1964). It is suitable for DC (static) and frequency analyses of networks. However, it is not applicable to time domain analysis, since the dynamics of a network are described by differential equations and not algebraic equations.

EXAMPLES OF LARGE NETWORKS

The class of network that will be considered consists of those that can be readily decomposed into identifiable parts that are "loosely" connected to other parts. Electric power system networks and electric filters belong to this class, as do also computer logic circuits and transistor amplifiers. Typically the order of the network is 200 to 1000. In Figure 2 are shown a band pass filter and a discretized model of a transformer winding, and in Figure 3 a gated full adder of order 78.

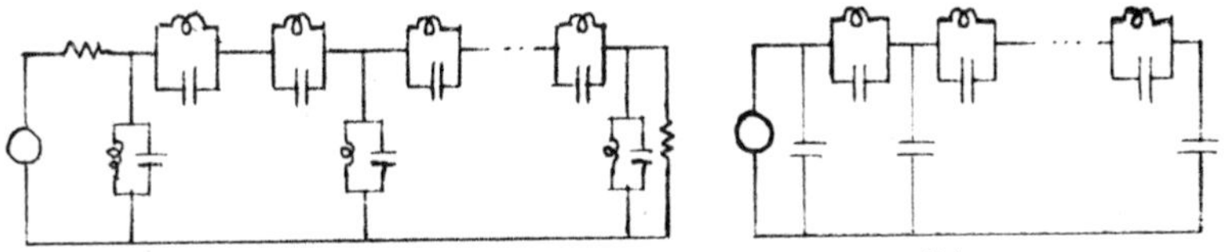

(a) (b)

Fig. 2 A band pass filter (a) and a transformer winding (b).

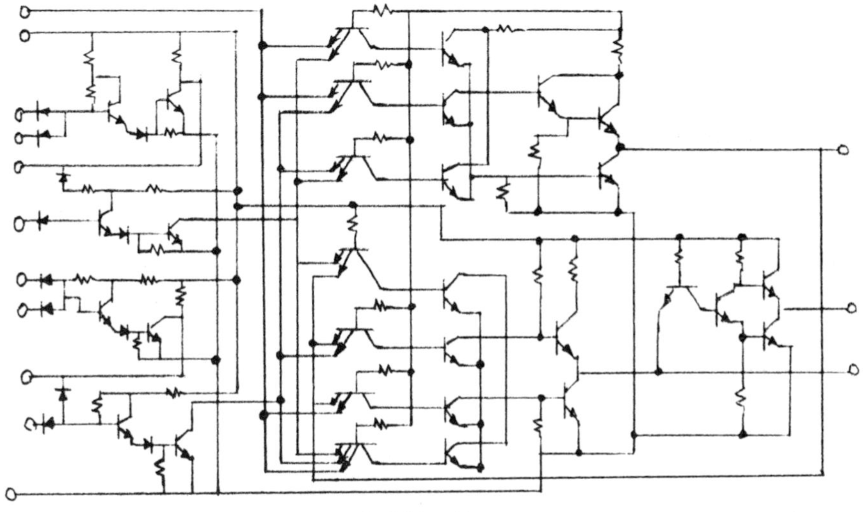

Fig. 3 A gated full adder.

ELECTRIC MODELS OF DIFFUSION AND WAVE EQUATIONS

The equations governing heat flow in a conducting medium is

$$\nabla^2 T = k \frac{\partial T}{\partial t} \tag{2}$$

where T is the temperature and k a constant depending on the thermal properties of the medium. It is easy to show that Eq. (2) can be modelled, for rectangular coordinates, by a network shown in Figure 4, where the voltages across the capacitors are analogous to temperature. The network is readily decomposed into cells. A typical one is shown in the figure.

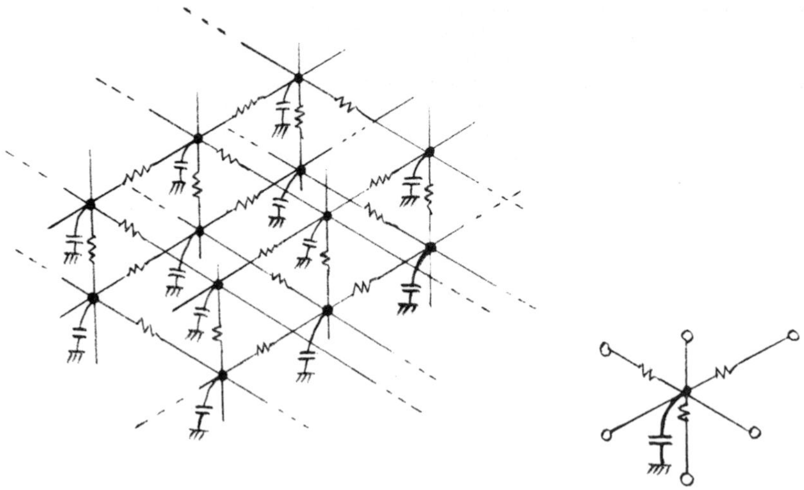

Fig. 4 Network model of a diffusion system.

For a longitudinal wave in a homogeneous elastic medium, the equation of motion is

$$\nabla^2 u = k \frac{\partial^2 u}{\partial t^2} \tag{3}$$

where u is a scalar potential function (Morse and Feshbach, 1953) such that its gradient is the displacement vector at a given point. It can be shown that Eq. (3) can be modelled by a network shown in Figure 5. The voltages on the capacitors represent the potential u.

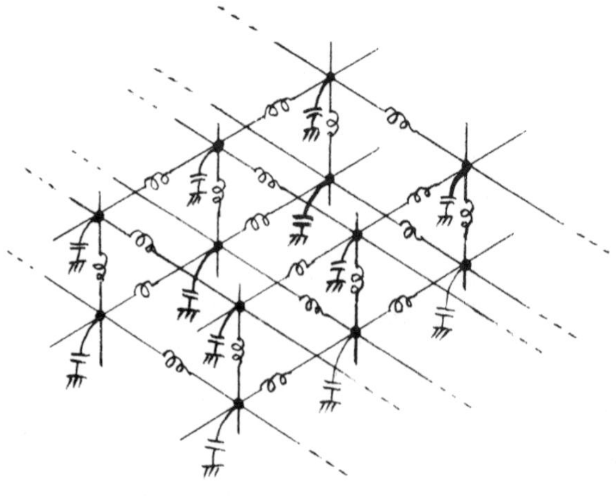

Fig.5 Network model of longitudinal wave motion in an elastic medium.

Equations (2) and (3) are usually solved numerically by using finite differences in time and in the space variables. The truncation error in time is of the order of h^2, where h is the time increment (Carnahan et al, 1969). On the other hand, if the network is decomposed into cells, and each is described by a differential equation that is integrated by using a numerical scheme such as Runge-Kutta, then the error can be made as small as of order h^5. The truncation errors in the space variables are the same in both approaches, however. Therefore, to obtain meaningful numerical results, the network must be decomposed into many cells and we have a large network.

FORMULATION OF DECOMPOSITION

Consider a network N composed of two subnetworks connected at two terminals A and B as shown in Figure 6. Assume there are no external excitations for the moment.

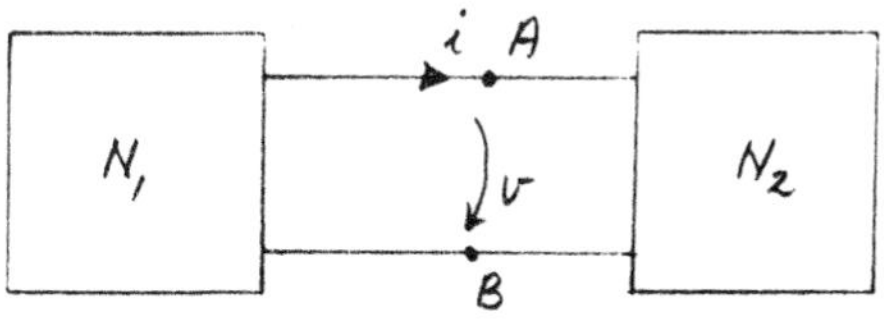

Fig.6 Two subnetworks connected at common terminals AB.

Let the voltage and current at the common terminals be v and i, respectively. The network can be decomposed in one or all of four ways shown in Figure 7, depending upon whether or not there is a capacitor loop spanning N_1 and N_2, an inductor cut set separating A and B, or a capacitor path from A to B in N_1 and N_2. In order to describe these situations, the notion of "proper partition" is introduced (Wing and Gielchinsky, 1972). A partition is said to be proper if the sum of the orders of the subnetworks including the sources assigned to each equals the order of the original network. In other words, the degree of freedom of the decomposed network must be the same as that of the network.

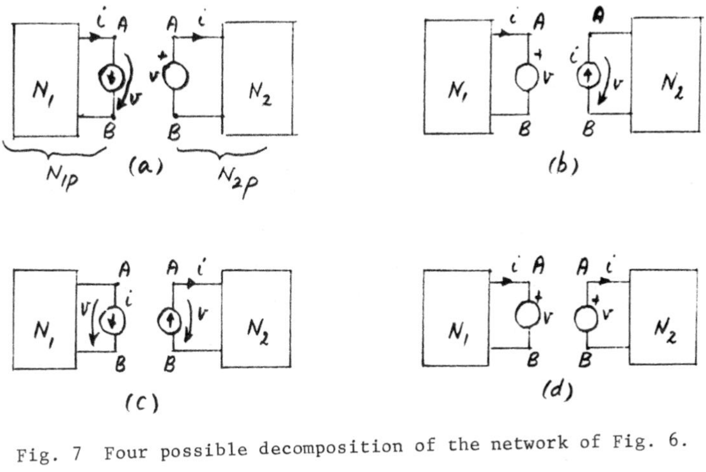

Fig. 7 Four possible decomposition of the network of Fig. 6.

For simplicity, we will assume in what follows that capacitor loops and inductor cut sets are absent, and that there does not exist an inductor cut set separating A and B in N_1, nor a capacitor path from A to B in N_2, so that partition (a) is proper. Electric models of the diffusion and wave equations are networks of this class. All the other contrary cases have been considered elsewhere (Wing and Gielchinsky, 1972; Gielchinsky, 1971) but are included in the Appendix for completeness.

For partition (a), the order of N_{1p} (N_{2p}) is the same as that of N_1 (N_2), and we can therefore choose the state variables of N_{1p} (N_{2p}) to coincide with those of N_1 (N_2). Let the state variables of N_1 and N_2 be $\underline{x}_1$ and $\underline{x}_2$, respectively. The state equations of the network are

$$\frac{d\underline{x}_1}{dt} = A_1 \underline{x}_1 + B_1 i \tag{4a}$$

$$\frac{d\underline{x}_2}{dt} = A_2 \underline{x}_2 + B_2 v \tag{4b}$$

with

$$v = C_1 \underline{x}_1 + r_1 i \tag{5a}$$

$$i = C_2 \underline{x}_2 + g_2 v \tag{5b}$$

where A_1, A_2, B_1, B_2, C_1 and C_2 are matrices and r_1 and g_2 are scalars. The last two equations are algebraic and are statements of the fact that the voltage v (current i) is a linear combination of the state variables $\underline{x}_1$ ($\underline{x}_2$) and the excitation i (v). The system of differential equations (4) together with the algebraic system (5) completely describe the behavior of the network. Notice that the two differential equations appear to be, but are not in fact, decoupled. The coupling is provided by the algebraic system.

EXISTENCE OF SOLUTION

Solving for i and v in (5), we obtain

$$v = \frac{1}{1 - r_1 g_2} (C_1 \underline{x}_1 + r_1 C_2 \underline{x}_2) \tag{6}$$

$$i = \frac{1}{1 - r_1 g_2} (g_2 C_1 \underline{x}_1 + C_2 \underline{x}_2) \tag{7}$$

Now from (5a) and Figure 7a, r_1 is seen to be the negative of the input resistance across AB in N_1 under the condition that all the capacitors are replaced by short circuits and inductors by open circuits. Therefore $r_1 \leq 0$ for a passive network. Moreover, $r_1 > -\infty$ since it is assumed that there does not exist an inductor cut set separating A and B in N_1. In a similar manner, g_2 is seen to be the input conductance across AB in N_2 under the same condition and we have $+\infty > g_2 \geq 0$. It follows that $-\infty < r_1 g_2 \leq 0$, and a solution to (6) and (7) exists.

If Eqs. (6) and (7) are substituted into (4a) and (4b), we obtain a system of differential equations in $\underline{x}_1$ and $\underline{x}_2$:

$$\frac{d}{dt}\begin{bmatrix}\underline{x}_1\\ \underline{x}_2\end{bmatrix} = \begin{bmatrix}A_{11} & A_{12}\\ A_{21} & A_{22}\end{bmatrix}\begin{bmatrix}\underline{x}_1\\ \underline{x}_2\end{bmatrix} \tag{8}$$

Since the order of the decomposed network is the same as the order of the original network, we can always choose the state variables of the latter to be $\underline{x}_1$ and $\underline{x}_2$. If a solution of the original network exists, then a solution to (8) exists, and by uniqueness of solution of a differential equation with constant coefficients, the solution in (8) is unique for a given set of initial conditions on $\underline{x}_1$ and $\underline{x}_2$.

SOLUTION ALGORITHM

Given the initial values of $\underline{x}_1$ and $\underline{x}_2$, we obtain the initial values of v and i from (6) and (7). With these, the next values of $\underline{x}_1$ and $\underline{x}_2$ are found from (4a) and (4b) separately by any numerical integration scheme. With the next values of $\underline{x}_1$ and $\underline{x}_2$ known, those of v and i are obtained from (6) and (7) and the process is repeated until the last time step is reached. Thus the solution at each time step requires the integration of a set of decoupled differential equations of low order and the solution of a set of algebraic equations.

CASCADE NETWORKS

The decomposition technique just outlined can now be applied to each of the subnetworks. For a network consisting of subnetworks connected in cascade, as shown in Figure 8, the extension of (4) and (5) is obvious.

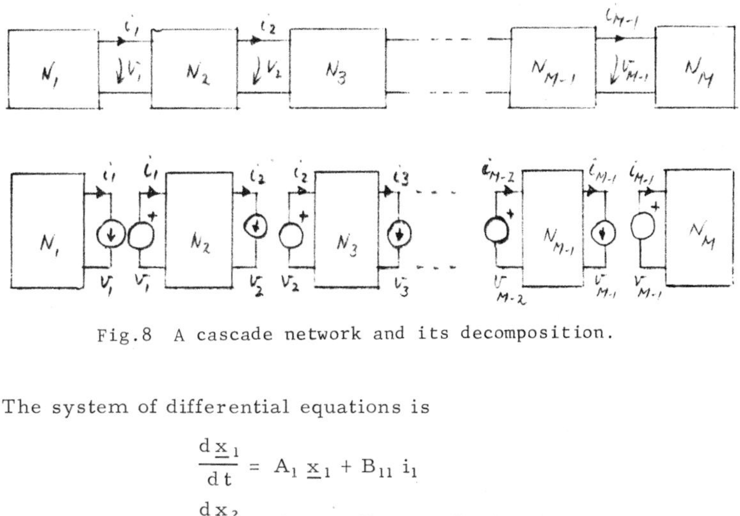

Fig.8 A cascade network and its decomposition.

The system of differential equations is

$$\frac{d\underline{x}_1}{dt} = A_1 \underline{x}_1 + B_{11} i_1$$

$$\frac{d\underline{x}_2}{dt} = A_2 \underline{x}_2 + B_{21} v_1 + B_{22} i_2$$

$$\frac{d\underline{x}_3}{dt} = A_3 \underline{x}_3 + B_{32} v_2 + B_{33} i_3 \qquad (9)$$

$$\cdots$$

$$\frac{d\underline{x}_M}{dt} = A_M \underline{x}_M + B_{M(m-1)} v_{M-1}$$

and the system of algebraic equations is

$$\begin{bmatrix} -r_{11} & 1 & & & & & \\ 1 & -g_{11} & -e_{12} & & & & \\ & -f_{21} & -r_{22} & 1 & & & \\ & & 1 & -g_{22} & -e_{23} & & \\ & & & -f_{32} & -r_{33} & & \\ & & & & \cdots & & \\ & & & & & 1 & -g_{(M-1)(M-1)} \end{bmatrix} \begin{bmatrix} i_1 \\ v_1 \\ i_2 \\ v_2 \\ i_3 \\ \vdots \\ v_{M-1} \end{bmatrix} = \begin{bmatrix} C_{11} & \underline{x}_1 \\ C_{12} & \underline{x}_2 \\ C_{22} & \underline{x}_2 \\ C_{23} & \underline{x}_3 \\ C_{33} & \underline{x}_3 \\ \vdots & \\ C_{(M-1)M} & \underline{x}_M \end{bmatrix} \quad (10)$$

The important point to note is that the system is tridiagonal and its solution requires 6(M-1) multiplications (Carnahan, 1969).

Let the order of subnetwork N_k be O_k. The integration of its corresponding differential equation in (9) requires $4\,O_k^2$ multiplications per time step if Runge-Kutta is used. The total number of multiplications per time step for the decomposed network is P_D given by

$$P_D = 4(O_1^2 + O_2^2 + \ldots + O_M^2) + 6(M-1) \quad (11)$$

On the other hand, if the network had not been decomposed, the number of multiplications per time step would be given by

$$P = 4(O_1 + O_2 + \ldots + O_M)^2 \quad (12)$$

For illustration, suppose that $O_1 = O_2 = \ldots = O_M$. Then $P_D = 4\,M\,O_1^2 + 6(M-1)$ whereas $P = 4\,M^2\,O_1^2$. The saving in computation time by using decomposition is now clear. In addition, the storage requirement for the decomposed network is of the order of

$$S_D = O_1^2 + O_2^2 + \ldots + O_M^2 + 6(M-1) \quad (13)$$

whereas for the original network

$$S = (O_1 + O_2 + \ldots + O_M)^2 \quad (14)$$

Again, the advantage of using decomposition is evident.

Example - Filter

In order to check for numerical accuracy, the decomposition technique was applied to the analysis of a constant resistance filter of Figure 9, whose step response is known to be $v_M(t) = \exp(-t)\, L_{M-1}(t)$, where $L_k(t)$ is the Laguerre polynomial of degree k (Gielchinsky, 1971).

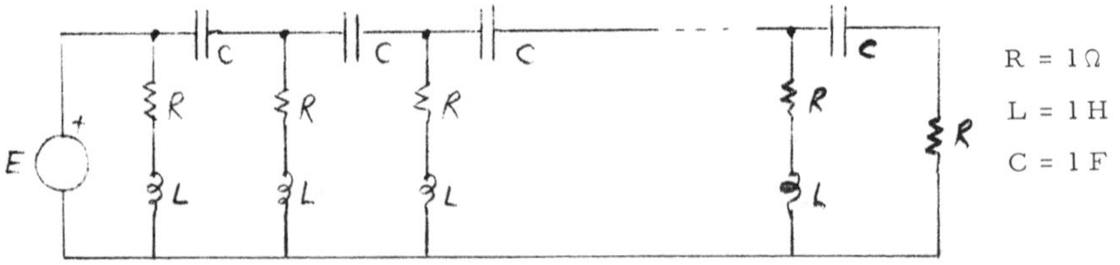

Fig.9 A constant resistance filter

The state equations are

$$
\begin{aligned}
\frac{dx_1}{dt} &= E - x_1 \\
\frac{dx_2}{dt} &= i_1 \\
\frac{dx_3}{dt} &= v_1 - x_3 \\
\frac{dx_4}{dt} &= i_2 \\
&\cdots \\
\frac{dx_{2M}}{dt} &= v_{M-1}
\end{aligned}
\tag{15}
$$

where x_{2k-1} is the inductor current and x_{2k} the capacitor voltage of the $k\underline{th}$ subnetwork. The algebraic system is

$$
\begin{bmatrix}
0 & 1 & & & & & & \\
1 & 0 & -1 & & & & & \\
 & -1 & 0 & 1 & & & & \\
 & & 1 & 0 & -1 & & & \\
 & & & -1 & 0 & 1 & & \\
 & & & & & \cdots & & \\
 & & & & & -1 & 0 & 1
\end{bmatrix}
\begin{bmatrix} i_1 \\ v_1 \\ i_2 \\ v_2 \\ i_3 \\ \vdots \\ v_M \end{bmatrix}
=
\begin{bmatrix} E - x_2 \\ x_3 \\ -x_4 \\ x_5 \\ -x_6 \\ \vdots \\ -x_{2M} \end{bmatrix}
\tag{16}
$$

The solution of the algebraic system is trivial. A variable step size Runge-Kutta was used in the integration of the differential equations. The numerical results agree with the exact values to within 0.01%. More importantly, the computation time per time step increases only about linearly with the order of the network as indicated in the table below.

M	Computation time per step (Sec)
50	0.02
100	0.04
200	0.04
500	0.29
1000	0.47

Note that if decomposition had not been used for the last case, we would have to solve Eq. (1) with matrix A being of order 2000×2000 and more than 50 per cent full.

EXAMPLE - LONGITUDINAL ELASTIC WAVE

Figure 10 shows a network model of a two-dimensional elastic medium. The voltage across the capacitor at x, y represents the scalar potential whose gradient is the displacement vector at that point. The network is decomposed into cells of order three as shown in the figure. The state equations of a typical cell are

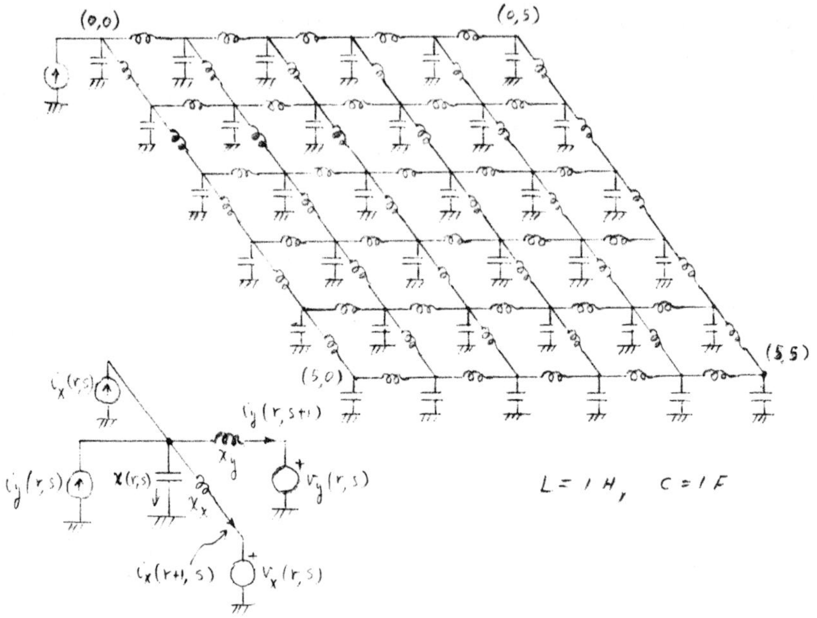

Fig.10 Network model of a 2-dimensional elastic wave

$$\frac{dx(r, s)}{dt} = -x_x(r, s) - x_y(r, s) + i_x(r, s) + i_y(r, s)$$

$$\frac{dx_x(r, s)}{dt} = x(r, s) - v_x(r, s) \tag{17}$$

$$\frac{dx_y(r, s)}{dt} = x(r, s) - v_y(r, s)$$

and the algebraic system is simply

$$\begin{aligned} v_x(r-1, s) &= x(r, s) \\ v_y(r, s-1) &= x(r, s) \\ i_x(r+1, s) &= x_x(r, s) \\ i_y(r, s+1) &= x_y(r, s) \end{aligned} \tag{18}$$

The differential equations for cells along $x = 6$ and $y = 6$ have to be appropriately modified. The boundary conditions are

$$\begin{aligned} x_x(0, 0) &= \text{unit step} \\ x_x(0, 1) &= x_x(0, 2) = \ldots = x_x(0, 5) = 0 \\ x_x(6, 1) &= x_x(6, 1) = \ldots = x_x(6, 5) = 0 \\ x_y(0, 0) &= x_y(1, 0) = \ldots = x_y(5, 0) = 0 \\ x_y(0, 6) &= x_y(1, 6) = \ldots = x_y(5, 6) = 0 \end{aligned} \tag{19}$$

The set of differential equations consists of thirty-six third order equations. Each was integrated by using a variable step, fourth order Runge-Kutta algorithm. The response v(0, 0) was computed to be as shown in Figure 11. As mentioned earlier, the truncation error in time is of the order h^5, which is much smaller than that if finite differences in time were used.

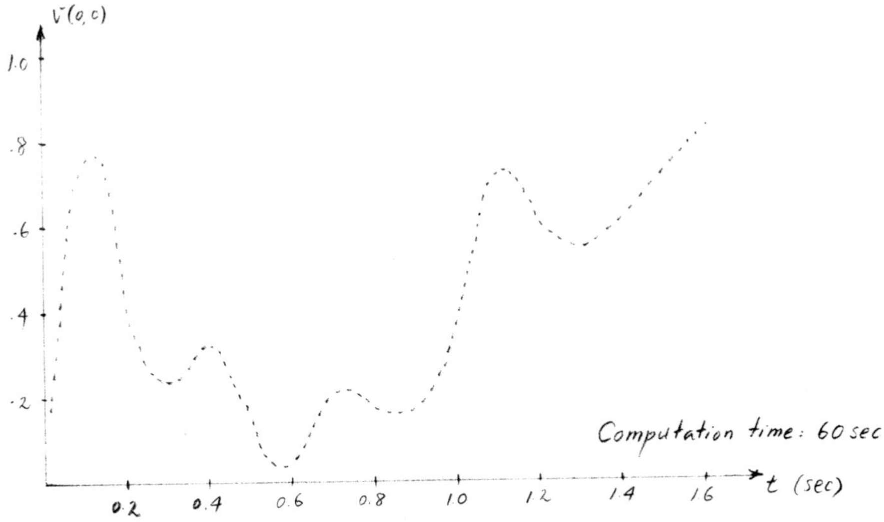

Fig.11 Step response of elastic wave at (0,0)

CONCLUDING REMARKS

We have presented a decomposition method for computing the response of a large network in the time domain. It is important to note that the decomposition is applied to the network, which is assumed to be composed of sparsely connected subnetworks, and not to its state equation, which may or may not be sparse. In place of a single state equation of large order, a system of decoupled state equations of low order and a system of sparse algebraic equations are formulated to describe the decomposed network. The variables of the algebraic system are the voltages and currents at the common terminals of the subnetworks. They may be regarded as auxiliary variables which serve to couple the state equations of the subnetworks.

The total computation time is the sum of the solution time of the system of decoupled differential equations and the time required to solve the system of algebraic equations. If the network is sparsely connected, significant savings in computation time and storage requirements can be realized, as a result of decomposition. If the network is not sparsely connected, then the system of algebraic equations is no longer sparse and the advantage of decomposition is lost.

APPENDIX

It shall be shown that a proper partition always exists for a network composed of two subnetworks connected at two common terminals, and that a solution to the equations that describe the decomposed network exists.

Case 1 There are no capacitor loops spanning N_1 and N_2 or inductor cut sets separating A and B in N.

Case 1a There is not any capacitor path from A to B in N_1 and no inductor cut set separating A and B in N_2. All four partitions are proper. The equations for partition (a) are given by Eqs. (4) and (5). Those for (b) are similar by symmetry. The equations for partition (c) are

$$\frac{d\underline{x}_1}{dt} = A_1 \underline{x}_1 + B_1 i \tag{A1}$$

$$\frac{d\underline{x}_2}{dt} = A_2 \underline{x}_2 + B_2 i \tag{A2}$$

$$v = C_1 \underline{x}_1 + r_1 i \tag{A3}$$

$$v = C_2 \underline{x}_2 + r_2 i \tag{A4}$$

Combining (A3) and (A4), we get

$$i = \frac{1}{r_2 - r_1} (C_1 \underline{x}_1 - C_2 \underline{x}_2) \tag{A5}$$

Using the same argument as in the text, one can show that $r_1 < 0$ and $r_2 > 0$. Hence a solution to (A5) and therefore to (A1) and (A2) exists.

The equations for partition (d) are deducible from those for (c) if duality is invoked, and they will be omitted.

Case 1b There is a capacitor-voltage-source path from A to B in N_1. Partitions (a) and (c) are proper if there is not any inductor cut set separating A and B in N_2. If there is, only partition (a) is proper. Existence of solution for (a) and (c) has been shown previously.

Case 1c There is an inductor-current-source cut set separating A and B in N_1. Partitions (b) and (d) are proper if there is not any capacitor path from A to B in N_2. If there is, only partition (b) is proper. Proof of existence of solution for (b) is the same as (a) by symmetry; and for (d), by duality of (c).

Case 2 There is a capacitor-voltage-source loop spanning N_1 and N_2. The order of N is therefore one less than the sum of the orders of N_1 and N_2. Partitions (a) and (b) are proper. Consider (b) only. In the decomposed network, the voltage source across AB in N_1 creates a capacitor-voltage-source loop. The order of N_{1p} is one less than the order of N_1. The order of the decomposed network is therefore one less than the sum of the orders of N_1 and N_2, and the partition is proper. The differential equations are

$$\frac{d\underline{x}_1}{dt} = A_1 \underline{x}_1 + B_1 v + E \frac{dv}{dt} \tag{A6}$$

$$\frac{d\underline{x}_2}{dt} = A_2 \underline{x}_2 + B_2 i \tag{A7}$$

$$i = C_1 \underline{x}_1 + g_1 v + f \frac{dv}{dt} \qquad (A8)$$

$$v = C_2 \underline{x}_2 \qquad (A9)$$

$$\frac{dv}{dt} = C_2 A_2 \underline{x}_2 + C_2 B_2 i \qquad (A10)$$

where $\underline{x}_1$ and $\underline{x}_2$ are linearly independent. Note that in (A9), the voltage v does not depend on i, since there is a capacitor path from A to B in N_2 and the voltage is simply a linear combination of the capacitor voltages along this path. Eliminating v and dv/dt, we get

$$(1 - f\ C_2 B_2) i = C_1 \underline{x}_1 + (g_1 C_2 + f\ C_2 A_2) \underline{x}_2$$

Since $\underline{x}_1$ and $\underline{x}_2$ are linearly independent, $1 - f\ C_2 B_2 \neq 0$ and a solution to (A8) - (A10) and hence to (A6) and (A7) exists.

Case 3 There is an inductor-current-source cut set separating A and B of N. The proper partitions are (a) and (b). The proof is entirely similar to that for case 2 if the principle of duality is used.

REFERENCES

Bashkow, T.R., (1957), The A matrix, a new network concept, IRE Trans. Circuit Theory, 4, 117.

Branin, F.H. Jr., (1962), DC and transient analysis of network using a digital computer, IRE Int. Conv. Rec., part 2, 236.

Branin, F.H. Jr., (1959), The relation between Kron's method and the classical methods of network analysis, IRE WESCON Conv. Rec., part 2, 3.

Brown, D.P., (1963), Derivative-explicit differential equations for RLC graphs, J. Franklin Inst., 275, 503.

Bryant, P.R., (1959), The order of complexity of electrical networks, Proc. IEE (London), Mono 335E, 174.

Bryant, P.R., (1962), The explicit form of Bashkow's A-matrix, IRE Trans. Circuit Theory, 9, 303.

Carnahan, R., Luther, H.A., and Wilkes, J.O., (1969), Applied Numerical Analysis, New York: J. Wiley, 363.

ibid, 442.

Dervisoglu, A., (1964), Bashkow's A-matrix for active RLC networks, IEEE Trans. Circuit Theory, 11, 404.

Gielchinsky, J., (1971), Computation of Transient Response of Large Networks by Decomposition, Ph.D. Thesis, Columbia University.

Householder, A.S., (1964), Theory of Matrices in Numerical Analysis, New York: Blaisdell, 123.

Kron, G., (1953), A set of principles of interconnecting the solutions of physical systems, J.A.P., 24, 965.

Morse and Feshbach, (1953), Methods of Theoretical Physics, New York: McGraw Hill, 142.

Pinel, J. F. and Blostein, M. L., (1967), Computer technique for the frequency analysis of linear electrical networks, Proc. IEE, 55, 1810.

Pottle, C., (1966), State-space techniques for general active network analysis, Chap. 3, System Analysis by Digital Computers, (ed. F. F. Kuo and J. F. Kaiser), New York: J. Wiley, 59.

Wing, O. and Gielchinsky, J., (1972), Computation of time response of large networks by partitioning, Proc. IEEE Int. Sym. Circuit Theory, 125.

ACKNOWLEDGEMENT

The author wishes to thank his former student, Dr. J. Gielchinsky, for providing the numerical examples.

ON THE SOLUTION OF THE SUBPROGRAMS IN THE DECOMPOSITION OF A MULTIFLOW PROBLEM

Jean François MAURRAS

Direction des Etudes et Recherches
d'Electricité de France
17, avenue du Général de Gaulle
92 - CLAMART (France)

Abstract : This paper presents a method for solving a multiflow problem issued from the problem of standardisation of a system.

A first model (3,000×8,000) has been solved; a second one (12,000×50,000) is almost solved. In order to cape with the difficulty of memory size the Dantzig and Wolfe method has been used with the reconstruction of matrices at each step and a quite efficient method of solving the flow subprograms.

Two main ideas lead to the method used to solve the later problem.

1) Characterization of the basic graph (with gains or not)
There is one and only one cycle for each connex component in the graph of the basis.

2) A full order corresponds to the subset order given by the above property.

The same simplex (or dual simplex) can then be applied but without the basis inversion. The number of operations is then proportional to m instead of m^2.

1. INTRODUCTION

Faced with a growing demand for low-voltage current, and a network having several medium voltages, Electricité de France has decided to normalize the medium voltage at 20 kV.

But what interests us is the MV/LV transformer inventory and in particular the 15 kV/LV transformers.

We thus have a line of equipment (the 15 KV/LV transformers) of which we have to make the best use. What is involved, in the presumably known future is determination of the most economical way of organizing the purchases, transformation, scrapping of this line.

Regarding purchases, we have two types of equipment to choose from, one more expensive but easily adapted to 20 kV, the other cheaper but costing more to transform.

This equipment, which has a given length of life, is spread over various areas which can make exchanges incurring transportation costs.

For a normalization period covering 40 years, a life-time of 30 years and only two areas, the model of this problem is a linear program (3,000 × 8,000); for eight areas (12,000 × 50,000).

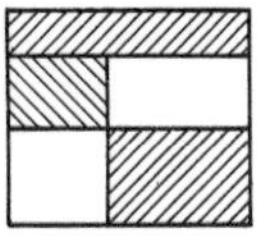

The structure of the matrix permits using the Dantzig and Wolfe decomposition method [3]. In fact, this matrix has the classical block angular structure. Various difficulties have been met with, those inherent to carrying out the decomposition method which we shall deal with in a first part, and those caused by the solving of subprograms which are, in this case, flow problems for which we have developed a specific adaptation of the simplex method [9].

1.1. PRESENTING THE PROBLEM

Two sets of equations are written; the first, which corresponds to the subprograms, expresses the inventory year by year, by area, and for each type and year equipment was purchased (e.g. the transformers 15 kV/LV purchased in 1965 is one of the type of equipment).

For a type of equipment purchased in the various areas during a given year, we express in each area and for each year the existing quantities of this equipment, and this for the entire life-time of the equipment. One must make a difference between the two types of equipment bougt in two different years because they don't have the same life.

The second set, which corresponds to the master program, expresses for each year and for each area that total existing equipment meets the needs of the area considered. We see that this set of equations relates the variables corresponding to each type of equipment.

1.2. DIFFICULTIES ARISING FROM DECOMPOSITION AND SIZE

We encountered four types of difficulties.

1.2.1. The first lay in choosing the number of subprograms. It was possible to have a variable number of subprograms appear, from 1 to 162 (60 of them corresponding to slack variables of the master program)while maintaining for them a flow problem structure. Our first choice was 2. The number of equations to be added to the matrix of the master program was thus minimized, one of the subprograms being a bounded polyhedron, the other a cone with the origine as vertex.

With this decomposition, we were never able to reach the optimum, in the limit of half hour of CDC 6600 computing time.

1.2.2. We then took the maximum number of subprograms (162). There being a large number of iterations, we were led to recalculate during the algorithm the inverse of the basis (reinversion) of the master program. At this stage we reinversed periodically. Noticing that the slack variables did not show optimum activity, we suppressed them and to our surprise solving time was increased by one-half. According to W. Orchard-Hays the elimination of the slack variables decreases the degree of freedom of the system.

1.2.3. We made various tests with different costs. In some cases we got no results, the recalculated inverse of the basis differing too greatly from the one obtained by the iterative method. The round-off arrow changed so much the inverse matrix of the master program, that the value of certain variables calculated after reinversion were negative. We sought another reinversion criterion : a reinversion before pivoting when the future pivot is below a given value. Furthermore, a very sure inversion method was chosen (that of Gauss with choice of the maximum pivot). This eliminated all our troubles, we were thus able to do without systematic periodic reinversion.

1.2.4. In the model with 8 areas (12,000 × 50,000) the size of the inverse of the basis is (250 × 250) and it occupies the greatest part of the storage we are allotted on the CDC 6600. From one iteration to the next we must therefore limit to a minimum the elements stored.

Given the structure of the matrix and the fact that a maximum number of subprograms is being used, we regenerate the vertical section of the constraint matrix containing the subprogram we are about to deal with.

As will be seen in the second part, thanks to the storage of only a single vector (12,000) characterizing the bases of the various subprograms, all the necessary elements of the subprogram can be recalculated quite successfully (solution vector, relative cost vector) and the successing iteration proceeded with.

2. SOLUTION OF THE SUBPROGRAMS

2.1. We are going to set forth our graph flow optimization method in the field of graphs with gains (a problem more complicated than the conventional one).

One method of solving the problem which derives from the Ford and Fulkerson algorithm was proposed in 1958 by Jewell [5] . A second method based on the theory of a general linear programming algorithm (simplex method) was proposed by Johnson [7] in 1965. It is the second type of method that we use here*.

Let us give a more accurate definition of a graph with gains.

A graph with gains is one in which the usual law of nodes (conservation of the flow at a node, i.e., the algebraic sum of flows arriving at a given node equal to zero) is replaced by a linear combination proper to each node.

The set of these linear combination coefficients forms the constraint matrix of the linear program representing the optimization problem. The i row of this matrix is $\sum_{j \in J} a^j x^j = b_i$ J is the set of the incident arcs to the i node, a, b, x will be defined later. This matrix is special in structure, its columns (corresponding to graph arcs) have at the most two non-zero elements.

In this linear program, the variables are the flows on the graph.

The procedure described here is effectively the simplex method, where at each iteration the calculations are made directly, i.e., without using the inverse matrix of the basis. Account can thus be taken of the very special structure of the problem, the direct calculations can be rapidly effected while making use of only reduced storage proportional to the number of arcs.

Two main ideas justify the use of this method :

1) the existence, in the graph of the basis, of one and only one cycle by connected component (§ 2.4.2.1.),

2) a special numbering of the graph (§ 2.4.2.2. and 2.4.2.5.).

2.2. NOTATION

A(mxn)	constraints matrix; this matrix has, in each column, at the most two non-null elements
I	set of m integers such that $I \subset N = (1, 2, \ldots, n)$ such that the m vectors A^i, $i \in I$ be linearly independent
A^I	reduction of A to its column vectors A^i, $i \in I$; the set of these vectors is called the A basis
c(1 × n)	cost vectors; c^I basis costs
b(m × 1)	second member vector
$G(\sigma, \xi)$	with σ set of nodes and ξ set of arcs, is the associate graph
u	vector 1 × m such that $u = c^I (A^I)^{-1}$ (u comes into the expression of relative costs $\bar{c}^j = c^j - u A^j$).
x	vector n × 1 each variable x_j represents the flow in the arc j.

2.3. THE PROBLEM POSED

Given the following linear program to be solved, written in its canonical form

$$(1) \quad \begin{cases} \min \ cx \\ Ax = b \\ 0 \leqslant x \leqslant a \end{cases}$$

The addition of artificial variables to the problem, the transformation of a problem given in the form of inequalities into its canonical form, do not alter

* Our article was already drafted when we heard of the work of Johnson, through M.L. Balinski.

the structure required by the following method. The slack variables or the artificial variables corresponds to A columns containing + 1 or - 1 in a given row.

Our method consists essentially in adapting the simplex method for the computations.

These changes can be made quite as well within the framework of dual-simplex method computations.

In what follows, we set forth our method within the framework of the simplex method only, as well as for the following simplified problem (1')

$$(1')\begin{cases}\min \ cx \\ Ax = b \\ x \geqslant 0\end{cases}$$

Remark

The introduction of upper bounds brings about a little difference in this presentation.

2.4. METHOD

The algorithm is developed in three steps in the same way as that of the simplex method called "explicit inverse of the basis" with the fundamental difference that the inverse of the basis is not available to us.

1) computation of the candidate column A^s
2) computation of the column to be taken out of the basis A^r
3) computation of the new u vector.

2.4.1. Computing the candidate column

s is determined such that $\bar{c}^s = \min\limits_{j=1,\ldots,n} (c^j - u A^j)$

If $\bar{c}^s \geqslant 0$ the algorithm is ended, the optimum is obtained

2.4.2. Search for the column to be taken out of the basis

It is here that the difficulties of the algorithm are encountered. First of all we shall prove a theorem that will enable us to label the arcs of the graph of the basis (graph whose arcs correspond to the A basis columns). This method will be defined subsequently. This numbering also applies to the columns of the basis matrix A^I. It will permit finding a simple solution to the linear system $A^I x = y$.

It is then shown, after exchanging the candidate column and the one from the basis, that the columns can be "re-labeled" simply.

2.4.2.1. Basic theorem

There is one cycle and only one by connected component in the graph of the basis.

Definition [2] A graph is connected if for any pair of distinct nodes there is a path from one to the other. A connected component piece of a graph is thus a connected part of the latter.

Proof [4] Let us isolate in the matrix of the basis a connected component; its matrix is square since it is of the basis and is therefore regular. Its cyclomatic number is $\nu(G_i) = m_i - m_i + 1 = 1$.

Corollary 2 of theorem 2 is applied [1] :

A graph G admits of a single cycle if and only if $\nu(G) = 1$ Q.E.D.

Remarks

In problems of conventional flows (gain equal to 1) the cycles are reduced to loops, effectively a matrix made up of columns each of which has two elements + 1 and - 1, and which is never regular.

For these problems, it is therefore advisable to particularize the algorithm in order to speed up code performance.

2.4.2.2. Consequence : numbering the graph of the basis

Remarks

As a result of the property referred to above, in the basis two cycles cannot be part of the same connecting component.

The arcs of the graph of the basis are numbered from 1 to m so that the numbering has the following property for the arcs not belonging to a cycle :

Property

Given a graph made up of arcs numbered from j to m : in this graph the arc j is loose. Further, for convenience sake, the loose end of the arc j is marked ① , the other end ② . In addition, the cycle arcs have either one of the possible numbers, satisfying the foregoing property, once they are opened at any one nod. Also marked ① and ② are the loose end and opposite end of each cycle arc, once a direction has been chosen for the numbering.

Remark

Johnson [6] proposes using Scoins' triple-label procedure.

Remark

The numbering can be done in more than one way.

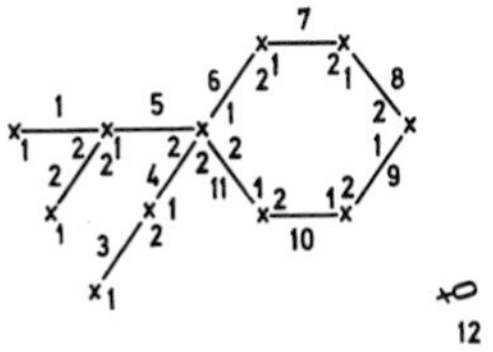

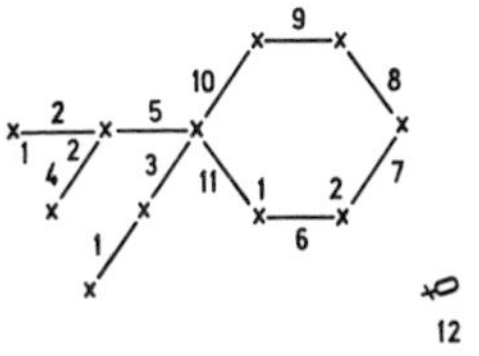

2.4.2.3. Solution of the linear system $B\,x = y$ $(B = A^I)$

Suppose that we have to calculate x_j, j being the j of the foregoing numbering.

a) the arc j does not belong to a cycle.

Let us assume that we know $x_1 \dots x_{j-1}$, only $x_j \dots x_m$ are unknown, the second member becomes $y - B^1_{x_1} \dots - B^{j-1}_{x_{j-1}}$

Now the first element as defined under 2.4.2.2., of the column B^j is alone in its row a^j_1 (i row).

There follows $x_j = \dfrac{y_i}{a^j_1}$. The routine can therefore proceed.

b) the arc j belongs to a cycle.

If this cycle is a loop, the result is trivial.

Let us suppose that the arc k is the first arc of this cycle and the arc ℓ the last.

An equation binds the "first" element of the column k to the matrix B and the "second" of the column ℓ to the said matrix.

Let us take the variable x_ℓ as paremeter, applying the previous method, we shall obtain in succession x_k, x_{k+1}, ..., x_ℓ as a linear function of this parameter, in particular $x_\ell = \lambda\, x_\ell + \mu$ which sets the value of

x_ℓ , $x_\ell = \frac{\mu}{1-\lambda}$; by replacing the parameter x_ℓ by its value in the expressions of $x_k, \ldots, x_{\ell-1}$, we obtain the values of $x_k, \ldots, x_{\ell-1}$.

Remark

In actual practice, it is advisable to take $x_\ell = 0$ as an initial operation, which determines μ, and compute independently $\lambda = \prod_{j=k}^{\ell} - \frac{a_2^j}{a_1^j}$, then restart the calculations with x_ℓ.

2.4.2.4. Determination of the column taken out of the basis

In the previous system, let us take as second member the candidate column A^s that was determined in § 2.4.1. The solution of this problem will be $\bar{A}^s = B^{-1} A^s$, and as in the simplex method, we can form the ratio $\Theta_i = \frac{\bar{b}_i}{\bar{a}_i^s}$ for $\bar{a}_i^s > 0$ with $\bar{b} = B^{-1} b$ ($\bar{b}$ is computed by modifying b at each iteration).

r is then determined so that $\Theta_r = \min \Theta_i$.

r is the index of the column coming out of the basis

Here the new $\bar{b}$' s are calculated, adding Θ_r to the variable x_s.

Remark

As in the simplex method, if the variables are not bounded, it may happen that the candidate variable can be given an infinite value, the optimum is then infinity.

2.4.2.5. Renumbering the columns

We have just found the column taken out of the basis, by means of very simple calculations and thanks to the numbering.

After the exchange of the candidate column and the column taken out of the basis, there is no reason for the numbering to satisfy the definition of 2.4.2.2.

We are going to show that systematic renumbering can be effected. Let us call "active graph" the graph of the arcs whose flows are modified by introduction of the candidate arc.

At stage 2.4.2.4. it was possible to mark this active graph.

Several cases can arise :

1) the candidate arc links two points of a connected component and here again we have two cases
   a) one of the cycles formed is broken
   b) the connected component splits up into two new ones

   } so as to remain in accordance with the basic theorem.

2) the candidate arc links two connected components, and we again find the same two sub-cases
   a) as above
   b) as above

We shall begin by numbering the loose paths not contained in the active graph and, in reverse order, starting with m, number the loops and the cycles not contained in the active graph.

We then number the active graph.

The different cases possible are given below :

ACTIVE GRAPH

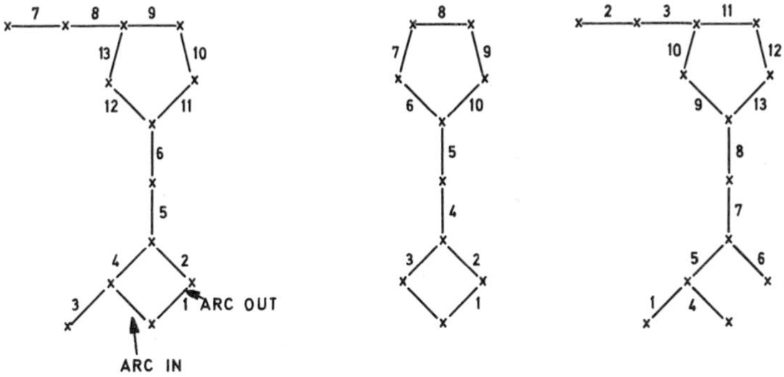

(1a)

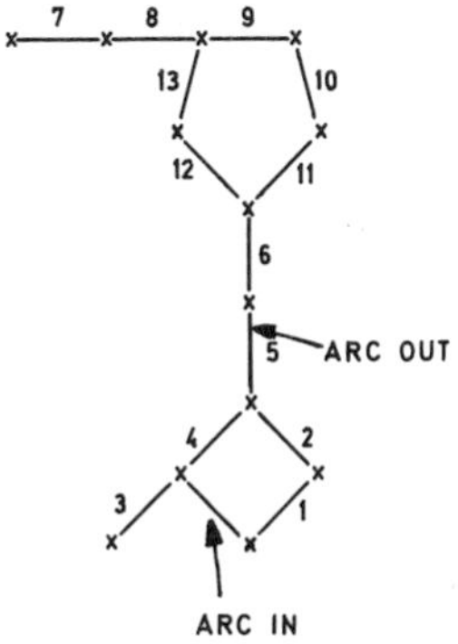

SAME ACTIVE GRAPH

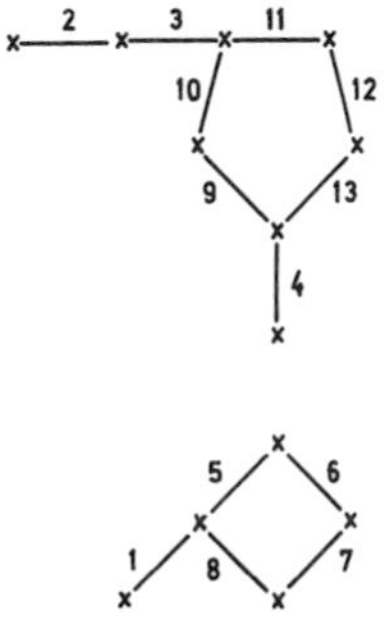

(1b)

ACTIVE GRAPH

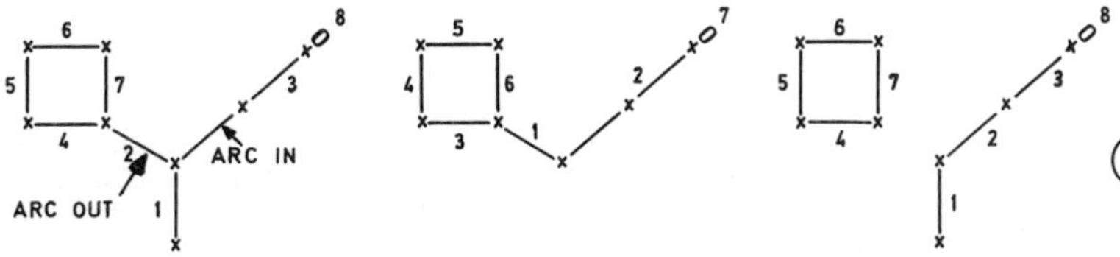

(2b)

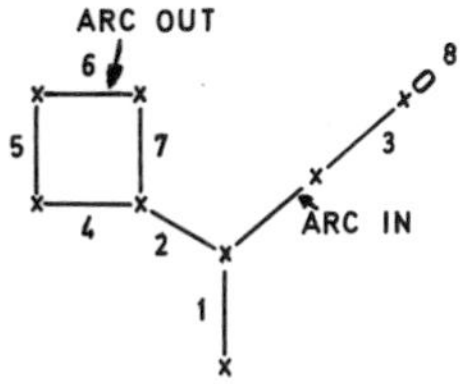

SAME ACTIVE GRAPH

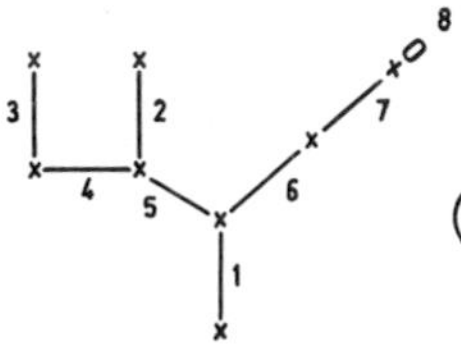

(2a)

2.4.3. Computing the new vector u

Given the linear problem $uB = c^I$ to be solved.

This problem is the transpose of the one solved in section 2.4.2.3.

It is solved by starting with the cycles, to which exactly the same method is applied.

Let j be an index that does not number a cycle arc, and let us consider as known all the u^t s that it was possible to compute for $i > j$.

Let ℓ be the number of the row of the "first" element of column j, k the row number of its "second" element.

We have the equation : $a_1^j u^\ell + a_2^j u^k = c^j$, now u^k is known by assumption, so we determine u^ℓ and proceed with the method until $j = 1$.

3. ACKNOWLEDGEMENTS

I wish to thank the Direction des Etudes et Recherches of Electricité de France who enabled me to carry out this work, and especially Mr HUARD who constantly gave me advise, help and encouragement.

4. REFERENCES

[1] C. Berge, Théorie des graphes et ses applications (Dunod éditeur, Paris, 1958, p 29).

[2] C. Berge et A. Gouilha-Houri, Programmes jeux et réseaux de transport (Dunod éditeur, Paris, 1962, p 122).

[3] P. Broise, P. Huard, J. Sentenac, Décomposition des programmes mathématiques (Monographies de recherche opérationnelle - Dunod éditeur, Paris 1967).

[4] M. Gondran, Oral communication.

[5] W.S. Jewell, Optimal flow through networks with gains (Interim technical report n° 8, Massachussetts Institute of technologie, Cambridge 1958 or Operation Research 1962 n° 10 p 476-499).

[6] E.L. Johnson, Networks and basic solutions (Operations Research Vol. 14 n° 4, 1966 p 619-623).

[7] E.L. Johnson, Programming in Networks and Graphs (ORC 65-1 University of Californie Berkeley).

[8] D. Lacaze, Flot optimal dans les réseaux avec multiplicateurs (Revue française de Recherche opérationnelle n° 37, 1965 p 271-293).

[9] J.F. Maurras, Méthode primale d'optimisation du flot dans un réseau avec multiplicateurs (Note interne EDF du 5.3.71), to be published in Mathematical Programming, titled "optimization of the flow through networks with gains".

ON THE SOLUTION OF LARGE-SCALE PROBLEMS BY DECOMPOSITION

Gábor E. Veress, Gedeon A. Almásy, Iván M. Pallai

Automation Research Institute Hungarian Academy of Sciences

Budapest, Hungary

Abstract: This paper deals with a general interpretation of the decomposition methods for solving large-scale problems, on the basis of the mathematical concept of problem decomposition. Problems are interpreted as mappings, thus composite problems can be considered as composite mappings. The solution of composite problems by structural decomposition is the decomposition of composite mappings. To illustrate these ideas some examples are given.

1. INTRODUCTION

Although the importance of solving large-scale problems is trivial and the decomposition methods have already been widely accepted, no precise definitions and interpretations thereof have been introduced so far. Generally methods, whereby the overall problem is transformed to a set of subproblems, are called decomposition methods.

Nowadays the decomposition methods can be classified according to the problems for which they are applicated. For example, Kron (1963) and Himmelblau (1966, 1967) deal with the decomposition of systems of equations in the first place, Kulikowski (1969) with that of optimization, Dantzig (1960), Nemhauser (1964), Lasdon (1970), Geoffrion (1970) and Mine (1970) with those relating to mathematical programming problems, Mesarovic (1970) with the decomposition of hierarchical systems in a general sense, while Vichenevetsky (1971) with differential equations.

This paper is an attempt for a general interpretation of decomposition methods. The basic idea is the interpretation of problems as mappings, so composite problems can be considered as composite mappings, while the structural decomposition is the decomposition of composite problems as composite mappings.

As preliminaries, the application of the concept of composite mappings for describing composite systems can be found in the work of Salovaara (1967), and the description of systems as mappings is given by Mesarovic (1970).

The interpretation of the problem composition and decomposition on the basis of the concept of composite mappings has been established in the authors' earlier papers (Almásy (1971), Pallai (1972)).

2. CHARACTERIZATION OF PROBLEMS

2.1. Interpretation of Problems

A problem can be regarded as a triplet: a problem variable set X, a problem solution set X^o and a problem solving operation P. Thus a problem can be interpreted as a mapping of the problem variable set onto the problem solution set i.e. a surjection (Polak 1971) that is

$$P : X \Longrightarrow X^o,$$

where $X^o \subseteq X$.

The following notation is introduced: $P(X)|X^o$.

Every problem can be interpreted in this way.

Example 1.

We first consider solving a set of simultaneous linear equations

$$Ax = b,$$

where $\underline{x},\underline{b} \in R^n$ (R^n is the n-dimensional Euclidean space) and A is a nonsingular matrix of order n.

Here the problem variable set X is R^n, and the problem solution set

$$X^o = \{ x:Ax = b \}.$$

Thus the solution of this equation system can be interpreted as a mapping of X onto X^o.

Example 2.

Let us take a mathematical programming (MP) problem interpreted in the usual way:

$$\max \{ f(x) \text{ subject to } g(x) \geq 0, \quad x \in X \subseteq R^n$$

Let us indicate the set of feasible solutions by X^F,

$$X^F = \{ x : g(x) \geq 0, \quad x \in X \}$$

and the set of the optimal solutions by X^o,

$$X^o = \{ x^o : f(x^o) \geq f(x) \quad \text{for all } x;\ x, x^o \in X^F \}$$

The mathematical programming problem can be interpreted as a mapping of X onto X^o.

2.2. Problem Transformation

The basis of problem solving is to transform the original problem into an equivalent problem easier to solve.

Two problems P' and P'' are <u>equivalent</u>, if their variable sets and solution sets are the same, that is, if

$$P' : X \Rightarrow X^o$$

$$P'' : X \Rightarrow X^o,$$

then $P'(X)|X^o = P''(X)|X^o$, or $P' = P''$.

The equivalence of problems is reflexive, symmetric and transitive.

A problem transformation resulting in a problem equivalent to the original one is <u>equivalent transformation</u>.

<u>Example</u>: <u>Transformation of mathematical programming problems</u>.

The first heuristic interpretation of MP problem transformation was given by Bellmore (1970), but it was not applicable in general. The problem transformation for decomposition was called "problem restructuring" by Brioschi (1967). The "problem manipulation" introduced by Geoffrion (1970) represented similar techniques although it applied to a somewhat wider field.

There are several types of MP problem transformation. Two of the most important types applicable for decomposition are the transformation of problems, introducing coupling variables, into a form in which the variables of the component problems are disjoint, and the other type is the transformation of problems by dualization.

Some other transformations, like inner linearization, outer linearization, introduction of slack variables to convert the inequality constraints into equalities, etc. result in easier solvability but they cannot be considered as tools of decomposition.

The proof of equivalence between the original and transformed problems is a very important question. For example, in the case of dualization the proof of equivalence is given by the well-known Duality Theorems.

3. COMPOSITE PROBLEMS

3.1. <u>Interpretation of Composite Problems</u>

As it is well-known, a mapping A

$$A : Y \Rightarrow Z$$

is a composite one, if there exist two mappings B and C, for which

$$B : V \Rightarrow Z$$

$$C : Y \Rightarrow V,$$

that is $A(Y)|Z = B(V)|Z \circ C(Y)|V$

or $A = B \circ C$

where the operation of the mapping composition is indicated by the sign "o".

A problem P^c is composite, if the mapping representing the problem is composite, that is

$$P^c = P \circ p,$$

where the problems P and p are called composing and component problems, respectively.

Example: Set of simultaneous linear equations.

Let us suppose, that the matrix A can be factored as a product of two matrices, i.e.

$$A = LU,$$

then the set of simultaneous linear equations (see Example 1 in Chapter 2.1.)

$$Ax = b$$

is equivalent to the sets

$$Ly = b$$

$$Ux = y.$$

The problem

$$P : X \Rightarrow X^o$$

where

$$X = R^n$$

$$X^o = \{x : Ax = b, x \in X\}$$

can be considered as a composite one, i.e.

$$P(X)|X^o = \Pi(Y)|X^o \circ p(X)|Y,$$

where

$$Y = \{y : Ux = y, x \in X\}$$

and the solution set X^o can be defined in the following form, also:

$$X^o = \{y : Ly = b, y \in Y\}.$$

3.2. Types of Composite Problems

A composite problem is simple, if it consists of one composing and one component problem only.

A composite problem is multicomponent, if it consists of one composing and at least two component problems, that is

$$P^c = P \circ \langle p_1, \ldots, p_n \rangle$$

or in detail:

$$P^c(X_1 \ldots X_n)|X^o = P(Y_1 \ldots Y_n)|X^o \circ \langle p_1(X_1)|Y_1, \ldots, p_n(X_n)|Y_n \rangle$$

where $X_i X_j$ is the union of sets X_i, X_j.

A composite problem can be considered as a component problem of another composite problem at a higher level, so a general multilevel composite problem is as follows:

$$P^c_{k-1} = P_{k-1} \circ \langle P_{k1}, \ldots, P_{kn_k} \rangle$$

On the basis of the definition of problems, it follows, that a composite problem as a mapping is a surjection, and the component problems are also surjections. On the other hand, the composing problem must be a surjection because the composite problem is a surjection.

A composite problem is regular, if both the composing and composite problems are surjections, otherwise it is irregular.

A composite problem is disjoint, if it consists of component problems with mutually disjoint variables. Without loss of generality we can suppose that, supplementing additional variables to those of the original problem, the variables of the component problems can be mutually disjoint.

A component problem of a disjoint composite problem is independent, if it has no common variables with the composing problem, otherwise it is dependent, while a composite problem is connected, if it has only dependent component problems, otherwise it is unconnected. The variables characterizing the connections between the composing and component problems are called a coordinating set.

3.3. Nomenclature

Altough the idea of composite problems has already been widely applied, no precise definitions and nomenclature thereof have been introduced so far. The nomenclature commonly used is essentially the following:

1. In the mathematical programming the component and composing problems are called subproblems and master problem, respectively.
2. In the hierarchical systems theory the component and composing problems are called subsystems and coordination, resp.
3. In the multilevel control the component and composing problems are called local and supervisory controllers, resp.
4. In the computer programming the component and composing problems are called subroutines and master routine, resp.

3.4. Decomposition of Problem

A problem P^* is decomposable, if there exists a transformation resulting in a composite problem P^c equivalent to the problem P^*, that is,

$$P^* = P^c$$

The decomposition of a composite problem is to decompose it into composing and component problems, i.e.,

$$P^c = P \circ p$$

where $P \circ p$ is the decomposed form of problem P^c.

A problem is separable (or multicomponent decomposable), if there exists a transformation resulting in a multicomponent composite problem equivalent to the original problem. A multicomponent decomposition is called separation. The decomposed form of a multicomponent composite problem is called separated form.

Assuming that a problem can be decomposed into given component problems, a separable problem is totally separated, if it is decomposed into all its component problems. A separable problem is partially separated (or a separable problem is partially composed), if the decomposition is only carried out for several component problems.

Example: Decomposition of Multilevel Systems

The Multilevel System Theory is well-known by the works of Mesarovic. Using his notations, a system can be represented as a map S,

$$S : MX \Rightarrow Y,$$

or

$$S(MX) | Y,$$

where M is the set of all possible inputs, X the set of initial conditions, while Y is the set of outputs.

Supposing that the system S can be decomposed into subsystems $S_1, \ldots, S_n$, first we must suppose, that the set MX is decomposable, that is, there exists a set decomposition map T_D

$$T_D : MX \Rightarrow M_1X_1$$
$$\vdots$$
$$MX \Rightarrow M_nX_n$$

or formally

$$T_D(MX) | \langle M_1X_1, \ldots, M_nX_n \rangle .$$

Let us suppose furthermore the existence of coordination maps $K_1, \ldots, K_n$,

$$K_i : MX \Rightarrow U_i \qquad i = 1, \ldots, n$$

A subsystem S_i can be considered as a map S_i

$$S_i : M_i X_i U_i \Rightarrow Y_i .$$

Finally we suppose, that there exists a set composition map T_c,

$$T_c : Y_1 \ldots Y_n \Rightarrow Y.$$

This way, Mesarovic's multilevel system decomposition can be represented as follows:

$$\begin{aligned} S(MX)|Y = T_c(Y_1,\ldots,Y_n)|Y \circ \\ \circ\langle S_1(M_1X_1U_1)|Y_1,\ldots,S_n(M_nX_nU_n)|Y_n\rangle \circ \\ \circ\langle K_1(MX)|U_1,\ldots,K_n(MX)|U_n\rangle \circ \\ \circ\langle T_D(MX)|\langle M_1X_1,\ldots,M_nX_n\rangle. \end{aligned}$$

4. PROBLEM SOLVING BY STRUCTURAL DECOMPOSITION

4.1. Methods for Solving Large-Scale Problems

The methods for solving large-scale problems can be classified into two groups:

1.) methods considering the large-scale problem as a whole, i.e. simultaneous solution;

2.) methods based on the character of the large-scale problem:
   2.1.) methods based on some special character of large-scale problems of a given type
   2.2.) methods whereby the overall problem is transformed to a set of subproblems, i.e. decomposition methods.

Decomposition methods can be divided into two groups:

a.) methods based on the decomposition of the problem variable set are called variable set decomposition methods:

b.) methods based on the decomposition of composite problems are called structural decomposition methods.

Example: Solution of Mathematical Programming Problem by decomposition.

Decomposition methods can be adapted to the mathematical programming problems.

a.) The variable set decomposition method is applied e.g. by the "branch and bound" method (for example Mitten (1970)).

b.) The structural decomposition methods are applied in a wide range, for example dynamic programming (Bellman 1957) and the decomposition principle (Dantzig 1960) are based on this idea, and its first general interpretation is given by Brioschi (1967) who has called it "structural programming".

4.2. Procedure of Problem Solving by Structural Decomposition

As it was mentioned earlier structural decomposition methods are based on the decomposition of composite problems.

The procedure of problem solving by structural decomposition consists of the following main steps:

a.) transformation of the original problem to an equivalent composite problem,

b.) decomposition of the composite problem,

c.) solving the decomposed problem.

According to the above procedure, there are several possible strategies for solving a composite problem by structural decomposition methods.

At the first step we should find composite problems equivalent to the original problem.

At the second step we decide the level of the decomposition depending on the given composite problem.

At the third step we choose a suitable method for solving the decomposed problem. This method is determined by the character of the given problem rather than by the way of decomposition.

When the composite problem equivalent to the original one is a general multilevel composite problem, these steps must be repeated at every level.

The purpose of problem solving by structural decomposition is to minimize the computational effort. Evidently, for a given problem one solving strategy is better than the other from the computational effort viewpoint. Special cases of the optimization of solving strategy can be found for example in the papers of Wilde (1965), Sargent (1967), Brioschi (1970), Almásy (1971) and of Westerberg (1971).

4.3. Examples for Solving MP Problems by Structural Decomposition

Depending on the type of the problem transformation, decomposition and solution, there are several structural decomposition methods in mathematical programming.

According to the type of problem transformation there are three types of decomposition methods. If the connection among the subproblems are in the variable, or in the model, or in the goal space, the decomposition method is non-disjoint, or primal, or dual, respectively.

On the basis of the problem decomposition, a decomposition method can be two or multilevel and two or multicomponent one.

In relation to the decomposed problem solution, there exist two essential types of decomposition methods: imbedding and coordinating iteration. In the case of imbedding the component problems are solved as functions of the coordinating variables, while in the case of coordinating iteration the component problems are solved at a fixed value of the coordinating variables.

Some examples for structural decomposition methods in MP are the following:

Discrete serial dynamic programming (Bellman 1957):
multilevel partially separated two-component composite problem solution by imbedding.

Discrete nonserial dynamic programming (Aris 1964):
multilevel partially separated multi-component composite problem solution by imbedding.

Method by Geoffrion (1967):
two-level not separated (two-component) composite problem solution by imbedding.

Method by Hocking-Shepard (1971):
two-level partially separated two-component composite problem solution by imbedding.

Dantzig-Wolfe decomposition method (Dantzig 1960):
two-level totally separated multi-component composite problem solution by coordinating iteration in the goal space.

Method by Kornai-Lipták (1965):
two-level totally separated multi-component composite problem solution by coordinating iteration in the model space.

REFERENCES

Almásy, G.A., Veress, G.E., Farkas, Z., Pallai, I.M.: Strategies of Optimizing Complex Steady-State Decision Systems IFAC Symposium on Digital Simulation of Continuous Processes, Győr, Hungary, 1971. Preprints I. A2-1--A2-10.

Aris, R., Nemhauser, G.L., Wilde, D.J.: Optimization of Multistage Cyclic and Branching Systems by Serial Procedures A.I.Ch.E. Journal 10(6), 913-919(1964).

Bellman, R.E.: Dynamic Programming Princeton University Press, Princeton, N.J., 1957.

Bellmore, M., Greenberg, H.J., Jarvis, J.J.: Generalized Penalty-Function Concepts in Mathematical Optimization. Operations Research, 18(2), 229-252(1970).

Brioschi, F., Locatelli, A.F.: Extremization of Constrained Multivariable Function: Structural Programming IEEE Trans. on Systems Science and Cybernetics, SSC-3(2), 105-111(1967).

Brioschi, F., Even, S.: Minimizing the Number of Operations in Certain Discrete-Variable Optimization Problems Operations Research, 18, 66-81(1970).

Dantzig, G.B., Wolfe, P.: Decomposition Principle for Linear Programs.
Operations Research, 8(1), 101-111(1960).

Geoffrion, A.M.: Solving Bicriterion Mathematical Programs
Operations Research, 15, 39-54(1967).

Geoffrion, A.M.: Elements of Large-Scale Mathematical Programming
Management Science 16(11), 652-691(1970).

Geoffrion, A.M.: Primal Resource-Directive Approaches for Optimizing Nonlinear Decomposable Systems.
Operations Research, 18(3), 375-403(1970).

Himmelblau, D.M.: Decomposition of Large-Scale Systems.
Chem.Eng.Science 21, 425-438(1966),
Chem.Eng.Science 22, 883-895(1967).

Hocking, R.R., Shepard, R.L.: Parametric Solution of a Class of Nonconvex Programs
Operations Research, 19, 1742-1747(1971).

Kornai, J., Lipták, T.: Two-level Planning
Econometrica 1965, 141-163.

Kron, G.: Diakoptics: The Piecewise Solution of Large-Scale Systems.
Macdonald, London, 1963.

Kulikowski, R.: Optimization of Large-Scale Systems.
Fourth Congress of IFAC, Warszawa, 1969, Survey Paper 16, 1-40 (1969).

Lasdon, L.S.: Optimization Theory for Large Systems
Macmillan Co., London, 1970.

Mesarovic, M.D., Macko, D., Takahara, Y.: Theory of Hierarchial, Multilevel Systems.
Acad.Press, New York, London, 1970.

Mine, H., Ohno, K.: Decomposition of Mathematical Programming Problems by Dynamic Programming.
Journal of Math.Anal. and Appl., 32, 370-385(1970)

Mitten, L.G.: Branch-and-Bound Methods: General Formulation and Properties.
Operations Research 18(1), 24-33(1970).

Nemhauser, G.L.: Decomposition of Linear Programs by Dynamic Programming.
Naval Res. Logist. Quart. 11, 191-195(1964).

Pallai, I.M., Almásy, G.A., Veress, G.E.: On the Optimal Strategy for Optimization of Chemical Plants.
Paper presented at the Fifth Congress of IFAC, Paris, 1972, Section 13.

Polak, E.: in Kuhn, H.W., Szego, G.P.: Differential Games and Related Topics
North-Holland, Amsterdam, etc., 1971.

Salovaara, S.: On Set Theoretical Foundations of System Theory
Acta Polytechnica Scandinavica, Mathematics and Computing Machinery Series No.15., Helsinki, 1967.

Sargent, R.W.H.: Integrated Design and Optimization of Processes.
Chemical Engineering Progress 63(9), 71-78(1967).

Vichenevetsky, R.: Serial Solution of Parabolic Partial Differential Equations.
Simulation, 13, 47-48(1969).

Westerberg, A.W., Edie, F.C.: Computer Aided Design
The Chemical Engineering Journal 2, 9-25, 114-123 (1971).

Wilde, D.J.: Strategies for Optimizing Macrosystems.
Chemical Engineering Progress 61(3), 86-93(1965).

PRACTICAL PROBLEMS IN LP DECOMPOSITION AND A STANDARDIZED PHASE I DECOMPOSITION AS A TOOL FOR SOLVING LARGE SCALE PROBLEMS

WILLIAM ORCHARD-HAYS

National Bureau of Economic Research, Inc.

Computer Research Center for Economics and Management Science

Cambridge, Massachusetts

Abstract: Automated LP capabilities have flourished for two decades in scope, power, generalization, and application. Yet decomposition, which is nearly as old in concept, remains more a principle than a practical method. Computerized techniques have been languishing on the shelf for years and use of general decomposition in a routine, production environment is virtually unknown. The reasons for this will be discussed as six major problem areas.

The second part will review work done during 1964-9 on standardizing decomposition formats for block-angular models and present a universal phase 1 procedure for several algorithmic approaches. This "basic gear" has been computerized but not used in a general way.

INTRODUCTION

The notion of applying decomposition techniques to a large LP model is older than the name, by about five years. [Orchard-Hays (1968)] In fact, if one accepts 1952 as the year in which the development of practical LP computational techniques began [Dantzig (1963)], then block partitioning is about two years younger. By 1955, rather elaborate schemes had actually been flow-charted. Yet, in two decades, while LP capabilities have flourished in scope, power, generalization and application, decomposition remains more a principle than a practical method. While it is true that the principle is quite often applied effectively in practical situations, and that the special case of Generalized Upper Bounding or GUB [Dantzig (1964)] has been highly developed, general-purpose decomposition procedures have all but been abandoned. Computerized systems have actually been languishing on the shelf for the last four years. Why?

They may be some slight consolation in recalling that as late as 1961 or 62, when general LP routines were already fairly powerful and reliable, the most widely used programs were for the transportation problem. One might say facetiously that, for several years, there appeared to be a widespread opinion that the transportation problem was the sum and substance of LP. Perhaps today's fabulously successful use of GUB--which is to decomposition what the transportation problem is to general LP--may portend a fuller appreciation of decomp. But the situations are not parallel. There were substantial and widely publicized efforts in decomposition eight years ago or more and the concept is inherent in much of the published research work of the past several years. Yet I know of not a single instance of the use of general decomposition in a routine, production environment such as has been commonplace with general LP for over a decade. Why?

There are a number of practical problems involved in the use of massive decomposition models. Our first task is to bring these out into the open and assess the possibilities of resolving them. Following that, a possible start on standardization of basic techniques will be presented.

A. The Problems

1. The Complexity of a Computerized System

The first problem has been with us from the outset. The 1955 flowcharts mentioned above were laid aside, not because there was any doubt about the mathematics, but becasue the computer programming problems were unmanageable at the time. Even now it is not clear that really large models of the generality then comtemplated could be handled adequately. The severest problem is the multiplicity and size of matrix files and the necessity for handling them both covariantly and contravariantly. However, we were then interested in block triangular models and in models with both linking constraints and linking variables. Had we stuck to simple block angular models, much more progress might have been made. Three or four years later, Dantzig and Wolfe used the simpler model as the basis for their decomposition principle.

Nevertheless, various implementations of algorithms for block angular models--both Dantzig-Wolfe and others--proved unsatisfactory in the long run, even though substantial amounts of time, money and talent were devoted to this area during the 1960's. This inadequacy was not entirely due to system problems but they aggravated other difficulties.

In 1968, S. T. (Shelly) Katz and I implemented the block-product algorithm [Orchard-Hays (1968)] in the "OPTIMA" system for Control Data Corpoation's (CDC) 6000 series of computers. This was the second implementation of the algorithm and the system had been designed from the outset with decomposition in mind. To my knowledge, this was the most comprehensive and smoothly-functioning system of automated procedures for decomposition which has been put together. Unfortunately, it has not seen much use and hence has not received the years of polishing which all systems require. It must be admitted, however, that even this system, which culminated several years effort, was not fully satisfactory in its data handling, and also lacked certain degrees of flexibility. This experience was very similar to that of other development groups. As a result, a general discouragement set in and algorithmic engineers turned their efforts toward other techniques such as GUB and mixed-integer programming or MIP [Balinski (1965)].

I now believe, though, that we gave up too soon. LP has always pushed computing systems to their limit. With decomposition, we simply got too far ahead of computer technology. It is only in the last very few years that the hardware has become adequate for massive problems and the basic software reliable enough for intricate control mechanisms. For example, we tend to think of the IBM 360 and its operating system (OS) as having been around since the mid-1960's. But that is only on paper. It was nearer 1970 before big systems were operating as intended. When we built "OPTIMA," we didn't have disk packs or extended core memory and had to rebuild the assembler and create our own system maintenance gear. I now think the system problems are manageable. Nevertheless, their complexity and the cost and time they will incur should not be underestimated. Nor are these the only problems.

2. The Competition of General LP Algorithms

Another factor that has impeded general development in decomposition has been the continuing improvement in general LP capability. This may seem like begging the question but diversion of interest and talent has a strong impact on any activity.

When we were thinking about decomposable models with perhaps a total of 200-300 constraints, suddenly general LP routines were solving 250-row models routinely. While "decomp" was being considered for 1,000-1,200 rows total,

"LP/90" and its derivatives were handling 1,024 rows quite nicely. Next Bonner and Moore's Univac package was attacking 2,000-order problems. The third-generation mathematical programming systems, called MPSs, promised 4,000 -constraint capability and actually began to solve 3,000-order problems and larger. Later systems and versions accept up to 8,000 rows and even 16,000, though whether any problems of this size have been successfully solved may be questioned. Models which can be formulated in GUB format may now go up to 50,000 constraints and more. Last year we successfully solved a model with over 49,000 GUB constraints, nearly 1,000 hard constraints and over 280,000 columns. Total solution time on a large IBM 370 Model 165 was finally reduced to 55 minutes. Other models approaching this size have now been formulated and are being successfully handled. So who needs decomposition?

This factor is closely related to the system problems and to the formulation problems discussed next. Had it been easier to experiment with mixtures of techniques, decomp would have been seen in a different light. D. M. (Dave) Carsters carried out a limited amount of such experimentation. He helped me perfect the block-product algorithm and was one of its chief implementers in "LP/600" for the General Electric (G.E.) 600 line. Later, he worked for G.E. for a while, before the merger with Honeywell. He cleaned up some of the problems that had remained with the first implementation and created interfacing gear so he could switch back and forth between decomp and general LP. First, he found that a large blocked problem, which could be solved as a general LP with "LP/600," could be solved in much less time with the decomp algorithm in the same system. But he also used general LP for phase 1 and decomp for phase 2, and then vice-versa. He found he could further reduce solution time. I believe the best result on his particular test case was using decomp for phase 1 and general LP for phase 2. The actual results are not so important as the concept and the ability. Of course, one has to be given a fairly open-ended budget for machine time and programming effort, a luxury not often available.

3. <u>Lack of Formulation and System Insight</u>

The third problem is simply the inability of most analysts and model formulators to adequately visualize the situation being modeled. This problem has been aggravated by the inability to experiment easily. Formulation is an art and most of us gain insights only when we actually see something. It is impossible to take advantage of a relationship which we have not yet grasped. Formulation involves a large measure of abstraction and an equally large measure of practical expediency. The number of people who can blend these two in proper proportion is limited.

Even where formulation insight is ample, it is not usually accompanied by system know-how. It is rare to find a team in a single department which includes both to a high degree. What is worse, this is not always realized. It is not uncommon to hear an analyst praising an assistant or co-worker as a system expert when he is really just a hard-working programmer who responds to the analyst's requests. The longer such a relationship exists, the less likelihood that anyone else's views will be considered. Most teams that function without friction tend to become myopic and inbred. That may seem a harsh judgement but new plateaus of achievement usually require new challenges and new ideas.

4. <u>The Inherent Uniqueness of Very Large Models</u>

The fourth problem subsumes in part, the first three. Suppose someone assembled a team of top-flight analysts and formulators, the most expert algorithmic designers, first-rate system implementers, and gave them the best available computing equipment. How much could they accomplish in developing practical approaches of wide utility for general decomposition, such as has been done with LP?

Before this team could accomplish much of anything, it seems safe to say that three things would be required:

(a) A difficult and important problem to attack, to give purpose and direction to the effort.

(b) A strong and knowledgable manager to resolve conflicts, establish priorities and deal effectively with the outside world.

(c) Continuing enthusiastic support accompanied by an eagerness to apply the system produced.

Already this begins to sound like a fairyland. Is all this really necessary? After all, other worthwhile developments have been made under considerably less than ideal conditions.

The difference lies in the complexity and variety of possibilities involved. General LP can be well defined. Someone can go off in a back room and develop a set of programs. As difficulties arise, these can be resolved by one or two people and improvements made. When enough experience is gained and enough demand is created, a team of half a dozen or so system programmers can put together an MPS and present it to the world. Its rules and capabilities can be taught in short seminars and a great variety of applications can be readily handled. As worthwhile new wrinkles evolve, the system can be upgraded by a small staff and the costs distributed over a large group of users.

The models for general LP are also relatively simple in concept, even though they may be large. LP models are large because they are combinatorial in nature. The basic data from which an LP model is implemented is usually relatively small, consisting of a few tables and lists which are exploded in various combinations to represent the interrelationship among various activities. In principle, of course, a large decomposable LP model is just more of the same. But in detail, "more of the same" is not so easily comprehended. Not only that, but a really comprehensive model begins to impinge on prerogatives which managers guard jealously. How many activities can one consider simultaneously before crossing departmental and divisional lines. When this happens, one must take account of speical constraints, policies, and interests.

Thus, really comprehensive models, to which decomposition is most meaningfully applied, tend to become unique. There is only one U.S. Navy Department, only one General Motors, only one NATO. Analytical questions are not the only basis for decisions. But even in handling the analytical relationships, simplifying assumptions must be made to bring the unmanageable complexity of the real world into a manageable model. The solution strategies then begin to rely on these assumptions which may not be valid in another situation. One can say that this is always true, but it is a matter of degree. In short, the world is simply not ready for the power which it is possible to assemble.

There are, however, important problems of large size but limited scope to which decomp can be applied. It will be wise to concentrate on these until proven methods are available. At best, it takes a lot of money to experiment in this area. The surest way to dry up sources of money is to create confusion at high levels.

5. The Data Management Problem

Another difficulty is related to both system file-handling problems and model scope but is really distinct. This is data management, per se. Had any important project required the use of decomposition and also been fully ready to supply data and utilize results, then some means for overcoming other difficulties would probably have been found. It is no accident that the petroleum industry was the first to make heavy use of LP and to put the results to practical use. It had classical process-flow problems, had accumulated a wealth of reliable technological data, some even in machine-readable form, and could make almost direct interpretation of results. An activity level for the fractionating tower, for example, meant to run that many thousands of barrels of crude through the equipment. The refinery manager could tell at a glance whether or not the figure made any kind of sense. As models become more abstract or encompass a range of possibilities which no single person has ever fully comprehended, it becomes more difficult to gain confidence in the model and to manipulate data in a meaningful way. Who knows what the effect is likely to be in year 5 of changing some input in year 1? How does one modify and keep track of a basic table of yield coefficients for multiple plants over multiple time periods? The nomenclature, identification and combinatorial problems become confusing enough without trying to interpret numerical values. In short, one loses all feel for reasonableness, not to speak of the greatly magnified possibility for data error.

The answer to this problem is a highly developed data management language and system. The management of data must itself be mechanized according to carefully worked out rules and procedures. Such a system is not cheap and its use requires special skills. Until very recently, an adequate system of this kind has simply not been available at any price. Fortunately, one now is but, like all powerful tools, it takes a significant effort to apply it to a particular environment.

6. Lack of Standardization

The final problem area I will mention is largely a consequence of the others but it contains additional ingredients and deserves to be recongnized independently. It is the lack of standardization in approaches to decomposition. One may argue that this is a feature of LP or mathematical programming in general, or of any similar technical field. However, there are usually de facto standards in a widely used technology. I don't place much reliance on standards committees or survey projects to make recommendations. In the long run, the essential fact is not whether the best of all possible arrangements has been achieved but whether a workable arrangement has been widely accepted. For example, it is now pointless to argue whether or not the IBM method for recording on magnetic tapes is the best one. Tens or hundreds of thousands of tape drives now exist which use that method. Other manufacturers conformed to it because it was predominant. As a result, one can put a reel of tape in his briefcase and travel over a good part of the world, reasonably confident that the computer he will use somewhere will be able to read the information recorded on it. This fact is far more important than whether or not some other recording mode would be better.

A similar thing happened in a smaller way with LP. In the early 1960's, the "LP/90" system became a de facto standard. Several other systems conformed to its formats. It is true it had the blessing of SHARE* but that was mostly a rubber stamp. Later, "MPS/360"** became a de facto standard. Both these systems had their faults, which I could easily recite. But such systems do have a standardizing influence which outweighs their own virtues and faults. When we built "LP/600" and "OPTIMA," we tried to correct one feature of "LP/90" which carried over to "MPS/360." In most contexts, one specifies a row first and a column second, e.g., in FORTRAN one writes A(I,J) which conforms to mathematical notation. So strong was the influence of "LP/90," which puts the

*The cooperative user's group for IBM computers.
**The standard large MPS for IBM 360 computers.

column name first, that the new format was not well accepted. "MPS/360" and its derivative systems continue to put column name first which is a de facto standard, whether aesthetically distressing or not. It is better to put the column name first than to be uncertain about its placement.

Fairly extensive efforts have been made to standardize decomposition formats but they have not taken hold. The second part of this discussion will review some of this work and present a phase 1 procedure which can be used with several decomposition algorithms for block angular models.

B. A Standard Phase 1 for Block Angular Models

1. A Typical Dantzig-Wolfe Algorithm

The most straightforward decomposition model has the following form:

$$\begin{bmatrix} A_0 & A_1 & A_2 & \cdots & A_p \\ & B_1 & & & \\ & & B_2 & & \\ & & & \ddots & \\ & & & & B_p \end{bmatrix} \begin{bmatrix} x_0 \\ x_1 \\ x_2 \\ \vdots \\ x_p \end{bmatrix} = \begin{bmatrix} b_0 \\ b_1 \\ b_2 \\ \vdots \\ b_p \end{bmatrix} \tag{1}$$

where A_0 and each B_p includes a full identity matrix for the corresponding logical (LGL) variables, i.e., the slacks. The objective form or functional is contained in the A_p blocks. For simplicity, we may as well regard it as the top row. Let

A_p be $m_0 x n_p$, $p = 0, 1, \cdots, P$

B_p be $m_p x n_p$, $p = 1, 2, \cdots, P$

x_p be $n_p x 1$, $p = 0, 1, \cdots, P$

b_p be $m_p x 1$, $p = 0, 1, \cdots, P$

Then, in detail, the problem can be written:

maximize x_0^1 subject to

$$\sum_{p=0}^{P} \sum_{j=1}^{n_p} a_{jp}^i x_p^j = b_0^i, \quad i = 1, \cdots, m_0 \tag{2a}$$

$$\sum_{j=1}^{n_p} b_{jp}^i x_p^j = b_p^i, \quad i = 1, \cdots, m_p; \; p = 1, \cdots, P \tag{2b}$$

and suitable ranges on all x_p^j.

We have not distinguished between LGL and structural (STR) variables, but have designated them all as x_p^j. In an actual implementation, these would have to be distinguished in standard fashion, e.g., as u_p^i and x_p^j, since their feasible ranges are treated and interpreted somewhat differently. However, we will simply assume all these details are taken care of. In the sequel, we will have occasion to introduce sets of dummy LGL variables which will be denoted by v_p^i.

We first note that the above model is not truly general. It does not, for example, accomodate time period models with multi-period lags--which are lower block triangular--or models with both linking constraints and linking variables. Perhaps the most important class it does not accomodate consists of the stair-step models which arise in time-staged production and inventory smoothing.* In these models, the B_p overlap in part, vertically. They arise frequently and are very difficult to solve. Nevertheless, the above model encompasses important classes of problems and is the only form for which highly developed solution techniques exist. The essential feature is that the B_p are independent, being related only through the A_p.

The equations (2a) are called the master problem and the set of equations (2b) are called the subproblems. A D-W (Dantzig-Wolfe) algorithm (i.e., one based on the D-W decomposition principle) never solves the master problem except as a post-optimal procedure. Instead, it solves the subproblems repeatedly to create a derived problem of the form

$$A_0 X_0 + \sum_{p=1}^{P} D_p Y_p = b_0 \qquad (3a)$$

$$\sum_{k=1}^{K_p} y_p^k = 1, \; p=1, \cdots, P$$
$$0 \leq y_p^k \leq 1, \text{ all } k, p \qquad (3b)$$

The columns constituting a D_p are mappings of feasible solutions to the p-th subproblem ("sub") into the master problem space. The equatons (3b) are convexity constraints on these sets of feasible solutions, thus insuring a feasible solution to each sub when a feasible solution to the derived problem is obtained. Certain y_p^k may be exempt from these conditions, being required only to be non-negative. This occurs when a mapping of an unbounded solution to a sub is generated. All decomposition techniques have to deal with the possibility that a subproblem treated separately may be unbounded in feasible solutions.

Much of the elegance and potential power of the D-W approach arises from the fact that subproblems need not be linear. All that is required is (i) that a set of Lagrange multipliers (or pricing vector) for the derived problem be interpretable, via A_p, meaningfully in the sub, and (ii) that a feasible solution to the sub can be meaningfully mapped into a D_p column in the linear derived problem. However, we shall here restrict ourselves to the case where all subs are linear.

The solution to a subproblem takes the following form:

*It is difficult to find a classification of model structures in the literature without an extensive search. See Dantzig (1963) for many examples and Orchard-Hays (1968) for some diagrammatic discussion.

Given a pricing vector π_o $(1 \times m_o)$, maximize the composite form $\pi_o \, v_p$ subject to

$$\begin{aligned} I_o \, v_p + A_p \, x_p &= 0 \\ B_p \, x_p &= b_p \end{aligned} \tag{4}$$

where x_p has the ranges originally specified and v_p is a $(m_p \times 1)$ column of free variables.

Since $A_p x_p = -v_p$, the column $-v_p$ is precisely a mapping of a feasible solution to the sub p into the derived problem. That is, each such $-v_p$ becomes some column D_{kp} when a 1.0 is appended in the appropriate row of (3b). If the form $\pi_o \, v^{kp}$ is driven to optimality, the sub p is optimal relative to the current solution to the derived problem. However, each such generated v_p enlarges the derived problem. The whole idea is to get the derived problem first feasible and then optimal and all subs relatively optimal to it, simultaneously. Consequently, it is usually considered desirable to produce several feasible solutions to each sub in the early stages--in order to create a meaningful derived problem. Driving each sub to relative optimality may be a waste of time until the total solution begins to stabilize.

In solving a sub, there can be three outcomes just as with any LP problems. First, if there is no feasible solution to (4) for any p, there is no solution to the entire problem, since (4) is less restrictive than (2a) and (2b). Second, if a feasible solution is found, there may be a finite maximum to $\pi_o \, v_p$.

Third, $\pi_o \, v_p$ may be unbounded subject to (4). This does not mean the entire problem is unbounded. The latter condition is difficult to determine unless $A_o \, x_o = b_o$ is already feasible and x^l is unbounded, a very unlikely situation. Note that, in many cases, A_o is simply I_o, i.e., the identity matrix of order m_o.

When one first starts to solve a problem with a D-W algorithm, the derived problem consists of only $A_o \, x_o = b_o$. If this has a feasible solution, an optimal solution may as well be obtained. If not, phase I on this problem will terminate with a "least infeasible" solution, in some sense. In either case, a π_o vector is produced, either an optimality or a feasibility form. Since the feasibility of all subs must be determined and feasible solutions found for them, and since D_{kp} columns are needed, each sub may as well be solved at least to feasibility using the π_o first obtained. Probably, a few feasible solutions with significant differences should be obtained for each sub, say a half dozen or so solumns D_{kp} for each p.

2. Setup and First Grand Iteration of Rosen's Algorithm

Before proceeding, let us look at two other algorithmic approaches for this model structure. D-W is a primal approach. The Rosen algorithm [Rosen (1963)] is a dual approach. It begins by forcing an artificial dual feasible solution. This is done as follows. An extra row is inserted at the bottom of the A_p blocks, i.e., m_o is increased by 1. This row has a 1.0 in its own LGL column in A_o, zero in all other LGL columns in all blocks and 1.0 in all their STR columns. The column b_o is extended with some large number Γ. Let u^{m_o} be the LGL in A_o for this row. Thus we have added the dummy constraint

$$u^{m_o} + \sum_{p=0}^{p} \sum_{j=m_p+1}^{n_p} x_p^j = \Gamma \tag{5}$$

(where it is assumed that x_p^j for j = 1, ···, m_p are LGLs).

Now let $\gamma = \max_{j,p} \{0, \max\{-a_{jp}^1\}\}$ where a_{jp}^1 are the cost elements. Then the π_o row

$$(1, 0, \cdots, 0, \gamma)$$

is dual feasible for the artificially constrained problem.

The subproblems are all solved to optimality with this π_o. There can be no unbounded feasible solutions since an upper bound was constructed artificially, namely $\gamma\Gamma$ for the whole problem. A derived problem is now constructed as follows:

Let $\begin{bmatrix} I_o & \underline{A}_p \\ & \underline{B}_p \end{bmatrix}$ be the optimal basis to sub p.

Let $\overline{A}_p$ be the upper part of the transformed nonbasic columns of sub p, ignoring any columns whose variables are fixed. That is, if $\begin{bmatrix} \mathring{A}_p \\ \mathring{B}_p \end{bmatrix}$ are the nonbasic columns,

$$\begin{bmatrix} \overline{A}_p \\ \overline{B}_p \end{bmatrix} = \begin{bmatrix} I_o & \underline{A}_p \\ & \underline{B}_p \end{bmatrix}^{-1} \begin{bmatrix} \mathring{A}_p \\ \mathring{B}_p \end{bmatrix} \tag{6}$$

as though a tableau procedure had been performed on the sub. Then the derived problem is

$$A_o\, x_o + \sum_{p=1}^{P} \overline{A}_p\, \overline{x}_p = b_o + \sum_{p=1}^{P} v_p \tag{7}$$

where the v_p are the same as in D-W.

We need not pursue the details and rationals of Rosen's algorithm; the above is sufficient for present purposes.

3. Phase I of the Block-Product Algorithm

The block-product algorithm is a parametric approach. One first attempts to solve the partial problem $A_o\, x_o = b_o$, expecting it to be infeasible. If and when this is determined, b_o is changed to $\overline{b}_o$ which is forced feasible. Then an optimal solution is obtained to the modified problem $A_o\, x_o = \overline{b}_o$, producing a π_o vector.* Now each subproblem is optimized with respect to this π_o. To control possible unbounded solutions in the subs, a bounding row is added to each sub in a manner similar to that for the entire problem in Rosen's algorithm. (In fact, were I implementing Rosen's algorithm, I would handle each sub individually.) To avoid the use of large numbers, which are always bad guesses, a special form of constraint was actually implemented. The row is treated as free unless no other constraint is binding. If and when this occurs, it is converted to a less-than-or-equal constraint with a good estimate of the proper upper limit which can be computed at that point. If any such constraints are binding in the final solution, an unbounded solution is

*It is conceivable that the modified problem might be unbounded. This would probably indicate that the entire problem, if feasible, is unbounded. This has not been investigated in detail and the situation has not arisen.

indicated. However, we need not pursue the details.

This algorithm does not use a derived problem but rather a modified RHS to the master problem which is then driven back to its original values, if possible, by a parametric technique. The modified master problem is

$$A_o\, x_o \;+\; \sum_{p=1}^{P} A_p\, x_p = \bar{b}_o - \sum_{p=1}^{P} v_p = \hat{b}_o$$

Letting $c_o = b_o - \hat{b}_o$, one has the parametric RHS problem,

$$A_o\, x_o \;+\; \sum_{p=1}^{P} A_p\, x_p = \hat{b}_o + \theta c_o \tag{8}$$

where the current solution is for $\theta = 0$ and $\theta = 1$ represents the original problem. Once this phase is complete, one has a primal and dual feasible solution to the modified problem (8) for $\theta = 0$. As θ is driven to 1.0, every solution is primal and dual feasible. If θ cannot be driven to 1.0, the entire problem is infeasible but one has a solution to a "nearest" feasible problem.

The algorithm derives its name from the fact that the inverse of the entire basis is maintained in product form where the elementary matrices are of block structure. The pivot element is a nonsingular sub-matrix. The parametric technique is relatively standard in itself and true parametric extensions to the model are readily handled.

4. The D-W Post-Iterative Solution Procedure

Both the Rosen and block-product algorithms produce values to the original variables directly. A D-W algorithm does not but rather produces answers to the derived problem. (The Rosen derived problem is in terms of the original variables.) A solution to (3a), (3b) gives values $X_o = \hat{X}_o$ and $Y_p = \hat{Y}_p$. The coefficient of a y_p^k is $D_{kp} = \sum_j A_{jp}\, \hat{X}^j_{pk}$ where $\hat{X}^j_{pk}$ are the values of the sub's variables in the k-th solution obtained and used in a mapping. Note that the ranges K_p are indefinite and unknown at the outset. (Indeed, one of the major difficulties in a D-W algorithm is in devising adequate indexing schemes.) It would be an intolerably enormous data processing task to keep track of the definitions of all D_{kp} in terms of X^j. Instead, one solves a final set of sets of simultaneous equations. Let[p]

$$b_{op} = D_p\, \hat{Y}_p,\; p = 1, \cdots, P$$

and resolve for X_p in the simultaneous equations

$$A_p X_p = b_{op}$$

$$B_p X_p = b_p$$

This can be done in matrix form as

$$\begin{bmatrix} I_o & A_p \\ & B_p \end{bmatrix} \begin{bmatrix} v \\ X_p \end{bmatrix} = \begin{bmatrix} b_{op} \\ b_p \end{bmatrix} \tag{9}$$

where the subcolumn v must be zero and X_p must be feasible. Such a solution is guaranteed to exist. It is a standard LP Phase I problem. The values obtained for X_p may be different from those implied by the $\hat{Y}_p$ and their D_p, i.e., an alternate solution. Note that the variables v are here considered fixed rather than free as were the original v_p.

5. The Common Solution Procedure

In all four of the foregoing situations, one has to solve a series of problems of the form of (4). Indeed, almost any imaginable approach to solving models of the form (1) will require solutions to LP problems of the form of (4). Let us list several known features of such submodels:

(a) The matrix consists of four parts:

(i) An identity always of order m_o in the upper left corner.
(ii) All zero block in lower left corner.
(iii) A matrix A_p in upper right corner whose m_o rows always have the same identification.
(iv) A matrix B_p in lower right corner which must contain a set of m_p linearly independent columns, possibly including LGL columns whose variables are fixed at zero.

(b) The maximum size of any such problem is m x n where

$$m = m_o + \max_p m_p$$

$$n = m_o + \max_p n_p$$

(c) If $\underline{B}_p$ is a nonsigular submatrix of B_p of order m_p, constituting the lower diagonal block of the current basis, and $\underline{A}_p$ is the part of A_p above $\underline{B}_p$, the basis inverse is

$$\begin{bmatrix} I_o & -\underline{A}_p\underline{B}_p^{-1} \\ 0 & \underline{B}_p^{-1} \end{bmatrix} \qquad (10)$$

This can be represented in standard product form with no more than m_p eta columns. Also the form (10) is itself a block elementary matrix. The product of several forms (10) for different p are commutative if they are expanded with diagonal identity matrices. For example, letting $-\underline{A}_p\underline{B}_p^{-1} = \hat{A}_p$ and setting p = 1 and 2,

$$\begin{bmatrix} I_o & \hat{A}_1 & \\ & \underline{B}_1^{-1} & \\ & & I_2 \end{bmatrix} \begin{bmatrix} I_o & & A_2 \\ & I_1 & \\ & & \underline{B}_2^{-1} \end{bmatrix} = \begin{bmatrix} I_o & \hat{A}_1 & \hat{A}_2 \\ & \underline{B}_1^{-1} & \\ & & \underline{B}_2^{-1} \end{bmatrix}$$

$$= \begin{bmatrix} I_o & & \hat{A}_2 \\ & I_1 & \\ & & \underline{B}_2^{-1} \end{bmatrix} \begin{bmatrix} I_o & \hat{A}_1 & \\ & B_1^{-1} & \\ & & I_2 \end{bmatrix}$$

and

$$\begin{bmatrix} I_o & \hat{A}_1 & \hat{A}_2 \\ & \underline{B}_1^{-1} & \\ & & \underline{B}_1^{-1} \end{bmatrix} = \begin{bmatrix} I_o & \underline{A}_1 & \underline{A}_2 \\ & \underline{B}_1 & \\ & & \underline{B}_2 \end{bmatrix}^{-1}$$

These latter forms are used extensively in the block product algorithm. They greatly facilitate transformations of segmented vectors. However, complications arise when the A_o block is involved unless it is the identity I_o.

(d) The variables associated with I_o are treated in various ways--sometimes as free, sometimes as fixed. In the A_o block, they have their own nominal feasible ranges.

(e) Either the last row of A_p or one row of B_p is special and its LGL variable may be subject to unorthodox conditions. (This is examined further below.)

(f) The master or derived problem's pricing vector, π_o, is regarded as the coefficients of a linear form in rows of A_p, p>o, and this linear form constitutes the objective function for the subproblem.

(g) The subproblems are treated independently but the upper part of various transformed columns are used separately for other purposes. (In the block product algorithm, the lower segments are also used separately.)

(h) The basis last obtained is a primal feasible basis to start the next grand iteration or to continue another phase.

These observations fairly well define a general decomposition computing system. It would not contain a complete algorithm but would provide most of the gear for implementing various algorithms. The chief ambiguity is in handling the various forms of bounding and convexity rows. This can be effectively resolved as follows.

D-W Iterations

It is unnecessary to include the convexity rows formally in the original problem. If a sub goes unbounded, create two D_{kp}. One is standard for the current feasible solution. The other is the upper part of the transformed column, i.e., $\hat{A}_{sp}$, where X_p^s is the variable which is unbounded. This D_{kp} is flagged in some way to show that it is not subject to the convexity condition. If such a column also goes unbounded in the derived problem, the whole problem is unbounded.

The D_{kp} are written out to a matrix file for the derived problem with some appropriate indexing scheme. Note that (3a), (3b) is a GUB model. If a GUB algorithm is used on it, the convexity rows are handled implicitly anyway. Otherwise, the algorithm can insert a 1.0 in the proper row on the basis of the index p. A D_{kp} from an unbounded solution must be treated specially in either event.

Rosen Iterations

When the model is set up, the extra row can be generated and γ calculated as the A_p are processed. The user should furnish a value for Γ. This is a standard inequality row. Alternatively, the same scheme may be used as in the block-product algorithm.

Block-Product Algorithm

An extra row is inserted at the top of each B_p as the problem is set up. This row then always has the same local index when handling subs. The LGL variable for this row has a special type. It is first a free variable. If a sub goes unbounded, it becomes a nonnegative slack with a calculated RHS value. In Phase II, these LGLs receive special treatment in pricing: essentially they are "encouraged" to come back into the basis and then revert to their original status. Sometimes, the calculated RHS value must be increased. If any of these constraints persist in binding, the entire model is unbounded. A byproduct is that these LGLs show the sum of values of the STRs in each sub.

Final D-W Solution Calculation

The Phase I calculations for (9) do not need the convexity rows. The b_{op} are computed columns for which feasible solutions to the subs in terms of the original variables are known to exist. The calculations are done with the original problem just as in iterating. The only change is that v is considered fixed instead of free.

The above scheme was implemented in 1968 for the "OPTIMA" MPS for CDC 6000 series computers. No provision was made for the Rosen algorithm but one was for Phase I of the block-product algorithm. A parameter caused D-W candidates to be produced (with the 1.0 or 0.0 in the p-th convexity row provided). Reinversion and restart facilities were provided and the whole arrangement worked beautifully.

Two important arrangements are required. First, the model must be input in the following order:

$$A_o, b_o, A_1, B_1, b_1, A_2, B_2, b_2, \cdots, A_P, B_P, b_P.$$

Second, P different eta files (product form of inverse) are used which must be switchable when changing subs. Solution bit maps are included. The complete solution vector is on a separate file, with a separate record for each segment.

C. Conclusion

The above arrangement appears simple and, hopefully, both useful and fairly general. However, both the simplicity and the generality are somewhat deceptive. The solution of each subproblem may require the use of an already elaborate and complicated MPS. The ability to handle a large number of big files and to switch them in and out is no easy system problem. The assumption that a decomposition model be submitted in a precise order and format may throw a road-block in front of some potential users.

Nevertheless, such a scheme is well within the present state-of-the-art of computing systems, even with more options than have been suggested. Until the math programming community is willing to accept conventions regarding input order, major operations and data handling, it cannot expect much further development of generalized capabilities. The truly interesting and important work lies beyond such a discipline, in the use of the facilities it makes available for new applications and projects. Existing MP technology is simply too complicated to be used in a casual or ad hoc manner on large and important tasks.

Bibliography

Balinski, M. L., Integer Programming Methods, Uses, Computation, Management Science, Vol. 12, pp. 253-313, November (1965).

Dantzig, G. B., "Linear Programming and Extensions," Princeton University Press, Princeton, N. J., (1963).

Dantzig, G. B., and R. M. VanSlyke, Generalized Upper Bounding Techniques for Linear Programming, pts. 1 and 2, Operations Research Center Reports, ORC 64-17 and ORC 64-18, University of California, Berkeley, August (1964) and February (1965).

Orchard-Hays, W., "Advanced Linear Programming Computing Techniques," McGraw-Hill, (1968).

Rosen, J. B., Convex Partition Programming, in Robert L. Graves and Philip Wolfe (eds.), "Recent Advances in Mathematical Programming," pp. 159-176, McGraw-Hill (1963).

UPPER-BOUNDING-TECHNIQUE GENERALIZED UPPER-BOUNDING-TECHNIQUE AND DIRECT DECOMPOSITION IN LINEAR PROGRAMMING: A SURVEY ON THEIR GENERAL PRINCIPLES INCLUDING A REPORT ABOUT NUMERICAL EXPERIENCE

HEINER MÜLLER-MERBACH
Fachgebiet Betriebswirtschaftslehre
Technische Hochschule Darmstadt
D-61 Darmstadt, Germany

Abstract: This paper will summarize the *upper-bounding-technique (UBT)*, the *generalized UBT*, and the method of *direct decomposition*. Furthermore, it shall show how direct decomposition can be adapted to any blockstructure of LP-problems.

The UBT was developed more than 15 years ago to treat *upper bounds on single variables* more efficiently. The generalized UBT treats *bounds on sets of variables*. The direct decomposition (first developed as an extension of the generalized UBT) is able to treat *blocks of constraints on sets of variables*.

The basic idea behind direct decomposition (and behind UBT which may be considered as its simplest version) is to leave the single blocks of constraints completely unchanged during the iterations unless the pivot-element lies within the block itself. In addition, those block areas of the coefficient matrix which consist of zeros in the beginning are kept unchanged throughout all the simplex iterations. If at any time any updated information (e.g. a row or a column) of any block is required, it can be generated easily.
Direct decomposition as well as the UBT can be inserted into the normal simplex method and into any of the revised simplex methods. For reasons of easier presentation, only the insertion into the normal simplex method will be considered in this paper.

Numerical experience with the block-angular type of problems and with different rules of pivot selection will be reported.

The aim of this survey is to provide a relatively easy access to the direct decomposition methods for those readers who are not as yet familiar with the details of the single principles of decomposition in linear programming.

1. THE GENERAL IDEA BEHIND UBT, GENERALIZED UBT, AND DIRECT DECOMPOSITION IN LINEAR PROGRAMMING

The upper-bounding-technique (UBT), the generalized UBT, and the method of direct decomposition (also called generalized generalized UBT) serve the purpose of accelerating the computation process of certain linear programming (LP) problems. The larger a problem is, the more significant this effect can be. The general principle behind UBT, generalized UBT, and direct decomposition is the same as that behind the revised simplex methods: In comparison to the normal simplex method, only a small part of the simplex tableau is updated in the single iterations, while the largest part remains unchanged; if any updated element or any row or column of the tableau is required it can be generated at any time.

LP-problems can be formulated as follows.

$$\begin{aligned} &\text{Maximize } z \\ &\text{s.t. } z + \underline{c}'\underline{x} = 0 \\ &\qquad \underline{A}\,\underline{x} \le \underline{b} \\ &\qquad \underline{x} \ge \underline{o} \end{aligned}$$

with $\underline{c}'$ as row vector with the coefficients of the objective function, $\underline{A}$ as coefficient matrix of the constraints, $\underline{b}$ as vector of the right-hand-side and $\underline{x}$ as vector of the variables.

The specific structure of the problems considered in this paper are characterized by matrices with block-structures like the following.

$$\underline{A} = \begin{array}{|c|c|c|c|}\hline \underline{A}_{o1} & \underline{A}_{o2} & \underline{A}_{o3} & \underline{A}_{o4} \\ \hline \underline{A}_{11} & \underline{0} & \underline{0} & \underline{0} \\ \hline \underline{0} & \underline{A}_{22} & \underline{0} & \underline{0} \\ \hline \underline{0} & \underline{0} & \underline{A}_{33} & \underline{0} \\ \hline \underline{0} & \underline{0} & \underline{0} & \underline{A}_{44} \\ \hline \end{array} \begin{array}{l} \} \text{ level 0} \\ \\ \left.\begin{array}{l} \\ \\ \\ \\ \end{array}\right\} \text{ level 1} \end{array}$$

The submatrices $\underline{A}_{ij}$ mean blocks with at least one non-zero element; the $\underline{0}$ indicate submatrices without any non-zero element. The submatrices $\underline{A}_{oj}$ which connect the single blocks of variables are defined as "level 0" or as the "master problem". The rows with disjunctive submatrices are defined as "level 1". Obviously, it can be rather arbitrary to draw the cutting lines between different levels.

In extension of the above two-level problem, problems with a hierarchy of three or more levels may arise in practice. And many different block-structures may occur.

If all the non-zero submatrices of level 1 consist only of one row and one column (and if no row and no column contains more than one non-zero element in level 1), the problem can be handled by the UBT as the simplest version of direct decomposition.

If the non-zero submatrices of level 1 consist only of one row each and if none of these submatrices has a column with a non-zero element in common, the generalized UBT (which is also a simple version of direct decomposition) can be applied.

The method of direct decomposition is based on the following technique: If in the course of the simplex iterations a pivot-element lies within the level 0, the pivot-operation is only carried out in the area of *this* level, while the areas of level 1 remain completely unchanged. If, however, a pivot-element lies within the level 1, the pivot-operation is carried out in the corresponding submatrix of level 1 as well as in the total area of level 0. The consequence is that the submatrices of level 1 are not kept updated. Hence, the necessary information of level 1 has to be generated as required. This can be carried out by very simple matrix multiplications. In addition, it may become necessary to exchange columns between the submatrices of level 1 which also is quite simple and requires little computation time.

Due to the fact that the level 1 is not at all affected by pivot-operations in level 0, the zero blocks of level 1 remain $\underline{0}$ throughout the whole course of iterations such that no storage space has to be provided for them. This is *one* advantage. The second advantage over the normal simplex method is that computation time is saved in that many iterations only work within level 0.

In the following sections, the mentioned techniques will be considered in more detail. For the sake of simplicity, these techniques will be combined with the normal simplex method. There exist no difficulties, however, to add these techniques to any of the revised simplex methods.

Starting with the simpliest case, the UBT will be considered in section 2. A bit more difficult is the generalized UBT, outlined in section 3. The actual method of direct decomposition will be presented in section 4. Numerical experience will be reported in section 5.

2. CHARACTERISTICS OF THE UBT

Upper bounds on single variables of LP-problems are the most simple case of decomposition structures. LP-problems with bounds on the single variables are of the type:

$$\begin{aligned} &\text{Maximize } z \\ &\text{s.t. } z + \textstyle\sum_{j=1}^{j=n} c_j\, x_j = 0 \end{aligned}$$

$$\textstyle\sum_{j=1}^{j=n} \underline{a}_{oj}\, x_j \le \underline{b}_o \qquad \text{(level 0)}$$

$$x_j \le b_j \ \forall\, j \ \text{(level 1, upper bounds)}$$

$$x_j \ge 0 \ \ \forall\, j$$

with $\underline{a}_{oj}$ and $\underline{b}_o$ being vectors, and z, c_j, b_j, and x_j being scalars.

Consider the following example for illustration.

$$\begin{aligned} &\text{Maximize } z \\ &\text{s.t. } z - 2x_1 - 4x_2 - 5x_3 - 3x_4 = 0 \\ &\qquad x_1 + 3x_2 + 6x_3 + 2x_4 \le 24 \\ &\qquad 5x_1 + 4x_2 + 4x_4 \le 20 \\ &x_1 \le 3 \quad x_2 \le 4 \quad x_3 \le 3 \quad x_4 \le 3 \\ &\qquad x_j \ge 0 \quad \forall\, j \end{aligned}$$

After adding slack variables, the following left tableau represents the initial solution. Notice that the unity matrix of the basic variables is not explicitly stated here and in the following tableaus. The following right tableau represents the optimal solution yielded by the normal simplex method without UBT or direct decomposition etc.

	x_1	x_2	x_3	x_4	RHS	
z(Max)	-2	-4	-5	-3	0	
x_5	1	3	6	2	24	level 0
x_6	5	4	0	4	20	
x_7	1				3	level 1; upper bounds
x_8		1			4	
x_9			1		3	
x_{10}				1	3	

		x_1	x_8	x_5	x_6	RHS
	z(Max)	1/2	1/6	5/6	1/3	82/3
level 0	x_3	-1/4	-1/6	1/6	-1/12	5/3
	x_{10}	-5/4	1	0	-1/4	2
level 1; upper bounds	x_7	1				3
	x_2		1			4
	x_9	1/4	1/6	-1/6	1/12	4/3
	x_4	5/4	-1	0	1/4	1

As far as the upper bounds are concerned the tableau of the solution shows:
(1) Since $x_1 = 0$, the slack variable $x_7 = 3$ remains at its upper bound.
(2) Since $x_2 = 4$ has reached its upper bound, the slack variable of this bound had to leave the basis ($x_8 = 0$). This required a simplex iteration with a pivot-element at level 1.
(3) The variable x_9 is the slack variable to the upper bound of x_3 (upper bound: $x_3 + x_9 = 3$). Since both variables, x_3 and x_9 , are basic variables and since the equation $x_3 + x_9 = 3$ must still hold, both rows must add up to $x_3 + x_9 = 3$. Therefore, the coefficients in the corresponding rows have to be the same, except for the sign. And the elements of both right-hand-sides (RHS) have to add up to the bound of 3.
(4) The equivalent is true for the rows of x_{10} and x_4 which have to add up to the upper bound $x_{10} + x_4 = 3$.

These oberservations can easily be generalized: Consider an upper bound $x_j \le b_j$. After adding the slack variable x_{n+m+j} the equation $x_{n+m+j} + x_j = b_j$ is yielded, the slack variable being the basic variable at the beginning. In the course of simplex iterations, this row can only be affected when x_j enters the basis. If x_{n+m+j} has to leave the basis for x_j , then this row will change into $x_j + x_{n+m+j} = b_j$, x_j becoming the basic variable (in this iteration only the RHS-vector will change its value while the coefficients of x_j only change their signs). If however, x_j enters the basis in any other row, then both, x_{n+m+j} and x_j, become basic variables of different rows. Since the equation $x_{n+m+j} + x_j = b_j$ must hold independently of the values of the non-basic variables, the equations represented by the two rows must add up exactly to the equation of the upper bound. This means that the coefficients of both rows must have the same values and differ only in their signs while both right-hand-sides must add up to the upper bound. From this follows that only one of the two rows has to be stored explicitly while the other can be simply generated from it.

Consequently, the following procedure seems to be advantageous: The upper bounds are kept in the level 1 area of the simplex tableau. This means that the basic variable of row j of level 1 is either x_{n+m+j} or x_j. If at any iteration x_j is to be exchanged against x_{n+m+j} , the pivot-operation will be carried out in level 1 and level 0. If, however, the pivot-element lies in level 0, the level 1 will not be affected at all.

In doing so, the information in those rows of level 1 is not adequately updated the corresponding variables of which are in the basis of level 0. But these rows can easily be generated as shown above, if necessary.

This modification of the simplex method is called *upper-bounding-technique* (UBT). The following tableau represents the optimal solution according to this technique.

	x_1	x_8	x_5	x_6	RHS	
z(Max)	1/2	1/6	5/6	1/3	82/3	
x_3	-1/4	-1/6	1/6	-1/12	5/3	level 0
x_{10}	-5/4	1	0	-1/4	2	
	x_1	x_8	x_3	x_{10}		
x_7	1				3	level 1; UBT applied
x_2		1			4	
x_9			1		3	
x_4				1	3	

Due to the fact that the coefficients and the RHS elements of level 1 remain unchanged, the UBT can be carried out by bookkeeping. This helps to accelerate the computation and to reduce the storage requirements.

Quite similarly a dual UBT can be applied in cases where the dual values of single variables have upper bounds.

3. CHARACTERISTICS OF THE GENERALIZED UBT

3.1. THE PRIMAL CASE OF GENERAL UPPER BOUNDS

If groups of variables instead of single variables have upper bounds, the generalized UBT applies instead of the simple UBT. Problems of this type have the following structure.

Maximize z

$$\text{s.t.} \quad z + \sum_{k=1}^{k=p} \underline{c}'_k \, \underline{x}_k = 0$$

$$\sum_{k=1}^{k=p} \underline{A}_{ok} \, \underline{x}_k \leq \underline{b}_o \qquad \text{(level 0)}$$

$$\underline{e}'_k \, \underline{x}_k \leq b_k \quad \forall\, k \qquad \text{(level 1)}$$

$$\underline{x}_k \geq \underline{o} \quad \forall\, k$$

with $\underline{A}_{ok}$ being the coefficient matrices of level 0, $\underline{x}_k$ and $\underline{b}_o$ being vectors, $\underline{c}'_k$ being row vectors, and b_k being scalars. The $\underline{e}'_k$ are row vectors consisting only of 1's ; they correspond in size with the vectors $\underline{x}_k$. The constraints of level 1 represent the generalized upper bounds, i. e. upper bounds on groups of variables. The index k (= 1, 2, ..., p) numbers these groups. The following example shall illustrate this case.

$$
\begin{aligned}
\text{Maximize } & z \\
\text{s.t. } & z - 7x_1 - 6x_2 - 8x_3 - 5x_4 = 0 \\
& x_1 + 5x_2 + 4x_3 + 3x_4 \le 18 \\
& 4x_1 + 2x_2 + 4x_3 + 2x_4 \le 12 \\
& x_1+x_2 \le 1 \qquad x_3+x_4 \le 9/2 \\
& x_j \ge 0 \quad \forall j
\end{aligned}
$$

After adding slack variables the following left tableau of the initial solution is yielded. The optimal solution obtained by the normal simplex method is shown in the following right tableau.

	x_1	x_2	x_3	x_4	RHS	
z(Max)	-7	-6	-8	-5	0	
x_5	1	5	4	3	18	level 0
x_6	4	2	4	2	12	level 0
x_7	1	1			1	level 1
x_8			1	1	9/2	level 1

		x_8	x_7	x_6	x_5	RHS
	z(Max)	6/5	7/5	13/10	2/5	148/5
level 0	x_1	2/5	4/5	1/10	-1/5	1/5
level 0	x_3	-7/5	-9/5	2/5	1/5	3/10
level 1	x_2	-2/5	1/5	-1/10	1/5	4/5
level 1	x_4	12/5	9/5	-2/5	-1/5	21/5

The following observations can be made from the optimal solution:
(1) Since the general upper bound $x_7 + x_1 + x_2 = 1$ must hold, the rows with the basic variables x_1 and x_2 must add up to the equation of the bound.
(2) The equivalent holds for the rows of x_3 and x_4 which must add up to the equation of the upper bound $x_8 + x_3 + x_4 = 9/2$.

Generalizing these observations, the following statement is obvious: Since the generalized upper bound $\underline{e}_k' \underline{x}_k \le b_k$ or (after the slack variable x_{m+n+k} is added) $x_{m+n+k} + \underline{e}_k' \underline{x}_k = b_k$ must hold throughout all the iterations, those rows must add up to the equation of the bound which have an element of the vector $\underline{x}_k$ as a basic variable. From this it follows that one of these rows need not be stored explicitly if the equation of the general upper bound is stored. Therefore, the upper bounds (equations of level 1) may remain unaffected by simplex iterations with a pivot-element in level 0. They can be updated at any time, if necessary. This is the basic principle behind the generalized UBT.

Applied to the above example, one yields the following tableau of the optimal solution.

	x_8	x_7	x_6	x_5	RHS	
z(Max)	6/5	7/5	13/10	2/5	148/5	
x_1	2/5	4/5	1/10	-1/5	1/5	level 0
x_3	-7/5	-9/5	2/5	1/5	3/10	level 0
	x_1	x_7	x_3	x_5		
x_2	1	1			1	level 1; generalized UBT applied
x_4			1	1	9/2	level 1; generalized UBT applied

Similar to the upper bounds on single variables (section 2) these upper bounds on groups of variables can also be handled by bookkeeping.

Similar to the UBT, there are so many alternatives to implement the generalized UBT that it does not seem to be useful to give algorithmic details here. The unique trick of UBT and generalized UBT is that the rows of the upper bounds are not altered in simplex-iterations with pivot-elements in level 0. If updated information of the rows of level 1 (upper bounds) is required, it can be generated by substracting the corresponding rows of level 0 from the equation of the upper bound.

If the non-zero coefficients of the generalized upper bounds are not equal to 1, either the coefficients are altered to 1 by scaling the columns of the simplex tableau or the generalized UBT has to be slightly modified.

If generalized upper bounds are not explicitly indicated in the course of constructing the initial tableau, it is still possible to define and handle some of the constraints as generalized upper bounds. The only condition is that no two or more of these constraints have a variable (with a non-zero coefficient) in common. In general, it is desirable to find the largest set of those independent constraints. This problem is identical with a *set covering problem* as well as with the problem of the *maximal internally stable graph* which can approximately be solved by several heuristic procedures, see MÜLLER-MERBACH (1971).

3.2. THE DUAL CASE OF GENERAL UPPER BOUNDS

A very frequent type of generalized upper bounds are dual generalized upper bounds. They appear in all multi-period problems where the coefficient columns of the stock-on-hand variables usually consist of one +1 and one −1. Problems of this type can be handled by the dual version of the generalized UBT, see MÜLLER-MERBACH (1971). One difficulty, however, arises from the fact that these dual generalized upper bounds are usually not independent from each other since two bounds always have one variable in common. This requires certain modifications of the dual generalized UBT.

4. CHARACTERISTICS OF DIRECT DECOMPOSITION

4.1. THE GENERAL PRINCIPLE, EXPLAINED BY THE BLOCK-ANGULAR CASE

The UBT and the generalized UBT are simple cases of the direct decomposition technique. This is due to the simple structure of the constraints of level 1. In this section 4, problems with more general block-structures will be considered. The first type of problem (section 4.1)is the following *block-angular case*.

Maximize z

$$\text{s.t.} \quad z + \sum_{k=1}^{k=p} \underline{c}'_k \underline{x}_k = 0$$

$$\sum_{k=1}^{k=p} \underline{A}_{ok}\underline{x}_k \le \underline{b}_o \qquad \text{(level 0)}$$

$$\underline{A}_{kk}\underline{x}_k \le \underline{b}_k \quad \forall\, k \qquad \text{(level 1)}$$

$$\underline{x}_k \ge \underline{o} \quad \forall\, k$$

where the $\underline{A}_{ok}$ and the $\underline{A}_{kk}$ are the coefficient matrices, the $\underline{x}_k$, $\underline{b}_o$, and $\underline{b}_k$ are vectors, and the $\underline{c}'_k$ are row vectors.

In addition to this block-angular case, some other structures shall be considered later on.

The following example is given for illustration.

Maximize z

$$\begin{aligned}
\text{s.t.} \quad z - 8x_1 - 3x_2 - 8x_3 - 6x_4 &= 0\\
4x_1 + 3x_2 + x_3 + 3x_4 &\le 16\\
4x_1 - x_2 + 3x_3 &\le 12\\
x_1 + 2x_2 &\le 8\\
3x_1 + x_2 &\le 10\\
2x_3 + 3x_4 &\le 9
\end{aligned}$$

$$4x_3 + x_4 \le 12$$
$$x_j \ge 0 \quad \forall j$$

After the slack variables x_5 to x_{10} have been added, the following left simplex tableau is yielded. The right tableau shows the optimal solution obtained by the normal simplex method.

	x_1	x_2	x_3	x_4	RHS	
z(Max)	-8	-3	-8	-6	0	
x_5	4	3	1	3	16	level 0
x_6	4	-1	3	0	12	level 0
x_7	1	2			8	level 1
x_8	3	1			10	level 1
x_9			2	3	9	level 1
x_{10}			4	1	12	level 1

	x_6	x_{10}	x_9	x_5	RHS
z(Max)	3/4	9/10	9/20	5/4	877/20
x_2	-1/4	3/10	-7/20	1/4	29/20
x_1	3/16	-3/20	-1/80	1/16	107/80
x_7	5/16	-9/20	57/80	-9/16	301/80
x_8	-5/16	3/20	31/80	-7/16	363/80
x_3	0	3/10	-1/10	0	27/10
x_4	0	-1/5	2/5	0	6/5

If the method of direct decomposition is applied to the same problem, the subproblems of level 1 are kept unchanged in those iterations with a pivot element in the area of level 0. If updated information of level 1 is required later on (e.g. the pivot column or the pivot row) it can be generated in a similar way as in the case of UBT and generalized UBT. The details shall be shown by means of the above example. The simplex tableaus of the four iterations are given in the following.

	x_6	x_2	x_3	x_4	RHS	
z	2	-5	-2	-6	24	
x_5	-1	4	-2	3	4	level 0
x_1	1/4	-1/4	3/4	0	3	level 0
	[x_1]	x_2	x_3	x_4		
x_7	1	2			8	level 1
x_8	3	1			10	level 1
x_9			2	3	9	level 1
x_{10}			4	1	12	level 1

Iteration 1 (x_1 vs. x_6, level 0)

	x_6	x_2	x_3	x_5	RHS	
z	0	3	-6	2	32	
x_4	-1/3	4/3	-2/3	1/3	4/3	level 0
x_1	1/4	-1/4	3/4	0	3	level 0
	[x_1]	x_2	x_3	[x_4]		
x_7	1	2			8	level 1
x_8	3	1			10	level 1
x_9			2	3	9	level 1
x_{10}			4	1	12	level 1
x_9	1	-4	4	-1	5	

Iteration 2 (x_4 vs. x_5, level 0)

	x_6	x_2	x_9	x_5	RHS	
z	3/2	-3	3/2	1/2	79/2	
x_4	-1/6	2/3	1/6	1/6	13/6	level 0
x_1	1/16	1/2	-3/16	3/16	33/16	level 0
	[x_1]	x_2	x_9	[x_4]		
x_7	1	2			8	level 1
x_8	3	1			10	level 1
x_3			1/2	3/2	9/2	level 1
x_{10}			-2	-5	-6	level 1
x_{10}	-5/6	10/3	-7/6	5/6	29/6	

Iteration 3 (x3 vs. x9, level 1)

	x_6	x_{10}	x_9	x_5	RHS	
z	3/4	9/10	9/20	5/4	877/20	
x_2	-1/4	3/10	-7/20	1/4	29/20	level 0
x_1	3/16	-3/20	-1/80	1/16	107/80	level 0
	[x_1]	[x_2]	x_9	x_{10}		
x_7	1	2			8	level 1
x_8	3	1			10	level 1
x_3			-1/10	3/10	27/10	level 1
x_4			2/5	-1/5	6/5	level 1

Iteration 4 (x_2 vs. x_{10}, level 1; two columns exchanged), optimum

In the first iteration, the column of x_1 is chosen as the pivot-column and the row of x_6 as the pivot-row, due to the standard selection rules. (All the pivot-elements are double underlined.) Since the pivot-element lies within level 0, the pivot operation is only carried out in this level. The consequence is that x_1 enters the basis of the master problem, but is still a non-basic variable of a subproblem (marked by a square).

In the second iteration, the column of x_4 is chosen as the pivot-column. To apply the standard rule for selecting the pivot-row, the current RHS RHS and the current pivot-column have to be generated in level 1. This can be done by simply substituting x_1 in the rows with x_7 and x_8 as basic variables. In terms of matrix operations this works as follows.

$$\begin{array}{c|cc|} & x_4 & \text{RHS} \\ \hline x_7 & 0 & 8 \\ x_8 & 0 & 10 \\ \hline \end{array} - \begin{array}{c|c|} & x_1 \\ \hline x_7 & 1 \\ x_8 & 3 \\ \hline \end{array} \bullet \begin{array}{c|cc|} & x_4 & \text{RHS} \\ \hline x_1 & 0 & 3 \\ \hline \end{array} = \begin{array}{c|cc|} & x_4 & \text{RHS} \\ \hline x_7 & 0 & 5 \\ x_8 & 0 & 1 \\ \hline \end{array}$$

The standard selection rule leads to the row of x_5 as pivot-row which again lies within level 0. This brings x_4 in the basis of level 0 which, at the same time, remains non-basic variable of subproblem 2 (also marked by a square).

In the third iteration, the column of x_3 is chosen as the pivot-column. The generation of the RHS and of the pivot-column of the level 1 can again be carried out by substitution of x_1 and x_4, respectively:

$$\begin{array}{c|cc|} & x_3 & \text{RHS} \\ \hline x_7 & 0 & 8 \\ x_8 & 0 & 10 \\ \hline \end{array} - \begin{array}{c|c|} & x_1 \\ \hline x_7 & 1 \\ x_8 & 3 \\ \hline \end{array} \bullet \begin{array}{c|cc|} & x_3 & \text{RHS} \\ \hline x_1 & 3/4 & 3 \\ \hline \end{array} = \begin{array}{c|cc|} & x_3 & \text{RHS} \\ \hline x_7 & -3/4 & 5 \\ x_8 & -9/4 & 1 \\ \hline \end{array}$$

$$\begin{array}{c|cc|} & x_3 & \text{RHS} \\ \hline x_9 & 2 & 9 \\ x_{10} & 4 & 12 \\ \hline \end{array} - \begin{array}{c|c|} & x_4 \\ \hline x_9 & 3 \\ x_{10} & 1 \\ \hline \end{array} \bullet \begin{array}{c|cc|} & x_3 & \text{RHS} \\ \hline x_4 & -2/3 & 4/3 \\ \hline \end{array} = \begin{array}{c|cc|} & x_3 & \text{RHS} \\ \hline x_9 & 4 & 5 \\ x_{10} & 14/3 & 32/3 \\ \hline \end{array}$$

The standard selection rule leads to the row of x_9 as pivot-row. This row lies within level 1. Therefore, the pivot operation is to be carried out in level 0 and also in that subproblem of level 1 that corresponds to the pivot-element. No difficulties arise while pivoting in subproblem 2 of level 1. But in order to pivot in level 0, the pivot-row has to be updated. This again can be done by substitution of x_4:

$$\begin{array}{c|ccccc|} & x_6 & x_2 & x_3 & x_5 & \text{RHS} \\ \hline x_9 & 0 & 0 & 2 & 0 & 9 \\ \hline \end{array} - \begin{array}{c|c|} & x_4 \\ \hline x_9 & 3 \\ \hline \end{array} \bullet \begin{array}{c|ccccc|} & x_6 & x_2 & x_3 & x_5 & \text{RHS} \\ \hline x_4 & -1/3 & 4/3 & -2/3 & 1/3 & 4/3 \\ \hline \end{array} = \begin{array}{c|ccccc|} & x_6 & x_2 & x_3 & x_5 & \text{RHS} \\ \hline x_9 & 1 & -4 & 4 & -1 & 5 \\ \hline \end{array}$$

This row has already been inserted in the corresponding tableau given on the preceding page. It can be erased after the iteration.

From now on, the whole problem can be considered as if x_3 and x_{10} were the basic variables of the subproblem 2 from the beginning on, while x_9 and x_4 would be the non-basic variables of this subproblem.

In the fourth iteration, the column of x_2 is selected as the pivot-column. The substitution of x_1 and x_4, respectively, yields the updated RHS and pivot-column:

$$\begin{array}{c|cc|} & x_2 & \text{RHS} \\ \hline x_7 & 2 & 8 \\ x_8 & 1 & 10 \\ \hline \end{array} - \begin{array}{c|c|} & x_1 \\ \hline x_7 & 1 \\ x_8 & 3 \\ \hline \end{array} \bullet \begin{array}{c|cc|} & x_2 & \text{RHS} \\ \hline x_1 & 1/2 & 33/16 \\ \hline \end{array} = \begin{array}{c|cc|} & x_2 & \text{RHS} \\ \hline x_7 & 3/2 & 95/16 \\ x_8 & -1/2 & 61/16 \\ \hline \end{array}$$

	x_2	RHS
x_3	0	9/2
x_{10}	0	-6

$-$

	x_4
x_3	3/2
x_{10}	-5

$\bullet$

	x_2	RHS
x_4	2/3	13/6

$=$

	x_2	RHS
x_3	-1	5/4
x_{10}	10/3	29/6

Alas, the standard rule selects the row of x_{10} as pivot-row. Therefore, the pivot-element lies in a zero submatrix. This would cause a connexion of the two subprob-lems of level 1 which should be avoided in order to keep the subproblems indepen-dent of each other. The latter can be achieved by a pivot operation in the subprob-lem 2. First of all, however, the pivot-row has to be updated:

	x_6	x_2	x_9	x_5	RHS
x_{10}	0	0	-2	0	-6

$-$

	x_4
x_{10}	-5

$\bullet$

	x_6	x_2	x_9	x_5	RHS
x_4	-1/6	2/3	1/6	1/6	13/6

$=$

	x_6	x_2	x_9	x_5	RHS
x_{10}	-5/6	10/3	-7/6	5/6	29/6

If the pivot operation in level 0 is carried out with this pivot-row (which is in-serted in the corresponding tableau, see above) and if the pivot-row replaces the row of x_4 in the master problem, then the level 0 tableau with x_2 and x_1 as basic variables is yielded as shown in the last tableau (see above). Now, x_{10} is still basic variable in the subproblem 2 while x_4 is only non-basic variable of subproblem 2, but is actually a basic variable of the current solution. A pivot operation in the subproblem 2, exchanging x_4 and x_{10}, brings this in order. This is identical with the exchange of the two columns of x_4 and x_{10} at the level 1.

Now the optimal solution is yielded. If one wishes, he may update the data of level 1 and obtain exactly the complete optimal tableau (as given above, next to the ini-tial tableau).

In generalization of this example, the steps of the direct decomposition method are the following.

(1) Selection of the pivot-column in regards to the current objective function.
(2) Generation of the RHS and the pivot-column of level 1.
(3) Selection of the pivot-row in regards to the quotients of the RHS coefficients and the pivot-column coefficients. If the pivot-row lies in level 0, go to step (5), otherwise to step (4).
(4a) Updating the pivot-row (in level 1) for pivoting in level 0. If the pivot-column hits the same subproblem as the pivot-row, then go to step (4c), otherwise to step (4b).
(4b) Selection of the pivot-column in the subproblem which is hit by the pivot-row. Any column can be taken which has a non-zero element in the pivot-row and which has such a non-basic variable that is also a basic variable in level 0. This row of level 0 will be replaced by the pivot-row as generated in step (4a).
(4c) Pivoting in that subproblem of level 1 which is hit by the pivot-row. Pivot-column either according to step (1) or (4b).
(5) Pivoting in level 0, using either the pivot-row of this level or the updated pivot-row of step (4a), respectively.

Many options exist to modify the single steps. But in any case, the basic operation of direct decomposition consists of leaving the matrices of level 1 less updated and to generate the specific updated information as necessary. In doing so, it is important that the single subproblems of level 1 remain independent of each other, i.e. it must not be allowed that initial variables of any sub- or masterproblem enter any other subproblem.

The following notation allows for a brief outline of the mathematical background:

B = set of basic variables
N = set of non-basic variables
0 = index of initial (original) tableau

C = index of current tableau of direct decomposition
o = index of master problem
k = index of k-th subproblem

The conditions of the independence of the subproblems are:

(1) $x_j \in B_k^C \cup N_k^C$ iff $x_j \in B_k^O \cup N_k^O$; $B_k^C \cup N_k^C = B_k^O \cup N_k^O \ \forall \ k$

(2) $x_j \in B_o^C$ iff $x_j \in B_o^O$ or $x_j \in N_k^C$

The general formula for generating the updated data (columns or rows) of level 1 is:

$$a_{ij} := a_{ij}^C - \sum_{x_l \in S_k} a_{il}^C \, a_{lj}^C$$

(a_{lj}^C → current column of master problem; a_{il}^C → current row of k-th subproblem)

with i indicating any row of the k-th subproblem, j indicating the column of any variable x_j, and S_k being the set of the variables x_l which - at the same time - are in the basis of level 0 ($x_l \in B_o^C$) and non-basic variable in the k-th subproblem of level 1 ($x_l \in N_k^C$).

After having considered the block-angular case, other types of decomposition structures shall be considered now.

4.2. THE STAIR-CASE TYPE

Less frequent than the block-angular case is the *stair-case* type of decomposition structure. Its shape is outlined in the following matrix.

$\underline{A}$ =
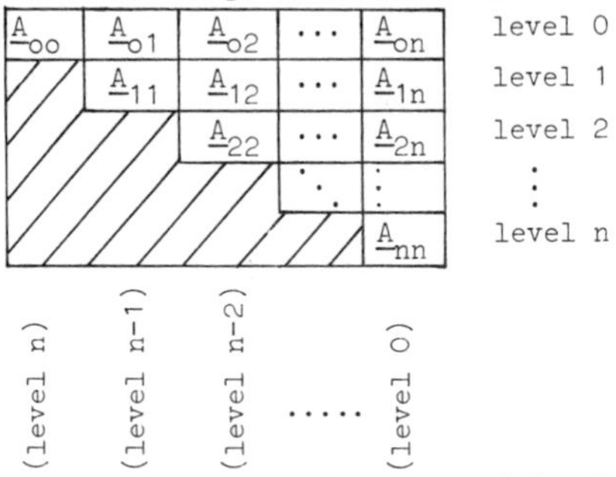

For solving problems of this type, the n level version of direct decomposition can be applied in two alternative ways. Either the matrices $\underline{A}_{oj}$ are considered as level 0, ..., and the matrix $\underline{A}_{nn}$ as level n, or - on the dual problem - the matrices $\underline{A}_{in}$ are considered as level 0, ..., and $\underline{A}_{oo}$ as level n.

The general procedure for this n level direct decomposition is that the pivot operations will be carried out only in the matrices of of level k, level $k-1$, ..., and level 0 if the pivot-element lies in level k. Consequently, updating a row or column in level k requires matrix multiplications through the levels 0 to k.

4.3. THE MULTI-PERIOD CASE

A third, very frequent type of decomposition structures is the *multi-period* type. The typical picture of the coefficient matrix is the following.

$\underline{A}$ =
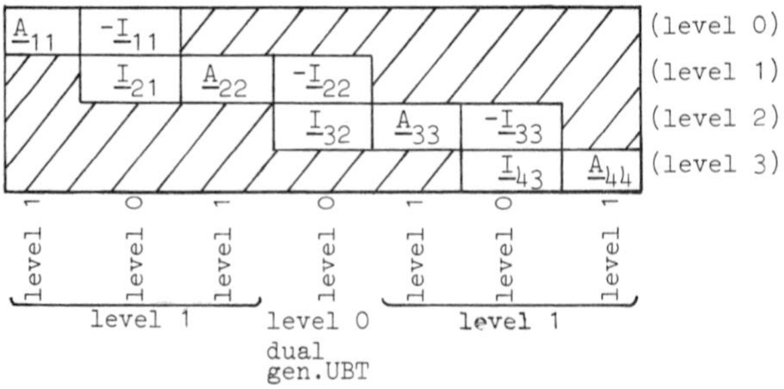

The matrices $-\underline{I}_{kk}$ and $\underline{I}_{k+1,k}$ usually correspond to the stock-on-hand between the periods k and $k+1$.

Problems of this kind can be handled by direct decomposition in many different ways. A first, although not very advantageous way, is to assign the k-th period to the k-th level, as indicated right of the matrix $\underline{A}$. If n periods are considered simultaneously, n levels are necessary. This is similar to the stair-case type.

A second and generally more advantageous cutting of the problem is that of assigning all the matrices $\underline{A}_{kk}$ to level 1 and all the unity matrices to level 0, as indicated below the matrix $\underline{A}$.

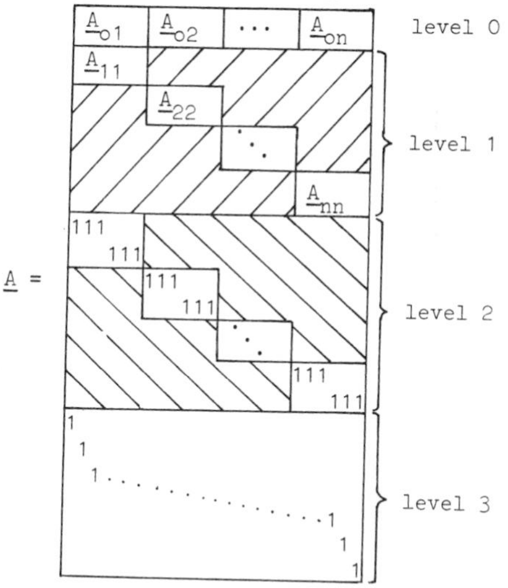

A third way is to combine pairs of two neighboured matrices $\underline{A}_{kk}$ and the enclosed unity matrices to subproblems of level 1 while the unity matrices between these subproblems are defined as level 0. Additionally, the dual generalized UBT can be applied to the problems of level 0. One will notice that all the subproblems of level 0 are independent of each other, as are the subproblems of level 1. Alternatively, one may also consider the subproblems of level 1 as of level 0 and vice versa.

Many more modifications are possible in the multi-period case.

4.4. TWO OTHER CASES OF BLOCK-STRUCTURES

It may be seen by the examples of section 4.1 to 4.3 that it is no problem to cut coefficient matrices with block-structures into a hierarchy of two or more levels. One can e. g. think of the case containing subproblems (level 1), generalized upper bounds (level 2), and single upper bounds (level 3), see above.

Another case of block-structure is the *bordered-angular* type, as given in the following matrix:

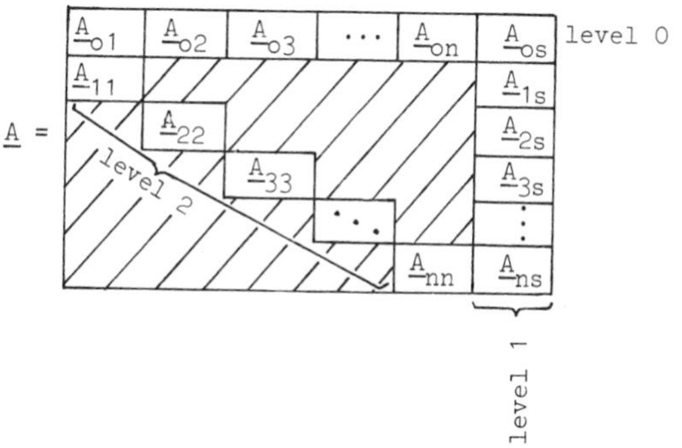

In this case it is possible to treat the matrices $\underline{A}_{oj}$ as level 0, $\underline{A}_{is}$ (i>0; s=n+1) as level 1, and all the $\underline{A}_{kk}$ as level 2.

4.5. GENERAL PRINCIPLES OF CONSTRUCTING A DIRECT DECOMPOSITION CODE

The principle of direct decomposition can be adapted to any block-structure. Only some cases were mentioned here.

The construction of a direct decomposition code for a specific block-structure requires considerations in three distinct sectors: (1) Structure analysis, (2) pivot selection rules, and (3) revised simplex methods.

Re (1): The structure of the coefficient matrix $\underline{A}$ has to be analyzed, the different levels and their subproblems have to be defined, and the correct mathematical relations between the submatrices of the different levels habe to be developed.

Re(2): Generally, many options exist for selecting the pivot-elements. The steepest ascent rule (as applied to the above example of section 4.1) e. g. does not take into account that the transfer of the single subproblems between the disc storage and the core storage takes a long time and should not be carried out more often than necessary. More specific pivot selection rules can be more efficient.

Re (3): The normal simplex method shall by no means be recommended to be applied in connexion with direct decomposition. Rather, some of the revised simplex methods have to be considered. Since - as seen above - mainly rows and columns have to be generated, the product form of the inverse or the symmetric revised simplex method, see MÜLLER-MERBACH (1965) will be advantageous. But still, if the subproblems are small and their matrices not sparse, the normal simplex method may be applied to the subproblems, and any of the revised simplex methods to the master problem.

It is a difficult task to find optimal combinations of the many choices in these three sectors for any specific structure. The only statement which seems to be true is that such an optimal combination has as yet been found for none of the structures mentioned.

5. NUMERICAL EXPERIENCE WITH THE BLOCK-ANGULAR CASE

5.1. THE STEEPEST UNIT ASCENT PIVOT SELECTION RULE

Finally, some numerical results shall be reported. In the author's first publication on direct decomposition, see MÜLLER-MERBACH (1965), he made an estimation on the reduction of operations, comparing the normal simplex method with the method of direct decomposition for the block-angular case. The result was that for certain structures the reduction of operations was in the neighbourhood of 70%. In numerical tests on different problems, reductions between 60 and 77% were counted, see MÜLLER-MERBACH (1966). OHSE (1967) reported similar results. The table on the following page is taken from MÜLLER-MERBACH (1966).

It is expected that the reduction of operations (and of computation time)will be much stronger if direct decomposition is combined with any of the revised simplex methods, particularly with the product form of the inverse. No experiments have been carried out in this direction so far.

5.2. A DUAL PIVOT SELECTION RULE

An interesting modification of the method of direct decomposition is the method suggested by OHSE (1971). He selects the pivot-elements in a different way. He starts - in the block-angular case - with pivoting in the matrices $\underline{A}_{kk}$ of level 1 and continues until all the subproblems are primal and dual feasible. In the second phase he applies the dual simplex method only to level 0. This brings him to a solution which is primal feasible in total. In the third phase he applies the dual simplex method to the single problems k consisting of the matrices $\underline{A}_{kk}$ and $\underline{A}_{ok}$. He keeps pivoting in these problems until all problems are primal feasible, too.

OHSE considers examples of the total size of up to 315 rows by 150 columns. He compares the performance of the classical decomposition approach of DANTZIG and WOLFE (1960) and (1961), of ROSEN (1964), of the direct decomposition method of MÜLLER-MERBACH (1964), (1965), and (1966), with his method of *dual direct decomposition.*

He found out that in his test problems the numbers of operations of the type a+b·c have the relationship 100:195:114:100 (DANTZIG/WOLFE:ROSEN:MÜLLER-MERBACH:OHSE). The numbers of transferring subproblems into the core storage have the relationship 277:382:2838:100.

The total computation times have the relationship 58:100:68:62, if all the data

No. of problem	Size of problem (rowsxcolumns)	Size of master problem (rowsxcolumns)	Number and size of subproblems (number x rows x columns)	Number of simplex iterations	Number of operations "a=b+cd" using direct decomposition	Number of operations "a=b+cd" using the normal simplex method	Reduction in operations by direct decomposition
1	70 x 100	20 x 100	5 x 10 x 20	38	79 089	202 234	60,9%
2	70 x 100	20 x 100	5 x 10 x 20	73	168 890	477 913	64,7%
3	70 x 150	20 x 150	5 x 10 x 30	63	202 523	582 607	65,2%
4	70 x 150	20 x 150	5 x 10 x 30	82	263 585	762 516	65,4%
5	100 x 120	16 x 120	6 x 14 x 20	77	178 550	694 966	74,3%
6	100 x 120	16 x 120	6 x 14 x 20	91	214 497	864 624	75,2%
7	100 x 120	16 x 120	6 x 14 x 20	117	275 730	1 145 938	75,9%
8	100 x 120	16 x 120	6 x 14 x 20	135	321 272	1 401 947	77,1%
9	120 x 150	20 x 150	5 x 20 x 30	197	762 446	3 359 291	77,3%
10	120 x 150	20 x 150	5 x 20 x 30	228	880 473	3 849 047	77,1%
11	120 x 200	20 x 200	10 x 10 x 20	184	828 844	3 566 208	76,8%
12	120 x 200	20 x 200	10 x 10 x 20	233	1 034 720	3 968 329	73,9%

are kept in the core storage throughout all iterations.

6. HISTORY OF DIRECT DECOMPOSITION

A brief historical synopsis will bring this paper to an end.

The method of direct decomposition has its root in the UBT. The development of UBT goes back as far as 1955 when DANTZIG (1955) and other authors published articles about it. Today the UBT is mentioned in almost all textbooks on linear programming and on operations research techniques. It is also implemented in most standard computer codes of linear programming.

Based on the simple UBT, the generalized UBT was developed in 1964 and published in 1967 by DANTZIG and VAN SLYKE (1967).

The first studies on direct decomposition as a further generalization of UBT were published by several authors, each of them approaching this technique differently. One of the first papers was that of ROSEN (1964) who called the method *primal partitioning*. Independently, the papers of MÜLLER-MERBACH (1964), (1965), and (1966) were published. Independently again, KAUL (1965) presented basically the same technique under the name of *generalized generalized upper-bounding-technique*. The method of BENNETT (1966) is similar. Direct decomposition and related methods are called *block-product* algorithms by ORCHARD-HAYS (1968) who applied these techniques since even longer. A major new contribution is that of OHSE (1971) and (1972). LASDON (1970) describes these methods in the context of large-scale programming. GRIGORIADIS (1972) works on an unification of these methods.

The numerical examples used in this paper are taken from MÜLLER-MERBACH (1964).

7. REFERENCES

BENNETT, J. M. (1966). An Approach to Some Structured Linear Programming Problems. Operations Research 14, p. 636

DANTZIG, G. B. (1955). Upper Bounds, Secondary Constraints, and Block Triangularity in Linear Programming. Econometrica 23, p. 174

DANTZIG, G. B. and P. WOLFE (1960). Decomposition Principle for Linear Programs. Operations Research 8, p. 101

DANTZIG, G. B. and P. Wolfe (1961). The Decomposition Algorithm for Linear Programs. Econometrica 29, p. 767

DANTZIG, G. B. and R. VAN SLYKE (1967). Generalized Upper Bounding Techniques. J. Comp. Sys. Sci. 1, p. 213

GRIGORIADIS, M. (1972) Unified Pivoting Procedures for Large Structured Linear Systems and Programs (in this volume)

KAUL, R. N. (1965). An Extension of Generalized Upper Bounded Techniques for Linear Programming. Operations Research Center Report 65-27, University of California, Berkeley

LASDON, L. (1970). Optimization Theory for Large Systems. New York

MÜLLER-MERBACH, H. (1964). Neue Entwicklungen in der linearen Optimierung. Darmstadt (Unpublished text for a course on recent developments in linear programming, held in the Deutsche Institut für Betriebswirtschaft, Frankfurt)

MÜLLER-MERBACH, H. (1965). Das Verfahren der direkten Dekomposition in der linearen Planungsrechnung. Ablauf- und Planungsforschung 6, p. 306

MÜLLER-MERBACH, H. (1965a). Die symmetrische revidierte Simplex-Methode der linearen Planungsrechnung. Elektronische Datenverarbeitung 7, p. 105

MÜLLER-MERBACH, H. (1966). Das Prinzip der "direkten Dekomposition" in der linearen Planungsrechnung. Zeitschrift für Angewandte Mathematik und Mechanik 46, p. T 102

MÜLLER-MERBACH, H. (1971). How to Apply the General Upper Bounding Technique to Any Linear Programming Problem. Forschungsbericht Nr. 32-1971, Mainz, (yet unplubished)

OHSE, D. (1967). Numerische Erfahrungen mit zwei Dekompositionsverfahren der linearen Planungsrechnung. Ablauf- und Planungsforschung 8, p. 289

OHSE, D. (1971). Ein dualer Dekompositions-Algorithmus zur Lösung block-angularer Probleme der linearen Planungsrechnung. Dissertation. Darmstadt

OHSE, D. (1972). A Dual Decomposition Method for Block-Diagonal Linear Programmes. Berkeley (yet unpublished)

ORCHARD-HAYS, W. (1968). Advanced Linear-Programming Computing Techniques. New York

ROSEN, J. B. (1964). Primal Partition Programming for Block Diagonal Matrices. Numerische Mathematik 6, p. 250

ON THE NUMBER OF ITERATIONS IN DANTZIG-WOLFE DECOMPOSITION ALGORITHM

by

Ilan Adler and Aydin Ülkücü
Department of Industrial Engineering
and Operations Research
University of California, Berkeley

May 1972

INTRODUCTION

Given a linear program in standard form, Dantzig-Wolfe decomposition algorithm replaces the original polytope by another one with many more variables but fewer equations. In this note we investigate the relations between the combinatorial structures of the two polytopes. In particular we show that, contrary to what one intuitively expects, the diameter of the original polytope may be smaller than the diameter of the decomposed one. Since the diameter of a polytope gives the maximal number of iterations taken by the "ideal" vertex-following algorithm, this observation may provide a clue for the slow convergence of the decomposition algorithm reported by some authors.

PRELIMINARY RESULTS

Given a polytope $\bar{\underline{X}} = \{x \mid Ax = b, x \geq 0\}$ where A is an $m \times n$ matrix, b is an m-vector and x is an n-vector. Let us partition A and b such that $A = \begin{bmatrix} A_0 \\ A_1 \end{bmatrix}$, $b = \begin{bmatrix} b_0 \\ b_1 \end{bmatrix}$ where A_0 is an $m_0 \times n$ matrix and b_0 is an m_0-vector. Define $\bar{\underline{X}}_1 = \{x \mid A_1 x = b_1, x \geq 0\}$; to simplify the discussion we assume that $\bar{\underline{X}}_1$ is bounded. Let $E = (x^1, \ldots, x^k)$

where $x^i (i=1, \ldots, k)$ are all the vertices of $\bar{\underline{X}}_1$ and define

$\bar{\underline{Y}} = \left\{\lambda \mid A_0 E\lambda = b_0 , \sum_{j=1}^{k} \lambda_j = 1 , \lambda \geq 0\right\}$ where λ is a k-vector. In the following proposition we present the well known correspondence between $\bar{\underline{X}}$ and $\bar{\underline{Y}}$ which is the basis for the Dantzig-Wolfe decomposition algorithm (See [1]).

Proposition 1

Let T be the linear mapping from $\bar{\underline{Y}}$ to $\bar{\underline{X}}$ defined by $T(\lambda) = E\lambda$. Then $T(\bar{\underline{Y}}) = \bar{\underline{X}}$ (i.e. T is *onto* $\bar{\underline{X}}$). The proof of Proposition 1 can be found in many linear programming textbooks and hence is omitted here.

The importance of Proposition 1 comes from its use in linear programming. Given a linear program P_0: min cx subject to $x \in \bar{\underline{X}}$, one instead can solve another linear program P_1: min $g\lambda$ subject to $\lambda \in \bar{\underline{Y}}$ where $g_i = cx^i (i=1, \ldots, k)$. It follows directly from Proposition 1 that if λ^* is an optimal solution of P_1 then $T(\lambda^*)$ is an optimal solution of P_0. (See [1]).

Since the simplex method, which is used to solve either problem, P_0 or P_1, is a vertex following algorithm (i.e. the algorithm moves from one vertex of the given polytope to another along the edges until the optimal solution is reached) it is interesting to find out the relations between the vertices and edges of $\bar{\underline{X}}$ and $\bar{\underline{Y}}$. Propositions 2, 3 and 4 shed some light on these relationships.

Although our main interest is the relation between vertices and edges, Proposition 2 is more general and contains some information about all faces of $\bar{\underline{X}}$ and $\bar{\underline{Y}}$. So let us first introduce the notion of faces of a polytope.

The *dimension* of a given polytope $\bar{\underline{X}}$ is the maximal number of affinely independent points contained in $\bar{\underline{X}}$. ($z^0, \ldots, z^\ell$ are affinely independent if and only if $z^1 - z^0, \ldots, z^\ell - z^0$ are linearly independent.) Given a polytope $\underline{\bar{X}} = \{x \mid Ax = b , x \geq 0\}$ let A' be a submatrix of A obtained by omitting some columns from A . If $\underline{\bar{X}}' = \{x' \mid A'x' = b , x' \geq 0\} \neq \phi$ then $\underline{\bar{X}}'$ is called a *d'-dimensional face* of $\underline{\bar{X}}$ where d' is the dimension of $\underline{\bar{X}}'$. Zero and one dimensional faces are called *vertices* (or *extreme points*) and *edges* respectively.

In the following lemma we give necessary and sufficient conditions for a subset of $\underline{\bar{X}}$ to be a face. This lemma is widely used and easy to prove, therefore we state it without a proof.

Lemma 1

Let $\underline{\bar{X}}$ be a polytope. A set F is a face of $\underline{\bar{X}}$ if and only if there exists a vector c such that $F = \{z \in \underline{\bar{X}} \mid cz = \min_{x \in \underline{\bar{X}}} cx\}$.

MAIN RESULTS

Given a polytope $\underline{\bar{X}}$ define $F_i(\underline{\bar{X}})$ as the set of all i-dimensional faces of $\underline{\bar{X}}$.

Proposition 2

(i) $F \in F_s(\underline{\bar{X}})$ implies that $T^{-1}(F) \in F_t(\underline{\bar{Y}})$ where $t \geq s$.

(ii) F is a face of $\underline{\bar{Y}}$ does not necessarily imply that $T(F)$ is a face of $\underline{\bar{X}}$.

Proof.

(i) By Lemma 1 there exists a vector c such that $F = \{z \in \underline{\bar{X}} \mid cz = \min_{x \in \underline{\bar{X}}} cx\}$. Let $cx^i = g_i\,(i=1, \ldots, k)$ and $g = (g_1, \ldots, g_k)$ then obviously, by Proposition 1, $T^{-1}(F) = \{\mu \in \underline{\bar{Y}} \mid g\mu = \min_{\lambda \in \underline{\bar{Y}}} g\lambda\}$. Hence, by Lemma 1, $T^{-1}(F)$

is a face of $\underline{\bar{Y}}$. Since F is of dimension s there exist $s + 1$ points $z^0, \ldots, z^s$ in F which are affinely independent, i.e. $z^i - z^0 (i=1, \ldots, s)$ are linearly independent or, equivalently, $\sum_{i=1}^{s} \mu_i (z^i - z^0) = 0$ implies $\mu_i = 0 (i=1, \ldots, s)$ where μ_i are real numbers.

For $i=1, \ldots, s$ let $\lambda^i \in T^{-1}(z^i)$ then

$0 = \sum_{i=1}^{s} \mu_i (E\lambda^i - E\lambda^0) = E \sum_{i=1}^{s} \mu_i (\lambda^i - \lambda^0)$ implies $\mu_i = 0 (i=1, \ldots, s)$. Hence, $\lambda^0, \ldots, \lambda^s$ are affinely independent and thus the dimension of $T^{-1}(F)$ is greater than or equal to s.

(ii) Consider the following example:

$$A = \begin{pmatrix} A_0 \\ A_1 \end{pmatrix} = \begin{pmatrix} 1 & 1 & 1 & 0 & 0 \\ \hline 2 & 1 & 1 & 1 & 0 \\ 0 & 1 & 0 & 0 & 1 \end{pmatrix} \quad \text{and} \quad b = \begin{pmatrix} b_0 \\ b_1 \end{pmatrix} = \begin{pmatrix} 3/2 \\ \hline 5/2 \\ 1 \end{pmatrix}.$$

It is easy to verify that the extreme points of $\underline{\bar{X}}_1$ are: (3/4,1,0,0,0); (5/4,0,0,0,1); (0,1,3/2,0,0); (0,1,0,3/2,0); (0,0,5/2,0,1); (0,0,0,5/2,1). So $\underline{\bar{Y}}$ is the set of all solutions to the following system:

$$7/4\lambda_1 + 5/4\lambda_2 + 5/2\lambda_3 + \lambda_4 + 5/2\lambda_5 = 3/2$$

$$\lambda_1 + \lambda_2 + \lambda_3 + \lambda_4 + \lambda_5 + \lambda_6 = 1$$

$$\lambda_j \geq 0 (j=1, \ldots, 5)$$

Moreover, $\bar{\lambda} = (6/7,0,0,0,0,1/7)$ is an extreme point of $\underline{\bar{Y}}$ while $T(\bar{\lambda}) = (9/14,6/7,0,5/14,1/7)$ is not an extreme point of $\underline{\bar{X}}$. ||

In Proposition 2 we proved that to every vertex of $\bar{\underline{X}}$ corresponds one or more vertices of $\bar{\underline{Y}}$. The effect of this fact on vertex following algorithms is twofold. The greater number of vertices in $\bar{\underline{Y}}$ suggests that such algorithms may take many more steps on $\bar{\underline{Y}}$ than on $\bar{\underline{X}}$. On the other hand a great number of vertices may present shorter routes on $\bar{\underline{Y}}$ (since every vertex has more neighbor vertices). In fact, it may be conjectured that if one uses an ideal vertex-following algorithm, then given any two vertices in $\bar{\underline{X}}$ and corresponding vertices in $\bar{\underline{Y}}$ the algorithms would take no more steps to connect those vertices in $\bar{\underline{Y}}$ than in $\bar{\underline{X}}$. In the following proposition we disprove this conjecture. This observation may partially explain the reported slow convergence of the Dantzig-Wolfe decomposition algorithm.

Proposition 3

Let $x^1 , x^2 \varepsilon \mathcal{F}_0(\bar{\underline{X}})$ such that x^1 , x^2 are joined by an edge of $\bar{\underline{X}}$. This does not necessarily imply the existence of $\lambda^i \varepsilon T^{-1}(x^i)$, i=1, 2 such that $\lambda^1 , \lambda^2 \varepsilon \mathcal{F}_0(\bar{\underline{Y}})$ and λ^1 and λ^2 are joined by an edge of $\bar{\underline{Y}}$.

Proof

Consider the following example:

$$A = \begin{pmatrix} A_0 \\ A_1 \end{pmatrix} = \begin{pmatrix} 1 & 0 & 0 & 1 & 1 & 0 \\ 4 & 2 & 3 & 4 & 5 & 6 \\ \hline 1 & 0 & 0 & 1 & 0 & 0 \\ 0 & 1 & 0 & 0 & 1 & 0 \\ 0 & 0 & 1 & 0 & 0 & 1 \end{pmatrix} \quad b = \begin{pmatrix} b_0 \\ b_1 \end{pmatrix} = \begin{pmatrix} 4/3 \\ 11 \\ \hline 1 \\ 1 \\ 1 \end{pmatrix}$$

Then $\bar{\underline{X}}_1$ is a 3-dimensional cube with 8 vertices and $\bar{\underline{Y}}$ is given by the set of all solutions to the following system:

$$2\lambda_1 + 2\lambda_2 + 2\lambda_3 + 2\lambda_4 + \lambda_5 + \lambda_6 + \lambda_7 + \lambda_8 = 4/3$$

$$15\lambda_1 + 15\lambda_2 + 12\lambda_3 + 12\lambda_4 + 12\lambda_5 + 12\lambda_6 + 9\lambda_7 + 9\lambda_8 = 11$$

$$\lambda_1 + \lambda_2 + \lambda_3 + \lambda_4 + \lambda_5 + \lambda_6 + \lambda_7 + \lambda_8 = 1$$

$$\lambda_i \geq 0 \ , \ i=1, \ldots, 8.$$

Take $x^1 = (0,2/3,2/3,1,1/3,1/3)$, $x^2 = (1,2/3,2/3,0,1/3,1/3)$ obviously x^1 and x^2 are vertices of $\bar{X}$ which share a common edge. However the two vertices of $\bar{Y}$ in $T^{-1}(x^1)$ are $(0,0,1/3,0,1/3,0,1/3,0)$ and $(1/3,0,0,0,0,0,2/3,0)$ while the two vertices of $\bar{Y}$ in $T^{-1}(x^2)$ are $(0,0,0,1/3,0,1/3,0,1/3)$ and $(0,1/3,0,0,0,0,0,2/3)$. Obviously no vertex of $T^{-1}(x^1)$ shares a common edge with some vertex of $T^{-1}(x^2)$ which proves the proposition.||

The significance of the last example becomes clearer when stated in terms of diameters of polytopes. The *diameter* of a given polytope $\bar{X}$ is defined as the smallest integer k such that any two vertices of $\bar{X}$ can be joined by a vertex-following path of length of less than or equal to k . In a sense, the diameter of a polytope gives the maximal number of iterations taken by the "ideal" vertex-following algorithm. The following proposition is a direct result of the example given in the proof of Proposition 3.

Proposition 4

Given polytopes $\bar{X}$ and $\bar{Y}$ as defined in this paper, the diameter of $\bar{Y}$ might be greater than or equal to the diameter of $\bar{X}$.

Remark:

The examples used in Propositions 2, 3 and 4 are by no means unique. In fact we can construct classes of such examples. In particular, it is interesting to note that these classes are constituted primarily of linear programs with bounded variables.

Reference

[1] Dantzig, G. B. "Linear Programming and Extensions," Princeton University Press, 1963.

SWITCHING BETWEEN BILL OF MATERIAL PROCESSING AND THE SIMPLEX METHOD IN CERTAIN LINEAR LARGE-SCALE INDUSTRIAL OPTIMIZATION PROBLEMS

HEINER MÜLLER-MERBACH

Fachgebiet Betriebswirtschaftslehre

Technische Hochschule Darmstadt

D-61 Darmstadt, Germany

Abstract: The computation of the optimal production program in large mechanical engineering companies suffers from the fact of the extremely large number of variables, i.e. products, assemblies, and parts. It will be demonstrated how this number can substantially be reduced by bill of material processing techniques. A typical case of an LP problem with 20,800 rows by 21,000 columns could be reduced to 300 rows by 500 columns. Particular attention will be paid to stock-on-hand of assemblies and parts which can be handled by an iterative technique switching between bill of material processing and the simplex method.

1. GENERAL DESCRIPTION OF THE PROBLEM

The typical structure of LP models for optimizing the production program in mechanical engineering companies is the following:

$$\text{Maximize } z$$

$$\text{s.t. } z - \underline{p}'\underline{y} + \underline{c}'\underline{x} = 0 \qquad (1)$$

$$\underline{y} + \underline{D}\,\underline{x} = \underline{x} \qquad (2)$$

$$\underline{A}\,\underline{x} \le \underline{b} \qquad (3)$$

$$\underline{x}, \underline{y} \ge \underline{o} \qquad (4)$$

with z being a scalar, $\underline{p}'$ and $\underline{c}'$ being row vectors of the objective function, $\underline{D}$ being a (*bill of material*) square matrix, $\underline{A}$ being a (*production coefficient*) matrix, $\underline{b}$ being a (*capacity*) vector, $\underline{y}$ being a vector of the *sold products*, and $\underline{x}$ being a vector of the *produced products*.

Due to the size of $\underline{D}$ (20,000 rows times 20,000 columns are quite common), problems of this type are difficult to solve with the standard LP codes. It shall be shown in this paper how problems of this type can be remarkably reduced in size.

The general idea is the following. The equation (2) can be rewritten by $\underline{y} = (\underline{I} - \underline{D})\underline{x}$, or after inverting $(\underline{I} - \underline{D})$ by $\underline{x} = (\underline{I} - \underline{D})^{-1}\,\underline{y}$. The substitution of $\underline{x}$ in the above problem leads to:

$$\text{Maximize } z$$

$$\text{s.t. } z - \{\underline{p}' - \underline{c}'\,(\underline{I} - \underline{D})^{-1}\}\underline{y} = 0 \qquad (5)$$

$$\underline{A}(\underline{I} - \underline{D})^{-1}\;\underline{y} \le \underline{b} \qquad (6)$$

$$\underline{y} \ge \underline{o} \qquad (7)$$

If $(\underline{I} - \underline{D})$ were not a matrix with extremely nice properties, its inversion would be quite a task. How its special structure can be taken advantage of shall be shown in this paper. The main points of interest are the computations of $(\underline{I} - \underline{D})^{-1}$ and $\underline{q}' = \underline{c}'(\underline{I} - \underline{D})^{-1}$ and $\underline{R} = \underline{A}\,(\underline{I} - \underline{D})^{-1}$ and $\underline{x} = (\underline{I} - \underline{D})^{-1}\,\underline{y}$. Some of these calculations can be carried out by bill of material processing related to shortest route calculations and to signal flow calculations in networks.

The example of the following section 2, partly taken from MÜLLER-MERBACH (1971b), shall be used for illustration.

2. A NUMERICAL EXAMPLE

A company produces the end products 1 and 2. For these it needs the assemblies 3, 4, 5, and 6. For the assemblies the parts 7, 8, 9, and 10 are required. The quantitative relations between these products are given in the following *Gozinto* graph and in the matrix $\underline{D}$. The term "Gozinto" graph is due to VAZSONYI (1958).

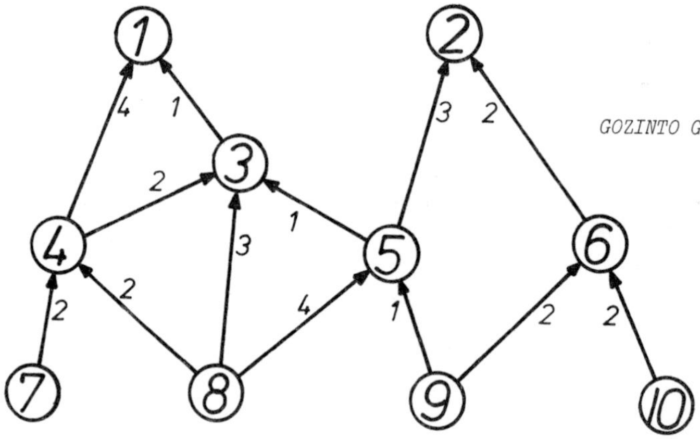

GOZINTO GRAPH

$\underline{D}$ =

	1	2	3	4	5	6	7	8	9	10
1										
2										
3	1									
4	4		2							
5		3	1							
6		2								
7				2						
8			3	2	4					
9					1	2				
10						2				

The single non-zero d_{ij} correspond to the arcs of the Gozinto graph. The *columns* of $\underline{D}$ represent the arcs entering a specific node; e. g. three pieces of 5 and two pieces of 6 are required for each piece of 2. The *rows* of $\underline{D}$ represent the arcs leaving a specific node; e. g. four pieces of 4 are required for each piece of 1 and two pieces of 4 are required for each piece of 3.

It is obvious that the total demand x_i of product i is equal to the sold quantity y_i of i plus the $\sum_j d_{ij}\, x_j$ pieces required for the other products. This is identical with the above eq. (2). Since end products are not required for other products, all the d_{ij} of the corresponding rows are zero. From this it follows that $x_i = y_i$ holds for the end products. Therefore, only one of both variables has to be explicitly used for the quantities of the end products.

In addition, the y_i have to be explicitly defined only for those products which are to be sold.

In all practical cases, the matrix $\underline{D}$ is very sparse. In addition, if the Gozinto graph does not contain circuits, $\underline{D}$ is always a triangular matrix. In the case of an appropriate order of the products, the triangularity is explicit (as in the above example). In general, $\underline{D}$ is only implicitly triangular.

After the discussion of the bill of material data, the production and capacity data shall be given now. The production is carried out in 6 "working units" (machines and workers). The matrix $\underline{A}$ contains the production times per piece of the products 1 to 10.

$\underline{A}$ =

	1	2	3	4	5	6	7	8	9	10
1							0.25	0.1	0.1	0.275
2								0.1	0.1	0.775
3			1			2				
4				0.5	1	0.5				
5	2	3								
6	4	1								

Each working unit has a capacity of 160 hours per month. Hence, the vector $\underline{b}$ consists of six elements of 160 each.

Only the end products 1 and 2 are to be sold. Therefore, only the prices of these products are of interest:

$\underline{p}'$ =

1	2
12,000	10,000

The direct production costs are given in the vector

$\underline{c}'$ =

1	2	3	4	5	6	7	8	9	10
550	450	100	150	150	100	160	150	260	210

.

The LP-model for maximizing the profit consists of 12 variables (2 y_i and 10 x_i) times 16 rows (10 according to eq. (2) and 6 according to eq. (3)). It would, of course, not be necessary to define the variables y_i *and* x_i for the end products, which reduces the model by two variables and two rows.

The optimal solution is: $y_1 = x_1 = 20$, $y_2 = x_2 = 20$, $x_3 = 20$, $x_4 = 120$, $x_5 = 80$, $x_6 = 40$, $x_7 = 240$, $x_8 = 620$, $x_9 = 160$, $x_{10} = 80$; and $z = 199{,}200$.

If the LP-model is reduced in size according to eq. (5) - (7) it reads:

Maximize z

$$\begin{aligned} \text{s.t.}\quad z - 5520\,y_1 - 4440\,y_2 &= 0 & 4\,y_1 + 4\,y_2 &\le 160 \\ 5\,y_1 + 3\,y_2 &\le 160 & 2\,y_1 + 3\,y_2 &\le 160 \\ 2\,y_1 + 5\,y_2 &\le 160 & 4\,y_1 + y_2 &\le 160 \\ y_1 + 4\,y_2 &\le 160 & y_1, y_2 &\ge 0 \end{aligned}$$

The optimal solution of this reduced problem is: $y_1 = 20$, $y_2 = 20$, $z = 199{,}200$. It is obvious that the handling of the reduced model requires much less computation time than that of the model first mentioned. On the other hand, the preparation of the data of the reduced model and the generation of the vector $\underline{x}$ from the resulting vector $\underline{y}$ requires additional computationr.How these additional computations can be carried out efficiently shall be shown in the following sections.

3. THE COMPUTATION OF $(\underline{I} - \underline{D})^{-1}$

The inversion of $(\underline{I} - \underline{D})$ would be a major task if standard codes were used. The following method for (even all-integer) computation of $(\underline{I} - \underline{D})^{-1} - \underline{I}$ is much more efficient. It was presented by MÜLLER-MERBACH (1969). It is based on the shortest route algorithm of FLOYD (1962).

This method for computing $(\underline{I} - \underline{D})^{-1} - \underline{I}$ consists of the formula

$$d_{ij}^{(k)} = d_{ij}^{(k-1)} + d_{ik}^{(k-1)} \cdot d_{kj}^{(k-1)} \quad \forall\, i, j \qquad (8)$$

which has to be carried out for $k = 1, 2, \ldots, n$ (n being the number of rows). The resulting $d_{ij}^{(n)}$ are the elements of $(\underline{I} - \underline{D})^{-1} - \underline{I}$. This can be shown from the close relation to the single steps of the inversion of $(\underline{I} - \underline{D})$ with pivot-elements in the diagonal, which is essentially identical.

One property of this algorithm is that no non-zero elements disappear during the single iterations. This makes it easy to store the non-zero elements d_{ij} in a one-dimensional list (instead of a matrix) and to link the elements of the single rows or columns by chains of addresses, see MÜLLER-MERBACH (1969). For each new non-zero element, this list grows by one item.

For the above example, the following matrix $(\underline{I} - \underline{D})^{-1} - \underline{I}$ is yielded.

$(\underline{I} - \underline{D})^{-1} - \underline{I}$ =

	1	2	3	4	5	6	7	8	9	10
1										
2										
3	1									
4	6		2							
5	1	3	1							
6		2								
7	12		4	2						
8	19	12	11	2	4					
9	1	7	1		1	2				
10		4				2				

This algorithm does not require that the rows and columns are ordered, i.e. the matrix $\underline{D}$ need not explicitly be a triangular matrix.

If only the columns of the end products are requested, the algorithm can be carried out only on these columns. In this case, the rows and columns must be ordered such that the matrix $\underline{D}$ is explicitly triangular.

Once the matrix $(\underline{I} - \underline{D})^{-1}$ has been computed, the products $\underline{q}' = \underline{c}'(\underline{I} - \underline{D})^{-1}$ and $\underline{R} = \underline{A}(\underline{I} - \underline{D})^{-1}$ and $\underline{x} = (\underline{I} - \underline{D})^{-1}\,\underline{y}$ can be generated. In the following sections, however, *bill of material processing* algorithms shall be shown which can be more efficient for the computation of $\underline{q}'$, $\underline{R}$, and $\underline{x}$.

4. THE COMPUTATION OF $\underline{q}' = \underline{c}'(\underline{I} - \underline{D})^{-1}$

The row vector $\underline{q}' = \underline{c}'(\underline{I} - \underline{D})^{-1}$ can be computed by multiplying $\underline{c}'$ and

$(\underline{I} - \underline{D})^{-1}$ or by solving the linear equation $\underline{c}' = \underline{q}'(\underline{I} - \underline{D})$. The latter can easily be carried out directly with the Gozinto graph. It starts with assigning the direct costs to the single nodes and continues with "exploding" the costs along the arcs from the parts to the end products. This is shown in the following Gozintograph:

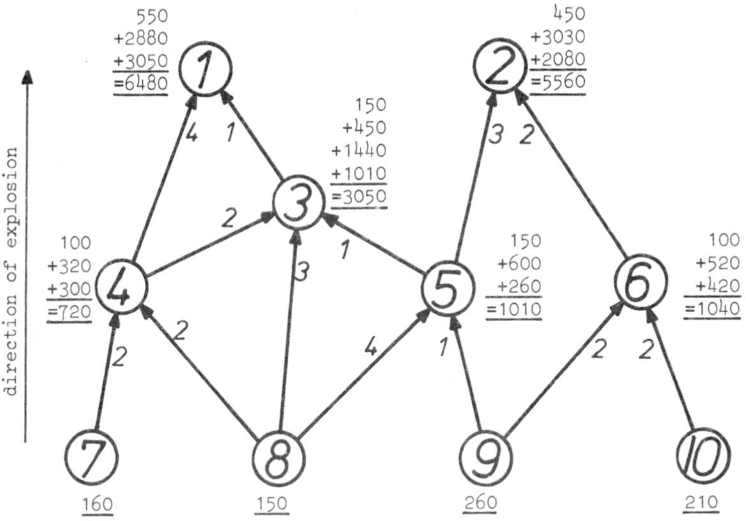

EXPLODING THE COSTS ALONG THE GOZINTO GRAPH

This explosion requires exactly as many operations as the number of arcs in the graph. If the number of arcs in the graph (or non-zero elements in $\underline{D}$) is smaller than the number of the required non-zero elements in $(\underline{I} - \underline{D})^{-1}$, this method works faster than the multiplication $\underline{q}' = \underline{c}'(\underline{I}-\underline{D})^{-1}$. "Required" are the elements of those columns of $(\underline{I} - \underline{D})^{-1}$ corresponding to products which are to be sold. If only a few elements of $\underline{q}'$ are requested, it can be still advantageous to multiply $\underline{c}'$ by the corresponding columns of $(\underline{I} - \underline{D})^{-1}$.

In the above example only the end products 1 and 2 are to be sold. Therefore, only q_1 and q_2 are requested. From the cost explosion it follows that $\underline{q}' = (q_1 \quad q_2) = (6480 \quad 5560)$. The corresponding elements of $\underline{p}' - \underline{c}'(\underline{I}-\underline{D})^{-1}$ as requested for the objective function are $(5520 \quad 4440)$, see section 2. Here the cost explosion required 13 operations while the corresponding vector-matrix-multiplication would have required 11 operations.

5. THE COMPUTATION OF $\underline{R} = \underline{A}(\underline{I} - \underline{D})^{-1}$

The computation of the coefficient matrix $\underline{R} = \underline{A}(\underline{I} - \underline{D})^{-1}$ can be computed in the same two ways as the objective function (section 4).

If it is done by explosion along the Gozinto graph, a vector has to be assigned to each node of the graph. This vector contains the required times at the single working units.

If only very few columns of $\underline{R} = \underline{A}(\underline{I} - \underline{D})^{-1}$ are requested, it is advantageous to multiply $\underline{A}$ by the corresponding columns of $(\underline{I} - \underline{D})^{-1}$.

Either method leads to the coefficients given in the reduced LP-model of section 2.

6. THE COMPUTATION OF $\underline{x} = (\underline{I} - \underline{D})^{-1}\ \underline{y}$

After the reduced problem has been solved, the vector $\underline{x}$ has to be generated from the solution vector $\underline{y}$. This can be carried out by multiplying the matrix $(\underline{I} - \underline{D})^{-1}$ by $\underline{y}$. It can also be done by exploding the quantities along the Gozinto graph, but now in the opposite direction. This is shown in the following Gozinto graph.

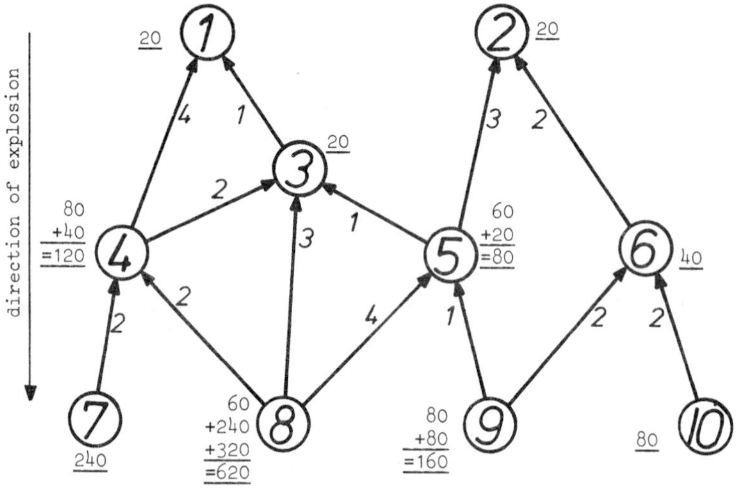

EXPLODING THE MATERIAL REQUIREMENTS ALONG THE GOZINTO GRAPH

This explosion, which corresponds to the solution of the equation $\underline{y} = (\underline{I} - \underline{D})\ \underline{x}$, is advantageous in the case of a large number of non-zero elements in the vector $\underline{y}$. If, however, this number is small it is more efficient to multiply the corresponding part of the matrix $(\underline{I} - \underline{D})^{-1}$ by $\underline{y}$.

7. ORGANIZATIONAL ASPECTS

A typical mechanical engineering company may have the following file sizes:

500 end products, 10,000 assemblies, 10,000 parts;
300 working units.

This means that $\underline{x}$ has 20,500 elements, $\underline{y}$ has 500 elements (if only the end products are to be sold; 20,000 elements of $\underline{y}$ are zero, due to definition), $\underline{D}$ is a 20,500 x 20,500 matrix, and $\underline{A}$ is a 300 x 20,500 matrix. The corresponding LP-model has the size of 20,800 rows and 21,000 columns!

The reduced model would have the size of only 300 rows and 500 columns. This is a reduction of size of 99.97 per cent.

To work with the reduced model requires the following steps if the inverse $(\underline{I} - \underline{D})^{-1}$ is used:

(0) Inversion of $(\underline{I} - \underline{D})$, see section 3.
(1) Computation of $\underline{p}' - \underline{q}'$ with $\underline{q}' = \underline{c}'(\underline{I} - \underline{D})^{-1}$ for those products which are to be sold, see section 4.

(2) Computation of $\underline{R} = \underline{A}(\underline{I} - \underline{D})^{-1}$ for those products which are to be sold, see section 5.
(3) Solving the problem by the simplex method, yielding the vector $\underline{y}$.
(4) Computation of $\underline{x} = (\underline{I} - \underline{D})^{-1}\underline{y}$, see section 6.

If the matrix $(\underline{I} - \underline{D})$ is not inverted, the following steps have to be carried out:

(1) Computation of $\underline{p}' - \underline{q}'$, $\underline{q}'$ by solving the equation $\underline{c}' = \underline{q}'(\underline{I} - \underline{D})$ by cost explosion along the Gozinto graph, see section 4.
(2) Computation of $\underline{R} = \underline{A}(\underline{I} - \underline{D})^{-1}$ by explosion along the Gozinto graph, see section 5.
(3) Solving the problem by the simplex method, yielding the vector $\underline{y}$.
(4) Computation of $\underline{x}$ by solving the equation $(\underline{I} - \underline{D})\underline{x} = \underline{y}$; this is again to be carried out by explosion along the Gozinto graph, see section 6.

All the explosion techniques can efficiently be carried out by bill of material processors which are provided as standard codes for many computers.

In a practical case that has been treated by the author and WIGGERT, the IBM processor "BOMP" and the "MPS" were used. Very few additional programs were necessary.

8. HISTORY

The combination of computing along the Gozinto graph and LP was first suggested by MÜLLER-MERBACH (1966) who considered chemical processes. The application to mechanical engineering companies was treated in the more recent and more specific publications by WIGGERT (1971a), (1971b),(1972a), and (1972b) and by MÜLLER-MERBACH (1971b). The advantage of FLOYD's (1962) algorithm for inverting $(\underline{I} - \underline{D})$ was explicitly described by MÜLLER-MERBACH (1969), (1971a). Most of the basic ideas concerning the Gozinto graph have their roots in VAZSONYI's book (1958).

The *explosion* method is closely related to the methods of *signal flow* as developed by MASON (1953). Due to the fact that Gozinto graphs do not have circuits, very simple versions of MASON's method can be applied.

9. HOW TO TREAT STOCK-ON-HAND

Up to here, this paper contained a systematic synopsis of the work carried out in the past. In addition, this section will report on some quite recent developments regarding the handling of stock-on-hand.

In practice, difficulties can arise if stocks of assemblies or parts have to be taken into consideration. Three ways of treating stocks shall be outlined here: (1) Extension of the model consisting of eq. (1) to (4). (2) Extension of the model consisting of the eq. (5) to (7). (3) Switching between solving the problem of eq. (5) to (7) and exploding the material requirements along the Gozinto graph.

9.1. FIRST APPROACH

If the model of eq. (1) to (4) is extended, the vectors $\underline{s}$ as stock-on-hand and $\underline{w}$ as the consumed stock quantity can be introduced as follows:

$$\text{Maximize } z$$

$$\text{s.t. } z - \underline{p}'\underline{y} + \underline{c}'(\underline{x} + \underline{w}) = 0 \qquad (9)$$

$$\underline{y} - \underline{w} + \underline{D}\,\underline{x} = \underline{x} \qquad (10)$$

$$\underline{A}\,\underline{x} \le \underline{b} \qquad (11)$$

$$\underline{w} \le \underline{s} \qquad (12)$$

$$\underline{x}, \underline{y}, \underline{w} \ge \underline{o} \qquad (13)$$

9.2. SECOND APPROACH

This first model is even bigger than the model of eq. (1) to (4). It can, however, be reduced in a similar way as shown in section 1. For this purpose, the set T is defined in addit.on to the previous notation. It contains all the products with stock-on-hand. Furthermore, a "partial" inverse $(\underline{I} - \underline{D})_t^{-1}$ of $(\underline{I} - \underline{D})$ is required for the reduction of the model. It is computed as shown in section 3, except for the fact that the eq. (8) is not carried out for the $k \in T$. This means that those quantities of the single products which are required for the products $i \in T$ are not taken as required for the products which the products $i \in T$ go into.

If in the above example the set T consists of the products 3, 5, and 7, the partial inverse reads as follows:

$(\underline{I} - \underline{D})_t^{-1} =$

	1	2	3	4	5	6	7	8	9	10
1	1									
2		1								
3	1		1							
4	4		2	1						
5		3	1		1					
6		2				1				
7	8		4	2			1			
8	8		7	2	4			1		
9		4			1	2			1	
10		4				2				1

In the reduced model has, according to WIGGERT (1972a) and (1972b), a variable y_i to be defined for each product which is to be sold, and the variables x_i and w_i are required for all the $i \in T$. Then the reduced model reads as follows:

Maximize z

$$\text{s.t.} \quad z - \{\underline{p}' - \underline{c}'(\underline{I} - \underline{D})^{-1}\}\,\underline{y} = 0 \qquad (14)$$

$$(\underline{I} - \underline{D})_t^{-1}(\underline{y} + \underline{x}) - \underline{w} = \underline{x} \quad \forall\, i \in T \qquad (15)$$

$$\underline{A}\,(\underline{I} - \underline{D})_t^{-1}\,(\underline{y} + \underline{x}) \le \underline{b} \qquad (16)$$

$$\underline{w} \le \underline{s} \quad \forall\, i \in T \qquad (17)$$

$$\underline{y}, \underline{x}, \underline{w} \ge \underline{o} \qquad (18)$$

If in the above example a stock-on-hand of $s_3 = 20$ pieces of product 3, of $s_5 = 20$ pieces of product 5, and of $s_7 = 500$ pieces of product 7 are available, the following reduced model is yielded:

Maximize z

$$\begin{array}{lrl}
\text{s.t.} & z - 5520\, y_1 - 4440\, y_2 & = 0 \\
& y_1 - x_3 - w_3 & = 0 \\
& 3\, y_2 + x_3 - x_5 - w_5 & = 0 \\
& 8\, y_1 + 4\, x_3 - x_7 - w_7 & = 0 \\
& 2.8\, y_1 + 1.5\, y_2 + 1.7\, x_3 + 0.5\, x_5 + 0.25\, x_7 & \le 160 \\
& 0.8\, y_1 + 3.5\, y_2 + 0.7\, x_3 + 0.5\, x_5 & \le 160 \\
& y_1 + 4\, y_2 + x_3 & \le 160 \\
& 2\, y_1 + 5\, y_2 + 2\, x_3 + x_5 & \le 160 \\
& 2\, y_1 + 3\, y_2 & \le 160
\end{array}$$

$$
\begin{array}{lr}
4y_1 + y_2 & \leq 160 \\
w_3 & \leq 20 \\
w_5 & \leq 20 \\
w_7 & \leq 500 \\
y_1, y_2, x_3, w_3, x_5, w_5, x_7, w_7 & \geq 0
\end{array}
$$

The optimal solution is: $y_1 = 35$, $y_2 = 20$, $x_3 = 15$, $w_3 = 20$, $x_5 = 55$, $w_5 = 20$, $x_7 = 0$, $w_7 = 340$; and $z = 282,000$.

9.3. THIRD APPROACH

Reduced models of the above type can still become quite large, due to a large number of products with stock-on-hand. Therefore, the following iterative approach which requires only a model of the same size as the model of eq. (5) to (7) shall be suggested here.

For this purpose, the set T will be subdivided into the two disjunct subsets T' and T''. T' contains those products the stock-on-hand of which is completely required for the computed production program; and T'' contains the remaining products whose stock-on-hand is only partly (or even not at all) required for the production program. Obviously, the subsets T' and T'' are not known definitively before the optimal production program is yielded. And this again depends upon the division of T into the two subsets. Therefore, the following procedure iterates between the determination of T' and T'' and the simplex method for computing optimal production programs.

In the following the matrix $(\underline{I} - \underline{D})$ is partially inverted according to section 3 such that eq. (8) is not carried out for the products $i \in T''$. The partial inverse is called $(\underline{I} - \underline{D})^{-1}_{t''}$. The stock-on-hand of these products lead to an alteration of the production coefficients in the same way as in eq. (16) compared to eq. (6).

On the other hand, the stock-on-hand of the products $i \in T'$ is handled by an alteration of the capacity vector of the right hand side. The vector $\underline{s}'$ corresponds to these products. It consists of their stock-on-hand quantities.

The model reads as follows:

Maximize z

$$\text{s.t.} \quad z - \underline{p}' - \underline{c}'(\underline{I} - \underline{D})^{-1}\underline{y} = 0 \tag{19}$$

$$\underline{A}\,(\underline{I} - \underline{D})^{-1}_{t''}\,\underline{y} \leq \underline{b} + \underline{A}\,(\underline{I} - \underline{D})^{-1}_{t''}\,\underline{s}' \tag{20}$$

$$\underline{y} \geq \underline{o} \tag{21}$$

If T' and T'' are properly defined, the solution of this problem (19) to (21) contains the optimal production program. This follows from the correctness of eq. (20). Its left hand side sums up the total production coefficients of the end products and their "predecessors" (assemblies and parts) except for those with an excessive stock-on-hand. Its right hand side adds to the capacity $\underline{b}$ the savings of capacity which result from the complete consumption of the available stock-on-hand of the products $i \in T'$.

Since only the set T is known in the beginning, but not T' or T'', the set T can be arbitrarily subdivided into T' and T''. Once the subsets are defined, the data of eq. (20) can be generated according to section 3 and 5. Then the problem (19) to (21) will be solved by means of the simplex method. The following computation of the required production quantities of the single products (according to section 6, extended by the rules for handling stocks) shows for each product $i \in T$ whether the available stock was not completely consumed (i enters T'') or was required in at least the available quantity (i enters T'). If the subsets T' and T'' remain stable, the optimal solution is yielded. Otherwise the eq. (20)

has to be redefined, and the cycle will be repeated as shown in the following figure.

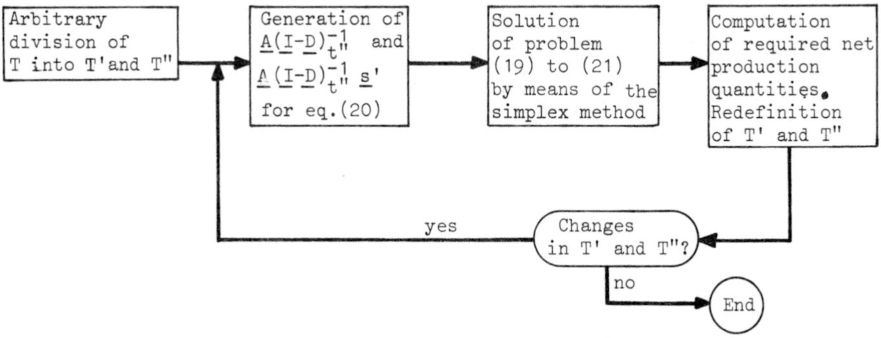

For the above example and for the stocks as given in section 9.2, the products 3 and 5 belong to the set T' while the product 7 belongs to the set T". These sets may be chosen by chance in the beginning or may be found iteratively. The corresponding LP problem reads as follows:

$$
\begin{array}{lrcrcr}
\text{Maximize} & z & & & & \\
\text{s.t.} & z - 5520\,y_1 & - & 4440\,y_2 & = & 0 \\
 & 2\,y_1 & + & 3\,y_2 & \le & 214 \\
 & 2\,y_1 & + & 5\,y_2 & \le & 194 \\
 & y_1 & + & 4\,y_2 & \le & 180 \\
 & 4\,y_1 & + & 4\,y_2 & \le & 220 \\
 & 2\,y_1 & + & 3\,y_2 & \le & 160 \\
 & 4\,y_1 & + & y_2 & \le & 160 \\
 & & & y_1,\, y_2 & \ge & 0
\end{array}
$$

The optimal solution is: $y_1 = 35$, $y_2 = 20$; and $z = 282{,}000$. The required net production quantities are: $x_1 = 35$, $x_2 = 20$, $x_3 = 35 - 20 = 15$, $x_4 = 170$, $x_5 = 75 - 20 = 55$, $x_6 = 40$, $x_7 = 340 - 340 = 0$ (160 remaining), $x_8 = 605$, $x_9 = 135$, and $x_{10} = 80$.

This iterative algorithm which switches between bill of material processing and the simplex method has not as yet been implemented in practice. Rather, it was only tested by small examples. The results were quite encouraging.

10. REFERENCES

FLOYD, R. W. (1962). Algorithm 97. Shortest Path. Comm. ACM 5, p. 345

MASON, S. J. (1953). Feedback Theory: Some Properties of Single Flow Graphs. Proc. IRE 41, p. 9

MÜLLER-MERBACH, H. (1966). Die Anwendung des Gozinto-Graphs zur Berechnung des Roh- und Zwischenproduktbedarfs in chemischen Betrieben. Ablauf- und Planungsforschung 7, p. 187

MÜLLER-MERBACH, H. (1969). Die Inversion von Gozinto-Matrizen mit einem graphenorientierten Verfahren. Elektronische Datenverarbeitung 11, p. 310

MÜLLER-MERBACH, H. (1971a). Operations Research (2nd ed.). München

MÜLLER-MERBACH, H. (1971b). Die Berechnung optimaler Produktionsprogramme bei mehrstufiger Fertigung. Werkstattstechnik 61, p. 19

VAZSONYI, A. (1958). Scientific Programming in Business and Industry. New York

WIGGERT, H. (1971a). Ermittlung der technologischen Matrix bei Anwendung der linearen Planungsrechnung in Betrieben mit mechanischer Fertigung. Betriebswirtschaftliche Forschung und Praxis 23, p. 158

WIGGERT, H. (1971b). Bestimmung des optimalen Fertigungsprogrammes in Betrieben mit mehrstufiger mechanischer Fertigung. Zeitschrift für wirtschaftliche Fertigung 66, p. 452

WIGGERT, H. (1972a). Kurzfristige Programmoptimierung mit Hilfe der linearen Planungsrechnung in Betrieben mit mehrteiliger und mehrstufiger mechanischer Fertigung. Dissertation. Frankfurt (to appear)

WIGGERT, H. (1972b). Programmoptimierung mit linearer Planungsrechnung. Berlin, Köln, Frankfurt (to appear)

DO-IT-YOURSELF-DECOMPOSITION

WILLIAM ORCHARD-HAYS

National Bureau of Economic Research, Inc.

Computer Research Center for Economics and Mangement Science

Cambridge, Massachusetts

Abstract: Specialized LP procedures recently incorporated in a powerful MPS can be used effectively as decomposition techniques. Strict partitioning is not necessary and, in many ways, the approach is more flexible and practical than true decomposition algorithms. These procedures are not practical for extremely large models but are useful for comprehensive industrial and similar models up to perhaps 4,000 constraints.

Intended originally to speed convergence when resolving revised models, they can be used to create submodels, auxiliary models and master problems. Masking and aggregation are used in conjunction with the idea of creating temporary bounds or fixed values for primal and dual variables. They utilize standard MPS procedures and algorithms, whose characteristics are well known, for actual optimization.

I. INTRODUCTION

About 18 months ago, a client of Management Science Systems, Inc., who developed MPS-III, asked them to develop an additional package of capabilities for inclusion in their system.* This package consists of several new routines and additional options for existing routines. There were two main purposes: (i) to attempt to minimize reoptimization time for revised models, and (ii) to permit fixed aggregations of parts of a model so that the remaining part or parts could be varied and studied while holding the aggregated conditions constant. It was soon realized that the second feature could be used as a form of decomposition. The package was developed on a short schedule and first installed in London. It has since been used in other locations and by other clients. Follow-on developments to the first feature are still going on and, in fact, are right now being tested in South America. This kind of technique may be of general interest and the main features and some of the difficulties will be discussed briefly.

II. The Package of Special Capabilities

A. Temporary Deletion of Unnecessary Model Elements

Before the request for the package, two important devices had already been incorporated in MPS-III. These are known as the NOFREE and NOFIX options which are invoked when a model is set up into work matrix format. It is common practice to include a number of free rows in a model for use as alternate objective forms or simply as accounting functions. These rows tend to be quite dense and, in extreme cases, may account for 50% or more of the volume of the matrix work file. Furthermore, they contribute heavily to the eta file of product-form columns. However, they do not affect the solution and are of no interest until an optimal solution is reached. The NOFREE option simply discards all elements in these rows. Since the set-up of a model takes no longer, substantially, that a major iteration with full pricing, deletion of

---

MPS-III is a mathematical programming system for IBM 360 and 370 computers which includes MPS/360 capabilities as a subset.

elements in these rows does not require much post-processing to restore the full model after the optimal basis is determined and before any reports requiring them are produced. A re-inversion is required but this is also very fast now and, likely as not, it will be time to do one anyway. NOFREE requires, of course, that free rows to be used be identified at set-up time.

The NOFIX option causes deletion of columns whose variables have fixed values and their aggregation into the RHS (right-hand-side). The increasing use of bound sets for LP variables increases the likelihood that a number of variables are fixed. In fact, some formulators do not use a RHS at all but construct one for each case by fixing certain variables via a bound set.

B. Reordering Matrix Columns

The use of a FLAGS procedure to delete selected LP columns had been common practice for many years. These columns are not necessarily fixed but are thought to be unlikely to have nonzero values for their variables in an optimal solution. Sometimes this medicine is too strong and what one really wants is something like the old curtaining techniques. Part of the package is a procedure called SEMIFLAG which moves selected columns from their normal position and puts them at the end of the work matrix behind a special marker. The original serialization is not changed but the columns are no longer in sequence on serial number. In pricing, the matrix is read only as far as the special marker except on occasional passes or when no other candidate can be found. Thus, the selected variables remain available but the coefficient columns do not contribute to file passing time except occasionally.

The procedure may be used repeatedly to reorder a matrix in some desired way. Any standard selection mechanism may be used, such as masks or first and last names of a set.

C. New SOLUTION Output Options

The first requirement for the main package was a new enfiling option for SOLUTION output. This is called RIMOUT since it writes the rim of the optimal tableau to a compact file. In other words, the LP row and column identifiers, status and primal and dual values are recorded. All the selection options of standard SOLUTION output are available plus a NOPRINT option which inhibits actual printing.

These files are recorded either as members of a partitioned data set or on a designated scratch file in a form very much like work matrix headers. They can be processed very quickly and form the main input to the modification and aggregation procedures.

D. Matching and Rectification Procedure

After a model has been optimized and a RIMOUT file produced, it will often be revised for some purpose. Since rows or columns may have been added or deleted, or even reordered, the first step is to match the old RIMOUT file with the revised model matrix and make it conform as nearly as possible. A new temporary RIMOUT file is produced on a scratch device and the old basis is honored so far as possible. The reinversion procedure is then executed to insure that a full, nonsingular basis matrix is defined. The actual inverse is not used and is usually discarded by later procedures.

E. Modification and Aggregation Procedure

This is the main procedure of the package. It reads the revised model matrix and the temporary RIMOUT file and produces a revision deck for input to REVISE. The purpose of this revision deck is to define a new row and/or column, or to redefine old ones, in such a way that the doubly revised model will be primal and dual feasible with certain standard parameter settings. Alternatively, the revision deck serves to aggregate selected columns of the matrix with auxiliary options to delete component columns and/or selected rows.

The three principle options will be discussed separately.

1. New change OBJ row and change RHS column.

For this option, the logical (LGL) variables are first processed and both primal and dual variables are feasibilized. This means that if the old values are within the new feasible ranges, they are honored. Otherwise, the old values are adjusted to the nearest feasible limit. In rare cases of ambiguity, a zero value is used. The feasibilized primal values are subtracted from an initially zero RHS change column. The feasibilized dual values constitute a working pi-vector.

Next, the structural (STR) variables are processed. The old primal values are feasibilized as for LGLs. Any new variables are set to feasible limits in standard fashion. The coefficient columns are multiplied by the feasibilized values and subtracted from the change RHS column.

The old dual values are first feasibilized and then new ones are computed using the working pi-vector. The difference is output as an element of a new OBJ change row. Finally, the specified RHS is added to the constructed change RHS column and the column is output.

When the model is revised and the basis re-inverted, the model will be primal and dual feasible if parameter values of -1.0 are used for both the change OBJ row and change RHS. These values may be parameterized or incrementally moved to a value of 0.0 at which point the originally revised model is solved. The rationale for this seemingly round-about method is to reduce the shock to a previous optimal solution which is engendered by a revision of the model. Moving parametrically or incrementally from the old simplex to the new one tends to keep the solution path in control and to avoid a long phase I followed by a long phase II which ends up relatively near to the starting solution.

2. New change OBJ row and adjusting STR column.

This option proceeds in almost identical fashion to option 1 except that the signs of the constructed column are reversed and it is considered a new STR column. The variable is bounded between -1.0 and 0.0 and forced to start at -1.0. Additionally, driver values for the OBJ and change OBJ row of the new column are calculated. There are two or three options for this, including a user-specified value. The purpose of the driver values is to cause the new STR variable to move from -1.0 to 0.0 as the RHS parameter is moved from -1.0 to 0.0. This avoids the complexity of double parameterization and allows the algorithms to vary the STR variable as required, not necessarily monotonically or in uniform steps.

Unfortunately, determination of the proper driver term is a difficult problem. A good value can be back-calculated after the desired solution is obtained but, of course, this begs the question. This difficulty is a major reason for continuing work on somewhat different techniques which make adjustments to feasible primal ranges. This work is being carried on by others and I am not prepared to discuss it.

3. New STR aggregation column.

In this option, dual values are ignored. Primal variables are feasibilized but, in most practical application, the old values are feasible.

The column is created as in option 2 but without subtracting the RHS or computing a driver term. The OBJ value in this column is simply whatever it aggregates to be. The new STR is given a fixed value of 1.0 in the current bound set.

This option may be used repeatedly to form several aggregations. The columns forming an aggregation may be optionally deleted. Some selection mechanism must be used for specifying these columns. If a bit map is used, then rows to be deleted may be optionally specified in the LGL section of the map. Thus one can, for example, delete all rows from the model which pertain only to the subproblem being aggregated. Clearly, this permits a form of user-directed decomposition, regardless of the original ordering of the LP rows and columns.

F. Auxiliary Utility Procedures

Three auxiliary procedures are provided. The first will output any RIMOUT or temporary RIMOUT file to print or punch format, either to standard output devices or to a specified file. The Matching and Rectification procedure has an option for reading a RIMOUT file in card-deck form. Thus RIMOUT files may be processed by other programs, such as FORTRAN routines, or by hand.

The second utility is called RIMERGE. It will merge two or more standard RIMOUT files with specified rules of survival for duplicate elements. The procedure is also a subroutine to the Matching and Rectification procedure where it may be required to reorder a RIMOUT file applied to a highly revised model.

The third utility revises revision decks and is intended to separate change rows and columns into partial vectors. It has not proved too useful and is sometimes replaced with special user-written routines which take other considerations into account and operate on a data base of revision files.

III. Experience with the Package

A. Temporary Deletion of Elements and Reordering Columns

Deletion of elements by NOFREE and NOFIX has shown the increased iteration speed expected. Increases of 50% in iterations per minute are not uncommon. The use of SEMIFLAG to move columns that are seldom in an optimal solution has also resulted in 40% to 50% reduction in run times. However, a couple of side effects should be noted. First, dramatic changes in iterating times may require readjustments in other run parameters for full effectiveness, for example, the cycling or partial pricing count and GUB pricing frequency. If reinversion is not under clock control, its frequency may also require adjustment.

Second, it might be thought that both those columns never in an optimal basis and those always in an optimal basis should be moved. The latter would be retrieved by the inversion procedure on its first execution. This does not necessarily work well. Sometimes these latter columns must be allowed to go in and out frequently while iterating.

The use of general bit maps and boolean operations on them is the most effective way to define the selection for SEMIFLAG. Without this capability the determination and specification of a set would be tedious.

B. Attempting to Maintain Primal/Dual Feasibility

There are actually several options available and several ways to manipulate a model using them. Since experience is still limited no one technique has been shown to be best. Nevertheless, impressive results have been experienced. Run time savings of 25% to 35% have occurred; the best report I have seen showed a 60% saving which resulted from splitting an STR column into many pieces. The use of an STR seems to be better than a change RHS. However, splitting a column can be overdone. If the number of STRs becomes large, many iterations are required just to get them all into the basis where they can vary continuously.

A number of methods of splitting the adjustment vectors have been tried, from merely separating negative and nonnegative values to pricing with a generated pi row. Much more investigation is needed before conclusive rules can be given and they may turn out to be model dependent. As indicated earlier, a similar situation exists with respect to driver values. Nevertheless, experience thus far clearly indicates that large differences in reoptimization times will occur with these techniques.

One problem occurs which is common to many convergence-acceleration approaches. The generated vectors, particularly the columns, tend to be unusually dense and to have large magnitudes. The number of iterations is usually reduced by keeping these in the basis as long as possible so continuous variations can occur. On the other hand, the presence of dense vectors reduces the iterating speed. In some trials, the net result was a wash in execution time. I ran into the same phenomenom many years ago with Dantzig-Wolfe algorithms.

C. Generation and Use of Aggregation Columns

Use of this option has been restricted to one or two installations and I have very little information on their experience. However, the matter of iteration speed is not the main concern here but rather flexibility in building, revising and studying a model. Suppose one has a typical block-angular model representing, say, manufacturing in three different areas and marketing throughout a combined region:

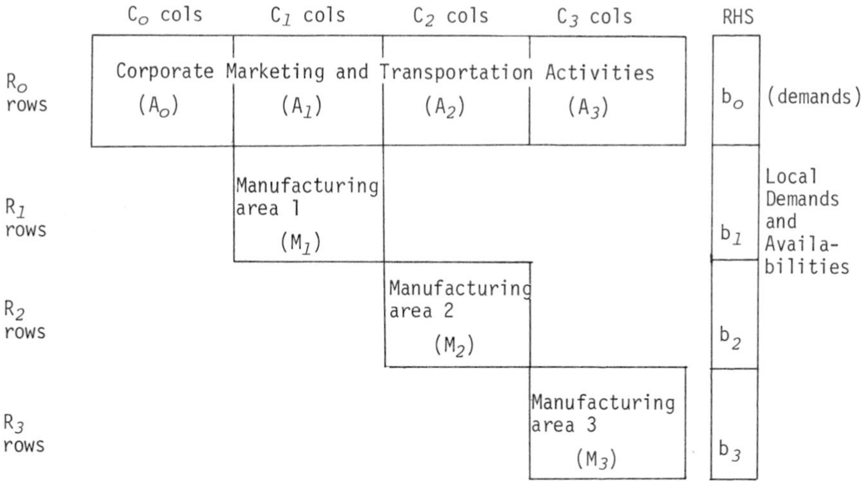

Suppose one has an old solution to the total model and that the model is being updated for the next planning period by several different groups. Suppose the corporate group wishes to study changes in area 1 and in corporate activities but new information is not yet available for areas 2 and 3. The best information available is the old solution. A RIMOUT for the old solution is filed and then blocks A_0, A_1, and M_1 are updated along with b_o and b_1. Then, in two passes, aggregation columns g_2 and g_3 are produced deleting rows in R_2 and R_3 and columns in C_2 and C_3. The following model is now available for study, where bars represent updated blocks and the columns g_2 and g_3 are fixed at level 1.0:

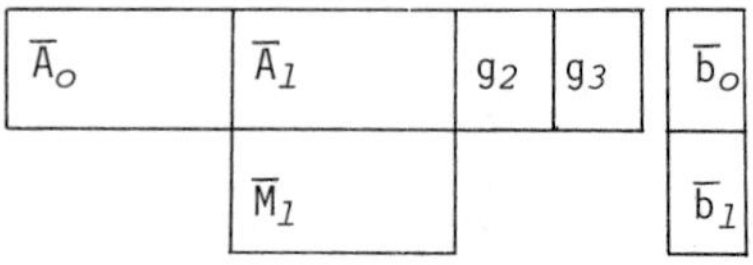

Suppose a satisfactory solution is obtained and new information on area 2 becomes available. Aggregating $\bar{A}_1$ into g_1 with the new solution and updating A_2, M_2 and b_2, the following model can be studied:

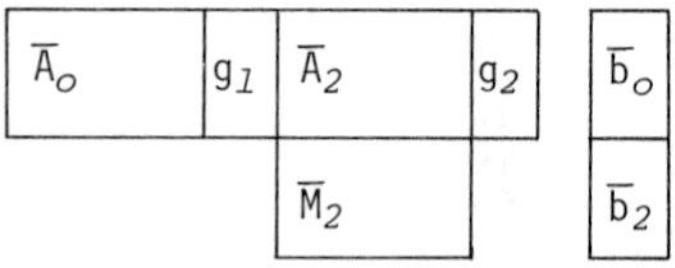

Area 3 can be treated similarly. Finally, the entire new model is solved using the combined solutions to $\bar{A}_1$, $\bar{A}_2$, $\bar{A}_3$, together with the last one obtained to $\bar{A}_o$, as a starting point.

Note that use of the other options and procedures can be used in any combination with this decomposition approach. The existence of such procedures and options in a powerful MPS gives an ingenious LP practitioner a handy bag of tricks to manipulate good-sized models and keep computer time to a minimum. Thus, he can effectively decompose models and study submodels in a fixed environment, with no special ordering of the input. The chief means for specifying parts of a model is through the user's own scheme of nomenclature in the LP identifiers and the use of corresponding name masks--thus, the term "do-it-yourself" decomposition.

DECOMPOSITION IN RESOURCE ALLOCATION

by L. S. Lasdon
Operations Research Department
Case Western Reserve University
Cleveland, Ohio 44106

1. Linear Programming Decomposition and Resource Allocation

1.1 General Formulation. A variety of resource-allocation problems, including multi-item production scheduling, may be formulated as follows. Let there be I activities and a single resource, to be allocated to these activities in each of T time periods; let x_{it} be the level of activity i in time period t and define

$$x_i = (x_{i1}, \ldots, x_{iT}) \tag{1}$$

The set of allowable activity levels is in part determined by the necessity of satisfying external demands and by the internal characterstics of the activity. For example, machine characteristics may limit the rate of production of a commodity. All such constraints, whether referring to demand or production, are called technological constraints and are symbolized by defining a constraint set S_i such that

$$x_i \epsilon S_i \tag{2}$$

Operating activity i incurs a cost and uses up the resource. Let

$c_i(x_i)$ = cost of operating activity i at levels x_i

$y_{it}(x_i)$ = amount of the resource used by activity i in time period t

b_t = resource availability in period t

$$Y_i(x_i) = (y_{i1}(x_i), \ldots, y_{iT}(x_i)) \tag{3}$$

$$b = (b_1, \ldots, b_T) \tag{4}$$

The problem of optimal resource allocation is to choose $x_1, \ldots, x_I$ to minimise the cost of operating all activities over time:

$$\text{minimize } z = \sum_{i=1}^{I} c_i(x_i) \tag{5}$$

subject to the resource limitations in all time periods:

$$\sum_{i=1}^{I} Y_i(x_i) \leq b \tag{6}$$

and the technological constraints

$$x_i \epsilon S_i, \quad i = 1, \ldots, I \tag{7}$$

The resource constraints (6) couple the activities together, forcing one to consider them all simultaneously when minimizing z. It is not difficult to extend this formulation to the case of more than one resource.

The problem of scheduling the production of I items over time may be included in this formulation by letting the activity level, x_{it}, be the quantity of item i produced in time period t. The resource could be labor or machinery, and $c_i(x_i)$ a production cost. Assuming that customer demands, r_{it}, are known for all items in all time periods, the constraints $x_i \in S_i$ could arise from upper and lower bounds on inventory levels, η_{it}, and production quantities, x_{it}; i.e.,

$$S_i = \left\{ x_i \,\middle|\, (x_{it})\min \leq x_{it} \leq (x_{it})\max,\ (\eta_{it})\min \leq \ _{it} \leq (\eta_{it})\max,\ t=1,\ldots,T \right\} \quad (8)$$

where the inventory levels are given by

$$\eta_{it} = \eta_{i,t-1} + x_{it} - r_{it}, \qquad t = 1,\ldots, T,\ \eta_{i0} \text{ given} \quad (9)$$

<u>Approximate Solution by Linear Programming</u>. The problem (5)-(7) may be difficult to solve if I is large, e.g., many items, or if the x_{it} are discrete, as when production levels can be varied only be allocating a limited number of machines. Manne [1] has shown that an approximate solution may be obtained for the case $I \gg T$ by recasting the problem as a large linear program as follows. Assume that, for each i, the set of all x_i that need be considered as candidates for optimality in (5)-(7) is finite. This is surely the case if S_i is finite, as when the x_{it} are discrete, and can be shown true in other situations, e.g., as in reference [2].

Let the jth such schedule from S_i be x_i^j. Define

$$c_{ij} = c_i(x_i^j) \quad (10)$$

$$y_{ij} = y_i(x_i^j) \quad (11)$$

and consider the following integer program:

$$\text{minimize} \sum_{i,j} c_{ij}\theta_{ij} \quad (12)$$

subject to

$$\sum_j \theta_{ij} = 1, \qquad i = 1, \ldots, I \quad (13)$$

$$\sum_{i,j} y_{ij}\theta_{ij} \leq b \quad (14)$$

$$\theta_{ij} \geq 0 \quad (15)$$

$$\theta_{ij} = \text{integer} \quad (16)$$

Constraints (13), (15), and (16) require that, for each i, exactly one $\theta_{ij}=1$, with all others zero. Thus θ_{ij} selects a particular schedule for item i, and (12) and (14) cause the schedules selected to satisfy the resource constraints and minimize the cost.

Solving (12)-(15) as an integer program would, however, be very difficult, owing to the potentially vast number of columns (one for each admissible schedule) and the many constraints if I or T is large. Fortunately, if $I \gg T$, the constraint θ_{ij} = integer may be dropped, and most of the variables in an optimal solution of the linear program (12)-(15) will still be integral. The precise result is this:

THEOREM. If $I > T$, then any basic feasible solution of the linear program (12)-(15) has the property that at least $I - T$ of the indices i have precisely one θ_{ij} positive (and hence unity).

PROOF. Any basic solution has at most $I + T$ variables positive. Each of the I constraints (13) requires at least one positive variable to satisfy it. The remaining T basic variables are then allocated to at most T different constraints of (13), leaving at least $(I - T)$ of these with exactly one positive variable.

There are many practical problems where $I \gg T$, for example in scheduling production of hundreds of items over a 12-month horizon. However, for such cases, solution of (12)-(15) is difficult, owing to the multiplicity of columns and many rows. Dzielinski and Gomory [2] have considered the application of the Dantzig-Wolfe decomposition principle [3] to this problem. This leads to a master program with only T+1 rows and even more columns which, however, may be dealt with by column generation. Thus an efficient solution is possible.

The constraint matrix of (12)-(15) has the following structure:

$$
\begin{array}{cccc}
\theta_{11}\cdots\theta_{1j_1} & \theta_{21}\cdots\theta_{2j_2} & \cdots & \theta_{I1}\cdots\theta_{Ij_I} \\[4pt]
\boxed{Y_{11}\cdots Y_{1j_1}} & \boxed{Y_{21}\ \cdots\ Y_{2j_2}} & \cdots & \boxed{Y_{I1}\cdots Y_{Ij_I}} \\[8pt]
\boxed{1\ 1\ \ldots\ 1} & & & \\[8pt]
& \boxed{1\ 1\ \ldots\ \ 1} & & \\[8pt]
& & \ddots & \\[8pt]
& & & \boxed{1\ 1\ldots\ \ 1}
\end{array}
$$

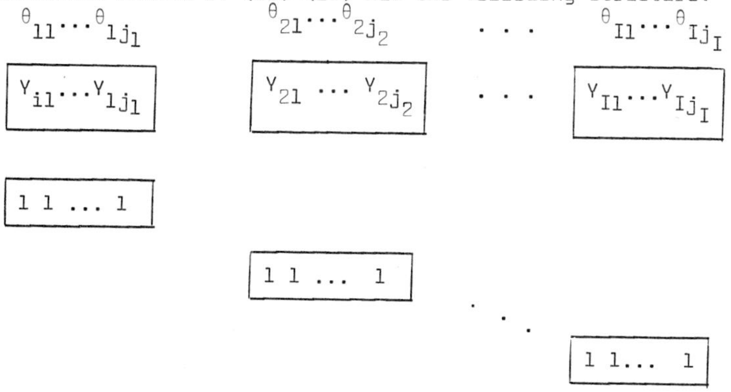

The constraints $\sum_{j}\theta_{ij} = 1$ obviously play the role of the subsystem constraints, while (14) couples the system together.

Defining,

$$e_i = (1, 1, \ldots, 1), \; j_i \text{ components} \tag{17}$$

$$\theta_i = (\theta_{i1}, \ldots, \theta_{ij_i}) \tag{18}$$

$$c_i = (c_{i1}, \ldots, c_{ij_i}) \tag{19}$$

$$\theta = (\theta_1, \ldots, \theta_I) \tag{20}$$

$$c = (c_1, \ldots, c_I) \tag{21}$$

$$L_i = [Y_{i1} \ldots Y_{ij_i}] \tag{22}$$

$$L = [L_1 L_2 \ldots L_I] \tag{23}$$

$$A = \begin{bmatrix} e_1 & & & \\ & e_2 & & \\ & & \ddots & \\ & & & e_I \end{bmatrix} \tag{24}$$

$$e = (1, 1, \ldots, 1), I \text{ components}$$

The linear program (12)-(15) may be more compactly written

$$\text{minimize } c\theta \tag{25}$$

subject to

$$L\theta + s = b \tag{26}$$

$$A\theta = e \tag{27}$$

$$\theta \geq 0, \; s \geq 0 \tag{28}$$

Since the set of all solutions to (27)-(28) is bounded, any solution may be written as a convex combination of extreme points, d_q:

$$\theta = \sum_q \lambda_q d_q, \qquad \lambda_q \geq 0, \qquad \sum_q \lambda_q = 1 \tag{29}$$

substituting (29) into (25)-(26) yields the following master program:

$$\text{minimize } \; z = \sum_q (cd_q)\lambda_q \tag{30}$$

subject to

$$\sum_q (Ld_q)\lambda_q + s = b \tag{31}$$

$$\sum_q \lambda_q = 1, \quad \lambda_q \geq 0, \; s \geq 0 \tag{32}$$

This program has T + 1 constraints and $\pi_i(j_i)$ columns, one for each exteme point of (27)-(28). To develop the subproblems, let (π, π_o) be a vector of simplex multipliers for (31)-(32). The relative cost coefficient for column q is

$$\bar{c}_q = cd_q - \pi L d_q - \pi_0 \tag{33}$$

To minimize $\bar{c}_q$, recall that d_q is an extreme point of the constraints

$$\left.\begin{aligned} e_i\theta_i &= 1 \\ \theta_i &\geq 0 \end{aligned}\right\} \quad i = 1, 2, \ldots, I \tag{34}$$

If d_q is partitioned as

$$d_q = (d_{q1}, \ldots, d_{qI}) \tag{35}$$

then it is easily verified that each d_{qi} has the form

$$d_{qi} = (0, 0, \ldots, 0, 1, 0, \ldots, 0) \tag{36}$$

The quantities cd_q and Ld_q may be similarly partitioned:

$$cd_q = \sum_{i=1}^{I} c_i d_{qi} \tag{37}$$

$$Ld_q = [L_1 \ldots L_I] \begin{bmatrix} d_{q1} \\ \vdots \\ d_{qI} \end{bmatrix} = \sum_{i=1}^{I} L_i d_{qi} \tag{38}$$

Thus the relative cost coefficient is

$$\bar{c}_q = \sum_{i=1}^{I} (c_i - \pi L_i) d_{qi} - \pi_0 \tag{39}$$

Since (39) is separable, minimizing $\bar{c}_q$ is equivalent to minimizing, for each i,

$$(c_i - \pi L_i) d_{qi} \tag{40}$$

over all d_{qi} of the form (36). The solution is obviously to put the "1" in the position where the minimal component of the vector $(c_i - \pi L_i)$ occurs. Using the definitions in (19), (10)-(11), and (22), this minimum is found by solving the subproblem

$$\text{minimize } (c_i(x_i^j) - \pi Y_i(x_i^j)) \tag{41}$$

Since $x_i^j \in S_i$, this is equivalent to

$$\text{minimize } c_i(x_i) - \pi Y_i(x_i) \tag{42}$$

subject to

$$x_i \in S_i \tag{43}$$

Note that this is a single item scheduling and inventory problem with a penalty term, πY_i, for using the common resource. The quantity $-\pi_t$ may be viewed as the price of this resource in period t.

Let x_i^0 solve subproblem i. By (39),

$$\min \bar{c}_q = \sum_{i=1}^{I} (c_i(x_i^0) - \pi Y_i(x_i^0)) - \pi_0 \tag{44}$$

Pricing out the slack columns, the relative cost coefficient is

$$\bar{c}_t = -\pi_t \tag{45}$$

If

$$\min \bar{c}_q < 0 \tag{46}$$

$$\min \bar{c}_q < \min \bar{c}_t \tag{47}$$

Then the column

$$\begin{bmatrix} Ld_q \\ 1 \end{bmatrix} = \begin{bmatrix} \sum_i Y_i(x_i^0) \\ 1 \end{bmatrix}$$

is formed and enters the current basis. If (47) does not hold, but $\min_t \bar{c}_t < 0$, then a slack column enters, while if all columns price out nonnegative the current basis is optimal.

1.2 Solution by Generalized Upper Bounding and Column Generation

The linear program in (12)-(15), section (1.1), may also be attacked directly by the simplex method, using column generation to ease the problem of many variables, and generalized upper boudning [4] (GUB) to reduce the effective number of rows. Let B be a basis matrix for (1.1-12) - (1.1-15) and let $\underline{\pi} = (\underline{\pi}_1, \underline{\pi}_2)$ be the set of simplex multipliers corresponding to this basis, with $\underline{\pi}_1$ associated with the constraints (1.1-13) and $\underline{\pi}_2$ with (1.1-14). Pricing out a non-basic column, the reduced cost coefficient is

$$\bar{c}_{ij} = c_{ij} - \underline{\pi}_2^T Y_{ij} - \pi_{1i} \tag{1}$$

To choose a column to enter the basis, we must find

$$\min_{i,j} \bar{c}_{ij} = \min_i \min_j \bar{c}_{ij} \tag{2}$$

The inner minimization in (2) may be accomplished by solving the subproblem

$$\underset{j}{\text{minimize}} [c_{ij} - \underline{\pi}_2^T Y_{ij}] \tag{3}$$

or, since all X_{ij} are elements of S_i

$$\text{minimize } c_i(X_i) - \underline{\pi}_2^T Y_i(X_i) \qquad (4)$$

$$\text{subject to } X_i \in S_i \qquad (5)$$

Note that these subproblems are single-activity problems with a penalty term $\underline{\pi}_2^T Y_i$ for use of the resource. Since S_i is assumed finite and the functions c_i and Y_i are assumed to be bounded below on S_i, the subproblems have finite solutions for all values of $\underline{\pi}_2$. To decide on which column is to enter the basis, subtract π_{1i} from the optimal objective value in (4)-(5) and find the minimum of these quantities over i.

The problem of many constraints still remains. However, due to the special form of the constraints (1.1-13), the problem can be solved by GUB, using a working basis whose dimension is the number of remaining constraints, (1.1-14), in our case (T X T). Since we are assuming that I is much greater than T, GUB greatly reduces the effort required to solve the problem.

1.3 Lot Size Problems with "Continuous" Set-Ups

A specialization of the model of the previous sections may be found in references [5] and [6]. This involves the production of I items over T time periods, with a specified number of machines available to produce these items. Demands for each item in each time period are assumed known. Costs are: (a) inventory holding and shortage costs and (b) setup costs. Setups occur when a given machine is changed from the production of one item to the production of another. We assume that any machine can, if suitably set up, produce any of the items, and that the setup cost is independent of which two items are involved in the changeover. A machine is set up to produce an item by installing in the machine a piece of equipment particular to the item. These pieces of equipment will be called dies. The problem is to allocate the machines to the items so that the sum of inventory and setup costs over all items is minimized.

This problem can be placed in the form of the resource allocation problem (1.1-5) - (1.1-7). Define

m_{it} = number of machines used to produce item i in time period t

M_i = $(m_{i1}, m_{i2}, \ldots, m_{iT})$

b_t = number of machines available in time period t

$\underline{b}$ = $(b_1, \ldots, b_T)$

d_{it} = demand for item i in time period t, assumed known

k_i = production rate of a machine producing item 1.

y_{it} = inventory of item i at the end of time period t.

= $y_{i,t-1} + k_i m_{it} - d_{it}$, $t=1,\ldots, T$, y_{io} given

p_s = cost of one setup

$(y_{it})_{max}, (y_{it})_{min}$ = given upper and lower boundson y_{it}

n_i = number of dies available for item i

Since the setup cost is assumed independent of the items involved, the total setup cost may be written

$$c_s = p_s \sum_{i=1}^{I} s_i(M_i)$$

where

$$s_i(M_i) = \sum_{t=1}^{T} (m_{it} - m_{i,t-1})_+, \; m_{io} \text{ given}$$

and where

$$(x)_+ = \begin{cases} x, & x \geq 0 \\ 0, & x < 0 \end{cases}$$

The inventory cost for item i is written

$$\gamma_i(M_i) = \sum_{t=1}^{T} \gamma_{it}(y_{it})$$

where γ_{it} represent holding costs for $y_{it} > 0$ and shortage costs for $y_{it} < 0$, and may have any convenient functional form. Let

$$c_i(M_i) = p_s \cdot s_i(M_i) + \gamma_i(M_i)$$

Then the problem is to choose $M_1 \ldots M_I$ to minimize

$$c = \sum_{j=1}^{I} c_j(M_j)$$

subject to

$$(y_{it})_{min} \leq y_{it} \leq (y_{it})_{max}, \text{ all } i, t \tag{1}$$

$$m_{it} \leq n_j, \text{ all } i, t \tag{2}$$

$$m_{it} = \text{nonnegative integer, all } i, t \tag{3}$$

and

$$\sum_{i=1}^{I} m_{it} \leq b_t, \; t = 1, \ldots, T \tag{4}$$

Let

$$S_j = \left\{ M_j \middle| M_j \text{ satisfies (1) - (3) for } i = j \right\}$$

since the n_i are finite, S_j is finite and is assumed nonempty. Let the elements of S_i be indexed by j, with M_{ij} the jth element. Defining

$$c_i(M_{ij}) \equiv c_{ij}$$

the linear program (1.1-12) - (1.1-15) - (2.11) assumes the form

$$\text{minimize} \quad \sum_{i,j} c_{ij}\, \theta_{ij} \tag{5}$$

subject to

$$\sum_{j} \theta_{ij} = 1, \qquad i = 1, \ldots, I \tag{6}$$

$$\sum_{i,j} M_{ij}\, \theta_{ij} \leq \underline{b} \tag{7}$$

and

$$\theta_{ij} \geq 0, \text{ all } i, j \tag{8}$$

The ith subproblem is:

$$\text{minimize} \quad p_s s_i(M_i) + \gamma_i(M_i) - \underline{\pi}_2^T M_i \tag{9}$$

subject to

$$(y_{it})_{min} \leq y_{it} \leq (y_{it})_{max}, \; t = 1, 2, \ldots, T \tag{10}$$

$$m_{it} \leq n_i \tag{11}$$

$$m_{it} = \text{nonnegative integer, all } t \tag{12}$$

Each subproblem may be solved by dynamic programming with two state variables, m_{it} and y_{it}. Since the initial states m_{i0} and y_{i0} are given, a forward recursion is used.

2. NONLINEAR RESOURCE ALLOCATION PROBLEMS AND LAGRANGIAN DECOMPOSITION

2.1 Decomposition Using a Pricing Mechanism

To introduce the basic ideas, consider the following simple example. Let there be n activities. Operating each activity incurs a cost and uses a certain amount of a single resource. Let

$c_i(x_i)$ = cost of operating activity i at level x_i

a_i = amount of resource used per unit of x_i, assumed constant

b = availability of resource

The problem of operating all activities at minimal cost while using no more of the resource than is available is

$$\text{minimize} \quad z = \sum_{i=1}^{n} c_i(x_i) \tag{1}$$

$$\text{subject to} \quad \sum_{i=1}^{n} a_i x_i \leq b, \qquad x_i \geq 0 \tag{2}$$

Let us consider a decentralized solution procedure for this problem, with a pricing system used as the coordinating mechanism. Let $u \geq 0$ be the price of the resource, with dimensions (units of z)/(unit of b). Given a value for u, a natural method of decomposition is to associate with each activity a "manager" who has full responsibility for choosing its level, x_i. The manager can purchase as much of the resource as he likes in order to run the activity but must pay for what he uses. A rational manager would choose that activity level which minimizes the total cost of operations, i.e. direct cost plus resource cost. This leads to the subproblem

$$\left.\begin{array}{ll} \text{minimize} & \tilde{f}_i(x_i,u) = c_i(x_i) + u(a_i x_i) \\ \text{subject to} & x_i \geq 0 \end{array}\right\} \text{subproblem } i \tag{3}$$

For fixed u, these subproblems are independent of one another. Let us (temporarily) assume that they have unique finite optimal solutions for any value of $u \geq 0$, denoted by $x_i(u)$. One feels that by increasing u, the managers can be induced to use less of the resource, and vice versa, so u provides some means of coordinating their actions. Natural questions are:

1. Does there exist a value of u, u^0 (called optimal), for which the subproblem solutions $\{x_1(u^0),\ldots,x_n(u^0)\}$ solve the original problem, (1)-(2)?
2. If the answer to question 1 is yes, how can optimality of $\{x_1(u^0),\ldots, x_n(u^0)\}$ be recognized?
3. How can an optimal value of u be computed, if it exists?

In the remainder of this section, partial answers to these questions will be given. The answers to 1 and 2 involve saddle points of Lagrangian functions, while the answer to 3 comes under the heading of duality. We note that the subproblems are being asked to provide the optimal solution directly. Thus the decision making may be said to be completely decentralized.

A clue to existence of optimal prices is provided by the Kuhn-Tucker theorem. Let the functions c_i be convex and differentiable. Form a Lagrangian function

$$L(x, u) = \sum_i c_i(x_i) + u\left(\sum_i a_i x_i - b\right) \tag{4}$$

$$L(x, u) = \sum_i \{c_i(x_i) + u a_i x_i\} - ub \tag{5}$$

Comparing (5) and (3), we see that the Lagrangian is additively separable and can be written

$$L(x,u) = \sum_i \tilde{f}_i(x_i,u) - ub \tag{6}$$

where $\tilde{f}_i$ is the subobjective function in (3). By the Kuhn-Tucker theorem, a vector $x^0 = (x_1^0, \ldots, x_n^0)$ solves (1)-(2) if and only if there is a multiplier $u^0 \geq 0$ such that x^0 satisfies

$$\frac{\partial L}{\partial x_i} = \frac{\partial \tilde{f}_i(x_i, u^0)}{\partial x_i} \geq 0, \qquad i = 1, \ldots, n \tag{7}$$

$$u^0 \left(\sum_i a_i x_i - b \right) = 0, \qquad \sum_i a_i x_i \leq b \tag{8}$$

$$x_i \frac{\partial \tilde{f}_i(x_i, u^0)}{\partial x_i} = 0, \qquad x_i \geq 0, \ i = 1, \ldots, n \tag{9}$$

Since $\tilde{f}_i$ is convex in x_i for any $u \geq 0$, (7) and (9) are necessary and sufficient that x_i^0 minimize $\tilde{f}_i(x_i, u^0)$ subject to $x_i \geq 0$, i.e., that x_i^0 solve the ith subproblem, (3), when $u = u^0$. Thus the Lagrange multiplier u^0 is the desired price. Conditions (8) say that u^0 is large enough to limit total resource consuption to the amount available and that, if optimal consumption is less than b units, the price u^0 is zero. Thus a resource in excess supply costs nothing.

2.2 Saddle Points of Lagrangian Functions

2.2.1 Basic Theorems. The previous discussion utilized the Kuhn-Tucker conditions to assert the existence of an optimal price. These are stationarity conditions, while the subproblems (3) are meaningfully stated only as minimizations. Stationarity and minimality are equivalent only for convex differentiable functions defined over convex sets. However, in many situations, these requirements of convexity and differentiability are not met, e.g., problems involving optimization over finite sets. Lagrange multipliers can still be useful in such problems. To see how, some theorems regarding saddle points of Lagrangian functions are needed.

$$\left.\begin{array}{ll} & \text{minimize } f(x) \qquad (1)\\ \text{subject to} & g_i(x) \leq 0, \quad i = 1, \ldots, m \qquad (2)\\ & x \in S \qquad (3)\end{array}\right\} \text{primal problem}$$

This is called the primal problem. The quantity x is an n vector, S in an arbitrary subset of E^n, and f and the g_i are arbitrary real-valued functions defined on S. The Lagrangian function associated with this problem is

$$L(x,u) = f(x) + \sum_{i=1}^{m} u_i g_i(x), \qquad u_i \geq 0 \tag{4}$$

DEFINITION. A point (x^0, u^0) with $u^0 \geq 0$ and $x^0 \in S$ is said to be a saddle point for L if it satifies

$$L(x^0, u^0) \leq L(x, u^0) \qquad \text{for all } x \in S \tag{5}$$

$$L(x^0, u^0) \geq L(x^0, u) \qquad \text{for all } u \geq 0 \tag{6}$$

The following theorem gives necessary and sufficient conditions for a saddle point of L.

THEOREM 1. Let $u^o \geq 0$ and $x^o \in S$. Then (x^o, u^o) is a saddle point for L if and only if

(a) x^o minimizes $L(x, u^o)$ over S (7)

(b) $g_i(x^o) \leq 0, \quad i = 1, \ldots, m$ (8)

(c) $u_i^o g_i(x^o) = 0, \quad i = 1, \ldots, m$ (9)

The usefulness of a saddle point is brought out in the following theorem.

THEOREM 2 (SUFFICIENCY OF SADDLE POINT). If (x^o, u^o) is a saddle point for L, then x^o solves the primal problem (1)-(3).

To relate Theorems 1 and 2 to our previous discussion regarding decomposition, let us apply these theorems to the allocation problem considered earlier. To place this in the general form of the primal problem (1)-(3), we let

$$S = \left\{ x \middle| x \geq 0 \right\} \tag{10}$$

Then the problem is rewritten

$$\text{minimize} \quad \sum_{i=1}^{n} c_i(x_i) \tag{11}$$

subject to

$$\sum_{i=1}^{n} a_i x_i - b \leq 0 \tag{12}$$

$$x \in S \tag{13}$$

The lagrangian function for this problem is

$$L(x,u) = \sum_{i=1}^{n} (c_i(x_i) + u a_i x_i) - ub \tag{14}$$

Condition (a) of Theorem 1 requires that $L(x, u^o)$ be minimized over S to obtain x^o. Since L is additively separable in the x_i for fixed u,

$$\min_{x_1 \geq 0, \ldots, x_n \geq 0} \sum_{i=1}^{n} (c_i(x_i) + u a_i x_i) = \sum_{i=1}^{n} \left\{ \min_{x_i \geq 0} (c_i(x_i) + u a_i x_i) \right\} \tag{15}$$

The quantity

$$\min_{x_i \geq 0} c_i(x_i) + u a_i x_i \tag{16}$$

is the optimal objective value for the ith subproblem (2.1-3). Thus these subproblems arise mathematically from trying to satisfy the first of the necessary and sufficient conditions for a saddle point. The second of these conditions [(b) of Theorem 1] says that the value of the optimal price, u^o, must be such that the subproblem solutions jointly satisfy the resource constraint. Condition (c) repeats our earlier remarks that a resource in excess supply has a price of zero.

The process of minimizing the Lagrangian $L(x,u)$ over $x \in S$ for fixed u leads to a set of independent subproblems whenever L is additively separable in x for fixed u and when S can be written as a Cartesian product. In the primal problem (1)-(3), let x be partitioned as

$$x = (x_1 \mid x_2 \mid \dots \mid x_p), \quad p \leq n \tag{17}$$

and assume that f, S, and the vector of constraint functions g may be written

$$f(x) = \sum_{i=1}^{p} f_i(x_i) \tag{18}$$

$$g(x) = \sum_{i=1}^{p} g^i(x_i) \tag{19}$$

$$S = S_i \times S_2 \times \dots \times S_p \tag{20}$$

where $\times$ denotes Cartesian product and each g^i is an m vector of functions. This is the important special case where the objective f and constraint functions g_i are additively separable, and when the constraints determining S consist of subsets of contraints involving the subvectors x_i separately. Then

$$L(x,u) = \sum_i f_i(x_i) + u \sum_i g^i(x_i) \tag{21}$$

$$L(x,u) = \sum_i (f_i(x_i) + ug^i(x_i)) \tag{22}$$

so the Lagrangian is additively separable in the x_i and

$$\min_{x_1 \in S_1, \dots, x_p \in S_p} L(x,u) = \sum_i \min_{x_i \in S_i} (f_i(x_i) + ug^i(x_i)) \tag{23}$$

Thus the Lagrangian can be minimized by solving the p independent subproblems

$$\left.\begin{array}{ll} \text{minimize} & f_i(x_i) + ug^i(x_i) \\ \text{subject to} & x_i \in S_i \end{array}\right\} \text{subproblem } i \tag{24}$$

These subproblems have the same pricing interpretation as in our earlier example, with u_k the price of the "resource" whose limited availability is represented by the kth constraint.

In choosing which constraints of the primal will be used to determine S and which written as $g_i \leq 0$, efforts should be made to achieve separability of L and S. That is, the choice should be made so that, if possible conditions (18)-(20) hold for some partition of $x, (x_1 \; x_2 \; \dots \; x_p)$.

All the previous results are easily extended to include equality constraints. If the kth constraint is $g_k(x) = 0$, then the multiplier u_k for this constraint may have any sign and condition (c) of Theorem 1 is superfluous, since $g_k(x^0) = 0$.

2.2.2 Everett's Theorem. Everett [7] has shown that if any set of multipliers $u \geq 0$ is chosen for which $\min_{x \in S} L(x,u)$ exists, any vector $x(u)$ which minimizes $L(x,u)$ solves a problem closely related to the primal problem. The result is as follows:

THEOREM 3 (EVERETT). If $x(u)$ solves the Lagrangian problem

$$\left.\begin{array}{ll} & \text{minimize } L(x,u) \\ \text{subject to} & x \in S \end{array}\right\} \text{Lagrangian problem} \tag{25}$$

with $u \geq 0$, then $x(u)$ solves the modified primal problem

$$\text{minimize } f(x) \tag{26}$$

$$\text{subject to} \quad g_i(x) \leq y_i, \quad i = 1, \dots, m \qquad \text{modified} \tag{27}$$

$$x \in S \qquad \text{primal}$$

$$\text{where} \quad \begin{array}{ll} y_i = g_i(x(u)) & \text{if } u_i > 0 \\ y_i \geq g_i(x(u)) & \text{if } u_i = 0 \end{array} \tag{28}$$

The modified primal has right-hand-side constraint constants either = or $\geq$ to $g_i(x(u))$. Let the constraints $g_i \leq 0$ be regarded as expressing resource limitations and the u_i be viewed as prices for these resources. Then the theorem says that any vector which minimizes $L(x,u)$ solves a primal problem which uses no more of the valuable resources than the vector itself, while availability of the free resources is at least as much as is used by the vector itself. The multipliers u thus convert a constrained problem to an unconstrained (except for the restriction $x \in S$) Lagrangian problem. As mentioned earlier, this is especially attractive if f, the g_i, and S have the separable forms in (18)-(20). Then the Lagrangian problem decomposes into the smaller independent subproblems (24). Note that no restrictions are posed on the functions f, g_i, or on the set S.

The proof of Theorem 3 follows immediately from the observation that, by Theorem 1, $(x(u),u)$ is a saddle point for the Lagrangian

$$L = f(x) + \sum_{i=1}^{m} u_i(g_i(x) - y_i) \tag{29}$$

One uses this theorem by choosing multipliers $u \geq 0$, solving the Lagrangian problem, and observing for what constraint constants y_i the primal problem has been solved. This procedure may be valuable if a family of primal problems is of interest, with different right-hand sides. Of course, not all right-hand-side vectors can be generated by this procedure. Sets of vectors y whose associated problems cannot be solved are termed "gaps" in [7]. A geometric characterization of these gaps is given in Section 2.3.

Existence of Saddle Points. Our questions regarding existence of optimal multipliers and of recognizing such multipliers if they exist are partially answered by the preceding results. Any vector u^0 which is the u component of a saddle point of L will serve as the desired multipliers, and conditions (a)-(c) of Theorem 1 show how such a point may be recognized. The following theorem deals with existence of a saddle point for problems with convex objective and constraint functions.

THEOREM 4 (KARLIN [8]). Let S be a convex subset of E^n, f a convex function defined on S, and $g(x) = (g_1(x), \ldots, g_m(x))$ a vector of convex functions defined on S. Assume that there exists a point $x \in S$ such that $g(x) > 0$. If x^0 is a point at which f(x) assumes its minimum subject to $g(x) \geq 0$, $x \in S$, then there is a vector of multipliers $u^0 \geq 0$ such that such that (x^0, u^0) is a constrained saddle point of

$$L(x,u) = f(x) + ug(x)$$

Conversely, if (x^0, u^0) is a constrained saddle point of L(x, u), then x^0 minimizes f(x) subject to $g(x) \leq 0$, $x \in S$.

2.3 Minimax Dual Problem

Although a number of results regarding existence of saddle points have been proved, nothing has yet been said as to how to find such a point. To answer this question, the behavior of the minimum value of L(x, u) must be studied as a function of u. That is, instead of focusing on a single set of Lagrange multipliers, u^0, we consider all $u \geq 0$.
Let

$$h(u) = \min_{x \in S} L(x,u) \tag{1}$$

$$X(u) = \{x \mid x \text{ minimizes } L(x, u) \text{ over } S\} \tag{2}$$

The function h is called the dual function. Its domain of definition is

$$D = \left\{u \mid u \geq 0, \min_{x \in S} L(x,u) \text{ exists}\right\} \tag{3}$$

That is, D is the set of nonnegative vectors u for which L(x,u) has a finite infimum over S which is attained at some point in S. If, for all $u > 0$, L(x,u) is continuous in x for $x \in S$ and if S is closed and bounded then, by the Weierstrass theorem [9],

$$D = (E^m)^+ = \{u \mid u \geq 0\} \tag{4}$$

However, D need not be convex and may be empty.

A primal dual pair of problems is now defined:

Primal

$$\text{minimize } f(x)$$

subject to (5)

$$g_i(x) \leq 0, \quad i = 1,\ldots,m$$

$$x \in S$$

Dual

$$\text{maximize } h(u)$$

subject to (6)

$$u \in D$$

We call this dual problem the minimax dual, since it may be phrased

$$\underset{u \in D}{\text{maximize}} \; \min_{x \in S} L(x,u) \tag{7}$$

Such dual problems have been studied by Falk [10], Mangasarian [11], Stoer [12], Rockafellar [13], and others. It is easily shown that if the primal is a linear program then (6) is its dual. Further, if $S = E^n$ and f and the g_i are convex and differentiable, the dual (6) is equivalent to the dual program of Wolfe [11].

We now give some important properties of the dual function h. Proofs and examples may be found in [15].

THEOREM 5

$$h(u) \leq f(x) \begin{cases} \text{for all } x \text{ satisfying the constraints of (5)} \\ \text{for all } u \in D \end{cases}$$

COROLLARY. If there exist primal feasible x^o and dual feasible u^o such that $f(x^o) = h(u^o)$, then x^o solves the primal and u^o solves the dual.

THEOREM 6. The dual function h is concave over any convex subset of its domain, D.

THEOREM 7. (x^o, u^o) is a saddle point for L if and only if x^o is primal feasible, u^o is dual feasible, and $f(x^o) = h(u^o)$.

Consider again primal problems for which L is separable in x, thus causing the Lagrangian problem to decompose into subproblems. The preceding results imply that one way to coordinate these subproblems so that their combined solutions solve the primal is to choose the multipliers, u, to solve the dual. The coordinator thus has the task of maximizing h subject to $u \in D$. Solution proceeds as outlined below.

1. Choose initial values $u \in D$ and solve the Lagrangian problem, (2.2-25). In the separable case this may be done by solving the subproblems, (2.2-24).
2. Evaluate the dual function h(u) and choose new multipliers so that it is increased.
3. Return to step 1, stopping when h is maximized.

If, upon termination of this procedure, u^o solves the dual, some $x^o \in X(u^o)$ is primal feasible, and $h(u^o) = f(x^o)$, then x^o solves the primal. How step 2 might be accomplished will be discussed shortly.

Geometric Interpretation   Added insight is obtained from the following geometric interpretation [16,17]. Imbed the primal problem in a family of perturbed problems, with perturbations y_i:

$$\text{minimize } f(x) \tag{8}$$

subject to

$$g_i(x) \leq y_i, \quad i = 1, \ldots, m \tag{9}$$

$$x \in S \tag{10}$$

That is, consider a family of problems with variable right-hand sides, $y = (y_1, \ldots, y_m)$. The primal problem corresponds to $y = 0$. Let

$$w(y) = \min \left\{ f(x) \middle| g(x) \leq y,\ x \in S \right\} \tag{11}$$

The function w may not exist for all y, owing either to the program (8)-(10) being infeasible or to the fact that the infimum of f subject to $g(x) \leq y$, $x \in S$, is not attained. To avoid this difficulty, assume that S is closed and bounded and that f and the g_i are continuous on S. Then the domain of w is

$$F = \left\{ y \,\middle|\, \text{there exists } x \in S \text{ such that } g(x) \leq y \right\} \tag{12}$$

The function w is nonincreasing over F. If f, g, and S are convex, then F can be shown to be a convex set and w is a convex function over F. Consider the set of points, R, in E^{m+1} on and above the graph of $w(y)$:

$$R = \left\{ (y_0, y) \middle| y \in F,\ y_0 \geq w(y) \right\} \tag{13}$$

If w and F are convex, this set is convex and might appear as shown in Figure 1. There is an intimate connection between duality and supporting hyperplanes for this set, as is shown by the following theorem.

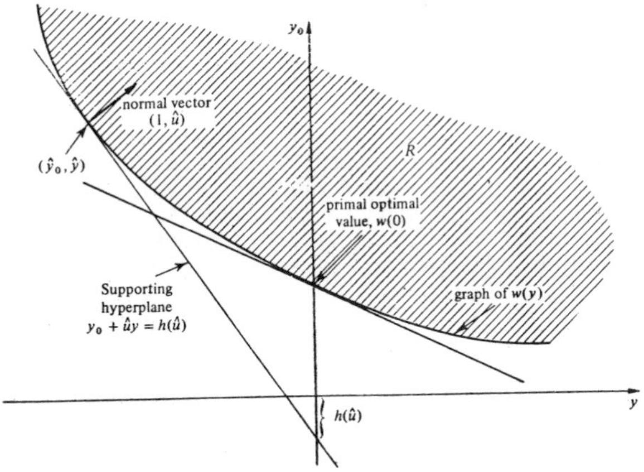

FIGURE 1   Supporting hyperplane for the set R.

THEOREM 8. Let $\hat{x} \in S$ and $\hat{u} \geq 0$. Then $\hat{x}$ minimizes $L(x,\hat{u})$ over S if and only if

$$H = \left\{(y_0, y) \middle| y_0 + \hat{u}y = L(\hat{x}, \hat{u})\right\} \tag{14}$$

is a supporting hyperplane for R at $(f(\hat{x}), g(\hat{x}))$.

According to the above result, the process of minimizing $L(x,\hat{u})$ over S is equivalent to finding a real number

$$h(\hat{u}) = \min_{x \in S} L(x,\hat{u}) = L(\hat{x},\hat{u})$$

such that the hyperplane H in (14) is a supporting hyperplane to the set R. The value of the dual function, $h(\hat{u})$, is the intercept of this support plane with the y_0 axis, and the slopes of the plane are the numbers $-\hat{u}$. The dual problem is that of finding the support plane with nonpositive slopes having maximal y_0 intercept, while the primal is that of finding a point $(y_0, 0)$ in R with minimal y_0 intercept, the number $w(0)$.

Since w is nonincreasing, all support planes to R have slopes $-u \leq 0$. It is evident from Figure 1 that any such support planes has y_0 intercept less than or equal to the primal optimum, $w(0)$. This statement corresponds to the inequality $h(u) \leq \min f$. If R is closed and convex, supporting hyperplanes exist at all boundary points, in particular at $(w(0),0)$. The intercept of this plane is, of course, equal to $w(0)$, yielding the relation $\min f = \max h$.

The geometric viewpoint also shows in what situations the Lagrangian approach can succeed. Consider a primal problem with optimal objective value $w(0) = f(x^0)$. The statement below follows directly from Theorem 8.

COROLLARY. Let x^0 solve the primal problem. There exist multipliers $u^0 \geq 0$ such that x^0 minimizes $L(x,u^0)$ over S if and only if the set R in (13) has a supporting hyperplane at the point $(f(x^0), g(x^0))$ with slopes $-u^0$.

This corollary applies to primal problems with arbitrary right-hand-side vector y. Thus we have a characterization of the "gaps" discussed in the previous section. Any vector y for which R has a supporting hyperplane at some point $(w(y),v)$ with $v \leq y$ can be solved by solving the Lagrangian problem for some $u \geq 0$. These are the vectors y for which $(w(y),v)$ lies in the intersection of the boundary of the convex hull of R with the boundary of R. Vectors y not having this property constitute the gaps. In Figures 2 and 3 the sets R are not convex, implying that the primal is not a convex program. Despite this, in Figure 2, R has a support plane at $(w(0),0)$, while there is no support plane at this point in Figure 3. There are a number of gaps, one of which is indicated in Figure 2.

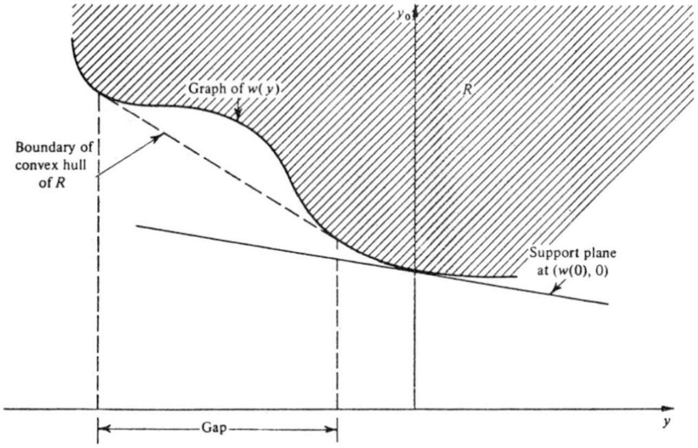

FIGURE 2 Support plane at optimal point.

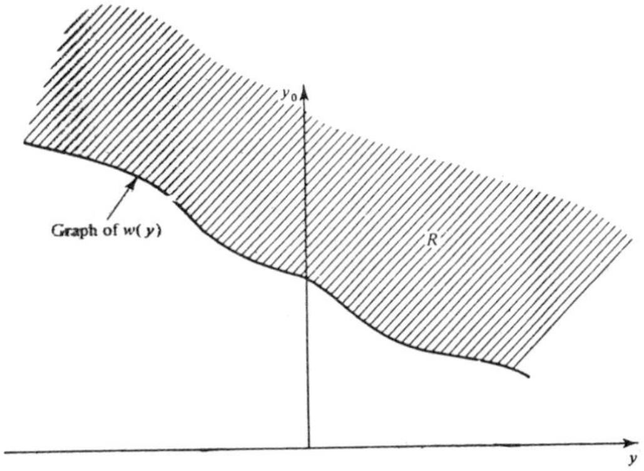

FIGURE 3 No support plane at optimal point.

2.4 Differentiability of the Dual Objective Function[1]

In constructing an algorithm to solve the dual problem, information regarding derivatives of the dual objective function, h, is essential. For example, when the feasible set of the dual, D, is $(E^m)^+$ and h is differentiable a gradient ascent method might well be used. A particularly nice property of the dual problem is that the partial derivatives $\partial h/\partial u_i$ are quite easy to obtain, when they exist. However, h is not, in general, differentiable at all points. Since h is concave over convex subsets of D, it is differentiable almost everywhere in the interior of such subsets [18] and possesses (one-sided) directional derivatives in all directions at all points in such subsets.

By analogy with the definition for convex functions, we define a subgradient of a concave function h at u as the vector of slopes of a supporting hyperplane to the graph of h at u. The set of all subgradients of h at u is $\partial h(u)$. Thus we have the following definition.

DEFINITION 1. x^* is a subgradient of h at u if

$$h(z) \leq h(u) + x^*(z - u) \qquad \text{for all } z \in E^m$$

The following shows that evaluation of $h(u)$ immediately yields an element of $\partial h(u)$.

THEOREM 9

$$g(x) \in \partial h(u) \qquad \text{for any } x \in X(u)$$

PROOF. By definition

$$\begin{aligned} h(z) &\leq f(x) + zg(x) && \text{for all } x \in S \\ h(u) &= f(x^0) + ug(x^0) && \text{for all } x^0 \in X(u) \end{aligned} \qquad (1)$$

Setting $x = x^0$ in (1) and subtracting yields

$$h(z) \leq h(u) + g(x^0)(z - u) \qquad \text{for all } z \in E^m$$

If h has only one subgradient at a point u^*, then that subgradient is its gradient and h is differentiable at u^*. Necessary and sufficient conditions for this to occur follow.

THEOREM 10. Assume S is a closed and bounded subset of E^n and f and all g_i are continuous on S. If $L(x, u^*)$ is minimized over S at a unique point $x(u^*)$, then h is differentiable at u^*, with partial derivations

$$\left.\frac{\partial h}{\partial u_i}\right|_{u = u^*} = g_i(x(u^*))$$

2.5 Computational Methods for Solving the Dual

Gradient Algorithms. Assuming, for the moment, that h is differentiable, a steepest-ascent algorithm, modified to handle the constraints $u \geq 0$ may be used to maximize h. This leads to a solution procedure which is a more complete version of that specified in Section 2.3:

1. Choose initial values $u^0 \geq 0$. Step i, $i = 0,1,2,\ldots$, proceeds as follows.
2. Solve the Lagrangian problem with $u = u^i$, obtaining a solution $x(u^i)$. In the separable case, this may be accomplished by solving the subproblems, (2.2-24).

3. Form the dual function $h(u^i) = L(x(u^i),u^i)$ and its gradient $\nabla h(u^i) = g(x(u^i))$.
4. Define a direction of search, s^i, by

$$s_k^i = \begin{cases} \left.\dfrac{\partial h}{\partial u_k}\right| u^i & \text{if } u_k^i > 0 \quad (1) \\ \max\left\{0, \left.\dfrac{\partial h}{\partial u_k}\right| u^i\right\} & \text{if } u_k^i = 0 \quad (2) \end{cases} \quad k = 1,\ldots,m$$

Choose a new vector u^{i+1} by

$$u^{i+1} = u^i + \alpha_i s^i \tag{3}$$

The step size α_i must be selected so that

$$h(u^{i+1}) > h(u^i) \tag{4}$$

If h is differentiable there exists $\alpha_i > 0$ satisfying the above unless u^i maximizes h. A common procedure is to choose α_i to maximize

$$r(\alpha) = h(u^i + \alpha s^i) \tag{5}$$

subject to the constraints $\alpha_i \geq 0$ and $u^{i+1} \geq 0$.

5. Return to step (2), stopping when $\alpha_i \approx 0$.

If $r(\alpha)$ is to be maximized, it must be evaluated a number of times, requiring a solution of the Lagrangian problem each time. Step 4 is simply Rosen's gradient projection method for linear constraints [19], specialized to the constraints $u \geq 0$. Convergence of this algorithm to a global solution has been proved in Part II of [19] for concave differentiable objectives.

This gradient procedure may be viewed as a coordination algorithm for a second-level coordinator, whose task is to solve the dual problem given values of h and ∇h. The first-level units solve the subproblems and provide these values. The algorithm has an interesting economic interpretation if the primal problem is viewed as one of minimal-cost resource allocation. Let the primal be

$$\text{minimize} \sum_j f_j(x_j) \tag{6}$$

subject to

$$\sum_j g_{ij}(x_j) \leq b_i, \qquad i = 1,\ldots,m \tag{7}$$

$$x_j \in S_j \tag{8}$$

where $f_j(x_j)$ is the cost of running activity j at levels x_j, $g_{ij}(x_j)$ is the quantity of resource i used by activity j at levels x_j, b_i is the amount of resource i available, and S_j is determined by technological constraints on each activity. Let $u^* > 0$ be viewed as a vector of prices for the resources, and assume that the Lagrangian problem has a unique solution when $u = u^*$,

$x(u^*)$. The subproblems are

$$\text{minimize } \tilde{f}_i(x_i,u) = f_i(x_i) + \sum_k u_k^* g_{ki}(x_i) \tag{9}$$

subject to $$x_i \in S_i \tag{10}$$

The subproblem objective represents the direct cost, f_i, plus the cost of resources used. The dual function is the sum of subproblem minima:

$$h(u^*) = \sum_i \tilde{f}_i(x_i(u^*),u^*) \tag{11}$$

The gradient of h at u* has components

$$\left.\frac{\partial h}{\partial u_i}\right|_{u^*} = \sum_j g_{ij}(x_j(u^*)) - b_i = e_i(u^*) \tag{12}$$

The summation term may be viewed as the total demand for resource i, while b_i is its supply. Thus $e_i(u^*)$ is the excess demand for this resource. By (1)-(3), the gradient algorithm requires that u_i^* be increased if excess demand is positive and decreased otherwise, unless $u_i^* = 0$. This is a familiar price-adjustment rule of classical economics. Exposing it as a gradient-ascent algorithm in this instance permits its convergence to be analyzed using established results.

Although the steepest-ascent approach is theoretically interesting, other more recent methods are recommended for computational purposes. If all constraints are equalities, implying that the multipliers u are unconstrained in sign, the variable metric method described by Fletcher and Powell [20], or the conjugate gradient method of Fletcher and Reeves [21] are recommended (see Section 1.1). A modification of the Fletcher-Powell method to handle linear constraints is described in [22] and is recommended when any constraints $u_i \geq 0$ occur. All these procedures require only function and gradient values, yet generally converge in far fewer iterations than the method of steepest ascent.

Methods for solving the dual which do not require that h be differentiable are found in reference [15], pp. 430-434.

2.6 Problems in which the Constraint Set is Finite: Multi-item Scheduling Problems. For separable problems involving optimization over a finite (but large) set, decomposition is attractive, since discrete optimization with only a few variables is easily accomplished by dynamic programming or some other combinatorial technique. An example of such a problem is the multi-item scheduling problem of Section 1.3. There, assuming that the number of items is much larger than the number of time periods, an approximate solution was found using linear programming. The algorithm involved solving a set of single-item scheduling problems, using prices provided by a "master" linear program. Here, using the minimax dual of nonlinear programming, the problem is attacked directly. The same subproblems are solved, but the prices are provided by maximizing the dual objective. Any solutions obtained are integral, independent of the number of items or time periods. However, optimal (or even feasible) solutions cannot always be obtained.

The Lagrangian function of the problem is

$$L(m,u) = \sum_{i=1}^{I} c_i(m_i) + \sum_{t=1}^{T} u_t \left(\sum_{i=1}^{I} m_{it} - b_t\right), \quad u_t \geq 0 \tag{1}$$

$$L(m,u) = \sum_{i=1}^{I} \left(c_i(m_i) + \sum_{t=1}^{T} u_t m_{it}\right) - \sum_{t=1}^{T} u_t b_t \tag{2}$$

The dual problem is

$$\underset{u \geq 0}{\text{maximize}} \quad \underset{m_1 \in S_1 \ldots m_I \in S_I}{\min} \quad L(m,u) \tag{3}$$

Again, since the Lagrangian is separable, the inner minimization yields the subproblems

$$\text{minimize} \left\{c_i(m_i) + u m_i\right\} \tag{4}$$

subject to $m_i \in S_i$ (5)

and the dual problem is

$$\underset{u \geq 0}{\text{maximize}}\ h(u)$$

where

$$h(u) = \sum_{i=1}^{I} h_i(u) - \sum_{t=1}^{T} u_t b_t \tag{6}$$

and where $h_i(u)$ is the minimum value attained in (4). The partial derivatives of the dual function (when they exist) are

$$\frac{\partial h}{\partial u_t} = \sum_{i=1}^{I} m_{it}(u) - b_t \tag{7}$$

Solution proceeds by choosing some initial $u_o \geq 0$ ($u_o = 0$ is a good choice) solving the subproblems for this u, evaluating the dual function and its gradient and, for example, changing u_o by moving in the gradient direction, thus increasing h. Note that the multipliers u_t appear in (4) as costs for the m_{it}. The gradient algorithm would increase the "cost" u_t if (7) were positive, i.e., if the number of machines used by all items in time period t were greater than b_t. Thus the subproblems are each penalized for collectively using more machines than are available.

Note that, since the sets S_i are finite, the minima in (4) exist, and thus the dual function exists (and is concave) for all $u \geq 0$. Additional properties of h follow from the finiteness of the sets S_i. Given any $u_o \geq 0$, we can imagine that the elements m_i of S_i are ordered along a line according to the value of the subproblem objective in (4) as shown in Figure 5. The lowest point represents the minimizing solution for $u = u_o$, and we

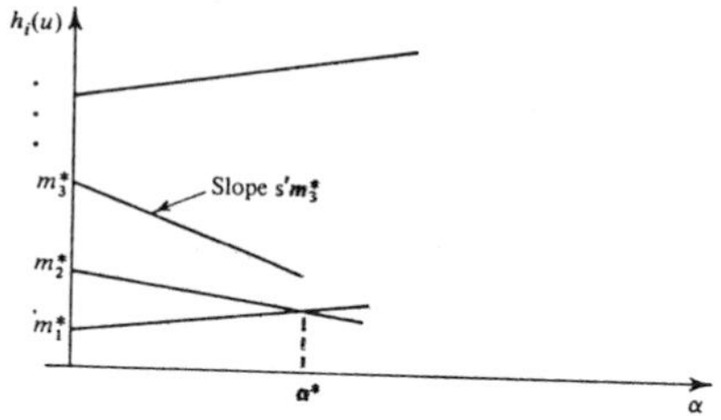

FIGURE 5

assume that each optimal solution, $m_i^*(u_o)$ is unique, implying that h is differentiable at u_o. Consider changing u_o by moving along a direction s; i.e.,

$$u = u_o + \alpha s, \qquad \alpha \geq 0 \tag{64}$$

As α increases, the function in (4) for each fixed m_i^* is given by

$$h_i^*(\alpha) = (c_i(m_i^*) + \alpha(s'm_i^*) \tag{65}$$

which is linear in α for fixed s, m_i^*. Since each of the h_i^* changes as shown in Figure 5, the optimal solution does not change until some point $a^* > 0$. Thus the subproblem objective function, $h_i(u(\alpha)$, is linear in α for $0 \leq \alpha \leq \alpha^*$ for any direction s;i.e., the dual function is linear as one travels along any direction in u space until one has traveled some positive distance, whereupon the optimal solution of some subproblem changes. This means the dual function is formed by sections of intersecting hyperplanes. One strategy for maximizing such a function is to follow the gradient along a given hyperplane until another hyperplane is reached, recompute the gradient, etc. (clearly the gradient exists except at the intersection of two hyperplanes, where it is discontinuous). The tangential approximation approach is perhaps even more appealing, since it does not depend on h being differentiable and since here h is piecewise linear. One might suspect that tangential approximation will find the dual optimum in a finite number of steps, although we have no proof of this.

Examples. There are no theoretical guarantees that discrete problems of the type considered here can be solved using duality. Thus solution of a number of sample problems was attempted to help assess the usefulness of the approach. Data for these problems is given in reference [15].

If each $c_i(m_i)$ is minimized separately subject to $m_i \in S_i$, and if the resultant solutions satisfy (1.3-4), these solutions are optimal for (1.3-1)-(1.3-4). The aggregate machine usage, $\sum_i m_{it}$, for these separate solutions is

Period	1	2	3	4	5	6
Total usage	27	28	31	33	36	28

The number of machines available in time period t, b_t, was chosen equal for all t(b_t = b for all t) and was varied to yield four different problems:

Problem	1	2	3	4
Machines available, b	35	34	33	32

Problem 1 is just barely constrained, as the total number of machines used by the unconstrained solutions is greater than 35 only in period 5. As b is decreased, the problem becomes more tightly constrained until eventually it becomes infeasible.

All problems were solved using a simple steepest ascent on h, modified to account for the constraints $u \geq 0$. Solution data are given in Tables 1-4. Optimal solutions were obtained for problems 1 and 3, as indicated by satisfaction of the machine availability constraints and complementary slackness conditions. With problems 2 and 4, the constraints could not be met in all time periods, but infeasibility occurs in only one time period, and its amount is small. Thus the schedules obtained are still useful.

TABLE 1 OPTIMAL SOLUTIONS

Problem	Dual optimum	Primal optimum (if attained)
1	2,877.40	2,877.40
2	2,885.14	—
3	2,910.88	2,910.88
4	2,947.60	—

TABLE 2 OPTIMAL MULTIPLIERS

Problem	u_1	u_2	u_3	u_4	u_5	u_6
1	0	0	0	0	6	0
2	0	0	0	1.2495	7.87495	0
3	0	0	0	3	9	0
4	0	0	0.045	6.78	1.3275	0

TABLE 3 MACHINE SLACK AT DUAL OPTIMUM: $\sum_i m_{it} - b$

	Time					
Problem	1	2	3	4	5	6
1	−8	−6	−4	−2	0	−7
2	−8	−6	−2	+2[a]	−2	−5
3	−6	−4	−2	0	0	−3
4	−4	−3	−1	0	+1[a]	−2

[a] Plus sign indicates constraint violation.

TABLE 4 ITERATION-BY-ITERATION SUMMARY FOR $b = 33$

	Multipliers						Machine slack						
Iteration	u_1	u_2	u_3	u_4	u_5	u_6	1	2	3	4	5	6	Dual objectives
1	0	0	0	0	0	0	−6	−5	−2	0	3	−5	2,867.14
2	0	0	0	0	3	0	−6	−5	−2	0	3	−5	2,885.14
3	0	0	0	0	6	0	−6	−4	−2	0	2	−5	2,901.40
4	0	0	0	0	8	0	−6	−4	−2	3	−1	−5	2,901.90
5	0	0	0	3	7	0	−6	−4	−2	0	2	−5	2,905.40
6	0	0	0	3	9	0	−6	−4	−2	0	0	−3	2,910.88

REFERENCES

1. Manne, A.S., "Programming of Economic Lot Sizes," Management Science, Vol. 4, (1958), pp. 115-135.

2. Dzielinski, B.P., and Gomory, R.E., "Optimal Programming of Lot Sizes, Inventory and Labor Allocations," Management Science, Vol. 11, No. 9, (1965), pp. 874-890.

3. L. S. Lasdon, "Optimization Theory for Large Systems," Macmillan, 1970, Chapter 3.

4. Ibid, Section 6.4.

5. Ibid, Section 4.2.

6. L. S. Lasdon and R.C. Terjung, "An Efficient Algorithm for Multi-Item Scheduling," Operations Research, May-June, 1971.

7. H. Everett, "Generalized Lagrange Multiplier Method for Solving Problems of Optimum Allocation of Resources," Operations Res., 11, 1963, pp. 399-417.

8. S. Karlin, Mathematical Methods and Theory in Games, Programming, and Economics, Vol. 1, Addison-Wesley Publishing Company, Inc., Reading, Mass., 1964.

9. G. Hadley, Nonlinear and Dynamic Programming, Addison-Wesley Publishing Company, Inc., Reading, Mass., 1964.

10. J. E. Falk, "Lagrange Multipliers and Nonlinear Programming," J. Math. Anal. Appl., 19, No. 1, 1967.

11. O. L. Mangasarian and J. Ponstein, "Minimax and Duality in Nonlinear Programming," J. Math. Anal. Appl., 11, 1965. pp. 504-518.

12. J. Stoer, "Duality in Nonlinear Programming and the Minimax Theorem," Numerische Mathematik, 5, 1963, pp. 371-379.

13. R. T. Rockafellar, "Duality and Stability in Extremum Problems Involving Convex Functions," Pacific J. Math., 21, 1967, pp. 167-187.

14. P. Wolfe, "A Duality Theorem for Nonlinear Programming," Quart.Appl. Math., 19, 1961, pp. 239-244.

15. L. S. Lasdon, "Optimization Theory for Large Systems," Macmillan, 1970, Chapter 8.

16. R. T. Rockafellar, "Nonlinear Programming," American Mathematical Society Summer Seminar on the Mathematics of the Decision Sciences, Stanford University, July-Aug. 1967.

17. D.G. Luenberger, "Convex Programming and Duality in Normed Space," Proceedings of the IEEE Systems Science and Cybernetics Conference, Boston, Oct. 11-13, 1967.

18. W. Fenchel, "Convex Cones, Sets, and Functions," mimeographed notes, Princeton University, 1951.

19. J. B. Rosen, "The Gradient Projection Method for Nonlinear Programming," Parts I and II, J. Soc. Ind. Appl. Math., 8, 1960, pp. 181-217;9,

1961, pp. 514-532.

20. R. Fletcher and M. J. D. Powell, "A Rapidly Convergent Descent Method for Minimization," Computer J., July 1963.

21. R. Fletcher and C. M. Reeves, "Function Minimization by Conjugate Gradients," Computer J., July 1964.

22. D. Goldfarb, "Extension of Davidon's Variable Metric Method to Maximization Under Linear Equality and Inequality Constraints", J.SIAM Appl. Math., 17, No. 4, 1969, pp. 739-764.

DECOMPOSITION OF A SHIP ROUTING PROBLEM

LEON S. LASDON
Department of Operations Research
Case Western Reserve University
Cleveland, Ohio 44106

Abstract: This paper describes a ship routing model used in applications on the Great Lakes. Some computational problems are discussed. Application of the Dantzig-Wolfe decomposition method is proposed to alleviate these problems.

1. PROBLEM FORMULATION

In this paper, we review a linear programming formulation of a ship routing problem, discuss some computational difficulties encountered in solving it, and consider the Dantzig-Wolfe decomposition principle, Lasdon (1970), as a means of easing these difficulties.

The formulation, originally presented in, Laderman (1966), deals with sets of supply and demand ports between which known quantities of goods are shipped. It is desired to route a given fleet of ships so that all goods are delivered, without using more than specified amounts of available time for each ship. The data of the problem are:

a_{ij} = amount to be shipped from origin i to destination j (tons), defined for (i,j) contained in a set S

t_k = available time for ship k (hrs)

e^k_{ij} = time for ship k to travel between i and j empty (hrs)

f^k_{ij} = time for ship k to travel between i and j full (including loading and unloading time) (hrs)

v^k_{ij} = capacity of ship k in going from i and j (tons)

For each pair $(i,j)\varepsilon S$, a unique commodity is assumed, although different a_{ij} may refer to different commodities. If there are $k\geq 2$ commodities to be shipped between some pair (i,j) then i may be broken into k pseudoports, $i_1,\ldots,i_k$, with each arc (i_m,j) associated with a single commodity. A similar comment applies if a port is an origin for one commodity and a destination for another. Then one pseudoport is an origin, the other is a destination, and travel time between these two pseudoports reflects average delay between unloading and loading. Of course, using these pseudoports increases the number of equations and variables in the model. One could eliminate pseudoports by introducing a commodity index. This would yield a smaller LP but could complicate the processing of the input and output data.

The decision variables are (all $\geq$ 0)

x^k_{ij} = number of full trips ship k makes from origin i to destination j

y^k_{ij} = number of empty trips ship k makes from j to i

The first set of constraints says that no more time must be used on each ship than is available. If A_k is the set of allowable* origin-destination pairs for ship k

*Port depth restrictions, among other conditions, may make it infeasible for a given ship to enter certain ports.

then these constraints are:

$$\sum_{(i,j)\varepsilon A_k} f_{ij}^k x_{ij}^k + \sum_{(i,j)\varepsilon A_k} e_{ij}^k y_{ij}^k + z_k = t_k \quad ,k=1,\ldots,K \tag{1}$$

where the z_k are slack variables. The next set of constraints specifies that all quantities a_{ij} shall be delivered. These are

$$\sum_k v_{ij}^k x_{ij}^k \geq a_{ij} \quad ,(i,j)\varepsilon S \tag{2}$$

The final set of constraints insures that each time a ship leaves an origin it returns to it, and each time it enters a destination it leaves it:

$$\sum_j x_{ij}^k - \sum_j y_{ij}^k = 0 \tag{3}$$

$$-\sum_i x_{ij}^k + \sum_i y_{ij}^k = 0 \tag{4}$$

Constraints (3) and (4) insure that the route for any ship can be broken into a set of simple cycles (round trips), between some sequence of ports. If it is desired to specify a starting or ending port, then the right hand side of the appropriate constraint in (3) or (4) can be changed to 1 or -1 respectively. We consider here only zero right hand sides.

The objective function given in Laderman (1966) is to maximize the total slack time

$$Z = \sum_k z_k \tag{5}$$

This yields a solution which delivers all goods in minimal total travel time, hence using the fleet most efficiently. The model may be resolved for various fleet sizes and compositions, to determine the minimal fleet size required to deliver all goods.

The structure of the constraint matrix for a 2 ship problem is shown in figure 1. The matrices T_k are transshipment matrices, with each column containing only 2 non zero entries, one +1 and -1. Each ship adds one such block to the problem. The number of rows in block k is the number of allowable ports for ship k, which can easily be in the hundreds. Hence with relatively few ships (say 10 or 15) problems with thousands of rows and many thousands of columns may be generated.

An additional computational problem arises because of the predominance of zeros on the right hand side. Basic solutions tend to be highly degenerate, i.e., more than 50 percent of the basic variables are at zero level. Hence the number of iterations required to reach optimality can be much larger than usual, e.g. 5 to 10 times the number of rows rather than 1 to 3 times.

The Chi Corporation* currently markets a software package, Chi Corporation (1971), built around this LP model. It includes (1) a matrix generator which reads the data on travel times, quantities to be shipped, allowable origins and destinations, etc., and composes the LP matrix (2) the UNIVAC 1108 LP system, which receives the LP matrix and applies the revised simplex method with basis inverse in product form to find an optimal solution (3) a report writer, which

*11000 Cedar Road, Cleveland, Ohio 44106

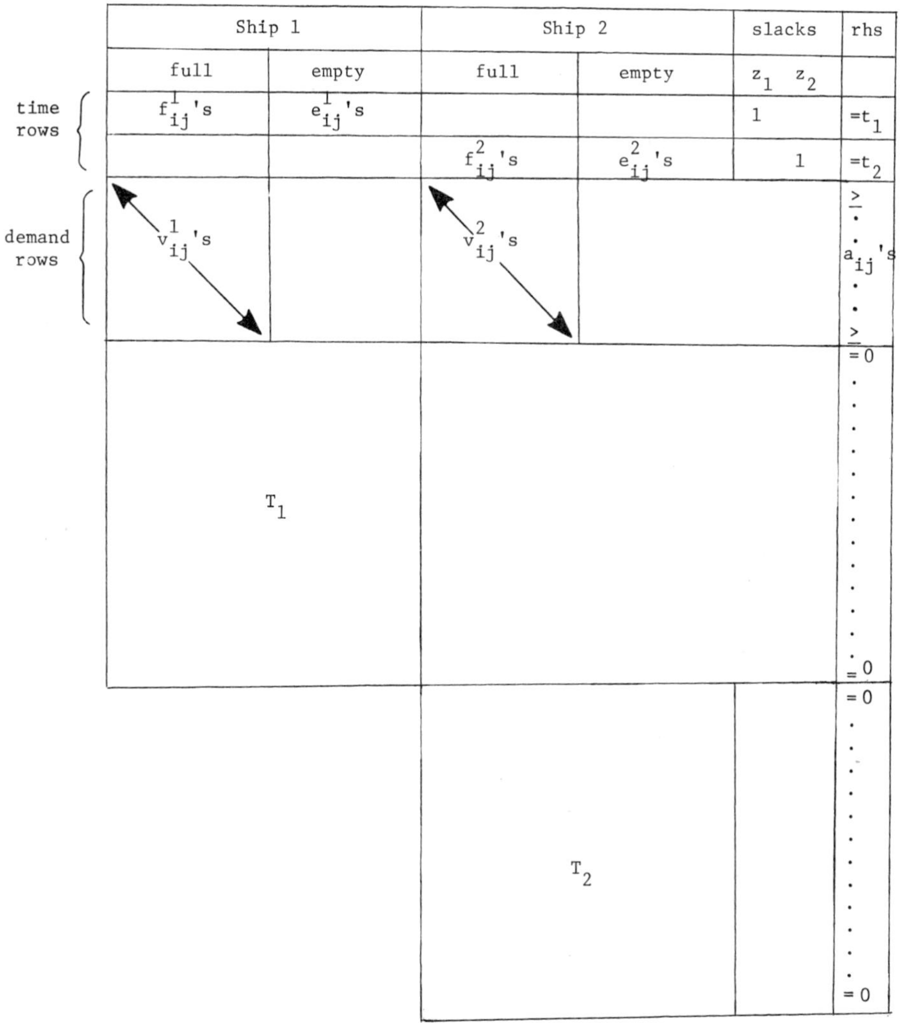

Figure 1 - Constraint Matrix for Ship Routing Problem

receives the optimal solution and outputs it in a form readable by management. A number of large problems generated by companies engaged in shipping on the Great Lakes have been solved. Run times for small problems (e.g. about 250 rows and 1000 or 1500 columns) are typically 5 minutes or less, but solution time increases repaidly with problem size. Models with 500 to 800 rows have run around an hour, and a model with about 2000 rows could not be solved in 24 hours. These times probably reflect the degeneracy problem mentioned above.

2. SOLUTION BY THE DANTZIG-WOLFE DECOMPOSITION METHOD

Since the problem has block diagonal structure with coupling rows, it may be attacked using the Dantzig-Wolfe decomposition method. The facts that relatively few of the rows are coupling rows and that each of the blocks is a transshipment matrix increase the attractiveness of this approach. Results similar to those presented here are found in, Rao (1968), although they are not derived using the Dantzig-Wolfe methodology.

To decompose this problem, let $(x^k y^k)$ be the vector of full and empty trip variables for ship k. The kth block of constraints (3)-(4) may then be written

$$T_k \begin{pmatrix} x^k \\ y^k \end{pmatrix} = 0 \tag{6}$$

$$x^k \geq 0, \; y^k \geq 0 \tag{7}$$

The set of all solutions to (6)-(7) is a convex polyhedral cone, and hence any solution can be written as a nonnegative linear combination of extreme rays of this cone. If the rth such ray is (x_k^r, y_k^r), then

$$(x_k, y_k) = \sum_r \lambda_k^r (x_k^r, y_k^r) \quad , \lambda_k^r \geq 0 \quad , \text{all } k,r \tag{8}$$

It can be shown that each extreme ray represents a simple cycle in the graph associated with T_k. That is, (x_k^r, y_k^r) contains a +1 for variables x_{ij}^k, y_{ij}^k whose arcs are in the rth cycle, and zeros elsewhere. Of course, since there are a great number of such rays, they are not tabulated in advance, but are generated during solution as needed.

Substituting (8) into (1),(2) and (5) yields the Dantzig-Wolfe master program:

$$\text{Maximize } \sum_k z_k$$

subject to

$$\sum_r \lambda_k^r t_k^r + z_k = t_k \quad , k=1,\ldots,K \tag{9}$$

$$\sum_{kr}\sum \lambda_k^r d_{ij}^{kr} \geq a_{ij} \quad , (i,j) \varepsilon S \tag{10}$$

$$\lambda_k^r \geq 0 \quad , \text{all } k,r$$

In the above, t_k^r is the time required for ship k to traverse the rth cycle, and d_{ij}^{kr} is the quantity of cargo carried by ship k between ports i and j while transversing cycle r. This is equal to v_{ij}^k if arc (i,j) is in the cycle and zero otherwise. This master program has far fewer rows than the original problem, usually at most a few hundred even for large problems. It has a variable for every possible simple cycle that each ship can traverse. To deal with this multitude of variables, assume we have a basic solution to (9)-(10) (perhaps with artifical variables included). Let (π, u) denote the vector of simplex multipliers associated with this solution, with π corresponding to the time rows (9) and u to the demand rows (10). Pricing out the column associated with λ_k^r yields the relative cost factor

$$\bar{c}_{kr} = -\pi_k t_k^r - \sum_{(i,j)\varepsilon S} d_{ij}^{kr} u_{ij} \tag{11}$$

The standard rule of the simplex method is to find the column which maximizes $\bar{c}_{kr}$, in order to locate an entering column which will improve the current solution. Doing the maximization over r first:

$$\max_{k,r}(\bar{c}_{kr}) = \max_{k} \max_{r}(-\pi_k t_k^r + \sum_{(i,j)\varepsilon S} d_{ij}^{kr}(-u_{ij})) \tag{12}$$

By bringing any profitable z_k columns or slack columns from a_{ij} rows into the basis before maximizing (12) we can insure that

$$\pi_k \geq 1 \quad ,\text{all } k \tag{13}$$

$$u_{ij} \leq 0 \quad ,(i,j)\varepsilon S \tag{14}$$

Hence the quantity to be maximized in (12) is a nonnegative linear combination of cargoes carried while traversing cycle r minus π_k times the time required to traverse the cycle. Intuitively, maximizing this should lead to cycles which carry much cargo in little time. In graph theoretic terms, we take the graph which connects allowable (i,j) pairs for ship k, and associate a weight with each arc. This weight is $-\pi_k f_{ij}^k - v_{ij}^k u_{ij}$ for an arc directed from i to j (full trip arc) and $-\pi_k e_{ij}^k$ for an arc directed from j to i (empty trip arc). Then, for each ship, we search for a simple cycle which maximizes the sum of the arc weights. This is the Dantzig-Wolfe subproblem. Of course, this is a traveling salesman problem, which is too difficult to solve in this context. However, a number of algorithms, e.g. shortest route algorithms, will yield a positive cycle if one exists or will show that none exists. If, for each ship, no positive cycle exists, the current basic solution is optimal. If any ships have positive cycles, we take these cycles, form the corresponding columns in (9)-(10), and solve a restricted master program. This program contains the above newly formed columns, and all slacks, as nonbasic columns, plus the current basic columns. Solving it yields an improved objective value, plus new simplex multipliers (π, u) with which we begin the next iteration. Since the procedure is just the simplex method, it converges in a finite number of cycles, at least if some anti-cycling rule is used.

A simple modification which avoids the traveling salesman problem is described in Dantzig (1966). One simply defines new variables $\bar{\lambda}_k^r$ satisfying

$$\lambda_k^r = \bar{\lambda}_k^r / t_k^r \tag{15}$$

and substitutes (15) into (9)-(10). The relative cost coefficient for the column associated with $\bar{\lambda}_k^r$ is

$$\bar{c}_k^r = -\pi_k + \sum_{(i,j)\varepsilon S} (-u_{ij}) d_{ij}^{kr} / t_k^r \tag{16}$$

For each k, maximizing $\bar{c}_k^r$ over r requires maximizing the sum in (16), which is the ratio of the (weighted) cargo carried in traversing cycle r to the time required to traverse it. A cycle which maximizes this ratio can be found using an algorithm based on the simplex method. The computation is much faster than solving a traveling salesman problem.

Solution via the Decomposition method appears particularly attractive here. There is no reason to expect the master program (9)-(10) to be degenerate, even though the original problem is highly degenerate. The subproblems may be solved by efficient graph theoretic methods. As more ships or ports are added, the number of rows in the master program increases only modestly, although the subproblems become more complex, and there are more of them. An easily computed upper bound, Lasdon (1970), is available to aid in terminating computations. Further, by compact storage of the basic cycles, it should be possible to solve large problems al in core. For these reasons, the Decomposition approach could well be significantly faster than direct solution via the simplex method. Unfortunately, verifying this conjecture is an expensive and time consuming task, involving much computer coding For this reason, computational work on the Decomposition approach has not yet begun. We hope to pursue this work in the future.

REFERENCES

The Chi Corporation, "Bulk Fleet Model Systems", 1971.

Dantzig, G. B. et. al., "Finding a Cycle in a Graph with Minimum Cost to Time Ratio with Application to a Ship Routing Problem", TR No. 66-1, Operations Research Department, Stanford University, November 1966.

Laderman, J., et.al., "Vessel Allocation by Linear Programming", Nav. Res. Log. Q. 13, 1966, pp. 315-320.

Lasdon, L. S., "Optimization Theory for Large Systems", The MacMillan Co., 1970, Chapter 3.

Rao, M. R. and S. Zionts, "Allocation of Transportation Units to Alternative Trips - A Column Generation Scheme with Out-of-Kilter Subproblems", Operations Research, 16, January-February 1968.

THE CONNECTION BETWEEN DECOMPOSITION ALGORITHMS AND OPTIMAL DEGREE OF DECOMPOSITION

OLI B. G. MADSEN

The Institute of Mathematical Statistics and Operations Research

The Technical University of Denmark
DK 2800 Lyngby - Denmark

Abstract: Large-scale mathematical programming problems can often be decomposed in many different ways. One can either solve a sequence of many smaller problems or a few larger problems. This paper describes an optimal method to decompose LP-problems with an angular structure. The objective is to minimize the total computing time and the method is based on information about the applied decomposition algorithm. A numerical example is demonstrated and it is shown how the optimal degree of decomposition is strongly influenced by the number of times the master problem has to be solved.

1. INTRODUCTION

Since the first decomposition algorithm appeared in 1958 (Dantzig, 1960) there has been developed about 50 - 75 different decomposition algorithms, see for instance the references in (Dantzig, 1967), but very little attention has been given to obtain practical experience with the proposed algorithms to find out how to decompose a given problem when a given algorithm is available.

2. THE PROBLEM

Let us assume that we have the angular structure shown in figure 1. Using a decomposition algorithm to solve the total problem one has several choices when the subproblems have to be solved:

a) One can solve each of the 8 subproblems separately.

b) One can solve 4 x 2 subproblems, for instance B1 and B2 as one larger subproblem, B3 and B4 as one larger subproblem and so on.

c) One can solve 2 x 4 subproblems, for instance B1, B2, B3 and B4 as one subproblem and so on.

d) One can solve B1, B2, ..., B8 as one larger subproblem.

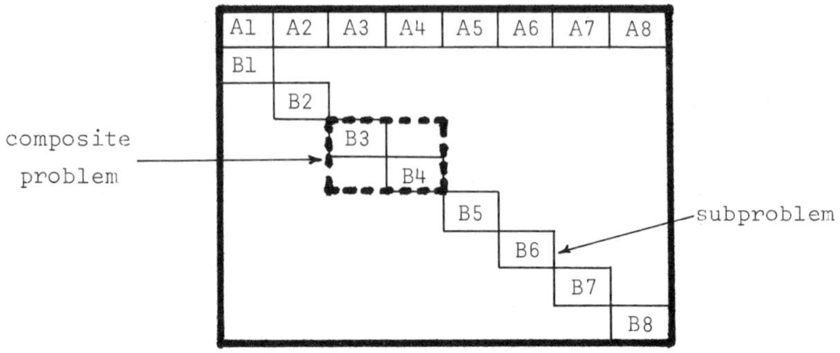

Fig. 1.

Let us call a tied group of subproblems, which is solved as one problem, a composite problem. The question now is how many composite problems is it reasonable to solve. That is which of the alternatives a, b, c or d gives the minimum total computing time. The optimization time for each of the composite problems leads to as many composite problems as possible while the problem of exchange of information and administrative difficulties leads to the smallest possible number of composite problems.

In his paper (Labro, 1964) Claude Labro has tried to compare different simple alternatives - as many composite problems as possible and only one large composite problem. Labro compares the total number of iterations in the two cases and shows with some simple examples that it is impossible to say anything in general about which alternative is the best. It depends on the actual coefficients in the problem. Labro compares only the number of master iterations and does not take into account the total computational effort.

In the following a very simple model to calculate the total computation time as a function of a varying number of composite problems is developed.

3. TERMINOLOGY AND BASIC ASSUMPTIONS

Terminology:

M The total number of subproblems

m The number of rows in each subproblem.

T_m The computation time used to find an optimal solution to a subproblem with m rows.

mo The number of common rows.

T_{mo} The computation time used to find an optimal solution to a problem with mo rows.

p The fraction of time it takes to reoptimize a modified problem from a given start solution in relation to the first optimization of the unmodified problem.

N The actual number of composite problems (the unknown).

D = N/M, the degree of decomposition. D=1 means that the problem is totally decomposed, D=1/M means that we only have one large composite problem. In the previous example we have: a) D = 1, b) D = 4/8=0.5, c) D = 2/8=0.25, d) = 1/8=0.125.

R The number of reoptimizations in connection with every composite problem that is necessary to carry out in order to reach an optimal solution, i.e. the number of master iterations.

S The time the computer uses to shift from one problem to another.

Comments and assumptions:

1) All subproblems have the same number of rows.

2) The computation time used to optimize a LP-problem depends on the cube of the number of rows. It is well known that the computation time is very difficult to estimate but the above mentioned expression is suggested in many books and articles, see for instance p. 152 in (Beale, 1968) and p. 195 in (Wolfe, 1963).

3) The time for reoptimization is proportional to the original optimization time, i.e. T_m (reopt) = pT_m. Practical experiments show that a good average estimate for p is between 0.05 and 0.25.

4) The number of reoptimizations R is very difficult to estimate. Let us assume that R depends on the number of composite problems, i.e. R = R(N). It is then possible to analyze the sensitivity of this dependence later on.

5) The shifting time S is assumed to be a constant.

6) The number of common rows is independent of N. This is not always true but it is an assumption which normally differs very little from reality.

7) The N composite problems contain the same number of rows.

4. THE MODEL

Using the terminology and assumptions from section 3 the total computing time can now be written as:

$$T_N = \underbrace{T_{mo} + N(M/N)^3 T_m + (N+1)S}_{\text{First optimization}} + R(N)\Big[\underbrace{pT_{mo} + pN(M/N)^3 T_m + (N+1)S}_{\text{Time for reoptimization}}\Big] \quad (1)$$

where:

M/N is the number of subproblems contained in each composite problem, i.e. the number of times there are m rows in a composite problem.

N+1 is the necessary number of shifts in one master cycle.

We can now formulate the following optimization problem:

$$\text{Minimize } T_N$$
$$\text{subject to } M/N \text{ and } N \text{ integers} \qquad (2)$$
$$\text{and} \quad N \geq 1$$

Taking into account that all the coefficients in T_N are nonnegative one can rewrite expression (1) and express T_N in terms of convexity:

$$T_N = \text{convex function} + R(N) \times \text{convex function} \qquad (3)$$

Unfortunately it is not possible to say anything about the convexity in connection with products of functions. T_N can therefore be a convex, a concave or a partly concave function of N. Figure 2 shows one example of expression (1) as a function of N.

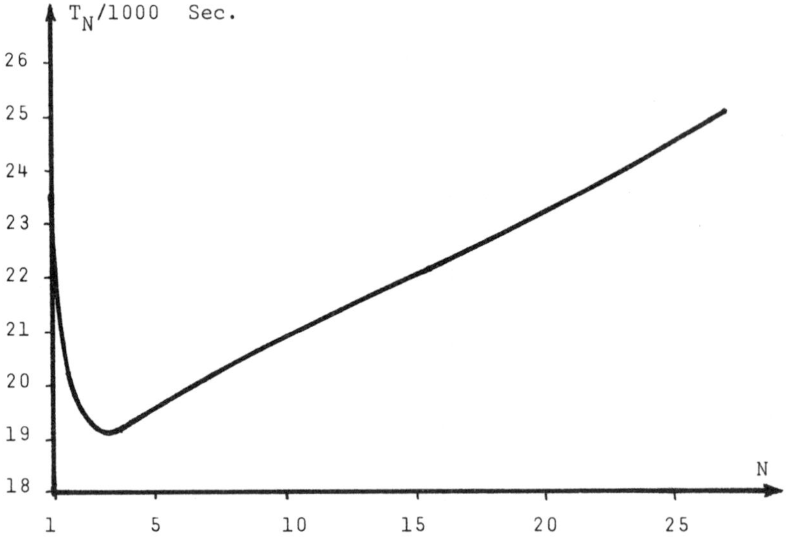

Fig. 2. $T_N = T_{mo} + N(M/N)^3 T_m + (N+1)S + R(N) \times \left[pT_{mo} + pN(M/N)^3 T_m + (N+1)S\right]$, where $T_{mo} = 0{,}45$ sec., $T_m = 0{,}25$ sec., $S = 0{,}005$ sec., M=32, p=0,2 and $R(N) = 10 \times N^{2,1}$.

It is neither possible - apart from very special cases - to find an analytic solution to problem (2). On the contrary, it is very easy to find a numerical solution to problem (2) when the coefficients are given. Problem (2) can be solved by simple enumeration because of the few possibilities left when both N and M/N are required to be integers.

5. NUMERICAL SOLUTION

Before doing further calculations let us assume that R(N) can be expressed as

$$R(N) = KN^q \tag{4}$$

where K and q are given constants (K > 0). The value of these constants depends on the actual decomposition algorithm used and it is therefore necessary to determine these quantities by practical experimentation with the algorithms.

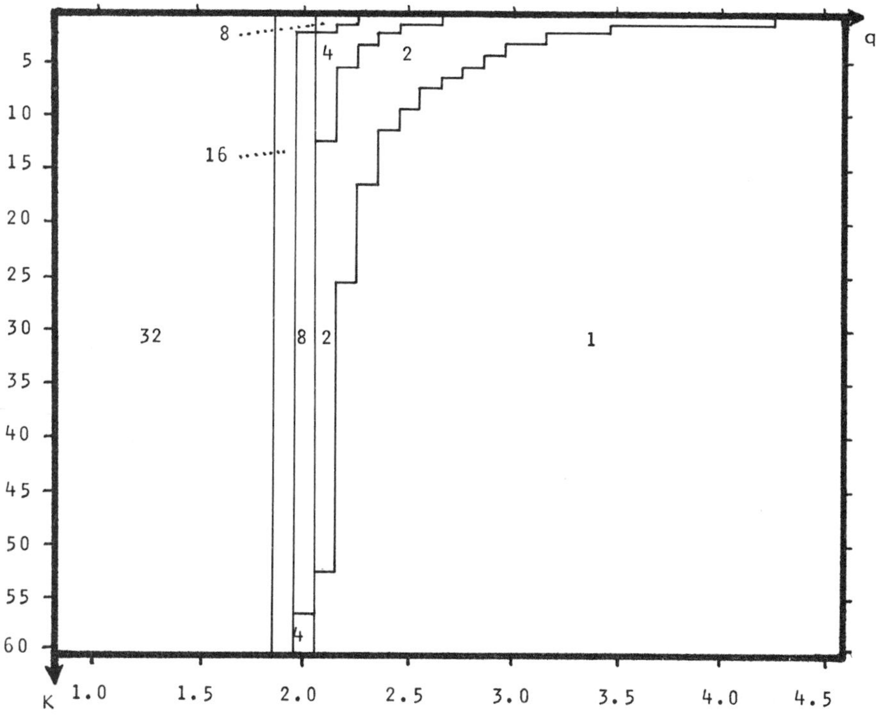

Fig. 3. The optimal number of composite problems as a function of K and q. $T_{mo} = 90$, $T_m = 50$, S = 1 and M = 32.

Using the usual theorems about convex functions it is now possible to say that T_N is convex if q complies with one of the following four conditions:

$$q \leq 1 \;, \quad q = 0 \;, \quad 1 \leq q \leq 2 \;, \quad q \geq 3 \tag{5}$$

There are still intervals where T_N can be nonconvex. It is now possible to calculate the optimal number of composite problems using a very simple computer program. Figure 3 shows the results as a function of K and q while T_{mo} , T_m , S, M and p are given constants.

Notice that the solution is not very sensitive with respect to changes in K.

Rewriting the expression for T_N as:

$$T_N/M = T'_{mo} + D^{-2}T_m + (D+1/M)S + KN^q\left[pT'_{mo} + pD^{-2}T_m + (D+1/M)S\right] \qquad (6)$$

where $T'_{mo} = T_{mo}/M$. It is easy to see that the optimal value for the degree of decomposition is almost independent of M, the number of subproblems. This can be verified by calculating the solution to problem (2) for various values of M. The result is shown in figure 4 where M has been given the values 16, 24, 30, 32, 36, 40 and 48.

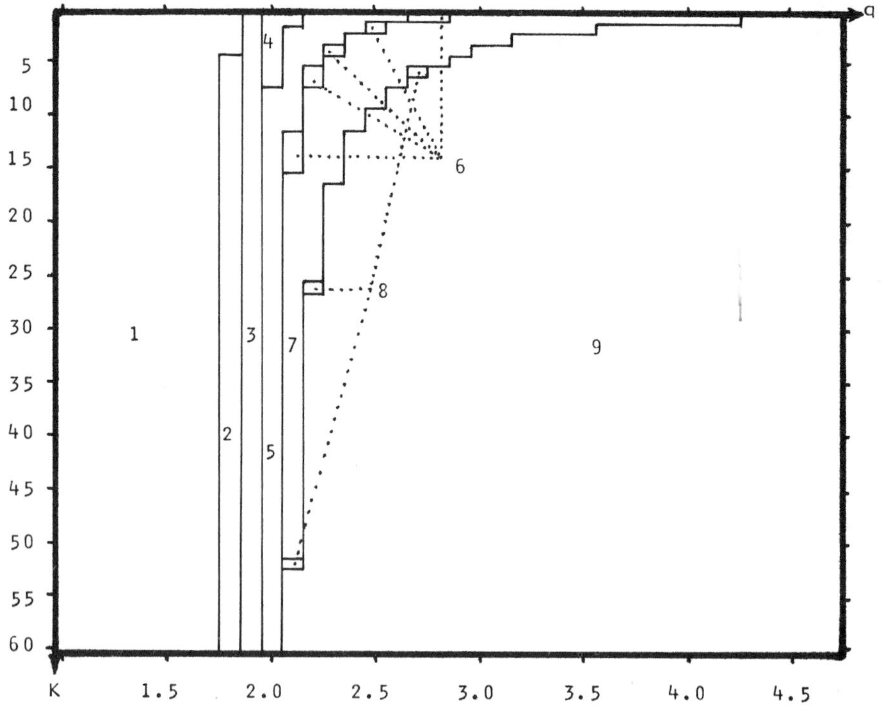

Fig. 4. Optimal degree of decomposition. Concentrated table of 7 computer runs with M = 16, 24, 30, 32, 36, 40 and 48. The code has to be interpreted as in (7). T_{mo} = 90, T_m = 50, S = 1 and p = 0.2.

The seven solution sets are summarized in one figure and the codes has to be interpreted as follows:

1 : D = 1 for all M

2 : D = 1 or D = 1/2 (7)

3 : D = 1/2 for all M

4 : $D = 1/2$ or $1/2 < D < 2/M$

5 : $1/2 < D < 2/M$ for all M

6 : $1/2 < D < 2/M$ or $D = 2/M$ (7)

7 : $D = 2/M$ for all M

8 : $D = 2/M$ or $D = 1/M$

9 : $D = 1/M$ for all M

i.e. the overlap for various M is indicated as even numbers.

To ensure that figure 4 also is fairly representative for other values of T_{mo}, T_m and S these constants are varied mutually on figure 5. Notice that inside a very extensive range of variation the relative dimensions of the constants is of no importance.

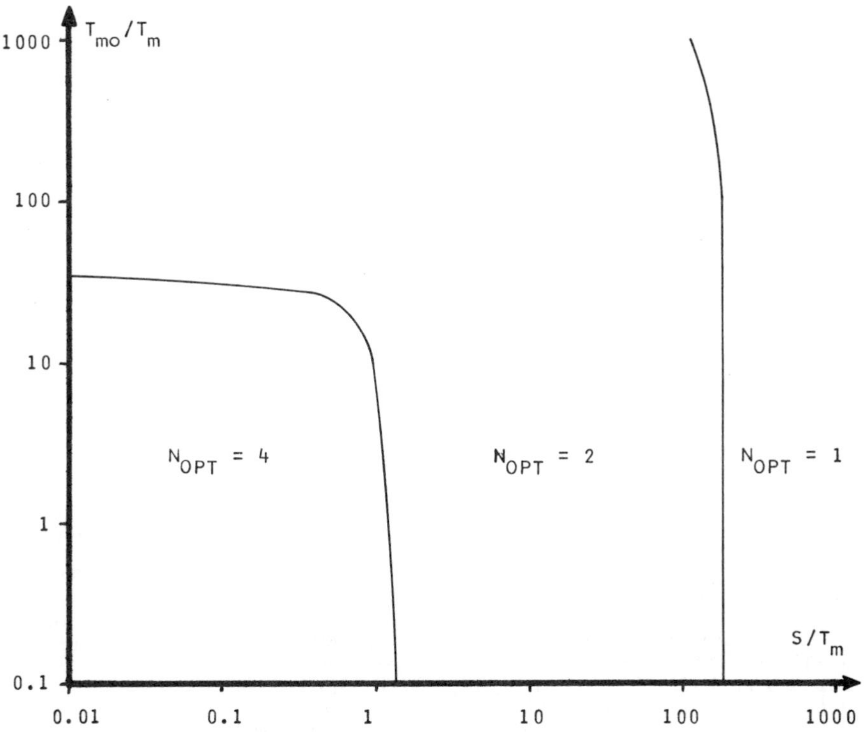

Fig. 5. The optimal number of composite problems as a function of T_{mo}, T_m and S. M = 32, p = 0.2, K = 10 and q = 2.1

As seen on the previous figures the change from complete decomposition to no decomposition occurs very suddenly when the exponent q has a value about 2. It is in a very narrow area a D-value between 1 and 1/M is realistic.

Due to uncertainty in the estimation of K and q it is tempting to reduce the area of the above mentioned partial decomposition by introducing a rule saying that if the difference between T_{NOPT} and $T_{N=1}$ or $T_{N=M}$ is less than a given tolerance then use N=1 or N=M respectively instead of the optimal value of N.

Figure 6 shows a concentrated calculation for 7 different values of M analoguous to figure 4 but using a tolerance of 20% in order to reduce the area of partial decomposition. The code is the same as in expression (7). The codes greater than 9 indicates a mixture of two codes in (7), for instance the code 19 indicates that for some M the code is 1 and for other M the code is 9.

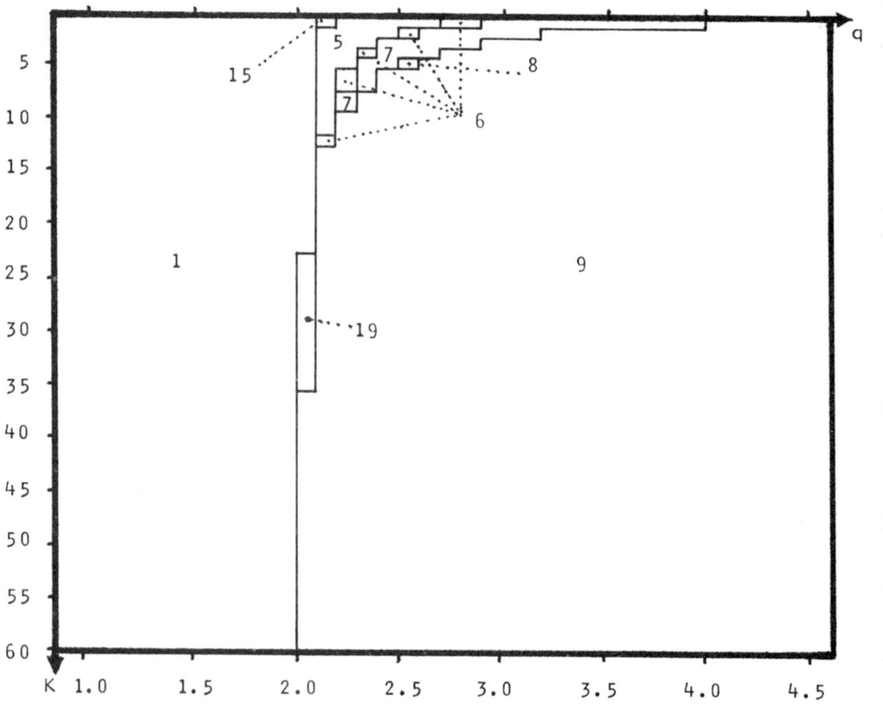

Fig. 6. Optimal degree of decomposition. Concentrated table of 7 computer runs with M = 16, 24, 30, 32, 36, 40 and 48. The code has to be interpreted as in (7). Tolerance = 20 pct. in favour of code 1 and 9. T_{mo} = 90, T_m = 50, S = 1 and p = 0.2

6. EXAMPLE - THE DANTZIG-WOLFE ALGORITHM

Some test examples using the Dantzig-Wolfe (D-W) decomposition algorithm with multiple pricing (Lasdon, 1970) have been run in order to find an estimate of K and q (Rasmussen, 1971). It is of course difficult to give an exact answer based on a few test examples but the lines in the result were quite clear.

$$K \sim 100 \ , \quad q \sim -0.75 \ \Rightarrow \ R(N) \sim 100 * N^{-0.75} \qquad (8)$$

From this it seems quite reasonable to stress that using the D-W decomposition algorithm with multiple pricing, then a given angular LP-problem has to be decomposed as much as possible.

7. SUPPLEMENTARY REMARKS

Having a given problem to be solved and a given algorithm to be used then one can either use the coarse rule with $q=2$ as mentioned in section 5 or one can with given coefficients optimize expression (1). The computer program to do this will only contain about 10 to 20 statements.

There might exist special circumstances which makes another approach necessary:

a) The total problem can be so extensive that one is forced to decompose due to the computer capacity.

b) There can be so special subproblems, for instance transportation-like subproblems, that it will be unreasonable to combine this with other problems.

c) In the previous mentioned examples only values of M which were divisible by 5 to 10 other numbers were used. Normally the M-values can be numbers as for instance 23 which is a prime. As above mentioned there was no strong correspondence between the optimal degree of decomposition and M. It will therefore be reasonable still to use the rule about total decomposition or no decomposition.

8. CONCLUSION

As a result of this work one can reach the following conclusion that it is possible to give a coarse rule on how to decompose a given angular LP-problem with subproblems of equal size when a given decomposition algorithm is available.

From experiences with the algorithm an estimate of K and q is found, where K and q form the expression $K(N) = K \cdot N^q$ - the necessary number of master iterations as a function of N, the number of composite problems. If $q > 2$ then it is normally not reasonable to use the algorithm at all. If $q < 2$ then the best thing to do is to decompose totally and use as many subproblems as possible. If q has a value just in the neighbourhood of 2 or if $K < 10$ then it might be useful to carry out a partial decomposition and solve composite problems consisting of more than one subproblem.

REFERENCES

Beale, E. M. L., (1968), Mathematical programming in practice, Cambridge University Press.

Dantzig, George B., (1967), Large-scale linear programming, Techn. Rep. No. 67-8, Department of OR, Stanford.

Dantzig, George B. and Philip Wolfe, (1960), Decomposition principle for linear programs, Operations Research 8, 101.

Labro, Claude, (1964), Efficiency and degrees of decomposition, Working paper No. 98, Center for Research in Management Science, Univ. of California, Berkeley.

Lasdon, Leon S., (1970), Optimization theory for large systems, Macmillan.

Rasmussen, Niels A., (1971), Decomposition in large-scale systems (in Danish), Master Thesis, The Institute of Mathematical Statistics and Operations Research, The Technical University of Denmark.

Wolfe, Philip and Leola Cutler, (1963), Experiments in linear programming, p. 177 in Graves and Wolfe (Eds.) Recent advances in mathematical programming, Mc Graw-Hill.

DECOMPOSITION AND THE OPTIMIZATION OF DISTRIBUTED PARAMETER SYSTEMS

DAVID A. WISMER

Systems Control, Inc.

Palo Alto, California, U.S.A.

Abstract: This paper reviews the problems of static and dynamic optimization when the model equality constraints are difference equations or differential - difference equations arising from the discretization of partial differential equations. It is shown that a wide class of these problems is particularly well-suited to solution by decomposition and hierarchical techniques. The dimensionality of these systems is proportional to the number of mesh points in the grid used in forming the difference approximations.

1. INTRODUCTION

It has long been recognized that a vast number of physical problems can be modeled using partial differential equations. Because of the importance associated with the solution of these equations and the fact that continuous mathematics provides only limited guidance to the solution of most practical problems, discrete approximation methods have long been the subject to intense investigation. Indeed, discrete approximations to distributed parameter systems can be fundamentally regarded as a means of decomposing these systems. Alternative schemes for solving these discrete systems of equations fall into several categories, including explicit, and implicit methods, as well as combinations or variations of these too numerous to mention. (Todd, 1962) The high dimensionality and the special structure resulting from the decomposed system of equations has prompted countless years of research in this area.

Recently, the already difficult task of solving partial differential equations has been compounded by attempts at optimization (Wang, 1964). The inherent computational difficulties of optimization together with the high dimensionality of discrete distributed parameter systems has stimulated additional research in this area. This research combines the decomposition methods of discrete approximation with the more recent results of hierarchical optimization theory (Wismer, 1971) into a unified approach to the problem. Several of the primary results of this research are reviewed here along with three examples which demonstrate the application of this approach to a variety of problem areas.

2. DISCRETE AND SEMIDISCRETE MODELS

Consider now the fairly general system equation

$$\frac{\gamma \underline{u}}{\gamma t}(\underline{X},t) = \mathcal{G}(\underline{u}(\underline{X},t),\ \underline{m}(\underline{X},t),\ \underline{X},\ t) \qquad \underline{X}\varepsilon\Omega \qquad t \geq t_o \tag{1}$$

with boundary conditions

$$\alpha(\underline{X},t)u_i(\underline{X},t) + \beta(\underline{X},t)\frac{\gamma u_i}{\gamma n}(\underline{X},t) = f_i(\underline{X},t) \qquad \underline{X}\varepsilon\Omega_b \qquad t \geq t_o \tag{2}$$

for each element i of the vector $\underline{u}$. The initial conditions are given by

$$\underline{u}(\underline{X},t_0) = \underline{u}_0(\underline{X}) \qquad \underline{X}\epsilon\Omega \tag{3}$$

The distributed parameter system described by Eqs. (1) to (3) is a multivariable system with a state vector $\underline{u}$ defined over an n-dimensional spatial domain Ω and ti The region Ω is assumed to be a given finite connected region in Euclidean n space and Ω_b is the boundary of Ω. The control vector $\underline{m}$ is similarly defined. The oper- ator $\mathscr{G}$ is a spatially varying vector differential operator on $\underline{u}$ which may include parameters which are functions of $\underline{u}$, $\underline{m}$, $\underline{X}$, or t. The functions α, β, and f_i are real-valued functions, piecewise C^1 on Ω_b and C^2 on t which satisfy the inequaliti

$$\begin{aligned} \alpha(\underline{X},t) &\geq 0 \\ \beta(\underline{X},t) &\geq 0 \qquad \underline{X}\epsilon\Omega_b \quad t \geq t_0 \\ \alpha(\underline{X},t) + \beta(\underline{X},t) &> 0 \end{aligned} \tag{4}$$

The symbol n indicates a direction normal to the boundary. In case the distribute control vector $\underline{m}$ is not present, the only means of controlling the system may be through the functions f_i defined on the boundary only. This problem is called the boundary control problem for partial differential equations. It possesses no analog for systems described by ordinary differential equations.

The performance functional to be minimized in our optimal control problem is denoted by

$$J(\underline{m}) = \int_{\overline{\Omega}} \int_{t_0}^{t_1} P(\underline{u}(\underline{X},t),\underline{m}(\underline{X},t),t)\,dt\,\underline{dX} \tag{5}$$

where P is a real-valued function of class C^2 on t and piecewise C^1 on Ω. The symbol $\overline{\Omega}$ represents the closure of Ω.

The control $\underline{m}$ may also be required to satisfy inequality constraints of the form

$$\underline{R}(\underline{u}(\underline{X},t),\underline{m}(\underline{X},t),\ \underline{X},t) \geq \underline{0} \qquad \underline{X}\epsilon\Omega \quad t \geq t_0 \tag{6}$$

and terminal constraints of the form

$$\begin{aligned} \psi_0(t_1) &= 0 \\ \underline{\psi}(\underline{u}(\underline{X},t_1),\underline{X}) &= \underline{0} \qquad \underline{X}\epsilon\Omega \end{aligned} \tag{7}$$

where $\underline{R}$ is a vector-valued function of dimension r with components R_i which are of class C^2 on t and piecewise C^1 on Ω. The vector-valued function $\underline{\psi}$ is of dimension q with components ψ_i having the same properties as R_i. The scalar function $\psi_0(t_1)$ specifies the final time t_1. We assume that the functions $\underline{R}$ and $\underline{\psi}$ are consistent with the boundary conditions in Eq. (2).

The system described by Eq. (1) is sufficiently general to include all the standard types of partial differential equations. For example, the hyperbolic and biharmonic classes of equations which involve higher-order time derivatives can be formulated in the notation of Eq. (1) by employing the notion of state variables.

The single exception is the class of elliptic partial differential equations which describes the steady-state behavior of the systems described by Eq. (1). These equations are time independent and can be represented by

$$\mathscr{G}(\underline{u}(\underline{X}), \underline{m}(\underline{X}), \underline{X}) = \underline{0} \qquad \underline{X}\varepsilon\Omega \tag{8}$$

with boundary conditions given by

$$\alpha_i(\underline{X})u_i(\underline{X}) + \beta_i(\underline{X}) \frac{\gamma u_i}{\gamma n} (X) = f_i(\underline{X}) \qquad \underline{X}\varepsilon\Omega \tag{9}$$

The criterion functional and inequality constraints then become

$$J(\underline{m}) = \int_{\bar{\Omega}} P(\underline{u}(\underline{X}),\underline{m}(\underline{X}), \underline{X})d\underline{X} \tag{10}$$

$$\underline{R}(\underline{u}(\underline{X}), \underline{m}(\underline{X}), \underline{X}) \geq \underline{0} \qquad \underline{X}\varepsilon\Omega \tag{11}$$

These continuous-space, continuous-time partial differential equations can now be approximated in several ways. By a discrete approximation we will mean replacing all derivatives by differences and all integrals by sums, thereby yielding a mathematical programming problem having a large number of algebraic or transcendental equality constraints. Alternatively, a semidiscrete approximation can be used on partial differential equations of the form of Eq. (1) where only the spatial derivatives are replaced by differences and the equality constraints then become sets of ordinary differential equations. Likewise, the criterion functional retains a single integral with respect to time and we have a problem in the calculus of variations. In this paper we will employ discretization in the space domain only, thus resulting in semidiscrete approximations of partial differential equations of the form of Eq. (1).

We discretize the space variables by defining a column vector

$$\underline{X}_{\underline{i}} = [i_1(\Delta x_1), i_2(\Delta x_2), \ldots, i_j(\Delta x_j), \ldots, i_n(\Delta x_n)]' \tag{12}$$

which in effect places a grid on the region Ω. The prime denotes the vector transpose. Here the elements of

$$\underline{i} = [i_1, i_2, \ldots, i_j, \ldots, i_n]'$$

are integers defined by $i_j = 0, 1, \ldots, N_j$ where

$$N_j = \frac{1}{\Delta x_j} \left[(x_j)_{max} - (x_j)_{min}\right]$$

Denoting the set of points defined by Eq. (12) by #, the following terms can be defined: mesh point--a point in #; interior point--a point in $\#\cap\Omega$; boundary point--a point in $\#\cap\Omega_b$; exterior point--a point belonging to $\#\cap C(\bar{\Omega})$ where $C(\cdot)$ represents the complement operator; regular point--a point belonging to $\#\cap\Omega$ and such that all adjacent points also belong to $\#\cap\Omega$; irregular point--a point belonging to # which is not a regular point.

For the case when $\beta_i = 0$ in Eq. (2), irregular points can be treated as regular points by appropriately defining a boundary (pseudo) mesh point at the intersection of the boundary and the line segment connecting the irregular point with each exterior point. In case $\beta_i > 0$, the boundary can be approximated by orthogonal line segments or by other suitable devices as discussed by Varga (1962).

For our purposes, it is sufficient to see that methods exist for handling any irregular boundary. For simplicity in the subsequent discussion, we will assume that $\beta_i = 0$ in Eq. (2) and that all mesh points are regular points.

The outcome of discretizing the operator $\mathcal{G}$ depends upon the specific form of $\mathcal{G}$ and upon the accuracy of the approximation. For example, if $\mathcal{G}$ contained no derivative terms, the discrete operator $\underline{G}_{\underline{i}}$ at $\underline{X}_{\underline{i}}$ would be a function of variables at $\underline{X}_{\underline{i}}$ only. If first derivative terms are present, at least one additional mesh point must be used in the approximation. For clarity of presentation, the operator $\mathcal{G}$ will be assumed to contain at most second-order space derivatives and the simplest possible approximation will be used. In this case we can express

$$\mathcal{G}(\underline{u}(\underline{X}_{\underline{i}},t),\underline{m}(\underline{X}_{\underline{i}},t),\underline{X}_{\underline{i}},t) \cong \underline{G}_{\underline{i}}(\underline{u}_{\underline{i}}(t),\underline{u}_{\underline{i}\pm\underline{I}_k}(t),\underline{m}_{\underline{i}}(t),t) \qquad k = 1,2,\ldots,n \tag{13}$$

where $\underline{I}_k = \{\underline{i}\,|\,\underline{i} = \underline{0}$ except for the kth element which equals 1$\}$, $\underline{i}$ ranges over all regular points, and the functions $\underline{G}_{\underline{i}}$ are assumed to be real valued and of class C^2.

Applying Eq. (12) to Eq. (1)-(7), the problem can now be posed as minimize $J(\underline{m}_{\underline{i}})$ where

$$J(\underline{m}_{\underline{i}}) = \sum_{\underline{i}\in\Omega} \int_{t_o}^{t_1} P_{\underline{i}}(\underline{U}_{\underline{i}}, \underline{m}_{\underline{i}}, t)dt \tag{14}$$

such that

$$\dot{\underline{u}}_{\underline{i}} = \underline{G}_{\underline{i}}(\underline{u}_{\underline{i}},\underline{u}_{\underline{i}\pm\underline{I}_k},\underline{m}_{\underline{i}},t) \tag{15}$$

with initial conditions

$$\underline{u}_{\underline{i}}(t_0) = \underline{u}_{0\underline{i}} \tag{16}$$

and final conditions

$$\underline{\psi}_{\underline{i}}(\underline{u}_{\underline{i}}(t_1)) = \underline{0} \tag{17}$$

and subject to the inequality constraints

$$\underline{R}_{\underline{i}}(\underline{u}_{\underline{i}}, \underline{m}_{\underline{i}}, t) \geq \underline{0} \tag{18}$$

Here $\underline{X}_{\underline{i}}\in\Omega$ and we are dealing with the distributed control problem in which one and only one control vector $\underline{m}_{\underline{i}}(t)$ is associated with each mesh point. We have assumed that the boundary conditions along with the inequality constraints completely determine the control on the boundaries and hence the boundary domain and the boundary conditions are excluded from the problem formulation. Furthermore, we assume that the final time t_1 is specified.

The optimization problem (14) to (18) is a straightforward problem in the calculus of variations or other relevant theory. The main potential difficulty comes from the problem size, which may be prohibitive for large space regions or small mesh spacings or both. The underlying computational problem involves the solution of a two-point boundary-value problem of a size at least twice as large as the number of internal mesh points. It is in the reduction of this computational problem to a manageable size that decomposition methods can possibly benefit.

3. DECOMPOSITION

Consider now the above problem and assume that we have arranged all elements of $\underline{u}_i$ in some convenient fashion. Now partition the resulting $\underline{U}$ vector consisting of all ordered elements into N subvectors, $\underline{u}_j$, j = 1, 2,...,N. Assuming that the elements of the vectors $\underline{u}_i$ and $\underline{m}_i$ are separable in the functions $\underline{P}_j$, $\underline{\psi}$, and $\underline{R}_j$ appearing below, we can express the problem (14) to (18) as N independent problems of the form

$$\min_{\underline{M}_j} \sum_{j=1}^{N} \int_{t_o}^{t_1} P_j(\underline{U}_j, \underline{M}_j, t)dt \tag{19}$$

such that

$$\dot{\underline{U}}_j = \underline{G}_j(\underline{U}_j, \underline{M}_j, \underline{S}_j, t) \tag{20}$$

with initial conditions

$$\underline{U}_j(t_o) = \underline{U}_{jo} \tag{21}$$

and final conditions

$$\underline{\psi}_j(\underline{U}_j(t_1)) = \underline{0} \tag{22}$$

and subject to the inequality constraints

$$\underline{R}_j(\underline{U}_j, \underline{M}_j, t) \geq \underline{0} \tag{23}$$

In Eq. (20) we have introduced a new vector $\underline{S}_j$ called a pseudocontrol vector in order to achieve the subsystem independence. For each element or function of elements $\underline{U}_i + \underline{I}_k$ in Eq. (15) which is not contained in $\underline{U}_j$ but represents a coupling between $\underline{U}_j$ and some $\underline{U}_k$, a corresponding element s_j is introduced on a one-for-one basis. The variable s_j may represent a single variable u_k, a function of u_k, or a function of several variables. By collecting all such interactions into a vector variable $\underline{S}_j$, the interaction between subsystems can be separated out and the N lower-dimensional subsystems can be treated independently. For clarity we consider here the simplest form of this constraint, namely

$$\underline{S}_j = \sum_{\substack{k=1 \\ k \neq j}}^{N} \underline{C}_{jk}\, \underline{U}_k \tag{24}$$

where $\underline{C}_{jk}$ is an $(m_j \times u_k)$-dimensional matrix of ones and zeros where a one is used to indicate a coupling constraint and a zero otherwise.

Employing the canonical form of the calculus of variations, we define the Hamiltonian function H as follows:

$$H = \sum_{j=1}^{N} \left\{ P_j(\underline{U}_j, \underline{M}_j, t) + \underline{\lambda}_j{}'\underline{G}_j(\underline{U}_j, \underline{M}_j, \underline{S}_j, t) + \underline{\rho}_j{}' \left[\sum_{k=1}^{N} \underline{C}_{jk}\, \underline{U}_k - \underline{S}_j \right] \right\} \tag{25}$$

where the vectors $\underline{\lambda}_j$ are adjoint variables of dimension n_j and the vectors $\underline{\rho}_j$ are Lagrange multipliers of dimension m_j. The necessary conditions for a minimum can now be expressed in terms of H.

$$\dot{\underline{U}}_j = \frac{\gamma H}{\gamma \underline{\lambda}_j} = \underline{G}_j(\underline{U}_j, \underline{M}_j, \underline{S}_j, t) \tag{26}$$

$$\dot{\underline{\lambda}}_j = \frac{\gamma H}{\gamma \underline{U}_j} = - \frac{\gamma P_j}{\gamma \underline{U}_j} - \left(\frac{\gamma \underline{G}_j}{\gamma \underline{U}_j}\right)' \underline{\lambda}_j - \sum_{\substack{k=1 \\ k \neq j}}^{N} \underline{C}_{jk}' \underline{\rho}_k \tag{27}$$

$$\frac{\gamma H}{\gamma \underline{M}_j} = \frac{\gamma P_j}{\gamma \underline{M}_j} + \left(\frac{\gamma \underline{G}_j}{\gamma \underline{M}_j}\right)' \underline{\lambda}_j = \underline{0} \tag{28}$$

$$\frac{\gamma H}{\gamma \underline{S}_j} = \left(\frac{\gamma \underline{G}_j}{\gamma \underline{S}_j}\right)' \underline{\lambda}_j - \underline{\rho}_j = \underline{0} \tag{29}$$

$$\frac{\gamma H}{\gamma \underline{\rho}_j} = \sum_{k=1}^{N} \underline{C}_{jk} \underline{U}_k - \underline{S}_j = \underline{0} \tag{30}$$

The coupling terms in the Hamiltonian can be rearranged using the adjoint relationship as

$$\sum_{j=1}^{N} \underline{\rho}_j' \sum_{\substack{k=1 \\ k \neq j}}^{N} \underline{C}_{jk} \underline{U}_k = \sum_{k=1}^{N} \underline{U}_k' \sum_{\substack{j=1 \\ k \neq j}}^{N} \underline{C}_{jk}' \underline{\rho}_j \tag{31}$$

Interchanging subscripts in Eq. (31), we can now write H in the form

$$H = \sum_{j=1}^{N} H_j \tag{32}$$

where

$$H_j = P_j(\underline{U}_j, \underline{M}_j, t) + \underline{\lambda}_j' \underline{G}_j(\underline{U}_j, \underline{M}_j, \underline{S}_j, t) + \underline{U}_j' \sum_{\substack{k=1 \\ k \neq 1}}^{N} C_{kj}' \underline{\rho}_k - \underline{\rho}_j' \underline{S}_j$$

The functionals H_j are called the subproblem Hamiltonian functions. The last two terms no longer have the form of a constraint attached to the subsystem criterion function P_j with Lagrange multipliers. These terms can, however, be regarded as penalty terms for noncoordination of the subproblems. By minimizing each H_j independently and then coordinating the solutions by choosing $\underline{\rho}_k$, the decomposed minimization problems taken together will under certain conditions solve the overall problem. This type of Hamiltonian decomposition is an extension of the Lagrangian decomposition methods of mathematical programming (Lasdon, 1970).

Having decomposed this problem, the difficult task of coordination remains.

4. COORDINATION

Several methods of coordination for this problem have been examined corresponding to the feasible and the dual feasible methods of Lagrangian decomposition (Wismer, 1971). However, a third method suggested by the Gauss-Seidel procedure for obtaining a numerical solution to elliptic partial differential equations has proved to be the most computationally efficient. For this reason, only this method, termed the Gauss-Seidel coordination algorithm, is discussed here. In this method, the subproblems are defined by Eqs. (26)-(28) while the coordination problem is given by Eqs. (29) and (30).

In the Gauss-Sendel method, Eqs. (26) to (28) are first solved for each subproblem using some initial estimates of $\underline{\rho}_k$ and $\underline{S}_j$. The resulting values of $\underline{\lambda}_k$ and $\underline{U}_k$ are then used in Eqs. (29) and (30) to obtain new values for $\underline{\rho}_k$ and $\underline{S}_j$ respectively. The interconnection between subproblems is provided by the coordination algorithm since the calculation of inputs $\underline{\rho}_k$ and $\underline{S}_j$ to the jth subproblem depends in general upon the solution of several other subproblems. Changing subscripts in Eq. (29) for notational clarity, the Gauss-Seidel coordination algorithm is given by

$$\underline{\rho}_k = \left(\frac{\gamma \underline{G}_k}{\gamma \underline{S}_k}\right)' \underline{\lambda}_k \tag{33}$$

$$\underline{S}_j = \sum_{k=1}^{N} \underline{C}_{jk} \underline{U}_k \tag{34}$$

The coordination algorithm is thus seen to consist of explicit relations which are quickly and easily solved on a digital computer. One method of solution is to solve all N subsystems using estimates for the vector inputs $\underline{\rho}_k$ and $\underline{S}_j$ to each subsystem. The outputs of these N subsytems are then combined by the coordination algorithm to form new input values. This approach is analogous to the Jacobi method for solving elliptic differential equations. An alternative approach, the Gauss-Seidel method, utilizes each new input value $\underline{\rho}_k$ or $\underline{S}_j$ for the jth subsystem as soon as it arises from the solution of some other subsystem. This method requires initial estimates of only half as many input values as the Jacobi method. A third method for solving partial differential equations, the method of overrelaxation, can be extended to this multilevel application. This technique requires the determination of a relaxation factor which, if properly chosen, speeds convergence over the Gauss-Seidel method but at the expense of additional computation. However, computational efficiency is of extreme importance in solving large problems, and the Gauss-Seidel method is considered perferable. Convergence conditions for this method require that Eqs. (26) and (27) satisfy a Lipschitz condition as discussed by Wismer (1971).

5. PROCESS CONTROL EXAMPLE

To demonstrate the method, we consider the problem of heating a long, thin rod with a given initial temperature distribution to some specified final temperature distribution with the minimum amount of energy. This problem can be posed as follows:

$$\min_{m} \int_0^1 \int_0^{t_1} m^2(x,t)dt\ dx \tag{35}$$

subject to the nonlinear diffusion equation

$$\frac{\gamma u}{\gamma t} = \alpha u \frac{\gamma^2 u}{\gamma x^2} + m \tag{36}$$

with boundary and initial conditions given by

$$u(0,x) = u(1,t) = 0$$

and

$$u(x,0) = u_0(x)$$

respectively. It is desired to attain the terminal state given by

$$u(x,t_1) = u_1(x)$$

at a specified time t_1.

The semidiscrete approximation to this problem is stated as

$$\min_{\underline{M}} \int_0^{t_1} \underline{M}'\underline{M}dt$$

subject to the differential equation

$$\dot{\underline{U}} = \underline{A(\underline{U})\underline{U}} + \underline{M}$$

$$\underline{U}(t_0) = \underline{U}_0$$

$$\underline{U}(t_1) = \underline{U}_1$$

where

$$\underline{U} = [u_1,u_2,\ldots,u_n]'$$

$$\underline{M} = [m_1,m_2,\ldots,m_n]'$$

If we partition $\underline{U}$ into two fourth-order subproblems, the state-dependent matrix $\underline{A}(\underline{U})$ becomes

$$\underline{A}(\underline{U}) = k \left[\begin{array}{cccc|cccc} -2u_1 & u_1 & & & & & & \\ u_2 & -2u_2 & u_2 & & & 0 & & \\ & u_3 & -2u_3 & u_3 & & & & \\ & & u_4 & -2u_4 & u_4 & & & \\ \hline & & & u_5 & -2u_5 & u_5 & & \\ & & & & u_6 & -2u_6 & u_6 & \\ & 0 & & & & u_7 & -2u_7 & u_7 \\ & & & & & & u_8 & -2u_8 \end{array}\right]$$

where $k = \alpha h^{-2}$ and h is the discretization interval. The Hamiltonian is is given by

$$H = \sum_{j=1}^{2} \left\{ \underline{M}_j{}'\underline{M}_j + \underline{\lambda}_j{}'[\underline{A}_{jj}(\underline{U}_j)\underline{U}_j + \underline{M}_j + \underline{D}_j(\underline{U}_j)\underline{S}_j] + \underline{\rho}_j'\left(\sum_{\substack{k=1 \\ k\neq j}}^{2} \underline{C}_{jk}\underline{U}_k - \underline{S}_j \right) \right\} \tag{37}$$

where

$$\underline{D}_1 = \underline{A}_{12} \qquad \underline{D}_2 = \underline{A}_{21}$$

and $\underline{C}_{12}$, $\underline{C}_{21}$ are 4 × 4 matrices of zeros except for a single one in the upper-left and lower-right corners respectively. In this problem, it is necessary to substitute the elements of $\underline{S}_j$ for the elements of $\underline{U}_k$ on a one for one basis because otherwise the multiplicative form of the nonlinearity couples the subproblems through the adjoint equation (27). The pseudocontrol variables and Lagrange multipliers are defined by

$$\underline{S}_1 = [s_1, s_2, s_3, s_4]' \qquad \underline{S}_2 = [s_5, s_6, s_7, s_8]'$$

$$\underline{\rho}_1 = [\rho_1, \rho_2, \rho_3, \rho_4]' \qquad \underline{\rho}_2 = [\rho_5, \rho_6, \rho_7, \rho_8]'$$

By applying the necessary conditions, the subsystem two-point boundary-value problems are readily seen to be

SUBPROBLEM 1:

$$\begin{aligned} \dot{u}_1 &= k[\qquad - 2u_1^2 + u_1u_2] - 2^{-1}\lambda_1 \\ \dot{u}_2 &= k[u_1u_2 - 2u_2^2 + u_2u_3] - 2^{-1}\lambda_2 \\ \dot{u}_3 &= k[u_2u_3 - 2u_3^2 + u_3u_4] - 2^{-1}\lambda_3 \\ \dot{u}_4 &= k[u_3u_4 - 2u_4^2 + u_4s_1] - 2^{-1}\lambda_4 \\ \lambda_1 &= -k[\qquad (-4u_1 + u_2)\lambda_1 + u_2\lambda_2] \\ \lambda_2 &= -k[u_1\lambda_1 + (u_1 - 4u_2 + u_3)\lambda_2 + u_3\lambda_3] \end{aligned} \tag{38}$$

$$\lambda_3 = -k[u_2\lambda_2 + (u_2 - 4u_3 + u_4)\lambda_3 + u_4\lambda_4]$$

$$\lambda_4 = -k[u_3\lambda_3 + (u_3 - 4u_4 + s_1)\lambda_4] - \rho_8$$

with boundary conditions

$$u_i(0) = u_{io} \qquad u_i(t_1) = u_{i1}$$

$$i = 1,2,3,4$$

SUBPROBLEM 2:

$$\begin{aligned}
\dot{u}_5 &= k[-2u_5^2 + u_5u_6 + u_5s_8] - 2^{-1}\lambda_5 \\
\dot{u}_6 &= k[u_5u_6 - 2u_6^2 + u_6u_7] - 2^{-1}\lambda_6 \\
\dot{u}_7 &= k[u_6u_7 - 2u_7^2 + u_7u_8] - 2^{-1}\lambda_7 \\
\dot{u}_8 &= k[u_7u_8 - 2u_8^2 \qquad] - 2^{-1}\lambda_8 \\
\dot{\lambda}_5 &= -k[(s_8 - 4u_5 + u_6)\lambda_5 + u_6\lambda_6] - \rho_1 \\
\dot{\lambda}_6 &= -k[u_5\lambda_5 + (u_5 - 4u_6 + u_7)\lambda_6 + u_7\lambda_7] \\
\dot{\lambda}_7 &= -k[u_6\lambda_6 + (u_6 - 4u_7 + u_8)\lambda_7 + u_8\lambda_8] \\
\dot{\lambda}_8 &= -k[u_7\lambda_7 + (u_7 - 4u_8)\lambda_8]
\end{aligned} \tag{39}$$

with boundary conditions

$$u_i(0) = u_{io} \qquad u_i(t_1) = u_{i1}$$

$$i = 5,6,7,8$$

In solving these subproblems, s_i and ρ_k are considered parameters which are determined by the coordination algorithm. Using Gauss-Seidel coordination, these values are determined by Eqs. (33) and (34) as

$$s_1 = u_5 \qquad \rho_8 = ku_5\lambda_5 \tag{40}$$

$$s_8 = u_4 \qquad \rho_1 = ku_4\lambda_4 \tag{41}$$

In practice, the solution to this problem is limited by the availability of a computational method for solving the nonlinear two-point boundary value problems given by (38) and (39). In this case the four mesh point subproblems require the solution of eighth order two-point boundary-value problems after the four adjoint equations are added. If this decomposition scheme were not used, the dimensionality of the problem would be sixteenth order.

The two level computational scheme requires an initial estimate of s_1 and ρ_8. The steps are given below:

1. Estimate $u_5(t)$, $\lambda_5(t)$ and set j=1 where j is the subsystem index.
2. If j=1, determine $s_1(t)$, $\rho_8(t)$ from (40).
   If j=2, determine $s_8(t)$, $\rho_1(t)$ from (41).
3. Solve subproblem j for $\underline{u}_j(t)$, $\underline{\lambda}_j(t)$

from Eqs. (38) or (39) using the latest information.

4. Is j = N?

   No - Set j = j+1 and go to step 2.

   Yes - Are Eqs. (40) and (41) satisfied to desired accuracy for j=1,2,...,N?

   No - Set j=j-1 and go to step 2.

   Yes - Stop.

This procedure has been found to have excellent convergence properties for both linear and nonlinear problems. The good convergence is attributed to the strongly diagonal character of the $\underline{A}$ matrix. This same factor speeds convergence of the subsystems over a wide range of initial estimates for $\underline{\rho}_k$ and $\underline{S}_j$. It is because of these convergence properties that this method of solution is attractive.

Numerical results have been obtained for this problem using the method of quasi-linearization to solve the two-point boundary-value problems. Taking k as 0.27, and $[0,t_1]$ as [0,5], the controls $\underline{M}_j$ and states $\underline{U}_j$ are given in Fig. 1 for the initial and final conditions shown. By symmetry, both subsystems have the same response. For initial estimates of

$$s_1(t) = 50 \qquad \rho_8(t) = 0$$

the convergence of the Gauss-Seidel algorithm was monitored by the norms

$$||e_1||_i = \int_0^5 \{|u_4 - s_8|_i + |ku_4\lambda_4 - \rho_1|_i\}dt \tag{42}$$

$$||e_2||_i = \int_0^5 \{|u_5 - s_1|_i + |ku_5\lambda_5 - \rho_8|_i\}dt$$

where i represents the iteration number. The values obtained for this example are shown in Fig. 2.

It is sometimes of interest to monitor the Hamiltonian function since for autonomous systems with prescribed terminal conditions H is constant at the optimum. Checking H for this condition serves as a convenient check of the numerical techniques. Of course the subystem Hamiltonians H_j of Eq. (37) are not necessarily constant since the optimum controls for the overall problem are not in general optimum for the subsystems taken separately.

6. AQUIFER IDENTIFICATION EXAMPLE

The pressure distribution in an underground reservoir is modeled by a linear parabolic partial differential equation of the form

$$\frac{\gamma}{\gamma x}\left(\tau \frac{\gamma P}{\gamma x}\right) + \frac{\gamma}{\gamma y}\left(\tau \frac{\gamma P}{\gamma y}\right) = \sigma \frac{\gamma P}{\gamma t} + q \tag{43}$$

Pressure is denoted by P and q represents a source strength (production rate per unit area) at some producing well. The coefficients in the equation characterize the porous medium. Transmissibility $\tau(x,y)$ is a measure of the ease with which fluid moves through the system and storage $\sigma(x,y)$ is a measure of system capacity. Vertical fluid flow is assumed to be negligible.

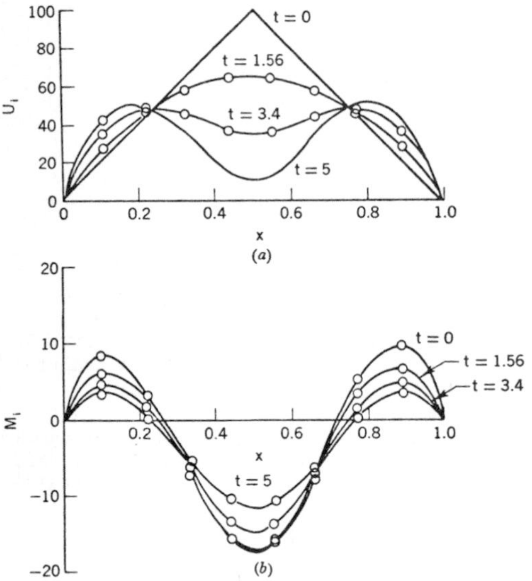

Fig. 1(a) Optimal State Response
(b) Optimal Distributed Controls

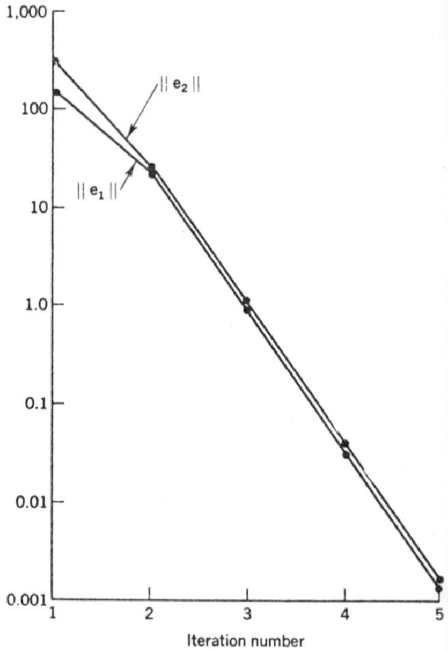

Fig. 2 Gauss-Seidel Convergence

We wish to determine the distributed parameters $\tau(x,y)$ and $\sigma(x,y)$ by fitting the model in a least-squares sense to observed pressure data obtained at a number of wells in the reservoir. When each well is producing, boundaries with zero potential gradient (no flow) must exist between wells. Fluid on opposite sides of these "boundaries" flows to opposite wells just as if the line of separation were impermeable. A convenient approximation in modeling the system is to separate regions containing a single well by straight-line boundaries. A pair of boundary lines on opposite sides of a well will intersect to form a wedge. As a result the model for N wells in a large reservoir is obtained by dividing the system into N wedge-shaped homogeneous regions, each radiating from a single arbitrarily selected origin and each enclosing a single well. We denote the ith well location by the angle θ_i. It is bounded by two boundaries at locations given by angles α_i and α_{i-1} where

$$\theta_i < \alpha_i < \theta_{i+1}$$

The distributed parameters $\tau(x,y)$ and $\sigma(x,y)$ are approximated by constant "average" values in each wedge-shaped region. The identification scheme must then determine values for the N-dimensional vectors $\underline{\tau}$ and $\underline{\sigma}$ composed of elements from each region along with the location of the N well boundaries $\underline{\alpha}$. Thus the geometry of the model is actually provided by the system behavior.

Indicating the model pressure by u to distinquish it from observed pressure values and employing the Boltzmann transformation, we can write Eq. (43) in a more convenient form

$$\lambda\frac{d^2u}{d\lambda^2} + (1 + \lambda)\,\frac{du}{d\lambda} = 0 \tag{44}$$

with boundary conditions

$$\lim_{\lambda\to\infty} u(\lambda) = P_1 \qquad \lim_{\lambda\to 0} 2\lambda\frac{du}{d\lambda} = \frac{q}{2\pi\tau}$$

where

$$\lambda = \frac{\sigma r^2}{4\tau t}$$

The solution of Eq. (44) is readily obtained in terms of the exponential integral $E(\xi)$:

$$u = P_1 - \frac{q}{4\pi\tau}\int_{\xi=\lambda}^{\infty} \xi^{-1}\exp(-\xi)d\xi = P_1 + \frac{q}{4\pi\tau} E\left(-\frac{\sigma r^2}{4\tau t}\right)$$

The method of images can be used to extend the solution to problems in a bounded, wedge-shaped, homogeneous region. (Dewiest, 1965). Suppose the ith actual image includes an angle of $2\pi/m_i$ radians between azimuths α_i and α_{i+1}, where m_i is an even integer. The image system then contains m_i wedges filling the entire plane. Each wedge has the same origin and contains a well which is the mirror image of the actual well and/or image wells across each of its adjacent boundaries. The angle between the ith well with azimuth θ_i and its kth image is given by $2\zeta_{ik}$ where

$$\zeta_{ik} = \left[\frac{k}{2}\right](\alpha_i - \theta_i) + \left[\frac{k-1}{2}\right](\theta_i - \alpha_{i-1} + 2\pi\delta_{i1}) \tag{45}$$

$$i = 1,2,\ldots,N \qquad k=1,2,\ldots,m_i$$

and δ_{ik} is the Kronecker delta. The notation $[k/2]$ and $[(k-1)/2]$ denotes that these quantities are truncated to integer values. The pressure computed at the ith well and the jth time corresponds to production q_i and is denoted by u_{ij}. Summing the pressure effects caused by m_i image wells for each of N actual wells yields

$$u_{ij} = P_1 + \frac{q_i}{4\pi\tau}\sum_{k=1}^{m_i} E_i\left(-\frac{\sigma_i r_i^2 \sin^2 \zeta_{ik}}{4\tau_i t_j}\right) \tag{46}$$

where r_i is the radial distance between the origin and the ith well. The trigonometric term arises directly from the model geometry.

We can now formulate this problem as a two-level optimization problem where each wedge-shaped region is considered as a subproblem. Since individual subproblems are coupled only by ζ_{ik}, the angle between the ith well and the kth image, we can regard these subsystems as independent after making the simple change of variables

$$\alpha_{i-1} = s_i \qquad i = 1,2,\ldots,N \tag{47}$$

where the s_i are the pseudocontrol variables. The decomposed static optimization problem can now be written as follows:

$$\min_{\tau_i,\sigma_i,\alpha_i,s_i}\left\{f = \sum_{i=1}^{N}\sum_{j=1}^{N}\left[P_1 + \frac{q_i}{4\pi\tau_i}\sum_{k=1}^{m_i} E_i\left(\frac{-\sigma_i r_i{}^2\sin^2\zeta_{ik}}{4\tau_i t_j}\right) - P_{ij}\right]^2\right\} \tag{48}$$

subject to the constraints

$$\underline{G}_i(\alpha_i,\tau_i,\sigma_i) = \begin{bmatrix} \theta_{i+1} - \varepsilon_i - \alpha_i \\ \alpha_i - \varepsilon_i - \theta_i \\ \tau_i - \beta_i \\ \sigma_i - \gamma_i \end{bmatrix} \geq \underline{0} \qquad i = 1,2,\ldots,N \tag{49}$$

where $m_i = 2\pi/(\alpha_i - s_i)$, ζ_{ik} is given by Eq. (45), and ε_i, β_i and γ_i are arbitrary small constants which insure that the inequality holds strictly in Eq. (49).

Equation (48) is separable and can be written

$$f(\underline{X};\ \underline{s}) = \sum_{i=1}^{N} f_i(\underline{X}_i;\ s_i)$$

$$X_i \triangleq (\tau_i,\ \sigma_i,\ \alpha_i)$$

$$\underline{X} = [\underline{X}_1,\ \underline{X}_2,\ \ldots,\ \underline{X}_N]'$$

Regarding s_i as a known parameter in the ith subsystem, the subsystems are uncoupled. Hence each subsystem optimization is performed by

$$\min_{\underline{X}_i} f_i(\underline{X}_i;\ s_i)$$

subject to the constraints (49).

In order to insure the minimization of Eq. (48) for N wells, we define the Lagrangian

$$L(\underline{X},\ \underline{\lambda},\ \underline{s},\ \underline{\mu}) = \sum_{i=1}^{N} L_i(\underline{X}_i,\ \underline{\lambda}_i;\ s_{i+1},\ s_i,\ \mu_i)$$

where

$$L_i = f_i(\underline{X}_i;\ s_i) + \underline{\lambda}_i'\underline{G}_i(\underline{X}_i) + \mu_i(\alpha_i - s_{i+1})$$

$$\underline{s} = [s_1,\ s_2,\ \ldots, s_N]'$$

$$\underline{\lambda} = [\lambda_1,\ \lambda_2,\ \ldots,\ \lambda_N]'$$

$\underline{\mu}$ = N-dimensional vector Lagrange multiplier

$$s_{N+1} \triangleq s_1$$

Assuming that the Kuhn-Tucker constraint qualification holds, the necessary conditions for a minimum of each subsystem are

$$\nabla_{\underline{X}_i} L_i(\underline{X}_i^*,\underline{\lambda}_i^*;\ s_i,\ s_{i+1},\ \mu_i) = \underline{0}$$

$$\lambda_i'\ \nabla_{\underline{\lambda}_i} L_i(\underline{X}_i^*,\ \underline{\lambda}_i^*;\ s_i,\ s_{i+1},\ \mu_i) = 0 \tag{50}$$

$$\lambda_{ij}^* \leq 0 \qquad i=1,2,\ldots,N \qquad j=1,2,3,4$$

The solution of Eq. (50) proceeds for given values of the parameters s_i, s_{i+1}, and μ_i by any of the standard nonlinear programming algorithms. It remains to determine these parameter values by satisfying the remaining necessary conditions for a minimum; namely,

$$\begin{aligned} \nabla_{s_{i+1}} L &= \nabla_{s_{i+1}} f_{i+1} - \mu_i^* = 0 \quad i = 1,2,\ldots,N \\ \nabla_{\mu_i} L &= \alpha_i - s_{i+1}^* = 0 \quad i = 1,2,\ldots,N \end{aligned} \tag{51}$$

Equations (51) are just the Gauss-Seidel coordination algorithm previously discussed. The solution proceeds iteratively until

$$||\underline{s}^{(k+1)} - \underline{s}^{(k)}|| \leq \delta_1$$

$$||\underline{\mu}^{(k+1)} - \underline{\mu}^{(k)}|| \leq \delta_2$$

where k is the iteration number and δ_1 and δ_2 are specified scalar values.

A numerical example has been solved using four wells and seven pressure observations for each well. A direct-search technique was found to be efficient for solving the subsystems and converged in six iterations. These results are given in detail by Wismer, et al. (1970).

7. GAS PIPELINE CONTROL EXAMPLE

The basic equations describing compressible flow in a pipeline have been discussed by Larson and Wismer (1971) and are summarized below:

Basic Equations for Transient Flow

$$\frac{\partial P(x,t)}{\partial t} = -C_1 \frac{\partial Q(x,t)}{\partial x} \tag{52}$$

$$\frac{\partial P^2(x,t)}{\partial x} = -C_2 |Q(x,t)| Q(x,t) \tag{53}$$

where t = time
x = distance along the pipeline
P(x,t) = pressure at distance x along the pipeline at time t
Q(x,t) = mass flow at distance x along the pipeline at time t
C_1, C_2 = pipeline constants.

The first equation expresses the conservation of mass. The second equation contains the dominant terms in Newton's second law. This equation defines the pressure gradient necessary to maintain a specified flow. Since the primary mechanism for flow in pipelines is diffusion, the momentum rate of change can be neglected by comparison with this diffusion effect.

The Nonlinear Diffusion Model

If Eqs. (52) and (53) are adopted as the fundamental equations for transient flow, the change of variable $U(x,t) = P^2(x,t)$ allows reformulation of the gas pipeline equations as a nonlinear diffusion process. The equations become

$$\frac{\partial U(x,t)}{\partial t} = \phi\left(U, \frac{\partial U}{\partial x}\right) \frac{\partial^2 U(x,t)}{\partial x^2} \tag{54}$$

where

$$\phi\left(U, \frac{\partial U}{\partial x}\right) = \frac{C_1}{\sqrt{C_2}} \left[\frac{U(x,t)}{\partial U(x,t)/\partial x}\right]^{\frac{1}{2}} \tag{55}$$

Using finite differences to express the variation in the space domain, these equations can be written

$$\dot{u}_i(t) = \phi_i(t)[u_{i+1}(t) + u_{i-1}(t) - 2u_i(t)] \tag{56}$$

where

$$\phi_i(t) = \frac{C_1}{h^2\sqrt{C_2}} \left[\frac{2h\ u_i(t)}{u_{i+1}(t) - u_{i-1}(t)}\right]^{1/2}$$

and h is the discretization interval.

Compressors

The basic control element in a pipeline network is the compressor. The jth compressor raises the inlet (suction) pressure, P_{sj}, to a higher outlet (discharge) pressure, P_{dj}. This higher outlet pressure then provides a pressure gradient to maintain flow in the next pipeline segment. Under the assumption of adiabatic compression, the horsepower required to maintain a specific compression ratio $(P_{dj}/P$ for a specified flow Q_{cj} is given by

$$(HP)_j = A_j Q_{cj}\left[\left(\frac{P_{dj}}{P_{sj}}\right)^{R_j} - B_j\right] \tag{57}$$

where

P_{dj} = discharge pressure for j^{th} compressor

P_{sj} = suction pressure for j^{th} compressor

Q_{cj} = flow through j^{th} compressor

$(HP)_j$ = horsepower required to achieve compression ratio P_{dj}/P_{sj} and flow Q_{cj} at compressor j

and A_j, B_j, R_j are constants for the jth compressor. The constant R_j is always smaller than 1, and it is typically between 0.2 and 0.3. The constant B_j is approximately 1.0, while the constant A_j depends on the compressor characteristics and the engineering units adopted.

If x_j is the distance coordinate of the jth compressor station, then the compressor imposes the following boundary conditions on the pipeline.

$$P(x_j^-, t) = P_{sj}(t) \tag{58}$$

$$P(x_j^+, t) = P_{dj}(t) \tag{59}$$

$$Q(x_j, t) = Q_{cj}(t) \tag{60}$$

The cost of running a compressor is obtained by integrating the horsepower over the control iterval. Thus

$$J_j = \int_{t_o}^{t_f} A_j Q_{cj}(t) \left[\left(\frac{P_{dj}(t)}{P_{sj}(t)} \right)^{R_j} - B_j \right] dt \tag{61}$$

Constraints on compressor operation include minimum and maximum values for flow, pressure, and compression ratio. These are expressed as

$$Q_{cj}^- \leq Q(x_j, t) \leq Q_{cj}^+ \tag{62}$$

$$P_{cj}^- \leq P(x_j^-, t) \leq P(x_j^+, t) \leq P_{cj}^+ \tag{63}$$

$$1 \leq r^- \leq \left(\frac{P(x_j^+, t)}{P(x_j^-, t)} \right) \leq r^+ \tag{64}$$

Junctions

A critical element in modeling large interconnected networks is the pipeline junction. Modeling of the junction depends on two physical principles: first, all pipelines meeting at the junction must have a common pressure at the junction point; and second, the instantaneous mass flow into the junction must be zero.

These principles can be used to establish a set of simultaneous nonlinear equations at these junctions. Thus, for a junction of N pipelines, if the junction pressure is denoted as $P_J(t)$ and the flows from each pipeline into the junction are denoted as $Q_{Ji}(t)$, i=1,2,...,N, then values for these (N+1) quantities are sought from the following (N+1) equations:

(i) N equations relating $P_J(t)$ and $Q_{Ji}(t)$ to the pressure values at sections further into the N pipelines

(ii) An additional equation for zero instantaneous mass flow into the junction, i.e.,

$$\sum_{j=1}^{N} Q_{Ji} = 0 \tag{65}$$

Delivery Points

The mth delivery point in a pipeline network, which has distance coordinate x_m, is characterized by the time-varying demand flow $D_m(t)$. This imposes a boundary condition of the form

$$Q(x_m^-, t) - Q(x_m^+, t) = D_m(t) \tag{66}$$

Typically, this flow must be delivered at a pressure greater than or equal to a specified contract pressure, P_{cm}. This imposes the constraint

$$P(x_m, t) \geq P_{cm} \, . \tag{67}$$

Sources

The ℓth source of gas, located at distance coordinate x_ℓ, is modeled in terms of its pressure and flow characteristics. For operation over a period of 24 hours or less, it is generally sufficient to assume that the source pressure is constant at $P_{w\ell}$ and that any desired flow, $Q_{w\ell}(t)$, can be furnished by the source provided this flow is between limits $Q^-_{w\ell}$ and $Q^+_{w\ell}$. The boundary conditions are thus

$$P(x_\ell, t) = P_{w\ell} \tag{68}$$

$$Q(x^+_\ell, t) - Q(x^-_\ell, t) = Q_{w\ell}(t) \tag{69}$$

where the input flow $Q_{w\ell}(t)$ is constrainted by

$$Q^-_{w\ell} \leq Q_{w\ell}(t) \leq Q^+_{w\ell} \tag{70}$$

Storage Fields

Storage fields are generally located close to demand points. They are filled during periods of low demand, and the gas in them is used during periods of high demand. For large storage fields a system of valves and compressors may be present at the inlet/outlet. Over a period of 24 hours or less the pressure at a large storage field can generally be assumed constant. Thus, the storage field behaves as either a constant pressure source or constant pressure demand point, depending on the relative pressure at other demand points.

Changes in Diameter and Pipeline Parameters

The pipeline parameters C_1 and C_2 in Eqs. (1) and (2) depend on specific characteristics of the pipeline, such as diameter, friction factor, etc. Whenever there is a change in one of these quantities, the parameters C_1 and C_2 must be modified. By choosing the boundaries of the discrete pipeline sections to lie at points where these parameters change, it is straightforward to simulate the changes.

CONTROL OF A NETWORK DRIVEN BY SINGLE COMPRESSOR

Formulation of the Control Problem

The basic unit for dynamic control is a single compressor and the network which it drives. As a minimum, this network will consist of a length of pipe leading to the next compressor, which acts as an equivalent demand point, and an equivalent source at the input to the compressor.

The dynamic control problem can be formulated as follows:

Given: (A) A compressor having operating costs over the control interval (on the order of 24 hours) given by Eq. (61); operating constraints given by Eqs. (62)-(64); and boundary conditions with the pipeline given by Eqs. (58)-(60).

(B) A pipeline network driven by the compressor and consisting of connected pipeline lengths and any or all of the elements discussed earlier, but including as a minimum the following:

(i) At least one demand point (possibly the equivalent demand point at the next compressor station) with a specified mass flow output as in Eq. (66) and a specified minimum delivery pressure as in Eq. (67).

(ii) At least one source (possibly the equivalent source at the input to the compressor) with a specified pressure, as in Eq. (68), and with mass flow input constraints as in Eq. (70).

(iii) At least one length of pipeline connecting the source and demand point in which the basic pipeline equations, Eqs. (52) and (53), are satisfied and for which maximum and (possibly) minimum values of pressure and flow are specified everywhere along the pipeline.

Find: The compressor operating policy over the control interval that minimizes compressor operating costs while satisfying all the constraints discussed above, including all constraints on compressor operation, demand points, sources, and pipeline lengths.

Solution: One control policy which yields a near-optimum solution has been obtained by Wong, et al. (1968). The policy can be stated as follows: Operate the compressor such that at all times the demanded flow is exactly furnished at all demand points and the minimum delivery pressure constraint is just met at one or more delivery points and exceeded at all others.

Control of Pipeline Networks

The optimal control problem is now formulated for a pipeline network consisting of N compressor-pipeline segments. For this problem, the objective is taken to be the minimization of energy input to the compressors.

$$\min_{U_{sj}(t),\ U_{dj}(t)} \sum_{j=1}^{N} \int_{t_o}^{t_t} A_j Q_{cj}(t) \left[\left(\frac{U_{dj}(t)}{U_{sj}(t)} \right)^{R_j/2} - B_j \right] dt \tag{71}$$

where from Eq. (53)

$$Q_{cj}(t) = +\left[\frac{1}{-C_2} \frac{\partial U_{dj}}{\partial x}(t) \right]^{1/2} = +\left[\frac{1}{-C_2} \frac{\partial U_{sj}}{\partial x}(t) \right]^{1/2} \tag{72}$$

If the jth pipeline segment is divided into n_j parts, the constraints can be written

$$\dot{U}_{ij}(t) = \phi_{ij}(t)[U_{i+1,j}(t) + U_{i-1,j}(t) - 2U_{ij}(t)] \tag{73}$$

where

$$\phi_{ij}(t) = \frac{C_1}{h^2\sqrt{C_2}} \left[\frac{2h\, U_{ij}(t)}{U_{i+1,j}(t) - U_{i-1,j}(t)} \right]^{1/2}$$

$$0 \leq U_{ij}(t_k) \leq U^+_{ij}(t_k)$$

$$0 \leq Q_j(t_k) \leq Q^+_j(t_k)$$

with initial conditions

$$U_{ij}(t_o) = U_{ijo} \qquad i=1,2,\ldots,n_j-1$$
$$j = 1,2,\ldots,N$$

The constraints at demand points are given by

$$U(x_{n_j}, t) \leq P_{cj}^2 \tag{74}$$
$$Q(x_{n_j}, t) = D_j(t)$$

where $j \varepsilon S_D$, the set of all segments leading to demand points. The constraints at supply points are given by

$$U(x_{oj}, t) = U_{oj}$$
$$Q_{oj}^- \leq Q(x_{oj}, t) \leq Q_{oj}^+ \tag{75}$$

where $j \varepsilon S_s$, the set of all segments emanating from supply points.

Considering now that the jth compressor pumps <u>into</u> the jth pipeline segment, the boundary conditions at each compressor are

$$U_{dj}(t) = U(x_{oj}, t) \tag{76}$$

$$U(x_{n_j}, t) = U_{sj}(t) \qquad j = 1,2,\ldots, N \tag{77}$$

$$Q(x_{n_j}, t) = Q(x_{oj}, t) \tag{78}$$

Substituting from Eq. (72)

$$\frac{\partial U}{\partial x}(x_{n_j}, t) = \frac{\partial U}{\partial x}(x_{oj}, t)$$

and using the backward and forward differences along with Eqs. (76) - (78), these boundary conditions can be replaced by

$$U_{sj}(t)-U(x_{n_{j-1}}, t) = U(x_{1j},t) - U_{dj}(t) \qquad j = 1,2,\ldots,N \tag{79}$$

If it is assumed (in keeping with good operating practice from the viewpoint of compressor maintenance) that U_{sj} should be kept constant, then, for $R_j < 1$, the criterion function is concave in $U_{dj}(t)$. In this case, optimizing each pipeline segment independently leads to the overall optimum if all boundary conditions and constraints are satisfield. The optimal policy of determining $U_{dj}(t)$ for each independent segment was discussed in the preceeding section. The task of the higher level optimizer is to select U_{sj}, $j=1,2,\ldots,N$, such that Eq. (79) is satisfied along with the demand and supply constraints, Eq. (74)-(75).

Clearly the minimum energy solution occurs when the suction and discharge pressures of each compressor are as close as possible consistent with constraints. Thus, one procedure for the second-level optimization is to choose U_{sj} to

$$\min_{U_{sj}} \; ||U_{dj}(t) - U_{sj}|| \qquad j = 1,2,\ldots,N \tag{80}$$

For example, one obvious norm is to make U_{sj} equal to the minimum value of $U_{dj}(t)$ over the time interval. By beginning the computation for all those segments leading to a demand point ($j \in S_D$), U_{sj} can be obtained from Eq. (80) explicitly. This value of U_{sj} in turn provides a known demand for all adjoining internal pipe segments. This procedure continues until all internal segments have been optimized. For those segments leading from a source to a compressor ($j \in S_s$), the procedure is to compute $U^+(x_{n_j}, t)$ and $U^-(x_{n_j}, t)$ using Eq. (73) corresponding to boundary conditions $Q(x_{oj}, t) = Q^+_{oj}$ and $Q(x_{oj}, t) = Q^-_{oj}$ respectively with $U(x_{oj}, t) = U_{oj}$ as in Eq. (75). If the suction pressures obtained by Eq. (80) satisfy

$$U^-(x_{n_j}, t) \le U_{s,j+1} \le U^+(x_{n_j}, t) \qquad (81)$$

for all t, and $j \in S_s$, then the solution is feasible. If Eq. (81) is not satisfied, increase U_{sj}, j = 1, 2,...,N by small increments until Eq. (81) is satisfied. This procedure will lead to a good feasible (although perhaps suboptimal) solution for the pipeline network.

8. CONCLUSION

This paper has reviewed an approach for solving the large sets of difference and differential-difference equations which result from the optimization of systems described by discrete approximations to partial differential equations. The approach suggested involves decomposition and a two-level structure for coordination. The method is computationally effective because of the diagonal dominance and sparsity resulting from the discrete problem formulation.

Although detailed computational results are given elsewhere (see references), three realistic example problems have been formulated to show how the proposed methods can be used in practice to solve complex problems.

9. REFERENCES

1. DeWiest, R. J. M., Geohydrology, Wiley, 1965.

2. Larson, R. L. and D. A. Wismer, "Hierarchical Control of Transient Flow in Natural Gas Pipeline Networks," IFAC Symposium on Distributed Parameter Systems, Banff, Alberta, Canada, June 1971.

3. Todd, John, Editor, Survey of Numerical Analysis, McGraw Hill, 1962.

4. Varga, R. S., Matrix Iterative Analysis, Prentice Hall, 1962.

5. Wang, P. K. C., "Control of Distributed Parameter Systems," Vol. 1, C. T. Leondes, Editor, Academic Press, 1964.

6. Wismer, D. A., "Distributed Multilevel Sysrtems," Ch. 6 in Optimization Methods for Large-scale Systems, D. A. Wismer, Editor, McGraw Hill, 1971.

7. Wismer, D. A., R. L. Perrine, and Y. Y. Haimes, "Modeling and Identification of Aquifer Systems of High Dimension," Automatica, Vol. 6, pp. 77-86, Pergamon Press, London, 1970.

AGGREGATIVE CONTROL OF CHEMICAL PLANTS

A. L. Diederich[1] and S. George Bankoff
Chemical Engineering Department, Northwestern University
Evanston, Illinois 60201

Abstract: Aggregative control theory, as developed by Aoki (Aoki, 1968a,b, 1971) for suboptimal control of large-scale linear systems, involves a decomposition into a dominant-eigenvalue subsystem and a subsystem with more rapid decay characteristics. An alternative approach employs an aggregation matrix dictated by the available measurements. Both methods are applied to the control of a stable gas absorber column, and of a stirred-tank chemical reactor with multiple consecutive reactions operated around the unstable steady-state point. As expected, the quality of control based on a decomposition dictated by measurements is inferior to the modal control, but is nevertheless quite acceptable.

1. INTRODUCTION

The vast majority of the flow processes which occur in chemical processing, as well as in many other applications, can be modeled by systems of hyperbolic or parabolic differential equations, or by sets of ordinary differential equations representing conservation equations for multiple-stage processes. In order to reduce the partial differential equations to sets of ordinary differential equations, it is common to introduce two- or three-point finite-difference approximations for the spatial derivatives. In either case, one is led to sets of ordinary differential equations governed by large, sparse matrices. For nearly all applications, these matrices are a specialized form of Jacobi matrices. After linearization and/or finite-differencing, if necessary, one arrives at a linear system of the form:

$$\dot{x}(t) = Ax(t) + Bu(t) \;;\; x(t_0) = x_0 \qquad (1)$$

where x is an n-dimensional state vector, u is an r-dimensional control vector, and A and B are constant matrices. Moreover, in many cases $A = (a_{ij})$, i, j = 1, ...,n is a stable Jacobi matrix, with entries of the form

$$a_{ij} = \begin{cases} \beta > 0 & i = j + 1 \\ = \gamma > 0 & i = j - 1 \\ = -\beta - \gamma & i = j \\ = 0 & \text{otherwise} \end{cases}$$

The eigenvalues and eigenvectors of these matrices are easy to calculate. Aggregation theory, due to Aoki and others,(Aoki, 1968a, 1968c, 1971, Chidambara, 1967, 1969, 1970, Davison, 1966, Fossard, 1970, Schainker, 1970), can then be applied to decouple the system into a subspace of significant modes, together with a subspace of response modes which can be safely neglected. This, however, requires that all the state variables be measured (rarely the case), or that estimates of the unmeasured variables be made available by means of a suitable observer (Luenberger, 1964). Alternatively, one may employ an aggregation matrix dictated purely by the measurements.

Other physical systems, such as a continuous stirred-tank reaction with multiple chemical reactions, may be operated around an unstable equilibrium point, but

[1] Currently with Procon, Inc., Des Plaines, Ill.

the matrix eigenvalue spectrum allows the system to be divided into two weakly-coupled subsystems. Both types of systems are considered in the present work.

2. FORMULATION OF THE PROBLEM

Consider the linear system (S_1) given by Eq. (1). If n is large, the modeling and optimal control of this system becomes a formidable task, and lower-order models are sought which reduce the computational burden. An obvious approach is the transformation to a lower-order system (S_2)

$$\bar{x} = Cx \tag{2}$$

where $\bar{x} \in E^m$, $m \ll n$, and C is a constant matrix, chosen to be full rank, called the aggregation matrix. The dynamics of this new state vector are given by

$$\dot{\bar{x}} = F\bar{x} + Gu \tag{3}$$

and it is readily verified that

$$CA = FC \tag{4}$$

$$G = CB \tag{5}$$

are necessary conditions for Eqs. (1), (2), and (3) to be consistent. The optimal proportional feedback control which minimizes the following quadratic objective function is well-known (Aoki, 1968b)

$$J = \int_0^\infty (x^TQx + u^TRu)dt \tag{6}$$

where $Q \geq 0$ and $R > 0$ are constant matrices chosen to reflect the allowable ranges of variation of both the state and control variables. The solution is

$$u(t) = -Kx(t) \tag{7}$$

where

$$K = R^{-1}B^TP \tag{8}$$

and P is the unique, positive-definite solution of the matrix Riccati equation:

$$A^TP + PA - PBR^{-1}B^TP + Q = 0 \tag{9}$$

For the reduced system, (S_2), one can similarly define a quadratic objective function

$$\bar{J} = \int_0^\infty (\bar{x}^T\bar{Q}\bar{x} + u^TRu)dt \tag{10}$$

with $\bar{Q} \geq 0$, which leads to the suboptimal feedback control

$$u(t) = -\bar{K}\bar{x}(t) = -R^{-1}G^T\bar{P}\bar{x}(t) \tag{11}$$

where $\bar{P}$ is obtained from

$$\bar{P}F + F^T\bar{P} - \bar{P}GR^{-1}G^T\bar{P} + \bar{Q} = 0 \tag{12}$$

The key to the problem is thus the specification of C and $\bar{Q}$, since F is determined from Eq. (4) by

$$F = CAC^T(CC^T)^{-1} \tag{13}$$

and G from Eq. (5). It is assumed that the pairs (A,B) and (F,G) are controllable. Aoki (1968b) shows that an appropriate choice for $\bar{Q}$ is

$$\bar{Q} = (CC^T)^{-1}CQC^T(CC^T)^{-1} \tag{14}$$

in which case $C^T\bar{P}C$ corresponds to P, so that a suboptimal feedback control is

$$u(t) = -R^{-1}B^TC^T\bar{P}Cx(t) \tag{15}$$

The choice of the aggregation matrix C is thus the principal problem facing the designer. One approach, which might be called "modal aggregation," involves a transformation to principal coordinates, which requires that the eigenvectors and eigenvalues of A be calculated. Suppose the eigenvalues are arranged in descending order of their real parts: $Re[\lambda_i] \geq Re[\lambda_2] \geq \ldots Re[\lambda_i] \geq \ldots Re[\lambda_n]$. Clearly, one cannot afford to discard unstable modes which persist for long periods of time; on the other hand, the response will essentially depend only on λ_i, $i \leq m$ after an initial period given by $t = 0(\frac{1}{\lambda_m})$. Thus, one approach is to choose C such that only the first m eigenvalues are retained in the aggregated system. The similarity to modal control (Chidambara, 1970, Davison, 1966, Schainker, 1970) is thus clear.

An alternative approach, which may be dictated by physical limitations of measurement devices and costs, might be called "physical aggregation" or lumping (Wei, 1969a, 1969b). This consists of defining C such that the new state vector components are weighted sums of the original vector components, and furthermore, such that stoichiometric relationships, such as mass conservation equations, are still obeyed. The new state vector components might thus be termed "pseudocomponents." An example is the use of boiling point fractions in complex reaction systems, such as petroleum mixtures. More generally, we may take C to be of the form $(E \mid 0)_{mxn}$, where E is a positive mxm matrix. In this case stoichiometry is preserved only if the columns of E all add up to the same (positive) constant, but in any case only the first m state variables will be involved in determining the control action.

To illustrate these procedures in a practical fashion, we discuss below the aggregative control of a CSTR operated near an unstable steady state, and of a five-plate gas absorber.

The controllability, stability, and loss of optimality of the aggregated control are all of considerable interest. The controllability is established by verifying that the controllability matrix is of maximum rank (Bryson, 1969).

In modal aggregation, n-m eigenvalues of A have been deleted from the original system, so that these modes will be uncontrollable. It is thus quite important to retain all unstable modes of A in order to stabilize them with the feedback control. From Eqs. (11) and (12) it is readily shown that

$$(F - G\overline{K})^T\overline{P} + \overline{P}(F - G\overline{K}) = -(\overline{K}^T R\overline{K} + \overline{Q}) \tag{24}$$

Since $R > 0$ and $\overline{Q} \geq 0$, it follows that $(\overline{K}^T R\overline{K} + \overline{Q}) > 0$. From Eq. (12) and a well-known theorem on the nature of the roots of equations of this form, due originally to Liapunov, $\overline{P} > 0$ (is positive-definite), and similarly from Eq. (4), that $(F - G\overline{K})$, which is the matrix of the closed-loop reduced system under suboptimal control, is also stable. On the other hand, since $C^T\overline{P}C$ is only positive semi-definite $(r(\overline{P}) = r(C^T\overline{P}C) < n)$, the closed-loop matrix of the original system, $(A-B\overline{K}C)$, will not necessarily be stable. However, Aoki (1968b) shows that $(A-B\overline{K}C)$ will be a stable matrix when $(F-G\overline{K})$ is stable, so long as the number of unstable modes of A is not greater than m.

The loss in optimality can also be estimated from upper and lower bounds provided by Aoki (1968c). The minimum or optimal value of the quadratic cost functional J is

$$J^*(x,u,t) = x_0^T P^* x_0 \tag{25}$$

On the other hand the suboptimal value of the quadratic cost functional resulting from implementation of the suboptimal policy

$$u = -R^{-1}B^T(C^T\overline{P}C)x = \overline{K}Cx \tag{26}$$

is given by

$$J(x,u,t) = x_0^T T x_0 \tag{27}$$

where T is the cost matrix associated with the suboptimal feedback gain matrix $\overline{K}C$. It can be shown that T is the solution of the equation

$$(A - B\overline{K}C)^T T + T(A - B\overline{K}C) + C^T\overline{K}^T R\overline{K}C + Q = 0 \tag{28}$$

Moreover, one can find upper and lower bounds of the Riccati equation, given by (2).

$$0 \le C^T\overline{P}C \le P^* \le T \tag{29}$$

3. SUBOPTIMAL CONTROL OF AN UNSTABLE SYSTEM

Consider a continuous-flow stirred-tank reactor (CFSTR) operated at an unstable steady state, with the homogeneous consecutive reactions: $A \xrightarrow{1} B \xrightarrow{2} C \xrightarrow{3} D \xrightarrow{4} E$. The feed is composed of pure A, possibly diluted with some inert materials. The dynamic behavior of the reactor can be studied via the transient heat and mass balances:

$$\rho V C_p \frac{dT}{dt} = \sum_{i=1}^{4} (-\Delta H_i) V k_i \exp\left(-\frac{E_i}{RT}\right) C_i - \rho C_p q (T - T_f) - U(T - T_c) \tag{30}$$

$$\begin{aligned}
V\frac{dC_1}{dt} &= qC_f - qC_1 - Vk_1 \exp\left(-\frac{E_1}{RT}\right) C_1 \\
V\frac{dC_2}{dt} &= -qC_2 + Vk_1 \exp\left(-\frac{E_1}{RT}\right) C_1 - Vk_2 \exp\left(-\frac{E_2}{RT}\right) C_2 \\
V\frac{dC_3}{dt} &= -qC_3 + Vk_2 \exp\left(-\frac{E_2}{RT}\right) C_2 - Vk_3 \exp\left(-\frac{E_3}{RT}\right) C_3 \\
V\frac{dC_4}{dt} &= -qC_4 + Vk_3 \exp\left(-\frac{E_3}{RT}\right) C_3 - Vk_3 \exp\left(-\frac{E_4}{RT}\right) C_4
\end{aligned} \tag{31}$$

The five balances are now expressed in dimensionless form:

$$\frac{d\tilde{T}}{d\theta} = \sum_{i=1}^{4} \alpha_i \beta_i e_i \tilde{C}_i - (\tilde{T} - 1) - \gamma(\tilde{T} - \tilde{T}_c) \tag{32}$$

$$\begin{aligned}
\frac{d\tilde{C}_1}{d\theta} &= \tilde{C}_f - \tilde{C}_1 - \alpha_1 e_1 \tilde{C}_1 \\
\frac{d\tilde{C}_2}{d\theta} &= -\tilde{C}_2 + \alpha_1 e_1 \tilde{C} - \alpha_2 e_2 \tilde{C}_2 \\
\frac{d\tilde{C}_3}{d\theta} &= -\tilde{C}_3 + \alpha_2 e_2 \tilde{C}_2 - \alpha_3 e_3 \tilde{C}_3 \\
\frac{dC_4}{d\theta} &= -\tilde{C}_4 - \alpha_3 e_3 \tilde{C}_3 - \alpha_4 e_4 \tilde{C}_4
\end{aligned} \tag{33}$$

Upon linearizing around the unstable steady-state $(\tilde{C}_{is}, \tilde{T}_s)$, one obtains:

$$\frac{dx_1}{d\theta} = \tilde{T}_s^{-2} \sum_{i=1}^{4} \alpha_i \beta_i \delta_i e_{is} x_i - (1 + \gamma) x_1 + \alpha_1 \beta_1 e_{1s} x_2 + \alpha_2 \beta_2 e_{2s} x_3$$

$$+ \alpha_3\beta_3 e_{3s}x_4 + \alpha_4\beta_4 e_{4s}x_5 + \gamma(\overline{T}_C - \overline{T}_{cs})$$

$$\frac{dx_2}{d\theta} = -(\widetilde{T}_s^{-2}\alpha_1\delta_1 e_{1s}\widetilde{C}_{1s})x_1 + (\widetilde{C}_f - \widetilde{C}_{fs}) - (1 + \alpha_1 e_{1s})x_2$$

$$\frac{dx_3}{d\theta} = \widetilde{T}_s^{-2}(\alpha_1\delta_1 e_{1s}\widetilde{C}_{1s} - \alpha_2\delta_2 e_{2s}\widetilde{C}_{2s})x_1 + \alpha_1 e_{1s}x_2 - (1 + \alpha_2 e_{2s})x_3 \qquad (34)$$

$$\frac{dx_4}{d\theta} = \widetilde{T}_s^{-2}(\alpha_2\delta_2 e_2\widetilde{C}_{2s} - \alpha_3\delta_3 e_3\widetilde{C}_{3s})x_1 + \alpha_2 e_{2s}x_3 - (1 + \alpha_3 e_{3s})x_4$$

$$\frac{dx_5}{d\theta} = \widetilde{T}_s^{-2}(\alpha_3\delta_3 e_{3s}\widetilde{C}_{3s} - \alpha_4\delta_4 e_{4s}\widetilde{C}_{4s})x_1 + \alpha_3 e_{3s}x_4 - (1 + \alpha_4 e_{4s})x_5$$

where the state variables, x_i $i = 1, \ldots, 5$ are now deviation variables. In matrix notation, one has

$$\dot{x}(t) = Ax(t) + Bu(t) \qquad (35)$$

$u^T = [u_1, u_2]$ is the control vector and A and B are defined by Eq. (34). The reactor is controlled by the temperature $\widetilde{T}_c$ of the coolant and the inlet concentration $\widetilde{C}_f$. Thus:

$$u_1 = \widetilde{T}_c - \widetilde{T}_{cs} \quad ; \quad u_2 = \widetilde{C}_f - \widetilde{C}_{fs} \qquad (36)$$

If, for simplicity, it is assumed that all deviations from the steady-state concentrations are to be weighted equally in the objective function, this implies that Q = I, the identity matrix. Other choices are, of course, possible so long as $Q \geq 0$. On the other hand, R must be chosen to reflect the control constraints and/or the costs of control. If the control costs are neglected, a suitable choice for the matrix, R, which will introduce penalties for high or low values of the control variables, will be

$$R = \begin{bmatrix} \frac{1}{\widetilde{\Delta T}_m^2} & 0 \\ 0 & \frac{1}{\widetilde{\Delta C}_m^2} \end{bmatrix} \qquad (37)$$

where ΔT_m and ΔC_m are proportional to the maximum allowable values of the control variables. The solution procedure consisted first of choosing physical parameters for the system (Diederich, 1971), and the dimension of steady-state temperature concentrations, and the components of the matrices A and B were then computed.

The solution of the time-invariant Riccati equation is based on a Newton-Raphson algorithm in function space. This involves the concept of Frechêt derivatives and is detailed elsewhere (Cuk, 1971, Kleinman, 1968, Krikelis, 1971).

In the present case there is one unstable mode, so that the dimension of the low-order system (S_2) can be as small as unity. A study was made of suboptimal control with 2, 3, and 4-dimensional reduced systems (S_2), using both modal and physical aggregation matrices. For modal aggregation, this implies that C takes the forms

$$C^T = (y_1, y_2) \qquad (I)$$

$$C^T = (y_1, y_2, y_3) \qquad (II)$$

$$C^T = (y_1, y_2, y_3, y_4) \qquad (III)$$

$$y_i^T A = \lambda_i y_i \; ; \; i = 1, \ldots, 5$$

where the eigenvalues have been arranged in descending order, beginning with the positive eigenvalue, λ_1. When dealing with physical lumping, on the other hand, the entries of C consist only of 1's and 0's, depending upon the components of the state vector which are measured:

$$C = \begin{bmatrix} 1 & 0 & 0 & 0 & 0 \\ 0 & 1 & 0 & 0 & 0 \end{bmatrix} \quad \text{(IV)}$$

$$C = \begin{bmatrix} 1 & 0 & 0 & 0 & 0 \\ 0 & 1 & 0 & 0 & 0 \\ 0 & 0 & 0 & 1 & 0 \end{bmatrix} \quad \text{(V)}$$

$$C = \begin{bmatrix} 1 & 0 & 0 & 0 & 0 \\ 0 & 1 & 0 & 0 & 0 \\ 0 & 0 & 0 & 1 & 0 \\ 0 & 0 & 0 & 0 & 1 \end{bmatrix} \quad \text{(VI)}$$

Table 1. Comparison of the Suboptimal and Optimal Control of a CFSTR with Various Aggregation Matrices

Matrix C	Controllability	$\|FC-CA\|$	J(x,u,t)	$(J-J^*)/J^*$%
III	yes	5.21	0.7518	0.2
II	yes	3.71	0.7869	5.0
I	yes	1.62	0.7752	3.4
VI	yes	7.70	0.8981	19.8
V	yes	7.70	0.9904	32.1
IV	yes	1.10	1.0659	42.1
Optimal control	yes		0.7497	

The controllability requirement for the optimal and suboptimal closed-loop systems was found, in every case, to be satisfied. One should note that, for the aggregation matrices (IV, V, and VI), dictated by physical measurements, this result is not guaranteed, since one cannot be sure that the unstable mode is included in the reduced system. Also, the consistency condition, Eq. (4), is only approximately satisfied, as may be observed from the table. Nevertheless, acceptable suboptimal controls are obtained, as shown in Figs. 1-4 for the temperature and product concentration curves. The apparent departure from optimality was even less for the other concentration variables.

4. SUBOPTIMAL CONTROL OF A STABLE, MULTIPLE-STAGE SYSTEM

The state matrix in a previous example was neither extremely sparse nor stable, and hence one is justified in some relief upon obtaining acceptable suboptimal control, based upon physical aggregation. The present example deals with a much more common example in practice, since it is typical of multiple-stage operations, such as distillation, absorption, and extraction, as well as the finite-difference representation of continuous transport processes, such as packed towers, heat exchangers, reboilers, condensers, and fluidized-bed systems. One obtains here (usually stable) Jacobi matrices, for which explicit formulae for the eigenvalues are available. Aggregative control would thus seem to be ideally suited to this class of problems; and indeed, modal aggregation is shown below to work very well. However, two problems arise: 1) physical aggregation does not guarantee stability of the closed-loop system, even when the plant is stable; 2) the eigenvalues of the Jacobi matrix lie in a relatively narrow range, so that separation in-

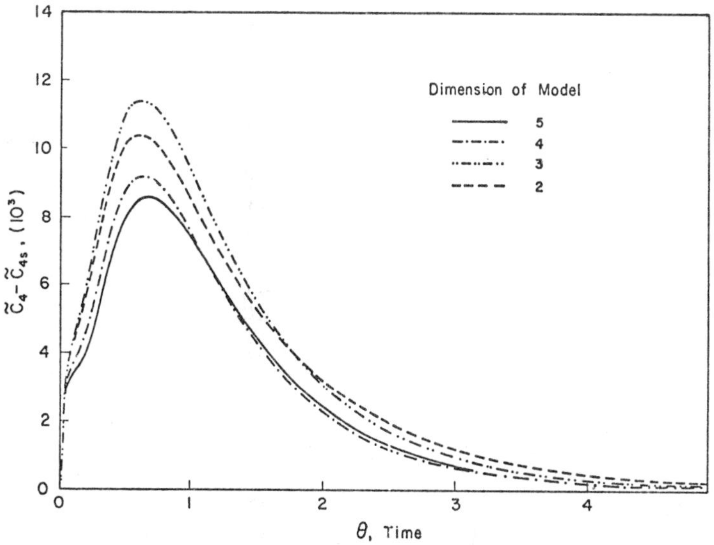

Fig. 1. Modal Aggregative Control of a Stirred-Tank Response of Dimensionless End-Product Concentration in Reactor Exit Stream; Initial Disturbance: $\Delta T = 5°C$; $\Delta C_1 = 1.0$ mole/liter. $\Delta\theta = 1$ corresponds to 1.1 sec. Intermediate-step concentration responses were much smaller, and showed little dependence on model dimension.

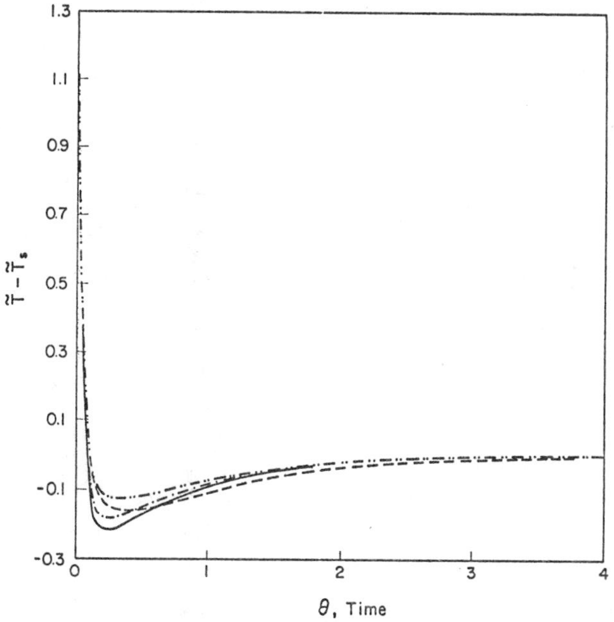

Fig. 2. Modal Aggregative Control of a Stirred-Tank Reactor; Response of Dimensionless Temperature. Disturbance and model dimension symbols same as in Fig. 1.

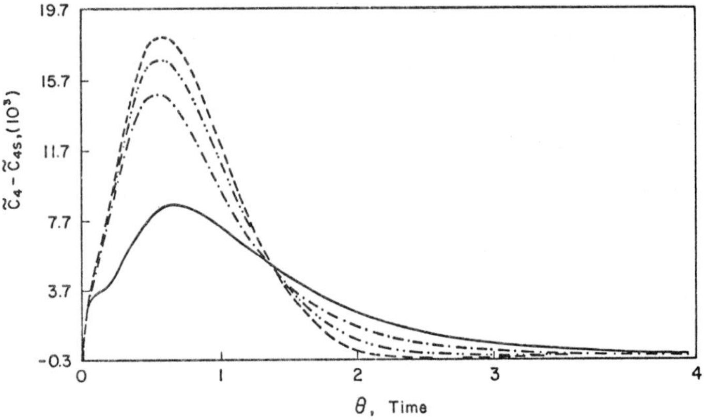

Fig. 3. Physical Aggregative Control of a Stirred-Tank Reactor; Response of Dimensionless End-Product Concentration. Disturbance and model dimension symbols same as in Fig. 1.

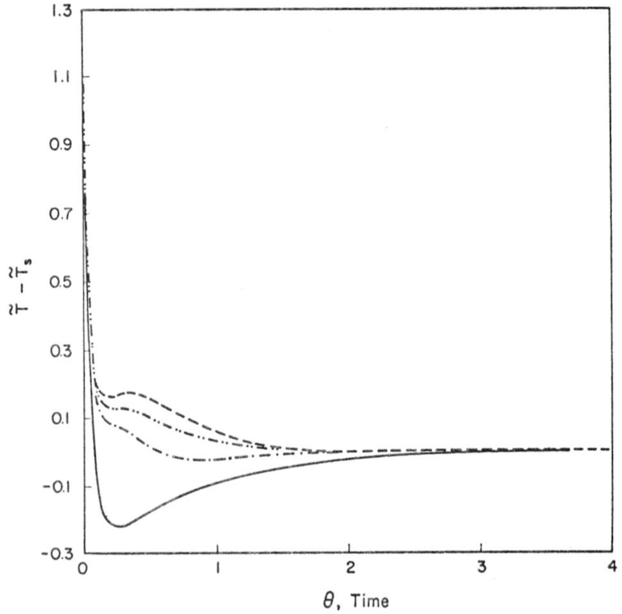

Fig. 4. Physical Aggregative Control of a Stirred-Tank Reactor; Response of Dimensionless Temperature. Disturbance and symbols as in Fig. 1

to weakly-coupled subsystems is not possible.

The system selected for study consists of a five-plate gas absorber (Lapidus, 1967) controlled by the compositions of the inlet liquid and gas streams. A material balance around the m-th plate leads to

$$H\frac{dy_m}{dt} + h\frac{dx_m}{dt} = L(x_{m-1} - x_m) + G(y_{m+1} - y_m) \tag{38}$$

where the compositions of the liquid and vapor leaving the m-th plate are governed by a linear equilibrium relationship:

$$y_m = ax_m + b \tag{39}$$

Upon substituting, one arrives at the set

$$\frac{dx_m}{dt} = \frac{d}{e}x_{m-1} - \frac{(d+1)}{e}x_m + \frac{1}{e}x_{m+1} \tag{40}$$

where

$$d = \frac{L}{Ga} \; ; \; e = \frac{Ha + h}{Ga}$$

This is clearly of the form of Eq. (1), with a stable Jacobi matrix. If the state variables are taken to be liquid concentration deviations from a specified steady-state set of conditions, and it is assumed that only x_5, the concentration in the liquid stream leaving the bottom plate, is initially non-zero (corresponding to a previous upset in the gas inlet stream concentration), one can compute optimal and sub-optimal controls, corresponding to a choice of identity matrices for Q and R in the objective function, and the following aggregation matrices:

$$C = \begin{bmatrix} y_2^T \\ y_5^T \end{bmatrix} \qquad \text{(I)}$$

$$C = \begin{bmatrix} y_2^T \\ y_5^T \\ y_3^T \end{bmatrix} \qquad \text{(II)}$$

$$C = \begin{bmatrix} y_2^T \\ y_5^T \\ y_3^T \\ y_4^T \end{bmatrix} \qquad \text{(III)}$$

for modal aggregation, with y_i^T the ith row eigenvector of A, where the eigenvalues of A were computed to be: $\lambda_1 = -2.1851$; $\lambda_2 = -0.1649$; $\lambda_3 = -1.1728$; $\lambda_4 = -1.7573$; $\lambda_5 = -0.5883$. For physical aggregation, the matrices were chosen to be:

$$C = \begin{bmatrix} 1 & 0 & 0 & 0 & 0 \\ 0 & 0 & 0 & 0 & 0 \end{bmatrix} \qquad \text{(IV)}$$

$$C = \begin{bmatrix} 1 & 0 & 0 & 0 & 0 \\ 0 & 0 & 0 & 0 & 1 \\ 0 & 1 & 0 & 0 & 0 \end{bmatrix} \qquad (\text{V})$$

$$C = \begin{bmatrix} 0 & 1 & 0 & 0 & 0 \\ 0 & 0 & 0 & 0 & 1 \\ 0 & 0 & 1 & 0 & 0 \end{bmatrix} \qquad (\text{VI})$$

$$C = \begin{bmatrix} 1 & 0 & 0 & 0 & 0 \\ 0 & 0 & 0 & 0 & 1 \\ 0 & 1 & 0 & 0 & 0 \\ 0 & 0 & 1 & 0 & 0 \end{bmatrix} \qquad (\text{VII})$$

The results are summarized in the following table:

Matrix C	Controllability	$\|FC-CA\|$	$J(x,u,t\|$	$(J-J^*)/J^*\%$
I	yes	10^{-9}	.01149	0.6
II	yes	0.94	.01171	2.5
III	yes	0.91	.01162	1.7
IV	yes	0.82	.01234	8.0
V	yes	0.82	.01226	7.2
VI	no			
VII	yes	0.82	.01219	6.7
Optimal Control	yes		.01143	

One can note several interesting results. The first is that the controllability criterion must be checked, even for a stable plant, when projecting the state vector into the output space which corresponds to physical aggregation. Another is that the loss of optimality is influenced heavily by the degree of approximation to the consistency condition, Eq. (4), as shown by the fact that system (I) had the best performance, despite the fact that its dimension was only two The performance of the modally-aggregated systems was clearly better (as would be expected) than those of the physically-aggregated systems, but in every case the performance was quite acceptable. The use of Luenberger observers (1964) to implement modal aggregation, even when the system is not decomposable into weakly-coupled subsystems, is therefore quite promising.

ACKNOWLEDGMENT

A portion of this work was done while one of the authors was studying in the United States under a grant from the French government. Partial support was also received from the Petroleum Research Fund, American Chemical Society, Grant 4666-AC

NOMENCLATURE

A	(n x n) constant matrix, Eq. (1)
B	(n x r) constant matrix, Eq. (1)
C	(m x n) aggregation matrix
C_{av}	reference feed concentration of A (moles/l)
C_f	concentration of A in the feed (control)
C_i	concentration of i-th component inside the reactor
$\tilde{C}_i$	$\frac{C_i}{C_{av}}$

C_p	heat capacity of reaction mixture (cal/g °K)
E_i	activation energy
e_i	$(\exp(-\delta_i/\tilde{T})$ (dimensionless)
F	(m x m) constant matrix, Eq. (3)
G	(m x r) constant matrix, Eq. (3); flow rate of inert gas
J	quadratic cost functional, Eq. (6)
J^*	optimal value of J
k_i	pre-exponential factor
P	(n x n) constant matrix, Eq. (9)
$\overline{P}$	(m x m) constant matrix, Eq. (12)
Q	(n x n) positive-semidefinite constant matrix, Eq. (6)
q	inlet flow rate (m^3/sec)
R	perfect gas constant
R	(r x r) positive-definite constant matrix, Eq. (6)
T	reactor temperature (°K)
$\tilde{T}$	$\frac{T}{T_f}$: dimensionless temperature
T_c	coolant temperature (°K)
$\tilde{T}_c$	$\frac{T_c}{T_f}$: dimensionless temperature of the coolant
T_f	inlet stream temperature (°K)
U	overall heat transfer coefficient (kcal/°K)
u(t)	control vector; r-dimensional
V	volume of CFSTR (m^3)
$\overline{x}$	aggregated state vector; m-dimensional
x_0	initial value of state vector
x(t)	state vector; n-dimensional
y_i	i-th row eigenvector of A; composition of vapor leaving ith plate
α_i	$\frac{V}{q}k_i$ (dimensionless)
β_i	$\frac{(-\Delta H_i)C_{av}}{\rho C_p T_f}$ (dimensionless)
γ	$\frac{u}{\rho C_p q}$ (dimensionless)
ΔH_i	heat of i-th reaction (cal/gmole)
δ_i	$\frac{E_i}{RT_f}$ (dimensionless)
θ	dimensionless time
λ_i	i-th eigenvalue of A
ρ	density of reaction mixture (kg/m^3)
τ	$\frac{V}{q}$, residence time (sec)
s	subscript denoting the unstable steady-state

*	superscript denoting optimal conditions
x_i	composition of liquid leaving ith plate
h	hold-up of inert liquid on each plate
H	hold-up of inert gas on each plate
L	flow rate of inert liquid
t	time

REFERENCES

1. Aoki, M.,(1968a), "Control of Dynamic Systems Containing a Small Parameter by Aggregation," Proc. IFAC Symp. Sensitivity, Dubrovnik, Yugoslavia.
2. Aoki, M. (1968b), "Control of Large Scale Dynamic Systems by Aggregation," IEEE Trans. Autom. Control, AC-13, 246.
3. Aoki, M. (1968c), "Note on Aggregation and Bounds for the Solution of the Matrix Riccati Equation," J. Math. Anal. Appli., 21, 377.
4. Aoki, M. (1971), chapter in Optimization Methods for Large-Scale Systems, D.A. Wismer, ed., pp. 191-232, McGraw Hill, N.Y.
5. Bellman, R. (1960), Introduction to Matrix Analysis, McGraw Hill, New York.
6. Bryson, A. E. and Ho. Y. C. (1969), Applied Optimal Control, Ginn & Co.
7. Chidambara, M. R. and Davison, E. J. (1967), "On a Method for Simplifying Linear Dynamic Systems," IEEE Trans. Auto. Contr., 12, 213.
8. Chidambara, M. R. (1969), "Two Simple Techniques for the Simplification of Large Dynamic Systems," Proc. JACC, 669.
9. Chidambara, M. R. and Schainker, R. B. (1970), "Lower Order Aggregated Model and Suboptimal Control," Proc. JACC, 842.
10. Cuk, N., Bingulac, S. P. and Stojic, M. R. (1971), "On an Iterative Solution of Time Invariant Riccati Equation," Proc. JACC, 178.
11. Davison, E. J. (1966), "A Method for Simplifying Linear Dynamic Systems," IEEE Trans. Aut. Contr., 11, 93.
12. Diederich, Alain L. (1971), "Aggregation of Complex Linear Systems," M. S. Thesis, Northwestern Univ., Evanston, Ill.
13. Fossard, A. (1970), "On a Method for Simplifying Linear Dynamic Systems," IEEE Trans. Aut. Contr., 15, 261.
14. Kantorovich, L. V. and Akilov, G. P. (1964), Functional Analysis in Normed Spaces, MacMillan.
15. Kleinman, D. L. (1968), "On an Iterative Technique for Riccati Equations Computations," IEEE Trans. Aut. Contr., 13, 114.
16. Krikelis, N. J. and Rekasius, Z. V. (1971), "On the Solution of the Optimal Linear Control Under Conflict of Interest,"IEEE Trans. Auto. Contr., 16, 140.
17. Lapidus, L. and Luus, R. (1967), Optimal Control of Engineering Processes, p. 49, Blaisdell Publ. Co., Waltham, Mass.
18. Luenberger, D. G. (1964), "Observing the State of a Linear System," IEEE Trans Mil. Electron., MIL-8, 74.
19. Schainker, R. B. (1970), "Suboptimal Control by Aggregation," Proc. JACC, 833.
20. Wei, J. and Kuo, J. C. W.,(1969a), "A Lumping Analysis in Monomolecular Reaction Systems," I&EC Fund., 8, 114.
21. Wei, J. and Kuo, J. C. W. (1969b), "A Lumping Analysis in Monomolecular Reaction Systems," I&EC Fund., 8, 124.

APPLICATIONS OF GRAPH THEORY TO DECOMPOSITION PROBLEMS OF STRUCTURES

William R. Spillers

Department of Civil Engineering
Columbia University
New York City

Abstract: Some decomposition problems of structures are discussed around the central theme of the decomposition of a graph into subgraphs which are serially connected. Three specific areas are studied: 1) Optimal methods for the solution of large structures. 2) Decomposition applied to structural design. 3) Problems of architectural layout.

1. INTRODUCTION

The range of decomposition techniques is so broad, even in structures, that it is virtually impossible in an article of this length to consider all existing applications. This situation is rather pleasant for an author since it allows him the freedom to discuss the aspects of a topic which are most interesting to him personally; the reader must of course beware since the material he receives is likely to be filtered more heavily than usual.

There is a thread which seems to recur in problems of structures and perhaps even all physical systems which are sparse. It is shown schematically in Fig. 1. The idea is that any graph can be decomposed into subgraphs which are themselves serially connected. Of course, when the graph is "complete", there is at most one subgraph in addition to the starting element but framed structures are characteristically sparse as are finite element or finite difference representations of continuous structures such as plates, shells, and solids. In these cases the subgraphs are "weakly connected," a fact which is used both directly and indirectly in constructing algorithms. It is about this thread that this paper is developed.

Three problem areas are included: structural analysis, structural design, and architectural layout. The first major section begins with the work of the late Gabriel Kron as it relates to the problem of solving structures in a manner which minimizes the computational effort required and traces its development through the past 10 - 15 years. Since structures applications commonly involve the solution of thousands of equations, this is a topic of considerable practical interest with a sizeable literature. The next section attempts to move smoothly into questions of optimization in structural design. In some cases there is really no transition at all to be made; in the case of generating new structures (graphs) from given examples there is a change of pace not only in terms of techniques used but also in terms of practical applicability since this area is only now under development. This second major sections also includes an unrelated but potentially powerful excursion into a decomposition used in truss design. The final section again returns to the central theme of the serial decomposition of a graph as it relate to architectural layout. Architectural layout problems in this case represent a complication of activity since they involve geometrical questions while the preceding sections are largely combinatorial as they relate only to the system graph. Here, experience with a heuristic algorithm for preliminary design is discussed.

While it is hoped that the emphasis taken in this paper is new, this is a review paper and with minor exceptions the results presented here are now available in the literature cited.

2. DECOMPOSITION IN STRUCTURAL ANALYSIS

This section is concerned with the question of solving efficiently large linear structures (systems of linear algebraic equations). In terms of the node method for skeletal structures (see e.g. Spillers (1972a)), Figure 2 indicates a structure, its graph, and the contribution of a typical member to the system matrix. From Figure 2 it can be concluded that most real structures correspond to sparse systems since it is uncommon for a single node to be connected by members to many other nodes.

Presuming that all solution methods are equally good, (which is clearly not the case), the method which costs the least to execute on a computer is considered to be the best. While the "best" method is therefore a machine dependent question it has become common to rate schemes machine independently by counting operations such as the required number of multiplications, disc I/O calls, etc. In these cases the concern is not with the specific value of a term in the system matrix but rather with the fact that is is or is not zero. For this reason the question of the optimal solution of a structure involves only the system graph and can be formulated as a graph problem.

Even more important for the solution of large structures is the fact that their system matrices are commonly well-conditioned. That aspect of structures will not be considered here.

2.2 Kron's Methods

Long before the advent of the digital computer, Kron showed how decomposition techniques could be used effectively to solve large systems. While Kron (1956) developed many techniques dealing not only with linear but also nonlinear systems, two of his techniques seems to predominate. They have been discussed in detail by Branin

(1959) and Roth (1959) for networks from which it is a simple matter to extend them to structures, given a network analogy.

Kron is probably best known for the technique of "tearing" which he derived from his orthogonal network concept. It was subsequently shown that tearing could be explained in terms of the method of modified matrices based upon the identity

$$(B - US\tilde{V})^{-1} = B^{-1} + B^{-1}U(S^{-1} - \tilde{V}B^{-1}U)^{-1}\tilde{V}B^{-1} \quad .$$

Branin (1959) has indicated that, applied properly, tearing can result in phenominal savings in terms of the required number of multiplications. To be brief, tearing requires the construction of an inverse system matrix (or subsystem matrix) at some level of computation. While a system matrix may be sparse, its inverse usually is not sparse. The result is that storage requirements for this inverse usually obviate the application of tearing to large structures today. A possible exception to this statement is the re-design situation (see e.g. Strauss (1969)).

Another of Kron's methods is called K-partitioning. It is based upon the fact that rather than solving the linear system $A_x = b$ directly as $x = A^{-1}b$, it may be partitioned

$$\begin{bmatrix} A_1 & A_2 \\ A_3 & A_4 \end{bmatrix} \begin{bmatrix} x_1 \\ x_2 \end{bmatrix} = \begin{bmatrix} b_1 \\ b_2 \end{bmatrix}$$

and solved

$$x_2 = (A_4 - A_3 A_1^{-1} A_2)^{-1} (b_2 - A_3 A_1^{-1} b_1)$$

$$x_1 = A_1^{-1} (b_1 - A_2 x_2)$$

assuming A_1 to be nonsingular. In order to include K-partitioning in schemes for the automatic analysis of structures, it is necessary to decide in advance how it is to be applied to an arbitrary structure. After some study (Spillers (1964)) it became clear that the computational efficiency improves with the number of substructures into which a structure is partitioned. Since K-partitioning is, in fact, a matrix form of Gaussian elimination, the question of optimal K-partitioning in the limit becomes a question of optimal Gaussian elimination.

In retrospect, the study of Kron's methods has led to techniqes which differ considerably from those used by Kron himself. This is probably due to the fact that the equipment available to Kron was much different than the computers which are available today.

2.3 Optimal Solution of Sparse Systems

The discussion of an optimal procedure for K-partitioning leads naturally into a discussion of optimal Gaussian elimination. For the case of the node method and stable structures, zeros do not appear along the diagonal and it is not necessary to interchange rows and columns during the solution process; that is, the ith row can be used to eliminate terms below it in the ith column of the system matrix. The elimination procedure is then completely specified when the joints of the structure or the nodes in the system graph have been numbered. An optimal Gaussain elimination then corresponds to an optimal numbering of the system graph.

From a practical point of view optimality is usually considered with respect to either required core storage or the required number of multiplications. While these two aspects of Gaussian elimination are essentially compatible, they are not entirely so as indicated in Figure 3. In the upper portion of this figure an ordering of the nodes which minimizes the required number of multiplications is indicated; in the lower portion an ordering which requires less storage is indicated. The node numberings are not the same.

Formally, the problems of minimizing the number of multiplications required by Gaussian elimination can be posed as follows (Spillers and Hickerson(1968)): Given a system graph and an arbitrary initial node ordering, find a permutation ξ of this ordering which minimizes $\Phi(\xi)$ where

$$\Phi(b) = \sum_{i=1}^{n} (b_i + 1)\, b_i - n$$

and b_i is the number of nonzero terms remaining in the ith row <u>after</u> the elimination phase has been completed. Let $W(G)$ be defined to be the number of multiplications required for the solution of a system whose graph is G and let G^i refer to the graph formed by the "elimination" of node i from G. Since W satisfies the functional equation

$$W(G) = \min_i [d_i(d_i + 1) - 1 + W(G^i)]$$

in which d_i is one plus the degree of node i in graph G, dynamic programming can be applied to the problem of a minimal ordering.

From a practical point of view, the dynamic programming solution for an optimal ordering is prohibitive in cost for large systems and it is necessary to resort to heuristics (Spillers and Hickerson (1968) and Jensen and Parks (1970)). Since the number of multiplications required depends upon the square of the number of nonzero terms in the pivotal row at any stage, a simple but effective scheme consists pivoting upon the row, at any stage, which contains the fewest number of nonzero terms. This algorithm is more in the spirit of a "shortest step" algorithm than it is in the spirit of Figure 1.

In most practical situations core storage requirements dominate in the solution procedure over other considerations such as the number of multiplications required. The primary reason for this situation is that when a problem becomes so large that its solution requires intermediate I/O on a scratch unit, the I/O dominates dwarfing the cost of the other operations. It is also common to pay disproportionately high charges for the use of large core at service centers. The result is that there seems to be a larger return for savings in required core storage than for savings in in-core machine operations.

When the nodes of the system graph are numbered properly, the non-zero terms become concentrated within a "band" about the diagonal (see Figure 4). The form of the band matrix is particularly useful since all the terms outside the band are known to be zero and are therefore neither stored nor operated upon.

In attempting to minimize core storage, it is common to attempt to find a node numbering for the system graph which minimizes the band width. This is equivalent to finding a node numbering which minimizes the maximum of the absolute value of the differences between the node numbers at the ends of each branch. Formally, given a graph which has been numbered arbitrarily, the optimal numbering scheme corresponds to a permutation ξ which minimizes

$$\Psi = \max_{[i,j] \in I_1} [|\xi_i - \xi_j|]$$

where I_1 is the set of node pairs of the given graph. The problem of minimizing the band width can also be attached using dynamic programming but again the possibilities for a simple, inexpensive solution are indeed bleak at the moment and it is again necessary to resort to heuristics in practical situations.

It seems that any commercial computer program for the automatic analysis of an arbitrary structure must provide for automatic node re-numbering since the cost of computation is sensitive to the band width and it is inconvenient for the user, even the sophisticated user, to concern himself directly with the question of band width reduction. There have been several schemes published for band width reduction in the past few years. Grooms (1972) reviews the state of the art and compares the relative efficiencies of these schemes. Basic to these algorithms is the procedure of first locating the members at which the maximum band width occurs and then attempting to reduce the band width by interchanging rows and columns. As Grooms indicates, this can be a very expensive procedure because the number of possible interchanges becomes very large for large systems.

The author has had excellent experience with a renumbering scheme based upon Figure 1. It proceeds as follows:

1. Number first the node with the smallest degree (number of branches incident upon it). This node constitutes the first subset.

2. Given the $(i-1)^{th}$ subset construct the ith subset to contain all nodes not previously numbered which are connected by members to nodes in subset $(i-1)$. Number consecutively the nodes in subset i next.

3. Stop when all nodes have been numbered.

The idea here, of course, is simply to start with the smallest possib band width and number next all the nodes, in general, which relate to the last node in a kind of shortest step procedure.

There are some simple bounds available. The band width associated with any ordering provides an upper bound on the optimal band width. A lower bound on D, the maximum difference in node numbers is

$$D \geq \begin{cases} d/2 & \text{when } d \text{ is even} \\ (d-1)/2+1 & \text{when } d \text{ is odd} \end{cases}$$

in which d is the maximum degree which occurs at any node.

2.4 The Utilization of Core Storage

It is convenient to define a system to be small if it can be solved on a given computer without the use of intermediate I/O. In terms of increasing size, there is an interesting class of problem which require a minimal use of this I/O. In this case the core is large enough to store the terms required for the elimination of any row without requiring information which has already been read onto disc or tape to be read back in. Figure 5 indicates this situation for Gaussian elimination in which it is required to keep b^2 terms in core. For systems in which there is not sufficient core to store these b^2 terms, the I/O dominates and therefore an optimal scheme becomes highly machine dependent. There appears to be little materia in the literature dealing with these extremely large systems.

2.5 Symmetric Systems

It has been convenient above to proceed directly from Kron's methods to Gaussian elimination neglecting the question of system symmetry. The system matrix of a structure is of course symmetric and it is important to use symmetry properly since there are savings of 50% to be had in almost every item to be considered:

1. It is only necessary to generate about half the coefficients in the system matrix.

2. Using symmetry, the number of multiplications required to solve an $n \times n$ system decrease from $n^3/3$ to $n^3/6$. For band matrices (Figure 5) the number of multiplications required decreases from nb^2 to $nb^2/2$.

3. The core storage required for the intermediate case of large systems decreases from b^2 to $b^2/2$.

The remarks made concerning the properties of the system graph are not affected by these considerations of system symmetry.

2.6 Reanalysis of Modified Structures

The recent work (Kirsch and Rubenstein (1972), Argyris and Roy (1972), Kavlie and Powell (1971)) concerned with questions of optimal methods for reanalysis is a logical extension of the work described above concerned with decomposition and the optimal solution of large systems. The problem here is that the analysis of a large system represents a considerable investment so that it is highly desirable not to have to repeat the original analysis when changes are made in a structure during the design process. Criteria concerning when it is economical to do a complete reanalysis and when it is not are problem dependent and rather complicated making it difficult to comment simply upon them here but it may be noted that Kavlie and Powell seem rather pessimistic concerning methods of modification.

3. STRUCTURAL DESIGN

3.1 Introduction

The state of the art of optimization in structural design has been described by Pope and Schmit (1971) and will not be repeated here. The attempt here will be to pursue this thread involving the serially connected subgraph (Figure 1) into the area of structural design. While the volume edited by Pope and Schmit is concerned largely with the application of mathematical programming to structural design, the emphasis here is more with some applications which are peculiar to structures.

This chapter begins with a discussion of cases in which analysis is essentially identical with design. It then moves into the area of mathematical programming and the decomposition of a truss problem is discussed. Finally some questions of total automation and the associated graph problems are presented.

3.2 Structural Design versus Structural Analysis

To the practical designer, design today depends strongly upon analysis. In most cases he has neither the tools nor the time to optimize a design carefully and spends most of his effort in the common design-analysis-redesign sequence. In the case of truss design it is common, e.g., to iterate

$$A_i^{(n+1)} = |F_i^{(n)}|/\sigma_i^{all}$$

in which

A_i - area of member i

F_i - force in member i

σ_i^{all} - "allowable" stress in member i

and the superscripts refer to the iteration number. In contrast to applications of mathematical programming which now present formidable difficulties, the commonly used heuristic design procedures such as the above iteration scheme have been highly successful over the years. This success has motivated a number of studies which will now be discussed. In these cases design, in terms of techniques, is simply analysis repeated a number of times; the decomposition techniques used are simply those of analysis.

In the iterative scheme just described, an analysis is followed by a selection of new member areas which is in turn followed by a reanalysis. The iterative scheme itself appears to deal with questions of safety as it attempts to satisfy allowable stress criteria. In order to motivate it in terms of optimization consider the following problem:

$$\text{minimize } \sum_i |F_i|L_i \qquad \text{subject to } \tilde{N}F = P$$

In terms of plastic design for trusses, the area of a member is proportional to the absolute value of the member force and the objective function above is thus proportional to the volume; the constraints represent joint equilibrium and P is the given joint load matrix. Other interpretations of the problem are possible (Spillers and Farrell (1969)).

In order to motivate the iterative design algorithm, it is convenient to first form the Lagrangian

$$\mathcal{L} = \frac{\sigma^{all}}{E} \sum_i |F_i|L_i + \tilde{\delta}(P - \tilde{N}F)$$

which yields

$$\partial\mathcal{L}/\partial F_j = 0 \quad \rightarrow \quad (N\delta)_j = L_j (\operatorname{sgn} F_j)\, \sigma^{all}/E \qquad \forall j$$

$$\partial\mathcal{L}/\partial \delta_j = 0 \quad \rightarrow \quad \tilde{N}F = P$$

The first of these equations yields the iterative scheme in terms of a Newtonian solution procedure for nonlinear systems since

$$(N\delta)_j = L_j (\operatorname{sgn} F_j)\, \sigma^{all}/E \quad \rightarrow \quad \frac{F_j L_j}{A_j E} = L_j \frac{F_j}{|F_j|} \frac{\sigma^{all}}{E}$$

using elastic analysis. Iteration may now proceed as defined above evaluating $|F_i|$ at the preceding step.

It is natural to attempt to extend this iterative procedure to classes of structures other than trusses. For frames and plates (Al Banna (1972)), the obvious generalizations of the iterative scheme used on trusses do not converge. In order to clarify this matter Al Banna considers the problem

$$\text{minimize} \quad t(F) \qquad \text{subject to} \quad \tilde{N}F = P$$

in which the function $t(F)$ is assumed to be convex, homogeneous, and separable. (The notation here follows the truss but only implies that the independent variable F must satisfy linear equality constraints.) He forms the Lagrangian and differentiates in the usual manner; linearizing using Newton's method he arrives at a linear system to be solved at each iteration. This system has the appearance of a structural analysis problem except that the stiffness matrix now comes from the Hessian of the objective function. Al Banna essentially indicates what stiffness matrix should be used in an analysis computer program in order to optimize.

3.3 A Decomposition Technique of Truss Design

In the case of truss design for two loading conditions, the following problem arises

$$\text{minimize} \quad \sum_i L_i \max\{|F_i^1|, |F_i^2|\} \qquad \text{subject to} \quad \tilde{N}F^1 = P^1,\ \tilde{N}F^2 = P^2$$

This problem is a rather obvious extension of the truss problem discussed above. The superscripts here refer to the two loading conditions.

It has been shown (Spillers and Lev (1971); Pope (1971)) actually attributes this result to Hemp (1968)) that using the identity

$$\max\{|x|,|y|\} = 1/2 \ \{|x+y| + |x-y|\}$$

that the two loading condition problem decomposes into two independen[t] "sum" and "difference" problems,

$$\text{minimize} \ \sum_i L_i |F_i^S| \qquad \text{subject to} \quad \tilde{N}F^S = P^S$$

$$\text{minimize} \ \sum_i L_i |F_i^D| \qquad \text{subject to} \quad \tilde{N}F^D = P^D$$

each of which has the appearance of a single loading condition problem. The sum and difference superscripts simply indicate

$$F^S = 1/2\,(F^1 + F^2) \qquad\qquad F^D = 1/2\,(F^1 - F^2)$$

$$P^S = 1/2\,(P^1 + P^2) \qquad\qquad P^D = 1/2\,(P^1 - P^2)$$

This decomposition is interesting both from the design point of view and the computational point of view.

Unfortunately, this decomposition does not extend either to more loading conditions or to other classes of structures. For the case of three or more loading conditions (Spillers (1972b)) it can be shown that the decomposition is a result of the description of an octahedron in n space in terms of its bounding planes. While a rectangular region can be defined by two sets of planes, the description of an octahedron, its three dimensional generalization, requires four sets of planes (Figure 6). The result is that for more than two loading conditions, the variables which correspond to the sum and difference variables for two loading conditions are no longer independent. If the decomposition is forced by ignoring the constraints between the variables a lower bound results. The application of these decompositions is essentially unexplored.

3.4 Questions of Complete Automation

Most of the literature concerning structural optimization deals with questions of sizing members once the structural configuration has been selected. That is, the concern is usually with the details of the members assuming that the geometrical location of the joints has been fixed. It has only been recently (e.g., Friedland (1971), Vanderplaats and Moses (1972)) that questions of optimal node location have been considered. This work is referred to as geometrical optimization.

The intension of optimization is in the long run complete automation. It is in fact not clear what complete automation means in practical terms, but the implication is that all existing processes will be automated. In view of the recent work on geometrical optimization, it appears that the complete automation of the selection of joint locations given a structure will shortly become a fact of life. It would appear that with the automation of geometrical optimization, design will become a graph problem, at least for framed structures.

3.5 The Graph Problem

There are many reasons for attempting to automate the selection of the graph of a structure. Most directly, it is one of the natural next steps to be taken. Another reason involves questions of artificial intelligence and the use of "examples". While it is now possible to program the responses of a good designer in fairly complex situations it is a rather expensive thing to do since the designer's decision tree is quite complicated. What appears now to be a more attractive alternative would be to program what the designer knew when he entered college and a learning algorithm. While there is considerable conjecture involved here since this learning algorithm is nonexistent, Figure 7 on the other hand, indicates a potential which now exists (Spillers and Friedland (1972)). In this figure the topmost truss represents an example; the other trusses represent improved designs generated automatically by computer. These designs are approaching Michell's classical result.

In switching theory it is now possible to realize designs directly from abstract problem statements. While it is now early, it seems likely that in the future a similar situation will develop for structures in which designs will be generated directly from abstract problem statements; an important role in the process will be played by graph theory. At the moment (Spillers (1971)) there are no uniformly accepted techniques available for dealing with graph problems but there are many possibilities.

One of the major problems is that there is no language or formalism now available in terms of which structures can be described. For example, while graph theory is adequate for describing the connectivity of a framed structure, there is no way available now to say that this framed structure is subjected to a uniform load which is to be divided up among the nodes depending upon their location. In switching theory it is common to begin with a truth table or even algebraic statements of the requirements a circuit must satisfy. Unfortunately, these representations do not at the moment seem to have any particular significance in structures.

One of the most promising representations of graphs is the binary vector (whose elements are 0,1) representation. Starting with the node-node matrix X (see, e.g., Grenander (1969)) whose elements $x_{ij} = 0,1$ if nodes i and j are not, are connected by a branch, it is a simple matter to map this array into a one-dimensional array to obtain a binary vector. If addition and multiplication are defined modulo 2, it is possible to manipulate these vectors (graphs) as matrices in an otherwise ordinary manner and also to bring to bear on these problems the considerable machinery of linear vector spaces.

Given a graph it is possible, for example, to generate a complete set of graphs using a nilpotent matrix T ($T^n = 0$ while $T^{n-1} \neq 0$). If the given graph is regarded as an example, the nilpotent matrix has the form of an operator operating on an example to produce an arbitrary configuration. From this simple point of view, a "learning" operator would modify the nilpotent matrix T.

In many ways these graph problems take the discussion back to theme indicated in Figure 1. In the first place it is somewhat natural to presume that optimal designs have sparse system graphs: that is certainly the present experience reinforced in theory by buckling problems and joint costs. In many cases modular construction tends to lead in this direction. With the exception of perhaps the formation of walls and solids by modules such as bricks, modular construction leads in many cases to systems in which modules or groups of modules are connected serially. While these questions are touched upon briefly in the references cited above, they are largely unexplored in terms of practical application.

4. ARCHITECTURAL LAYOUT

4.1 Introduction

In this section the thread indicated in Figure 1 is pursued into the problem of architectural layout. This is the problem the designer must face in the early stages of a design project when he has only the relatively small amount of information given him by the client and perhaps some preliminary site information and must begin to make trial layouts. In spite of its inherent vagueness, several warehouse type models have been set up to formalize the design problem at this early state. (An extensive bibliography on problems of this type has been prepared by Miller (1971).) In one of these models (Spillers and Wiedlinger (1970)) the designer is given a list of department areas A_i and an adjacency matrix which describes quantitatively or qualitatively the relative affinities of the departments for each other and is asked to produce a layout which minimizes the weighted walking distance

$$\Phi = \sum_{i,j} A_i A_j \omega_{ij} r_{ij}$$

Here r_{ij} is distance between the centroids of departments i and j and ω_{ij} is a weighting factor based upon the given adjacency matrix While perhaps not identical, other mathematical models also convert the rather informal architectural design problem into a mathematical programming problem in terms of some generalized walking distance. Once again, there are no available algorithms and it is necessary to resort to heuristics.

4.2 An Algorithm

For the architectural space allocation problem it is again convenient to think in terms of Figure 1. In this problem, most of the time conflicting interests develop as various departments vie to be as close to each other as possible. The usual difficulties arise: for problems of meaningful size, the number of possible layouts is enormous and it is necessary to eliminate many of them in order to produce a useable computational scheme. In the algorithm to be described the number of trials is cut down in two ways. First, after a layout has been divided into modules which are subsequently placed serially, a module is only considered or tried in the available spaces adjacent to modules which have already been placed. Second, in spite of the fact that modules placed early must be placed on the basis of incomplete information, once a module has been placed it is not moved.

There is next the question of a criterion to use in placing a module. That is, granted that each available module will be tried in each available place in some subset of the actual space of available positions, which of these modules should actually be placed. On the basis of some numerical experiements it becomes clear that several obvious decision rules don't work. For example, if the module which adds the least to the objective function is positioned, modules which are unrelated are placed side by side in a very uneconomical layout. A decision rule based upon a kind of gravitational attraction seems to work relatively well. Under this rule the module is placed which maximizes the function

$$W = \sum_{j \in I_1} \frac{A_j' \omega_{ij}}{(r_{ij}')^n}$$

over the available locations. Here,

- I_1 - set of all modules which have already been placed
- A_j' - area of the modules of department j which have already been placed
- r_{ij}' - distance between point i and the centroid of the area A_j'.
- n - an integer which can be used to exercise some control over department shape

The relationship of this algorithm to Figure 1 is a little more difficult than the preceding cases for a number of reasons. They were, first of all, graph problems while the layout problem is highly geometrical; while the relationships between nodes come directly from the system graph in these cases, in architectural space allocation the relationships between the departments come from the adjacency matrix and are not immediately of the either - or type. The serial decomposition is none the less important as it indicates a coarse hierarchy of departments for the layout procedure.

As in the case of the band width problem simple bounds are available. Any layout provides an upper bound on an optimal layout; a lower bound can be achieved by assuming that all departments are square and can be placed directly adjacent to each other.

5. CONCLUDING REMARKS

While this article is something of a potpourri in terms of topic there is an underlying commonality which the author has tried to represent in Figure 1. Probably the most important characteristic shared by the problems discussed is sparseness: In terms of its system graph structures are usually sparse while the space allocation problem is sparse in terms of information content (i.e. each department is usually only related to a few other departments). In these cases Figure 1 can be used to construct algorithms; it is only for sparse systems that this serial decomposition has significances.

It is possible to view these algorithms in terms of "shortest step" procedures. In the case of minimizing the band width of a system graph the information available says that the absolute value of the difference between the node numbers of a member is to be minimized which motivates algorithms which attempt to do so at every step. The situation is similar in the other cases discussed. In all these cases a shortest step logic is being executed. Motzkin and Straus (1956) have attempted to go beyond this type of procedure in the case of a particular type of extremum problem which is both topological and geometrical. On the whole, this area should be particularly active in the near future as more sophisticated techniques are developed.

Acknowledgement

This work has been supported by the National Science Foundation.

REFERENCES

Sami Al Banna (1972).
"Iterative Structural Design", Ph.D. Thesis, Columbia University, New York, New York 10027.

John H. Argyris and John R. Roy (1972).
"General Treatment of Structural Modifications", Proc. Amer. Soc. Civil Engineers, ST2, pp. 465-492.

Franklin H. Branin (1959).
"The Relation Between Kron's Method and the Classical Methods of Network Analysis", IRE-WESCON Convention Record, Part 2, pp. 1-29.

Lewis R. Friedland (1971).
"Geometric Structural Design", Ph.D. Thesis, Columbia University, New York, New York 10027.

U. Grenander (1969).
"Foundations of Pattern Analysis", Quart. Appl. Math., 27, pp. 1-55.

Henry R. Grooms (1972).
"Algorithm for Matrix Bandwidth Reduction", Proc. Amer. Soc. Civil Engineers, ST1, pp. 203-214.

W. S. Hemp (1968).
"Abstract of Lecture Course "Optimum Structures", 2nd ed., Engr. Lab., University of Oxford.

H. Gordon Jensen and Gary A. Parks (1970).
"Efficient Solutions for Linear Matrix Equations", Proc. Amer. Soc. Civil Engineers, ST1, pp. 49 -

Dag Kavlie and Graham H. Powell (1971).
"Efficient Reanalysis of Modified Structures", Proc. Amer. Soc. Civil Engineers, ST1, pp. 377-392.

Uri Krisch and Moshe F. Rubinstein(1972).
"Reanalysis for Limited Structural Design Modifications", Proc. Amer. Society of Civil Engineers, EM 1, pp. 61-70.

Gabriel Kron (1956).
Diakoptics, Macdonald, London. (Reference to much of Kron's work can be found in this book).

William R. Miller (1971).
"Bibliography: Computer-Aided Space Planning", 8th Annual Design Automation Workshop, SHARE, ACM, IEEE, Atlantic City.

T.S. Motzkin and E. G.Straus (1956).
"Some Combinatorial Extremum Problems", Proc. Amer. Math.Soc., 7, pp. 1014-1021.

G. G. Pope (1971)
"Classical Optimization Theory Relevant to the Design of Aerospace Structures", loc. cit., Pope and Schmit (1971).

G. G. Pope and L. A. Schmit (eds.) (1971).
Structural Design Applications of Mathematical Programming Techniques, AGARD ograph No. 149, AGARD-NATO, Technical Editing and Reproduction Ltd., Hartford House, 7-9, Charlotte St., London.

J. Paul Roth (1959).
"An Application of Algebraic Topology: Kron's Method of Tearing", Quart. Appl. Math., 17, 1.

W. R. Spillers (1964).
"Network Techniques Applied to Structures", The Matrix and Tensor Quarterly, 15, 2, pp. 31-41.

W. R. Spillers (1971).
"Graph Theory, Switching Theory, and Structural Design", Appl. Methcnics Reviews, pp. 501 - 504.

William R. Spillers (1972a).
Automated Structural Analysis: An Introduction, Pergamon Publishing Co., New York City.

William R. Spillers (1972b).
"A Note on the Decomposition of an Absolute-Value Linear Programming Problem", Quart. Appl. Math., pp. 541-544.

W. R. Spillers and John Farrell (1969).
"An Absolute-Value Linear Programming Problem", J. Math. Analysis and Appl., 28, 1, pp. 153-158.

W. R. Spillers and Lewis Friedland (1972).
"An Adaptive Structural Design", to appear in Proc. Amer. Soc. Civil Engineers.

W. R. Spillers and Norris Hickerson (1968).
"Optimal Elimination for Sparse Symmetric Systems as a Graph Problem", Quart. Appl.Math., 16, 3, pp. 425-432.

W. R. Spillers and O. Lev (1971).
"Design for Two Loading Conditions", Int. J. Solids Structures, 7, pp. 1261-1267.

W. R. Spillers and Paul Weidlinger (1970).
"An Algorithm for Space Allocation", 5th Annual ACM Urban Symposium, Assoc. for Computing Machinery, pp. 142-157.

Charles Michael Strauss (1969).
"3DPDP - A Three-Dimensional Piping Design Program", Ph.D. Thesis, Brown University.

Garret M. Vanderplaats and Fred Moses (1972).
"Automated Design of Trusses for Optimum Geometry", Proc. Amer. Soc. Civil Engineers, ST 3, pp. 671 - 690.

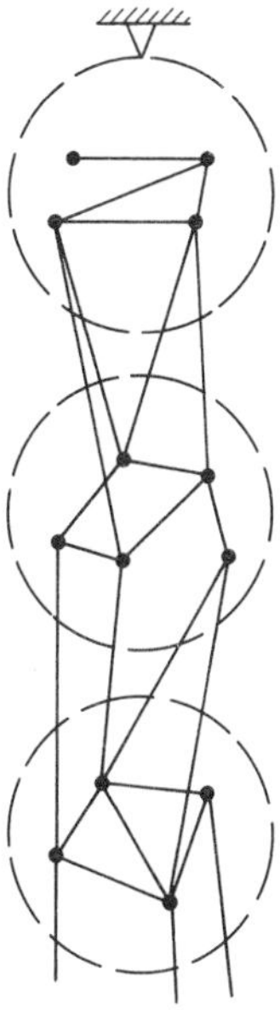

System Graph

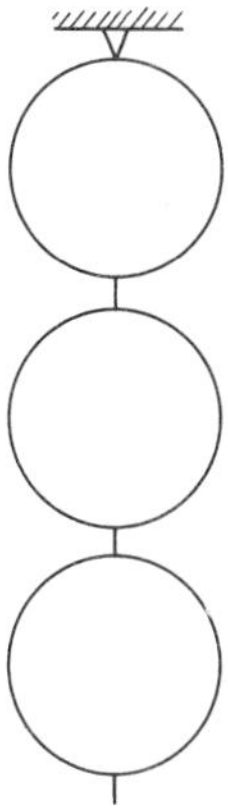

Decomposed System Graph

Figure 1.

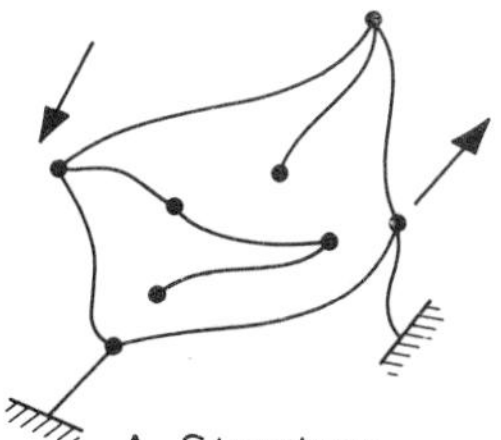

A Structure

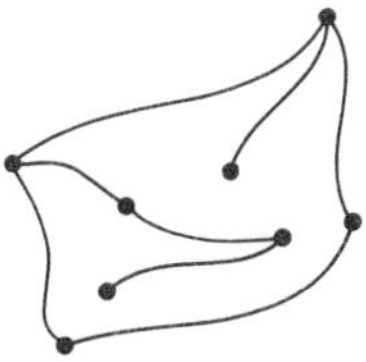

Its Graph

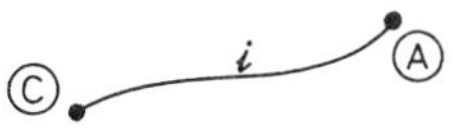

A Typical Member

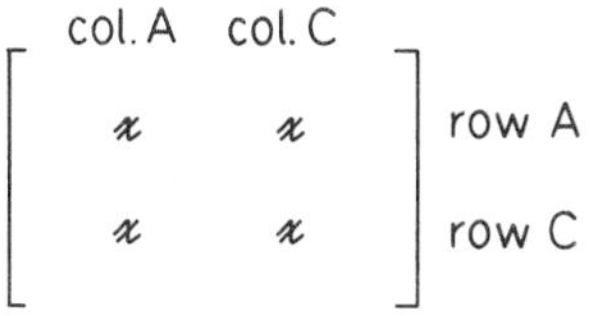

Contribution of Member *i* to the System Matrix

Figure 2.

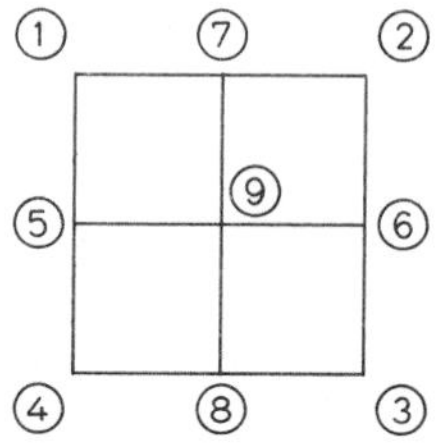

SYSTEM GRAPH

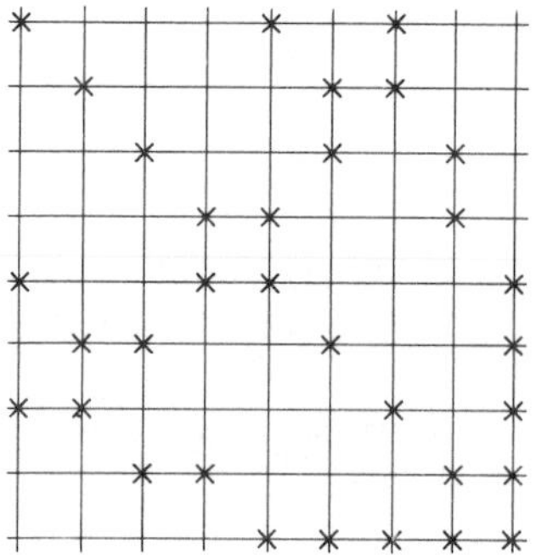

i	b_i	storage req'd at i^{th} step
1	3	0
2	3	3
3	3	6
4	3	9
5	4	12
6	4	16
7	3	20
8	2	17
9	1	13

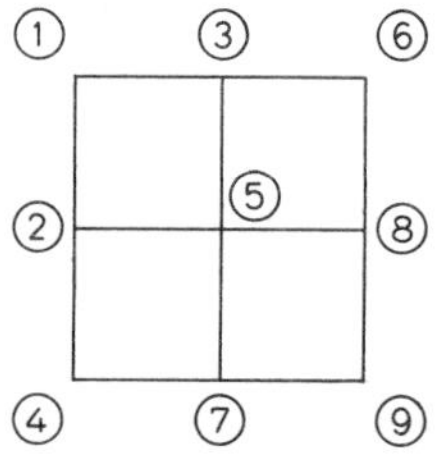

SYSTEM GRAPH

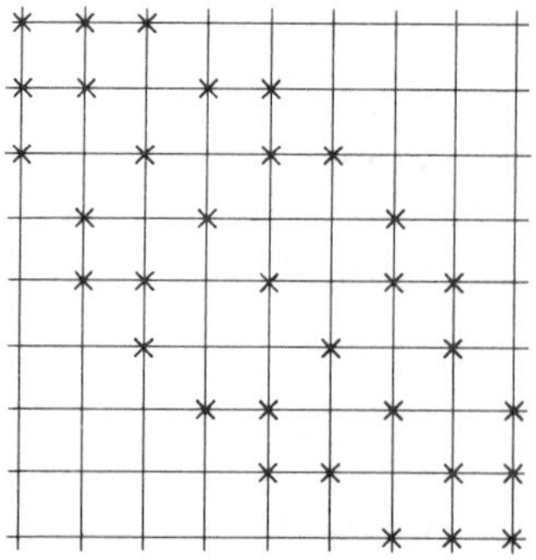

i	b_i	storage req'd at i^{th} step
1	3	0
2	4	3
3	4	7
4	4	8
5	4	12
6	3	12
7	3	11
8	2	10
9	1	5

Figure 3.

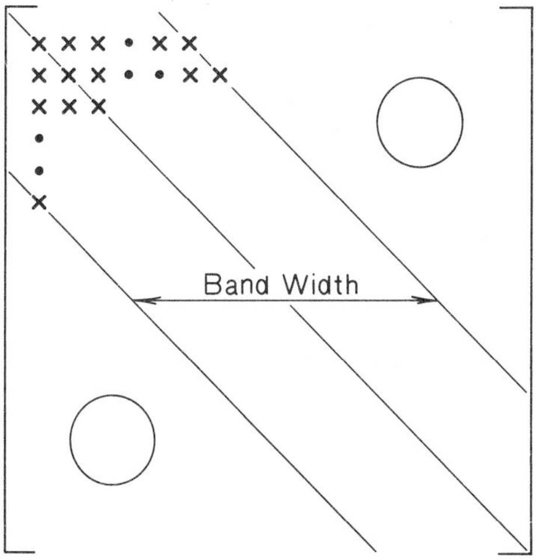
Band Width

Figure 4.

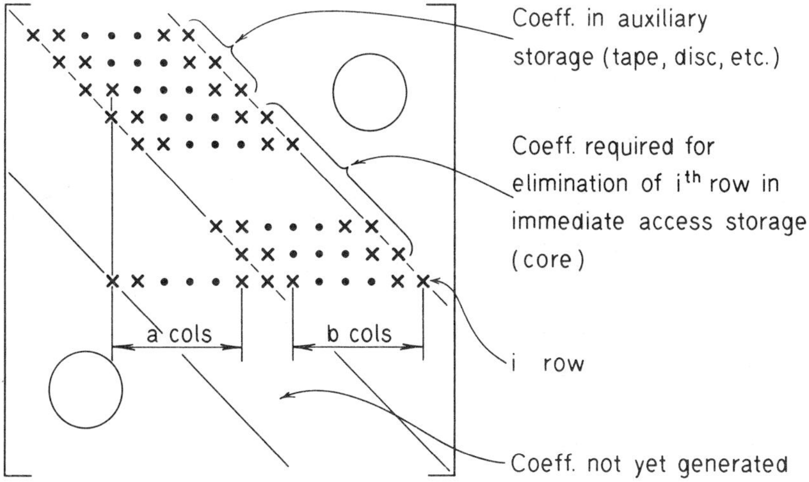
Coeff. in auxiliary storage (tape, disc, etc.)
Coeff. required for elimination of i th row in immediate access storage (core)
i row
a cols
b cols
Coeff. not yet generated

Figure 5.

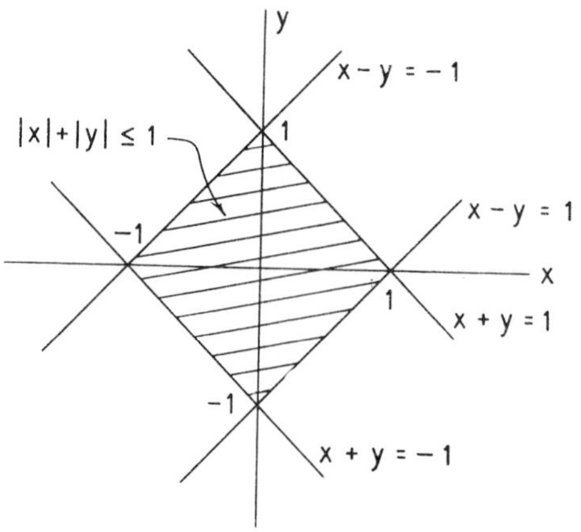

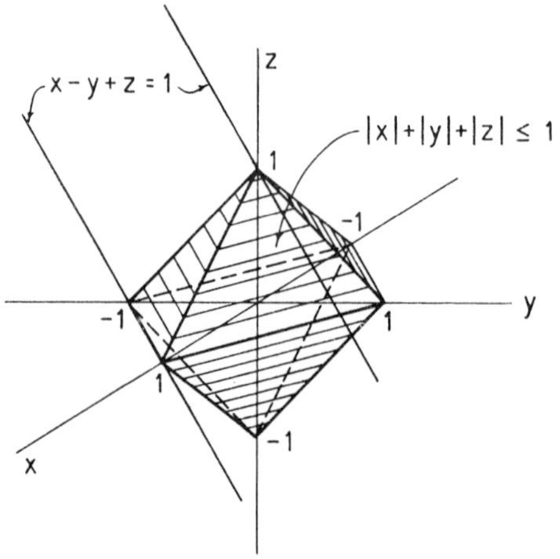

Figure 6

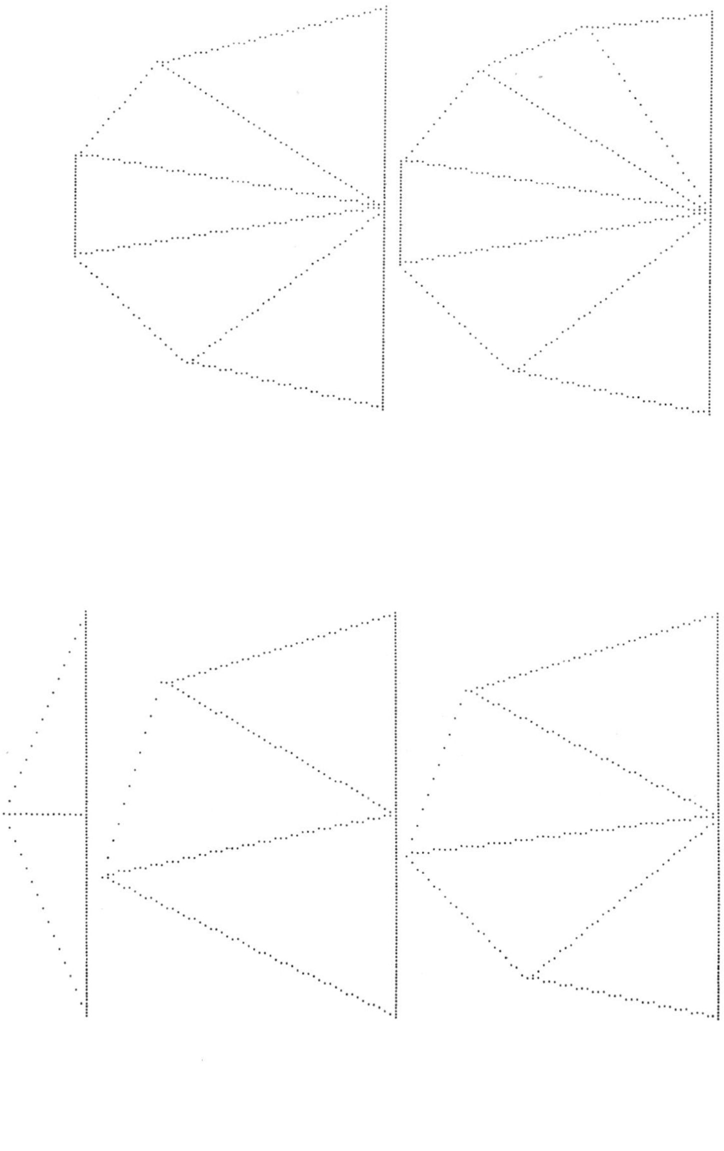

Figure 7

GRAPH-THEORETIC DECOMPOSITION OF LARGE STRUCTURAL NETWORK SYSTEMS

JOHN P. BATY and GWYN WILLIAMS

Department of Civil Engineering and Building Technology
The University of Wales Institute of Science and Technology
Cardiff, South Wales

Abstract: The use of a mixed method of analysis, incorporating both the usual nodal and loop methods, is applied to the compilation of the system equations of large decomposed structural networks. In particular degenerate forms of decomposition are discussed where the decomposition is achieved by tearing at nodes. Bases transformations are also discussed for the modification of the cutset bases in the subnetworks to remove singularities occuring in floating subnetwork stiffness (nodal) equations. A method of network condensation is proposed to determine the form of the bases transformation and the coupling matrices arising as a result of decomposition. An alternative method of subnetworks is suggested for structural problems where bases transformations fail to remove the floating subnetwork singularities. The mixed method equations are also derived for non-linear network systems.

1 INTRODUCTION

This paper describes methods of analysis, based on graph theoretic decomposition, for the solution of large structural systems. All the systems considered here are those which have a diagrammatic representation or model in the form of a network consisting of m-node ($m \geq 2$) elements connected at a finite number of points. The network model is not exclusive to structural engineering, as many types of engineering systems have network models, so that the techniques described in this paper may be applied outside the structural field where appropriate. The main factor distinguishing structural networks from electrical and hydraulic networks for example, is the consideration of the geometrical properties of the network from a practical computation viewpoint, the topological properties being common for all these various network systems.

The theoretical aspects of tearing, dissection or decomposition of networks based on mixed methods of analysis have been well documented for structural systems by Stewart (1964), Baty (1967a, 1969, 1971), Wiberg (1967, 1970) and for electrical networks by Amari (1962), Branin (1962, 1967) and Riaz (1970). Further background material on the applications of algebraic topology to networks will be found in Langefors (1959, 1961) and Samuelsson (1962). Kron's (1967) work on diakoptics was based on his orthogonal analysis which was shown to be related to the mixed methods of analysis for specialised decompositions by Baty (1967a). Happ (1965, 1971) has continued Kron's work and extended it by developing a general contour theory for the analysis of electrical networks. Baty (1967b) indicated how general coutour systems could be related to standard tree-link contours in structural networks and these relations are described in terms of bases transformations.

The only practical implementation of the mixed method of analysis in the structural field known to the authors is that by Wiberg (1970). This paper will also concentrate on practical techniques for solving decomposed structural networks emphasising degenerate forms of the mixed method system equations. The use of graph-theoretical or topological concepts facilitates the administration and strategy of compiling and solving the equations of state of decomposed structural network systems. In this context a method of condensation is proposed for a simple representation of decomposed networks in order to determine the form of a cutset bases transformation matrix T_A and a topological matrix G describing the coupling between the subnetwork system equations. The details of the storage schemes and solution techniques adopted for deriving the solution of equations of state of decomposed networks is described in some detail.

Certain structural systems such as pin-jointed trusses, ball-jointed space frames and finite element models of plane stress or strain models do not lend themselves to the mixed method of analysis particularly when floating networks with singular equations of state occur as a result of decomposition as the singularities cannot be removed by a simple cutset basis transformation. An alternative solution technique based on Przemieniecki's (1963) substructure method is suggested in these cases. (For this type of network the grounding of a single node does not prevent a rigid body rotation about that point). A network description and comparison of Przemieniecki's (1963) method is made and its dual loop analysis substructure method is evident from the network notation used.

Finally there exists a large interest in non-linear network systems from many fields of engineering and so a non-linear form of the mixed method of analysis is given for decomposed non-linear network systems. The advantage of maintaining the network formulation for non-linear network problems means that the expertise and techniques adopted for compiling and solving linear network systems can be utilised immediately for non-linear networks.

2 NETWORK ANALYSIS PRELIMINARIES

The topology or manner in which the elements are inter-connected to form a given structural system is represented by the line diagram of the network, called a graph (directed graph), when the elements are given an orientation. The theoretical aspects of graph and digraph theory have been given by Harary (1970, 1965). The digraph representation for oriented network systems will be used in this paper. The foundation nodes are merged into a single ground node and the remaining nodes are said to be independent.

If a digraph of a structural network has b branches, n independent nodes and l independent loops (b = n + l) then it is possible to establish two dual pairs of independent contours on the digraph, a path contour pair (μ^o, μ^c) consisting of n open and l closed paths and a cutset contour pair (ω^c, ω^o) consisting of n cutsets (closed) and l chords (open cutsets) respectively. The information describing these contours in terms of the branches is conveniently summarised in the four connection matrices B, C, A andD (Baty 1967b, 1971) of order b x n, b x l, b x n and b x l, where:

$$\left.\begin{matrix} B_{ij} \\ C_{ij} \\ A_{ij} \\ D_{ij} \end{matrix}\right\} = \left\{\begin{matrix} 1 \\ -1 \\ 0 \end{matrix}\right\}, \text{ if branch } i \text{ is } \left\{\begin{matrix} \text{positively} \\ \text{negatively} \\ \text{not} \end{matrix}\right\} \text{ incident to } \left\{\begin{matrix} \text{open path } \mu^o_j \\ \text{closed path } \mu^c_j \\ \text{cutset } \omega^c_j \\ \text{chord } \omega^o_j \end{matrix}\right\} \text{ respectively}$$

The matrix pairs X = (B,C) and Y = (A,D) are non-singular square matrices of order b and the duality relations existing between the contour pairs

$$\begin{bmatrix} \mu^o \\ \mu^c \end{bmatrix} = \begin{bmatrix} B^t \\ C^t \end{bmatrix} \underline{b} = X^t\ \underline{b}, \qquad \begin{bmatrix} \omega^c \\ \omega^o \end{bmatrix} = \begin{bmatrix} A^t \\ D^t \end{bmatrix} \underline{b} = Y^t\ \underline{b} \qquad \text{.......................(1)}$$

is expressed by the equation:

$$X^tY = (B,C)^t\ (A,D) = I_b \qquad \text{...(2)}$$

or its alternate forms $YX^t = XY^t = Y^tX = I_b$, and when equation (2) is multiplied out one has:

$$B^tA = I_n, \quad B^tD = 0_{nl}, \quad C^tA = 0_{ln}, \quad C^tD = I_l \quad \text{.......................(3)}$$

Normally a tree-link partition of the branches of the digraph is selected and the associated dual contour pairs (μ^o, μ^c), (ω^c, ω^o) are taken as the n independent node-to-datum node tree paths, l link determined closed paths (loops), n independent node-star cutsets and l individual links respectively when the connection matrices have the partitioned forms:

$$(B_s, C_s) = \begin{bmatrix} B_T & C_T \\ 0 & I \end{bmatrix}, \quad (A_s, D_s) = \begin{bmatrix} A_T & 0 \\ A_L & I \end{bmatrix} \quad \text{................................(4)}$$

and this type of selection for the contours will be called standard bases. If general contour bases $(\mu^o, \mu^c)g$, $(\omega^c, \omega^o)g$ are selected in the digraph then they may be related to a standard bases contour selection by dual matrix transformations of the form (Baty 1967a):

$$\begin{bmatrix} \mu^o \\ \mu^c \end{bmatrix}_g = \begin{bmatrix} T_B^t & L_{CB}^t \\ 0 & T_c^t \end{bmatrix} \begin{bmatrix} \mu^o \\ \mu^c \end{bmatrix}_s = M^t \begin{bmatrix} \mu^o \\ \mu^c \end{bmatrix}_s, \quad \begin{bmatrix} \omega^c \\ \omega^o \end{bmatrix}_g = \begin{bmatrix} T_A^t & 0 \\ L_{AD}^t & T_D^t \end{bmatrix} \begin{bmatrix} \omega^c \\ \omega^o \end{bmatrix}_s = Q^t \begin{bmatrix} \omega^c \\ \omega^o \end{bmatrix}_s \quad \text{...............(5)}$$

The corresponding relations between the connection matrices follow from equations (1) and (5) and are:

$$\begin{bmatrix} \mu^o \\ \mu^c \end{bmatrix}_g = X_g^t \underline{b} = M^t \begin{bmatrix} \mu^o \\ \mu^c \end{bmatrix}_s = M^t X_s^t \underline{b}, \quad \text{or} \quad X_g = X_s M \quad \text{.........................(6)}$$

$$\begin{bmatrix} \omega^c \\ \omega^o \end{bmatrix}_g = Y_g^t \underline{b} = Q^t \begin{bmatrix} \omega^c \\ \omega^o \end{bmatrix}_s = Q^t Y_s^t \underline{b}, \quad \text{or} \quad Y_g = Y_s Q \quad \text{.........................(7)}$$

and using equation (2) the duality relation between M and Q is:

$$X_g^t Y_g = M^t X_s^t Y_s Q = M^t Q = I, \quad \text{or} \quad Q = (M^t)^{-1}, \quad M = (Q^t)^{-1} \quad \text{................(8)}$$

Figure 1 illustrates these topological relations for a digraph associated with a simple structural network.

The network analysis is completed physically by superimposing an algebraic structure on the network consisting of the interrelations between the load (flow) and displacement (potential) variables associated with the branches and contour pair bases (μ^o, μ^c), (ω^c, ω^o) selected in the digraph of the network. N, P and R will be used to denote the branch, open path and closed path load variables and n, p and r the corresponding branch, cutset and chord displacement variables associated with a standard bases situation respectively. P^o, R^c, p^c, r^o will be used to denote the general bases form of the load and displacement variables, the branch variables N and n being invariant with respect to bases transformations.

The load-displacement (Hooke's Law) relations for the b branches of the disconnected or primitive network are specified and given in either the matrix stiffness form:

$$N_t = N + N_p = K(n + n_p) = Kn_t \quad \text{...................................(9)}$$

or the inverse flexibility form:

$$n_t = n + n_p = F(N + N_p) = FN_t \quad \text{...................................(10)}$$

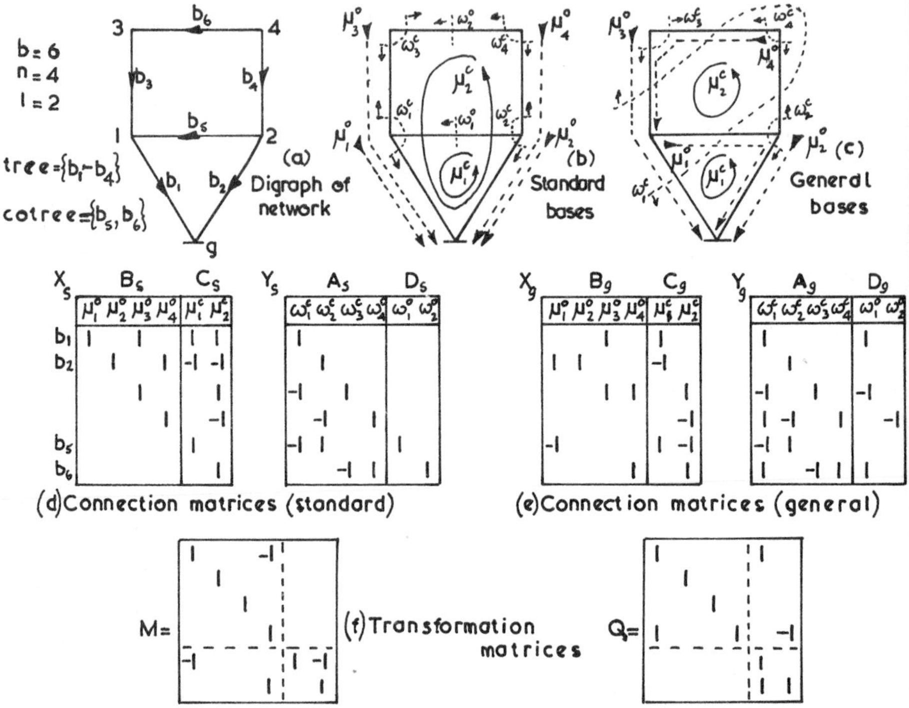

FIGURE 1

where K and F $(=K^{-1})$ are diagonal or block diagonal matrices of order b. The vector pairs N_t and n_t, N and n, N_p and n_p represent the total, induced and prescribed (due to internal sources on the branch) branch loads and displacements respectively. The sign convention for these variables is shown in figure 2.

The act of interconnecting the branch elements to form the structural network imposes constraints on the physical variables (Kirchoff Laws) which are known as the dual load equilibrium and displacement compatibility relations in structural analysis. These dual relations are summarised with the aid of the topological connection matrices and are:-

Load equilibrium	Displacement compatibility	
$N = CR$,	$n = Ap$	(11)
$N_p = BP$,	$n_p = Dr$	(12)
$N_t = (B,C)\begin{bmatrix} P \\ R \end{bmatrix} = X\begin{bmatrix} P \\ R \end{bmatrix}$,	$n_t = (A,D)\begin{bmatrix} p \\ r \end{bmatrix} = Y\begin{bmatrix} p \\ r \end{bmatrix}$	(13)
$A^t N_t = P$,	$B^t n_t = p$	(14)
$D^t N_t = R$,	$C^t n_t = r$	(15)

Algebraic manipulation of equations (9) to (15) produces the well-known stiffness (nodal) and flexibility (loop) system equations for a network and the variable transformation diagram in figure 3 (Roth, 1955) provides a neat summary of these fundamental relations between the network variables.

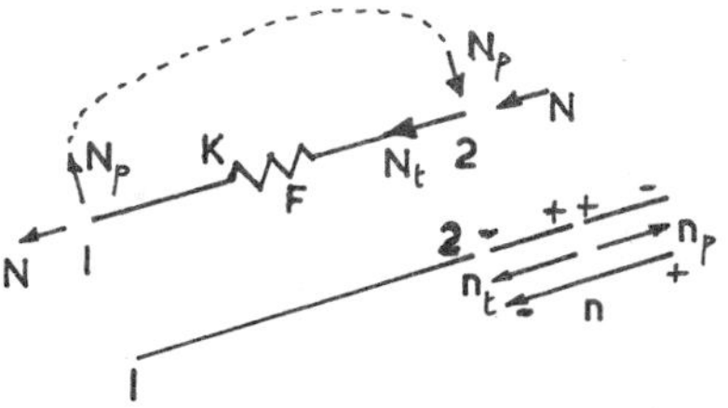

Branch variables-sign convention
FIGURE 2

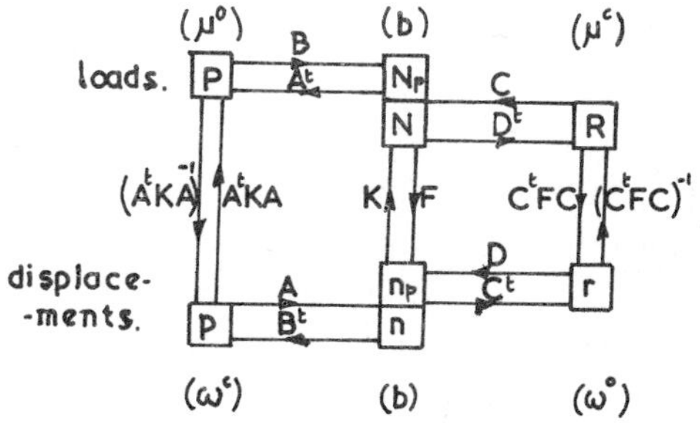

Branch variables-transformation diagram
FIGURE 3

Equations (9) to (15) hold for the standard or general bases form of the structural variables and connection matrices and the relations between them can be deduced from equations (6), (7) and (15) as follows:

$$N_t = X_g\begin{bmatrix}P^o\\R^c\end{bmatrix} = X_sM\begin{bmatrix}P^o\\R^c\end{bmatrix} = X_s\begin{bmatrix}P\\R\end{bmatrix}, \quad \text{or} \quad M\begin{bmatrix}P^o\\R^c\end{bmatrix} = \begin{bmatrix}P\\R\end{bmatrix} \quad \ldots\ldots\ldots\ldots\ldots\ldots\ldots(16)$$

$$n_t = Y_g\begin{bmatrix}p^c\\r^o\end{bmatrix} = Y_sQ\begin{bmatrix}p^c\\r^o\end{bmatrix} = Y_s\begin{bmatrix}p\\r\end{bmatrix}, \quad \text{or} \quad Q\begin{bmatrix}p^c\\r^o\end{bmatrix} = \begin{bmatrix}p\\r\end{bmatrix} \quad \ldots\ldots\ldots\ldots\ldots\ldots\ldots(17)$$

In structural network problems either K or F and the variables P and n_p (or N_p) are specified and N and p are required from the analysis. If general bases are used then a transformation is required to convert P to P^o and then p^c back to p using equations (16) and (17) or their inverses.

The components of the load and displacement vectors are themselves often vectors in structural analysis with as many components as there are degrees of freedom associated with the structural elements. The entries of the connection matrices are then unit matrices of the appropriate order provided a global co-ordinate system is used for the structural variables, see Fenves (1963, 1966), Spillers (1965). There is a choice of placing all the geometrical information either in the Hooke's Law equations (9) and (10) or in the connection matrices. In the latter case the connection matrices would need redefining but their structure does not change.

3 MIXED METHOD OF ANALYSIS

The mixed method of analysis is a method for establishing the equations of state of decomposed network systems in mixed form incorporating both the pure cutset (nodal) and closed path (loop) methods of analysis simultaneously. The branch set S{b} of a network S are partitioned by the decomposition into two subsets labelled $S_1\{b_1\}$ and $S_o\{b_o\}$ respectively. Cutset analysis is applied to the S_1 type subnetworks and closed path analysis to S_o type subnetworks. The decomposition is performed in such a way that the sets of closed paths $\{\mu^c\}$ and cutsets $\{\omega^c\}$ are also partitioned into two subsets $\{\mu_1^c, \mu_o^c\}$ and $\{\omega_1^c, \omega_o^c\}$ with the property that closed paths $\{\mu_1^c\}$ consist of branches from $\{b_1\}$ only and cutsets $\{\omega_o^c\}$ consist of branches from $\{b_o\}$ only. This partitioning of the closed path and cutset sets induces a corresponding partitioning of the network variables $P = (P_1, P_o)$, $p = (p_1, p_o)$, $R = (R_1, R_o)$ and $r = (r_1, r_o)$. The connection matrices are also partitioned into the forms:

$$\begin{matrix} & n_1 & n_o \\ b_1 & B_{11} & B_{10} \\ b_o & B_{01} & B_{00} \end{matrix} \quad \begin{matrix} l_1 & l_o \\ C_{11} & C_{10} \\ 0 & C_{00} \end{matrix} \quad \begin{matrix} n_1 & n_o \\ A_{11} & 0 \\ A_{01} & A_{00} \end{matrix} \quad \begin{matrix} l_1 & l_o \\ D_{11} & D_{10} \\ D_{01} & D_{00} \end{matrix} \qquad (18)$$

and this leads to the mixed method system equations (see Branin 1967, Baty 1967a, 1971, Riaz 1970, Wiberg 1967, 1971) given here in general bases form:

$$\begin{bmatrix} K_1 & G \\ -G^t & F_o \end{bmatrix} \begin{bmatrix} p_1^c \\ R_o^c \end{bmatrix} = \begin{bmatrix} A_{11}^t & \{(B_{11}P_1^o + B_{10}P_o^o) - K_{11}n_{pl}\} \\ C_{oo}^t & \{n_{po} - F_{oo}(B_{o1}P_1^o + B_{oo}P_o^o)\} \end{bmatrix} \qquad (19)$$

where $K_1 = A_{11}^t K_{11} A_{11}$, K_{11} = diagonal matrix of branch stiffnesses in S_1,

$F_o = C_{oo}^t F_{oo} C_{oo}$, F_{oo} = diagonal matrix of branch flexibilities in S_o,

$G = A_{o1}^t C_{oo} = -A_{11}^t C_{10}$.

The compilation of the flexibility part F_o in equation (19) associated with the S_o subnetwork can be avoided if degenerate decompositions are used. These are achieved by splitting nodes of the digraph of network S and inserting fictitious branches with zero flexibility across the split nodes. If the decomposition is then made with the fictitious branches in S_o and the real branches in S_1, equation (19) has the special form:

$$\begin{bmatrix} K_1 & G \\ G^t & 0 \end{bmatrix} \begin{bmatrix} p_1^c \\ R_o^c \end{bmatrix} = \begin{bmatrix} P_1^o - A_{11}^t K_{11} n_{pl} \\ 0 \end{bmatrix} \qquad (20)$$

(symmetry is achieved by negating the S_o rows of equation (20)) when the partitioned form of the connection matrices are:

$$\begin{matrix} & n_1 \\ b_1 & B_{11} \\ b_o & 0 \end{matrix}, \quad \begin{matrix} l_1 & l_o \\ C_{11} & C_{10} \\ 0 & I \end{matrix}, \quad \begin{matrix} n_1 \\ A_{11} \\ A_{01} \end{matrix}, \quad \begin{matrix} l_1 & l_o \\ I & 0 \\ 0 & I \end{matrix} \qquad (21)$$

Equation (20) is used as the basis of a solution algorithm described in the next section for solving decomposed structural systems. For these degenerate types of decomposition the number of variables in equation (20) is $(n + 2n_s)$ -(n_s = number of split nodes)- compared to n for a straight stiffness analysis. The main aim of degenerate decomposition is to induce a block diagonal form on the K_1 part of equation (20) by partitioning the real branches of S into convenient subsets S_1 as a result of splitting certain nodes.

The pattern of elements in the K_1 $(= A^t_{11}K_{11}A_{11})$ block is determined by the product $A_{11}^t A_{11}$ and a standard bases selection may be unsatisfactory under decomposition because: (a) A_{s11} cannot be partitioned into block diagonal form and hence K_1; (b) the stiffness sub-blocks associated with floating subnetworks are singular. Both these difficulties can be removed by using a cutset transformation to produce both a cutset block diagonal matrix A_{g11} and non-singular stiffness equations for the floating sub-networks. These new general bases are related to standard bases by a transformation of the form (see equation (8)):

$$Q = \begin{bmatrix} T_A & 0 \\ 0 & I \end{bmatrix}, \qquad M = \begin{bmatrix} (T_A^t)^{-1} & 0 \\ 0 & I \end{bmatrix} \qquad (22)$$

which produces the following relations between the connection matrices from equations (6) and (7):

$$B_g = B_s(T_A^t)^{-1}, \quad C_g = C_s, \quad A_g = A_sT_A, \quad D_g = D_s \qquad (23)$$

The new cutset displacements and open path load relations, using equation (17) and the inverse of equation (16), are:

$$p = T_Ap^c, \qquad P^o = T_A^tP \qquad (24)$$

and then the right-hand side of equation (20) is determined from the second part of equation (24) and the solution p^c is transformed by the first part of equation (24) to produce the node displacements p relative to the ground node.

The coupling matrix G in equation (20) is also affected by this cutset bases transformation and is now given by:

$$G = A^t_{g01}C_{00} = A^t_{g01} = T^t_AA^t_{s01} \qquad (25)$$

as $C_{00} = I$. The pattern of elements in G is more complicated with the new bases and its compilation is discussed in detail later. Figure 4 shows a simple network which has been decomposed into two subnetworks by splitting node 2. The connection matrices in standard and general bases form, the transformation matrix T_A and the coefficients of the mixed method equation (20) are also shown in scalar form assuming unit stiffnesses for the branches. Of the boundary nodes in the floating subnetwork $S_{1.a}$, node 1 has been selected as a ground node for this subnetwork and node 2 is torn to produce an appropriate decomposition.

When the decomposition technique is applied to more complicated networks it is necessary to select one boundary node in each floating subnetwork as a ground node so as to connect it to a neighbouring subnetwork which may be truly grounded or not. If that neighbouring subnetwork is floating then it must be grounded itself to another neighbour and so on until a simple path to a true ground is achieved. The remaining non-grounded boundary nodes are torn. This process of linking the subnetworks is helped conceptually by performing a condensation of the decomposed network.

The torn network is condensed by contracting each S_1 subnetwork to a node, coloured black if that S_1 is grounded and white if it is floating. The condensation diagram is completed by connecting tree branches (solid lines): (a) from each grounded S_1 subnetwork node to the datum node z; (b) from each floating S_1 subnetwork node to its neighbour defined by the choice of boundary ground node in the floating S_1, and link branches (dashed lines); (c) between each pair of subnetwork nodes which share a torn node of the original network.

The solid line part of the condensation diagram should form a tree with datum node z as root. For consistency the nodes of the condensed network diagram are labelled from the top of the tree down to the root. (See figure 4i and 7c).

The mixed method equations for such degenerate decompositions, with the basis transformation matrix T_A defined, then produce a block diagonal form as shown in figure 5, where to take full advantage of decomposition as many as possible of the K subnetwork stiffness blocks would be made identical.

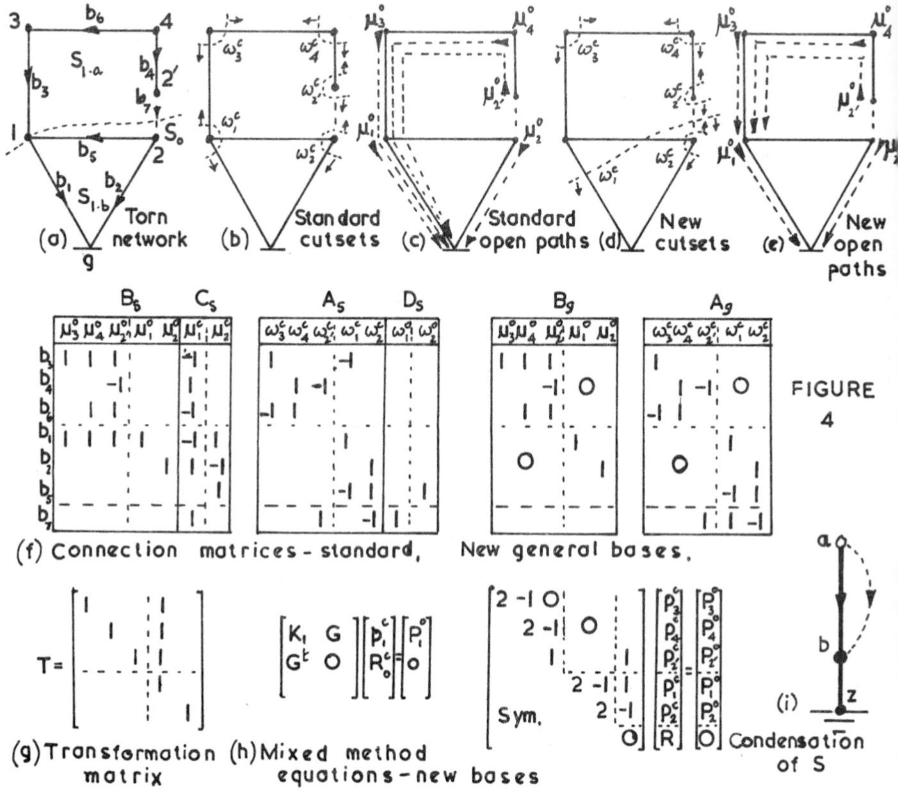

FIGURE 4

A fundamental cutset and open path bases selection would also produce block diagonal and non-singular forms of equation (20) but the simplicity of compiling the triple products A^tKA is then lost compared to either standard bases or the transformed bases suggested here. The maximum number of elements per row in the A_s or A_g matrices is two and this feature is the main reason for the popularity of the stiffness method in structural analysis.

The topological aspects of decomposition of structural networks has been discussed so far with all the structural variables expressed in a global co-ordinate form. For numerical reasons it is usual to use a main co-ordinate representation of the variables and the mixed method equation (20) is transformed from global to main co-ordinate form by using relations of the form:

$$\begin{bmatrix} p_1^c \\ R_o^c \end{bmatrix}_G = \begin{bmatrix} (H_1^t)^{-1} & 0 \\ 0 & H_o \end{bmatrix} = \begin{bmatrix} p_1^c \\ R_o^c \end{bmatrix}_M, \qquad \begin{bmatrix} P_1^o \\ 0 \end{bmatrix}_G = \begin{bmatrix} H_1 & 0 \\ 0 & (H_o^t)^{-1} \end{bmatrix} \begin{bmatrix} P_1^o \\ 0 \end{bmatrix}_M \quad \ldots\ldots(26)$$

where the transfer matrices H_1 and H_o are block diagonal matrices with elements of the form, see Fenves (1963, 1966), Spillers (1965):

$$H = \begin{bmatrix} I & 0 \\ V & I \end{bmatrix}, \qquad V = \begin{bmatrix} 0 & -z & y \\ z & 0 & -x \\ -y & x & 0 \end{bmatrix} \quad \ldots\ldots(27)$$

$$\begin{bmatrix} K_{1.1} & & & & G_1 \\ & K_{1.2} & & & G_2 \\ & & \ddots & & \vdots \\ & & & K_{1.m} & G_m \\ G_1^t & G_2^t & \cdots & G_m^t & O \end{bmatrix}$$

FIGURE 5

$$\begin{bmatrix} N_1 \\ N_2 \\ \vdots \\ N_m \end{bmatrix} = \begin{bmatrix} K_{11} & K_{12} & \cdots & K_{1m} \\ K_{21} & K_{22} & \cdots & K_{2m} \\ \vdots & \vdots & \ddots & \vdots \\ K_{m1} & K_{m2} & \cdots & K_{mm} \end{bmatrix} \begin{bmatrix} n_1 \\ n_2 \\ \vdots \\ n_m \end{bmatrix}$$

FIGURE 6

Algebraic manipulation of equations (20) and (26) produces the main co-ordinate form of the degenerate mixed method equation, which is:

$$\begin{bmatrix} H_1^{-1} & 0 \\ 0 & H_o^t \end{bmatrix} \begin{bmatrix} K_{1G} & G_G \\ G_G^t & 0 \end{bmatrix} \begin{bmatrix} (H_1^t)^{-1} & 0 \\ 0 & H_o \end{bmatrix} \begin{bmatrix} p_1^c \\ R_o^c \end{bmatrix}_M = \begin{bmatrix} K_{1M} & G_M \\ G_M^t & 0 \end{bmatrix} \begin{bmatrix} p_1^c \\ R_o^c \end{bmatrix}_M = \begin{bmatrix} p_1^o \\ 0 \end{bmatrix}_M \quad \ldots\ldots\ldots\ldots\ldots\ldots (28)$$

where $K_{1M} = H_1^{-1}K_{1G}(H_1^t)^{-1}$ and $G_M = H_1^{-1}G_GH_o$.

The main co-ordinate representation affects the elements of the coupling matrix G and transformation matrix T_A but not their position, the unit matrices are replaced by transfer matrices. K_{1M} is the stiffness form commonly used in structural analysis and consequently it is possible to extend the analysis to networks composed of higher order elements, i.e. finite element systems. The load-displacement relation for a typical m-node element has the main co-ordinate form shown in figure 6 and is an alternate form of equation (9). The contribution of an m-node element i to the subnetwork stiffness equation to which it belongs is evaluated by a triple matrix product $\alpha_i^t.K_i.\alpha_i$ where α_i is the appropriate row partition of a co-incidence matrix (Langefors 1959) which lists the subnetwork nodes to which the element is connected.

It should be emphasised that the analysis described in this section is only applicable to structural network systems for which the selection of a single ground node in the network S or the subnetworks $S_{1.i}$ prevents rigid body displacements, i.e. rigid plane, grid or space frames, plate bending finite element models. Systems such as trusses, ball-jointed grillages or space frames, plane stress or strain finite element models are excluded as the cutset basis transformation described in this section does not eliminate the singularity associated with floating subnetwork stiffness equations arising from decomposition. Such systems can be solved by an alternative decomposition technique due to Przemieniecki (1963) outlined in section 6.

4 PRACTICAL COMPILATION OF MIXED METHOD EQUATIONS

For practical implementation of the degenerate decomposition method outlined in section 3, the main co-ordinate form equation (28) is used for numerical conditioning reasons rather than the global form equation (20). The compilation of the stiffness blocks of K_{1M} is straightforward given the element-node connection properties of the subnetworks. For a given decomposition of a network the structure of the matrices T_A^t and G can be obtained in: (a) block; (b) global co-ordinate, or (c) main co-ordinate form by setting up a tree-path matrix B* for the condensation diagrams S* of the decomposed network S. The B* matrix is, in fact, see second part of equation (24), the block form of the matrix T_A^t. The global form of T_A^t is then obtained by detailing the nodes of the subnetwork blocks and entering unit matrices in the leading diagonal positions and in the selected ground node rows of the non-zero blocks below the diagonal as indicated by the non-zero contributions in B*. The main co-ordinate form T_{AM}^t follows by replacing the unit matrices below the leading diagonal by transfer matrices H_{ij} in position (i,j) where the V matrix of equation (27) now has elements of the form $x = x_i - x_j$, $y = y_i - y_j$ and $z = z_i - z_j$. The elements of the coupling matrix G follow from equation (25) as A_{s01} is the incidence matrix of the fictitious branches inserted at the split nodes and therefore the columns of G are the difference of the torn node rows of T_A^t. The contents of these various matrices is given in figure 7 for a simple degenerate decomposition.

The condensation technique allows a simple representation of large decomposed structural network systems and the structure of the T_A^t and G matrices are more easily discerned. The development of the algorithm for compiling these matrices is simplified as the condensation highlights the topological or connection properties of the torn systems. For problems with a large number of torn nodes the G and F blocks become quite large also but with the compact storage scheme outlined in section 5 only the non-zero parts of these matrices are stored.

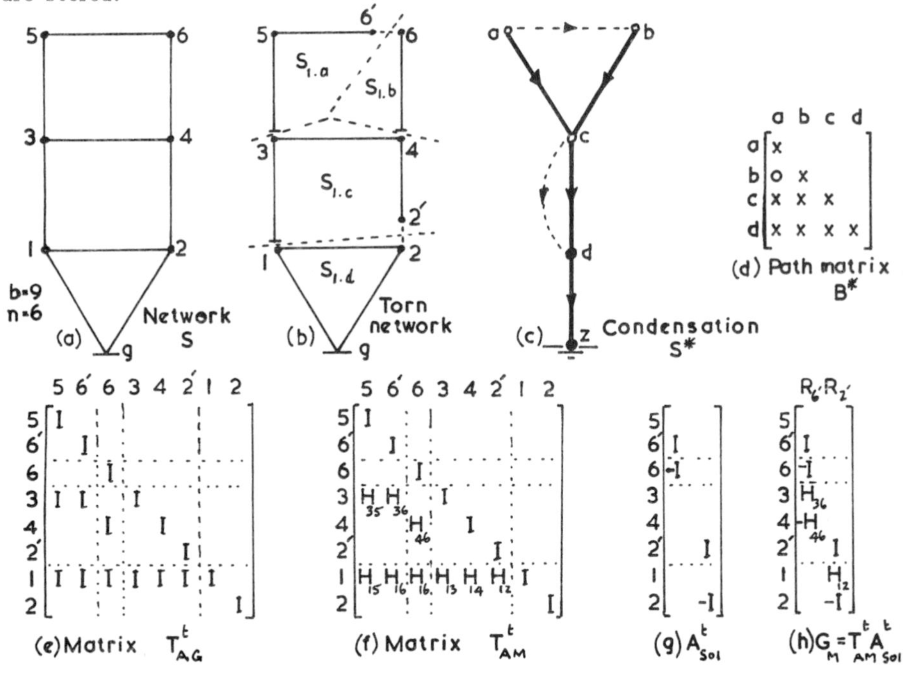

FIGURE 7

5 STORAGE SCHEME AND SOLUTION TECHNIQUE

To minimise the storage required for the symmetric form of the mixed method equation (29) only the upper triangular part of the K and F diagonal blocks and the non-zero columns of the G blocks are stored by columns in single dimension arrays. A variable band width is used for the individual K and G subnetwork blocks and the F block is stored in full upper triangular form as this corner, initially zero, fills as the solution proceeds. The variable bandwidth technique requires an additional vector JB to define the shape of the variable band and to address the elements within the single dimension array. The component JB(j) holds the row number of the first non-zero element in column j and the addresses of the diagonal elements are then evaluated from the sequence:

$$DG(1) = 1, \quad DG(j) = DG(j-1) + JB(j) - j + 1, \quad j = 2, 3, \ldots n;$$

and the address of a general element (i,j), $i \leq j$, is given by the expression $(DG(j) + i - j)$.

Random access disc is the backing store used on the ICL 4/70 by the authors and the maximum size of a disc record is 896 single words and these can be linked to form larger logical records, if necessary. If there are m sub-networks and n_s split nodes arising from a decomposition then the 2m blocks $K_{1.1}$, $G_{1.1}$, $K_{1.2}$, $G_{1.2}$, ... $K_{1.m}$, $G_{1.m}$ and the n_s individual columns F_1, ...F_{ns} are stored as single logical records on disc. The schematic form of this storage technique is shown in figure 8 (see also Jennings (1971), Wiberg (1970).)

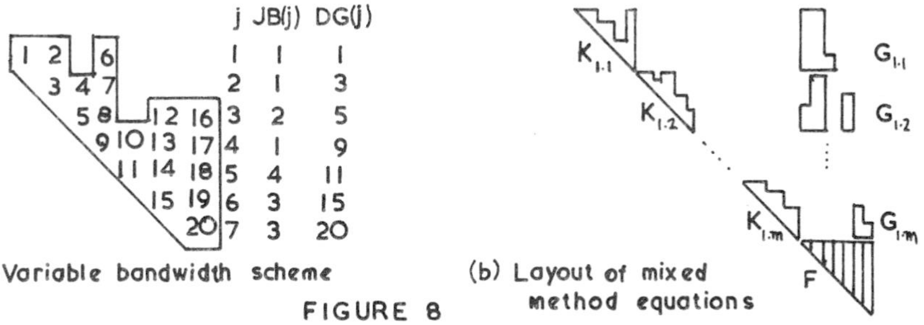

(a) Variable bandwidth scheme (b) Layout of mixed method equations

FIGURE 8

The shape of the K blocks is a function of the node numbering scheme adopted in each subnetwork. The node numbering scheme adopted by the authors is to number: (a) the true ground nodes -1, -2, ...; (b) the selected ground node of a floating subnetwork o; (c) the internal nodes 1, 2, 3, ...J; and (d) the boundary nodes J+1, J+2, ... $J+n_s$ in each subnetwork. Numerical information describing the linkage of the boundary nodes to the neighbouring subnetworks and the type of linkage is also given with the data for each sub-network. Although not incorporated in this program, King (1970) gives an algorithm for minimising a variable bandwidth storage scheme such as used here.

Compact elimination is performed within the area of the variable bandwidth scheme shown in figure 8 and is modified to cope with the block form of the equations. For a typical subnetwork $S_{1.i}$ only the $K_{1.i}$, $G_{1.i}$ and the appropriate columns of the F block, as determined by the non-zero columns of $G_{1.i}$, are in core at any one time.

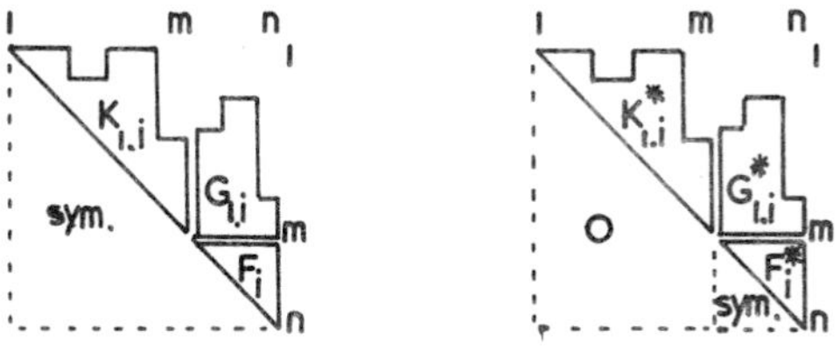

FIGURE 9

The elimination is carried out within the $K_{1.i}$ and $G_{1.i}$ blocks as far as row M in figure 9 and the modified $K_{1.i}$, $G_{1.i}$ and F blocks are then returned to disc store. Each subnetwork is dealt with in a similar manner until finally the complete F block is compiled and then the elimination is continued on the F block itself. The back substitution process follows after the right-hand side vector has first been modified by a forward substitution process on account of the triangularisation. The solution consists of the values of the cutset displacements p_1^c and loads R_o^c. The first part of equation (24) is then used to transform the cutset displacements p_1^c to the actual node displacements p (relative to ground). The member forces N are evaluated from equation (9) using the compatibility equation (11).

The compact elimination for the variable bandwidth form is based on the formula:

$$a_{ij} = a_{ij} - \sum_{k=q}^{i-1} \frac{a_{ki}\, a_{kj}}{a_{kk}}, \qquad \begin{aligned} i &= JB(j) + 1, \ldots\ldots, j; \\ j &= 2, \ldots\ldots, n; \\ q &= \max(JB(i), JB(j)), \end{aligned}$$

where only non-zero contributions under the summation sign are evaluated for efficiency; this is determined by the contents of the JB vector. It is also possible to incorporate further constraints on the variables in the mixed method equations by modifying the JB vector as suggested by Wiberg (1970).

The logic of the algorithm described above has been checked and solutions of decomposed rigid structural network systems have been obtained. Large scale tests have not yet been carried out or a comparison made with other techniques such as Przemieniecki's (1963) sub-structure method or Irons (1970) frontal solution method. It's main advantage would be for large regular systems where the elimination work can be reduced when there is a repetition of identical K blocks. Although not tested yet the conditioning of the mixed method equations may be better than the normal stiffness equations of large systems.

6 PRZEMIENIECKI'S SUBSTRUCTURE METHOD

Przemieniecki's (1963) method of substructure analysis is quoted here in our notation to give a comparison and because the network formalism of this paper immediately suggests the dual flexibility (loop) application of his technique. In this method the network is decomposed into subnetworks by tearing or splitting nodes. The stiffness equations are established in partitioned form for each subnetwork, the partitioning dividing these equations into the internal and boundary node components. This type of decomposition corresponds to a simple row partition of the A connection or α co-incidence matrix. The internal node variables p_I are removed by elimination from each subnetwork equation system to give a modified stiffness matrix for the boundary node part of each subnetwork. These modified boundary stiffnesses could be thought of as load-displacement relations for a finite element as defined by the boundary nodes and shown in figure 6. This interpretation of the modified boundary equations suggests that Irons (1970) frontal solution technique could be used to obtain the solution of the interconnected boundary network system for the boundary node variables p_B. In Przemieniecki's paper the boundary network system equations are compiled using the load equilibrium and displacement compatibility relations at the boundary nodes. The remaining displacements p_I are determined by back substitution in each of the subnetwork blocks once the displacements p_B are known. The essence of Przemieniecki's method and pertinent equations are given in figure 10a.

The dual method follows by analogy, the network systems are now decomposed by tearing or releasing loops and the flexibility equations are established in internal and boundary loop partitioned form for each subnetwork. The decomposition this time corresponds to a row partitioning of the branch-loop connection matrix C. The remainder of the dual analysis then follows the same path as the stiffness method and is shown figuratively in figure 10b.

The advantage of Przemieniecki's method is that it is not restricted to

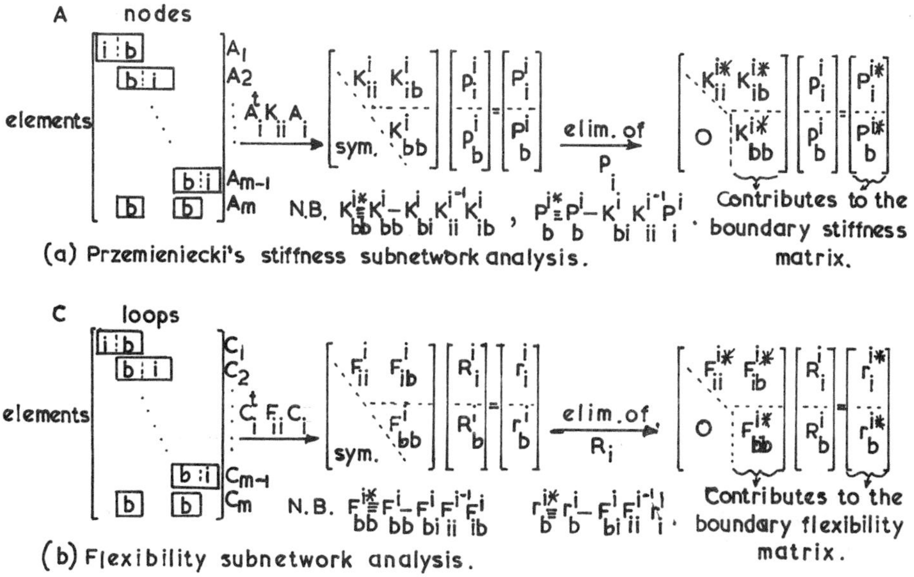

FIGURE 10

particular structural problem types and does not involve basis transformations. The technique may present problems however if the boundary network equations are too large to go in core unless the frontal solution method is incorporated.

7 NON-LINEAR NETWORK SYSTEMS

Many practical problems in civil engineering give rise to network models with non-linear characteristics. Structural analysis provides many such examples and hydraulic network system analysis is another area. A review of non-linear structural theory has been given by Åkesson (1967) in network notation with an extensive bibliography. Baumann (1971) has presented a non-linear network formulation of electrical systems for both the nodal and loop methods of analysis and with a suitable reinterpretation of the variables his theory could apply equally well to structural and hydraulic networks. The use of network theory suggests an extension of Baumann's equation to derive the non-linear form of the mixed method equations for decomposed network systems and this is now presented.

A decomposition of a network S is performed so that the connection matrices associated with the decomposition have the partitioned form of equation 18. If the Hooke's Law equations (9) and (10) are written in non-linear form for the branches in S_1 and S_o then:

$$N_{t1} = K_{11}f(n_{t1}) \qquad (29)$$

$$n_{to} = F_{oo}g(N_{to}) \qquad (30)$$

and the load equilibrium and displacement compatibility relations for the S_1 and S_o subnetworks are expressed in the partitioned form of equations (11), (12), (14) and (15) as:

$$n_{t1} = A_{11}p_1 + n_{p1} \qquad (31)$$

$$N_{to} = C_{oo}R_o + N_{po}, \quad (\text{N.B. } N_{po} = B_{ol}P_1 + B_{oo}P_o) \quad \ldots\ldots(32)$$

$$P_1 = A^t_{11}N_{t1} + A^t_{ol}N_{to} = A^t_{11}K_{11}f(n_{t1}) + A^t_{ol}N_{to} \quad \ldots\ldots(33)$$

$$r_o = C^t_{1o}n_{t1} + C^t_{oo}n_{to} = C^t_{1o}n_{t1} + C^t_{oo}F_{oo}g(N_{to}) \quad \ldots\ldots(34)$$

using equations (29) and (30).

A vector function is defined as follows:

$$\Phi(\underline{x}) = \Phi\begin{bmatrix} n_{t1} \\ N_{to} \end{bmatrix} = \begin{bmatrix} A^t_{11}K_{11}f(n_{t1}) + A^t_{ol}N_{to} - P_1 \\ C^t_{1o}n_{t1} + C^t_{oo}F_{oo}g(N_{to}) - r_o \end{bmatrix} = 0 \quad \ldots\ldots(35)$$

where $\underline{x} = (n_{t1}, N_{to})^t$ and Φ represents the vector of out of balance node loads in S_1 and loop displacements in S_o respectively.

Equation (35) can be solved numerically by successive approximation where from a starting vector $\underline{x}^o$ a sequence $\underline{x}^i$ is generated by successive corrections:

$$\underline{x}^{i+1} = \underline{x}^i + \Delta\underline{x}^i \quad \ldots\ldots(36)$$

where the corrections are calculated from the truncated Taylor's expansion:

$$\Phi(\underline{x}^{i+1}) = \Phi(\underline{x}^i) + \left\{\frac{\partial\Phi_i(\underline{x}^i)}{\partial x_j}\right\}\Delta\underline{x}^i = \Phi_i + J_i\Delta\underline{x}^i = 0 \quad \ldots\ldots(37)$$

By solving:

$$J_i \, \Delta\underline{x}^i = -\Phi_i \quad \ldots\ldots(38)$$

In our notation, since $\Delta n_{ti} = A_{11}\Delta p_1$ and $\Delta N_{to} = C_{oo}\Delta R_o$ from equations (31) and (32), one obtains the mixed method form of equation (38) as:

$$\begin{bmatrix} K_1 & G \\ -G^t & F_o \end{bmatrix}_i \begin{bmatrix} \Delta p_1 \\ \Delta R_o \end{bmatrix}_i = \begin{bmatrix} P_1 - A^t_{11}K_{11}f(n_{t1}) - A^t_{ol}N_{to} \\ r_o - C^t_{1o}n_{t1} - C^t_{oo}F_{oo}g(N_{to}) \end{bmatrix}_i \quad \ldots\ldots(39)$$

where $K_1 = A^t_{11}\left(K_{11}\dfrac{\partial f}{\partial n_{t1}}\right)A_{11}$, $\quad F_o = C^t_{oo}\left(F_{oo}\dfrac{\partial g}{\partial N_{to}}\right)C_{oo}$,

and $\quad G = A^t_{ol}C_{oo} = -A^t_{11}C_{1o}$.

The nature of the mixed method equation (39) implies that any special techniques developed for the linear network problem which take account of symmetry, bandwidth storage schemes and use sophisticated elimination methods may be applied now to the non-linear problem. As non-linear problems require iterative solution techniques the saving of time and effort in obtaining solutions is multiplied by the number of iterations.

The non-linear version of the algorithm has not yet been developed for the solution of decomposed non-linear network systems but the logic has been tested with small networks by hand.

8 CONCLUSIONS

One aim of this paper has been to indicate how graph theoretic concepts can be applied to the analysis and development of solution algorithms for decomposed network systems. By concentrating on the connection properties or topology of network models it is possible to establish various strategies for the solution of large networks based on the mixed methods of analysis which highlight the duality between the traditional nodal and loop methods of analysis and incorporate both.

REFERENCES

1 AMARI, S. (1962) 'Topological foundations of Kron's tearing of electrical networks', RAAG Memoirs 3, Div. F-VI, p 88.

2 BATY, J. P. and STEWART, K. (1967a) 'Dissection of structures', JASCE, 93 No. ST5, p 217.

3 BATY, J. P. and STEWART, K. (1967b) 'Orthogonal formulation of structural network problems using the concepts of open and closed paths', Proc. 4th IKM, Weimar, 1967, p 26.

4 BATY, J. P. and STEWART, K. (1969) 'Some applications of dissection theory', Matrix and Tensor Quarterly, 19, p 81.

5 BATY, J. P. and STEWART, K. (1971) 'Organisation of network equations using dissection theory', in Large sparse sets of linear equations, Academic Press, 1971, p 169.

6 BAUMANN, R. (1971) 'Sparseness in power system equations', ibid, reference 5, p 105.

7 BRANIN, F. H. (1962) 'Machine analysis of networks and its applications', IBM Technical report TR 00.855.

8 BRANIN, F. H. (1967) 'Computer methods of network analysis', IBM Technical report 00.1562.

9 FENVES, S. and BRANIN, F. H. (1963) 'Network-topological formation of structural analysis', JASCE, 89 No. ST4, p 483.

10 FENVES, S. (1966) 'Structural analysis by network, matrices and computers', JASCE, 92 No. ST1, p 199.

11 HAPP, H. H. (1965) 'The structure of orthogonal electric networks with conventional networks as special cases', Matrix and Tensor Quarterly, 15, p 99 and 16, p 35.

12 HAPP, H. H. (1971) 'Diakoptics and networks', Academic Press, 1971.

13 HARARY, F., NORMAN, R. I. and CARTWRIGHT, D. (1965) 'Structural models', John Wiley and Sons, 1965.

14 HARARY, F. (1970) 'Graph theory', Addison Wesley, 1970.

15 IRONS, B. M. (1970) 'A frontal solution program for finite element analysis', Int. J. Num. Meth. in Eng., 2, p 5.

16 JENNINGS, A. and TUFF, A. D. (1971) 'A direct method for the solution of large sparse symmetric simultantous equations', ibid, reference 5, p 97.

17 KING, I. P. (1970) 'Automatic re-ordering scheme for simultaneous equations derived from network systems', Int. J. Num. Meth. in Eng., 2, p 523.

18 KRON, G. (1963) 'Diakoptics', MacDonald, London, 1963.

19 LANGEFORS, B. (1959) 'Algebraic topology of networks', SAAB, Technical Note 43.

20 LANGEFORS, B. (1961) 'Algebraic topology for elastic networks', SAAB, Technical Note 49.

21 PRZEMIENIECKI, (1963) 'Matrix structural analysis of substructures', JAIAA, 1, No. 1, p 138.

22 RIAZ, M. (1970) 'Piecewise solutions of electrical networks with coupling elements', Journal of Franklin Inst., 289, p 1.

23 ROTH, J. P. (1955) 'An application of algebraic topology to numerical analysis: On the existence of a solution to the network problem', Proc. Nat. Acad. Sciences, 41, p 518.

24 SAMUELSSON, A. (1962) 'Linear analysis of frame structures by use of algebraic topology', Doctor's thesis, Chalmers Tekniska Hogskola, Goteborg.

25 SPILLERS, W. R. and DIMAGGIO, F. L. (1965) 'Network analysis of structures', JASCE, 91 No. EM3, p 169.

26 STEWART, K. (1964) 'Some notes on theory of diakoptics', Matrix and Tensor Quarterly, 14, p 42 and 15, p 84.

27 WIBERG, N-E, (1967) 'Diakoptics and Codiakoptics', Chalmers Tekniska Hogskola, Publ. 67:8, Goteborg.

28 WIBERG, N-E (1970) 'System analysis in structural mechanics', Doctor's thesis, Chalmers Tekniska Hogskola, Goteborg.

29 ÅKESSON, B. (1967) 'Non-linear structures, the Maxwell-Mohr theorem, and the displacement method', Trans. of Chalmers Tekniska Hogskola, No. 316, Goteborg.

ON VERTICAL DECOMPOSITION AND MODELING OF LARGE SCALE SYSTEMS

MIHAJLO D. MESAROVIC
Systems Research Center
Case Western Reserve University
Cleveland, Ohio, USA

1. INTRODUCTION

We shall consider in this paper the so-called vertical decomposition because of its considerable importance in the analysis and synthesis of the large scale systems and also because all other presentations at this Symposium are dealing with the alternative, horizontal, decomposition. We shall start with some general comments on decomposition and then present basic concepts of the multilevel systems theory. In conclusion we shall give two examples of the application of vertical decomposition.

2. COORDINATION, RETICULATION AND DECOMPOSITION

(a) Classification of Decomposition-type Concepts

We shall start from a map $S: M \times X \to Y$ which we shall take to represent a system. It can be shown quite readily that all kinds of systems (e.g. differential equations, algebraic-type systems, etc.) can be viewed as a realization of such a map. M is the set of all possible inputs, X the set of initial conditions while Y is the set of outputs; $\times$ denotes the Cartesian product.

Assume that M, X and Y can be represented in terms of the component sets

$$M = M_1 \times \ldots \times M_n$$
$$X = X_1 \times \ldots \times X_n$$
$$Y = Y_1 \times \ldots \times Y_n$$

so that $m = (m_1, \ldots, m_n)$ where $m \in M$, $m_i \in M_i$ and similarly for $x \in X$ and $y \in Y$.

We can now represent subsystems of S on the component sets; i.e., S_i on M_i, X_i, Y_i, $i=1,\ldots,n$. To account for interaction between such subsystems additional sets have to be introduced, U_i for S_i, etc. The subsystems are then given by

$$S_i: M_i \times X_i \times U_i \to Y_i$$

where U_i is the interaction set. To specify U_i the interaction maps have to be introduced, namely

$$K_i: M \times X \to U_i$$

Finally, P_i and K_i have to be such that $P(m) = (y_1, \ldots, y_n) = (P_1(m_1, x_1, K_1(m,x)), \ldots, P_n(m_n, x_n, K_n(m,x)))$.

The system S can be now viewed as decomposed into a family of subsystems $\overline{S} = \{S_1, \ldots, S_n\}$ and the interaction map $K: M \times X \to U$ where

$$K(m,x) = (K_1(m,x), \ldots, K_n(m,x))$$

To talk about decomposition we need three items, therefore: S which we shall refer as the *overall system*; $\overline{S} = \{S_1, \ldots, S_n\}$ which we shall refer to as the *subsystems*; and, K which we shall refer to as the *interaction map* or the *coordination map*.

Three types of problems can be now defined:

(1) Given S and $\overline{S} = \{S_1,\ldots,S_n\}$, find K; we shall refer to this problem as *coordination*.

(2) Given S and K, find $\overline{S} = \{S_1,\ldots,S_n\}$; we shall refer to this problem as *reticulation*.

(3) Given S, find $\overline{S} = \{S_1,\ldots,S_n\}$, and K. We shall refer to this problem as *decomposition*.

Coordination encompasses various kinds of interaction changes, such as the problem of decoupling in classical control theory. Reticulation includes such approaches as various tearing methods in the network theory. "True decomposition" as defined in (3) is performed in two stages performing first the reticulation and then the coordination. For example, the subsystems can be the ones obtained by the projection (properly defined) of S on M_i, X_i and Y_i and then the coordination problem is how to link them together in a proper way.

The mappings S and $\overline{S}$ can be the input-output representations of systems in which case no additional interpretations of the associated sets is needed, or they might be decision-making systems (as e.g. in the consideration of the decomposition of an optimization procedure) in which case additional descriptions are needed. We shall consider the second case in some more details because of its practical importance.

When S is a decision procedure it is defined in terms of three maps

$$P: M \times X \to Y$$
$$G: M \times Y \to V$$
$$D: X \to M$$

where P is the *constraint map*, G the *evaluation map*, D the *decision map* while V is the *value set*. The map D is such that for any $x \in X$ it selects an element $D(x) \in M$ so that

$$G(D(x),P(D(x),x)) \in V$$

has a given property, e.g., is a minimum. Obviously an optimization algorithm for the problem with equality constraints is a realization of such a system. For simplicity we shall not consider the inequality constraints here. The subsystems represent then algorithms for the subproblems defined by

$$P_i: M_i \times X_i \times A_i \to Y_i$$
$$G_i: M_i \times Y_i \times B_i \to U$$
$$D_i: X_i \times A_i \times B_i \to M_i$$

and D_i acts so that $D_i(x: \alpha_i,\beta_i) \in M_i$ has a special property, e.g., is minimal for G_i while satisficing the constraint defined by P_i.

The coordinator now is defined by a map

$$K: X \times A \to B$$

where $A = A_1\times\ldots\times A_n$, $B = B_1\times\ldots\times B_n$ and is such that it selects (α,β) so that the local decisions selected by $D_1,\ldots,D_n$ satisfy the overall objective which satisficing the local objectives.

Various decomposition techniques for optimization problems such as Dantzig-Wald decomposition principle belong to this type of problems. However, it should again be mentioned that the dominant problem in decomposition is coordination. The reticulation is usually done in reference to the special properties of the evaluation map and the constraint map (e.g. the overall evaluation map is a sum of the corresponding local maps and the local constraint is simply the projection - appropriately defined - of the overall constraint on the local variables) and the real problem is how to specify an efficient coordination algorithm. *Coordination appears, therefore, as the central problem in decomposition.*

3. MULTILEVEL SYSTEMS CONCEPTS†

For a system to belong to the class of multilevel systems it is necessary at least that:

(i) The system consists of two or more (explicitly recognized) interacting subsystems.

(ii) The subsystems are arranged vertically, i.e., there exists a way to identify the subsystems as being on different levels; in such an arrangement the behavior of the subsystems on the lower levels is conditioned (determined, specified) by the inputs from the higher levels while the performance (success) of the subsystems on the higher level depends upon the behavior and performance of the lower levels (Fig. 1).

The inputs from the higher to the lower levels are referred to as *interventions* (or *coordination parameters*) while the inputs in the other direction are referred to as the *performance feedbacks*. Implicit in the concept of intervention is the priority of actions or goals of the higher level units.

(a) *Strata - Levels of Description or Abstraction*

Truly complex systems almost by definition evade complete and detailed descriptions. The dilemma in description is basically between simplicity, one of the prerequisites for understanding, and the need to take into account numerous behavioral aspects. A resolution of this dilemma is sought through a multilevel, description. One describes the system by a family of models each concerned with the behavior of the system as viewed from a different level of abstraction. For each level there is a set of relevant features, variables, laws and principles in terms of which the system's behavior is described. For such a hierarchy to be effective it is necessary that the description on any level be considered independent of the description on other levels. We shall refer to this type of systems as *stratified descriptions* of a system. The levels of abstraction involved in a stratified description will be referred to as *strata*.

Examples of stratified descriptions in the natural sciences are abundant. One can study living organisms on molecular, cellular, organ and total organism levels. Although a complete decoupling of levels cannot be fully justified, the assumed decoupling of levels enables the study of system behavior on any *stratum* to be performed in considerable detail. Neglecting the cross-strata interdependence leads to an incomplete understanding of the whole system behavior. Indeed, a restriction to biological-type inquiries represents in itself an isolation since apparently the system under consideration can also be described on the stratum of concern to chemistry or physics on one hand and ecology and perhaps social science on another.

One can take an electronic computer as an example of a stratified description of a man-made system (Fig. 2). Its functioning is customarily described at least on two strata: in terms of physical laws describing the functioning and interconnection of the constituent parts and in terms of processing abstract non-physical entities, such as digits or information sequences. On the stratum of information-processing one is concerned with such problems as computation, programming, etc.,

---

† M.D. Mesarovic, D. Macko, Y. Takahara, "Theory of Hierarchical, Multilevel Systems", Academic Press, 1970.

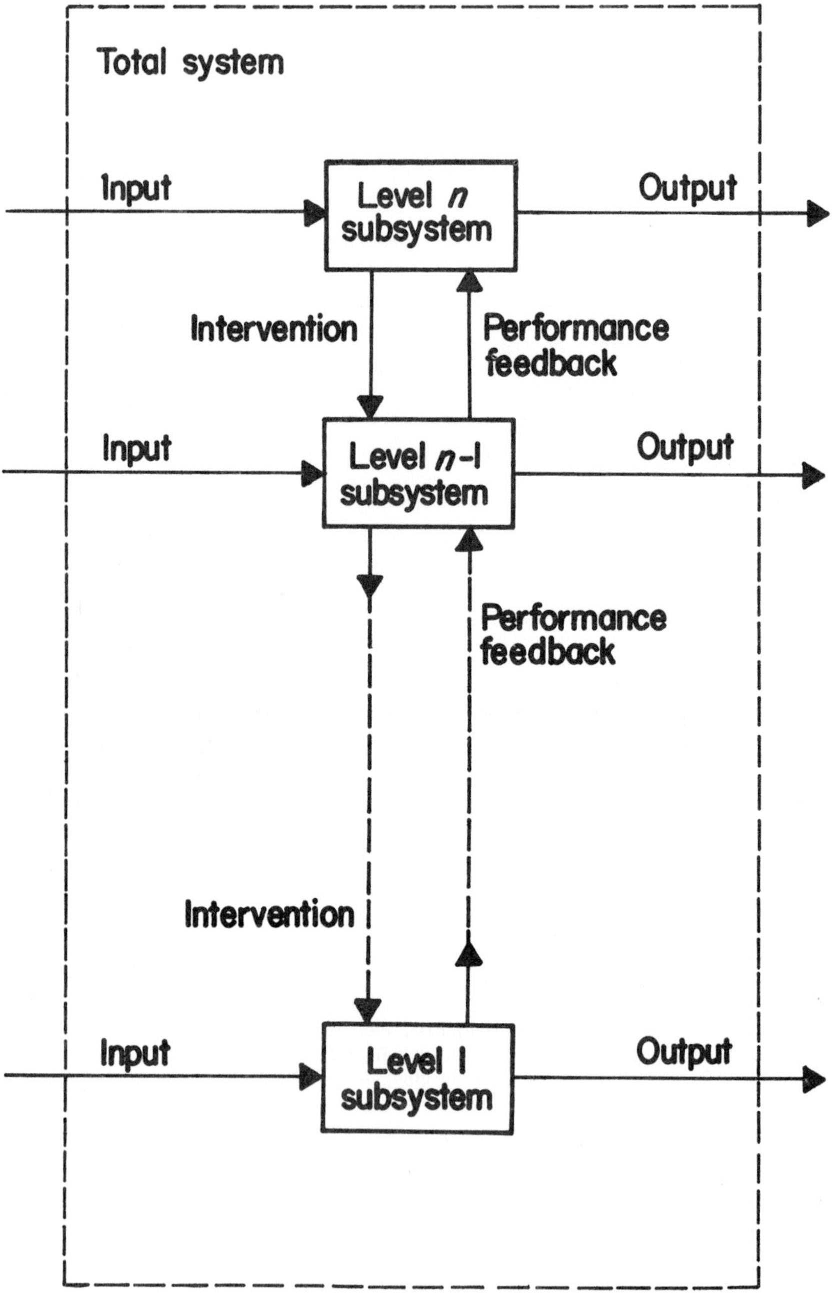

FIGURE 1

GENERAL DIAGRAM OF A MULTILEVEL SYSTEM

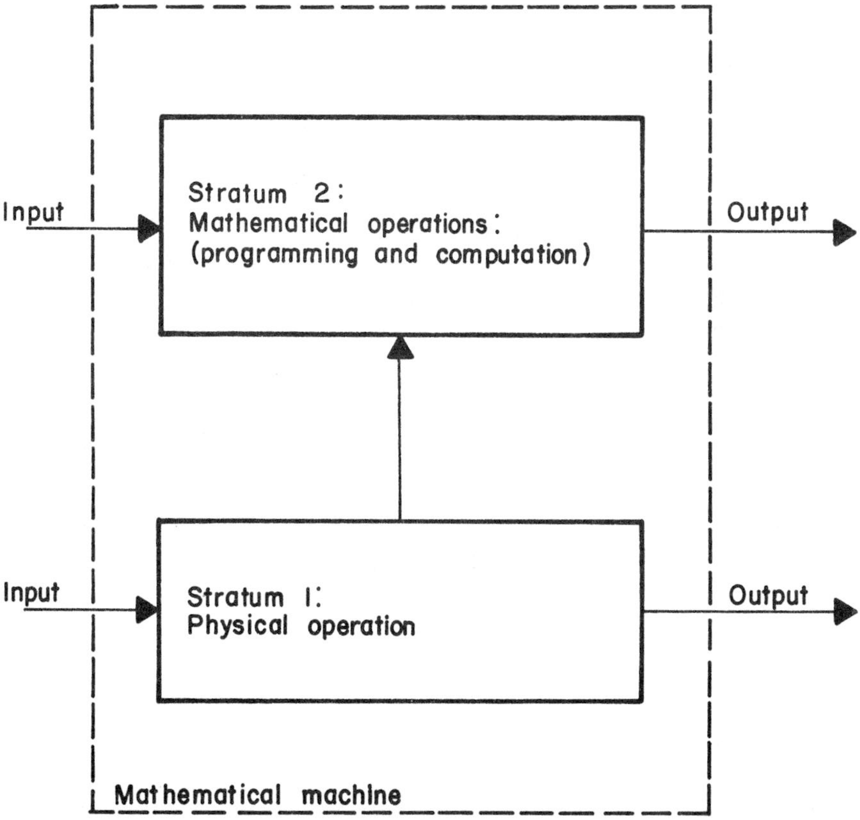

FIGURE 2

TWO-STRATA REPRESENTATION OF A COMPUTER SYSTEM

and the underlying physical basis of operation is not explicitly considered. Of course, a description of the system or some of its subsystems on other strata may also be of interest; the stratum of atomic physics for some component design or the so-called systems stratum for problems such as time sharing. Another example of stratification relevant for industrial automation is shown in Figure 3. On the first stratum the system is described in terms of the laws of physics and chemistry. On the second stratum one is dealing with signals and information and the principles of decision making (control) and information processing are used. On the third stratum the systems operation is described in economic terms.

Let us now briefly present a formalization of the concept of a stratified system.

The starting point for stratification of a given system $S: X \to Y$ is the assumption that the set X of outside stimuli and the set Y of responses are both representable as cartesian products; it is assumed there are given two families of sets X_i, $1 \leq i \leq n$, and Y_i, $1 \leq i \leq n$, such that

$$X = X_1 \times \ldots \times X_n, \quad \text{and} \quad Y = Y_1 \times \ldots \times Y_n . \tag{1}$$

This assumption corresponds to the ability of partitioning the input stimuli and responses into components.

Given that the sets X and Y are representable as in (1), each pair (X_i, Y_i), $1 \leq i \leq n$, is assigned to a particular stratum. The *ith stratum* of the system S is a system represented as a mapping S_i:

(i) $S_i: X_i \times W_i \to Y_i$ if $i=n$,

(ii) $S_i: X_i \times C_i \times W_i \to Y_i$ if $1 < i < n$,

(iii) $S_i: X_i \times C_i \to Y_i$ if $i=1$.

A family of such systems S_i, $1 \leq i \leq n$, is a *stratification* of S if there exist two families of mappings $h_i: Y_i \to W_{i+1}$, $1 \leq i < n$, and $c_i: Y_i \to C_{i-1}$, $1 < i \leq n$, such that for each x in X, and $y = S(x)$:

$$\begin{aligned} &\text{(i)} \quad y_n = S_n(x_n, h_{n-1}(y_{n-1})), \\ &\text{(ii)} \quad y_i = S_i(x_i, c_{i+1}(y_{i+1}), h_{i-1}(y_{i-1})), \; 1 < i < n, \\ &\text{(iii)} \quad y_1 = S_1(x_1, c_2(y_2)). \end{aligned} \tag{2}$$

The image set Y_i represents the responses of the ith stratum S_i. The sets C_i and W_i represent the sets of stimuli respectively from the strata immediately above and below the ith stratum. The mappings h_i and c_i are referred to as the ith stratum *information function* and *decision function* respectively; they tie the strata together as in (2) to form the system S.

In reference to the mappings h_i and c_i, we can formally describe what is meant by a successful stratification; in so doing, we recognize several degrees of stratification.

The system S is *completely stratified* if any stratum S_i, $1 \leq i \leq n$, is such that for any pair (y_i, w_i) in $C_i \times W_i$, and any two elements x_i and x_i' in X_i:

$$\begin{aligned} h_i(S_i(x_i, y_i, w_i)) &= h_i(S_i(x_i', y_i, w_i)), \\ c_i(S_i(x_i, y_i, w_i)) &= c_i(S_i(x_i', y_i, w_i)). \end{aligned} \tag{3}$$

This simply means that for a given intervention y_i and feedback w_i, the response of the subsystem S_i to any change in the stimuli x_i will be such that there is no change either in intervention y_{i-1} or feedback w_{i+1}; in other words, the response

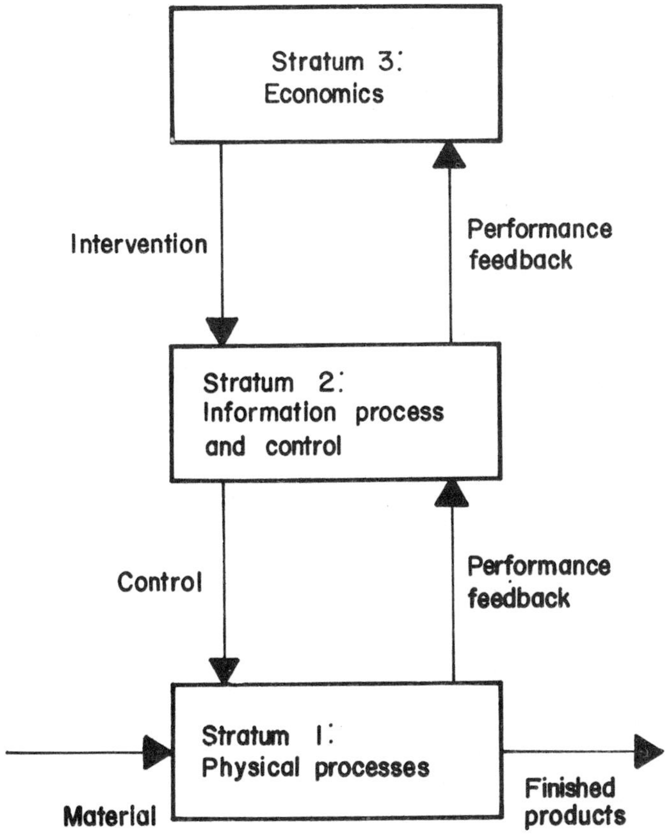

FIGURE 3

THREE-STRATA MODEL USED IN INDUSTRIAL CONTROL

is confined to the ith stratum only. Notice that complete stratifiecation depends not only on the transformation S_i but upon the mappings c_i and h_i as well; the stratification depends upon interstrata relationships as well as the strata themselves.

To require that the response of each stratum be completely localized is, of course, a strong condition. A weaker concept is that of a stable stratification in which localization is achieved for some but not all intervention-feedback pairs. In this, the prominence of the top stratum is acknowledged. The topmost stratum has its own set X_n of outside stimuli, and its response depends upon the entire hierarchy below. The requirement which the top level places upon the hierarchy can then be represented in terms of the feedback information w_n, since this is the only input from below.

A *stable stratification* is characterized, therefore, by the following condition: there exists some x in X and y = S(x), and for each i, $1 \leq i \leq n$, a pair (y_i, w_i) such that

$$\begin{aligned} w_i &= h_{i-1}(y_{i-1}), & 1 < i \leq n, \\ y_i &= c_{i+1}(y_{i+1}), & 1 \leq i < n, \\ y_i &= S_i(x_i, y_i, w_i), & 1 \leq i \leq n, \end{aligned}$$

and, moreover, for all x' in X and i, $1 \leq i \leq n$, condition (3) holds.

Distinction between complete and stable stratification is that, in the latter, it is not required that the strata be independent for every intervention-feedback pair, but rather only that there exist such "states of the entire system," so to speak, in which the responses are localized. Of course, there is the question of how a stable state in a hierarchy can be reached, but this is a question which needs deeper structure for the analysis.

Both complete and stable stratifications represent only idealized cases and, therefore, approximations to the hierarchies in the actual real-life systems. There are a number of ways in which the conditions can be weakened leading to less completely stratified systems. We shall not proceed to formalize such weaker, but more realistic, notions but shall restrict ourselves to several comments.

Condition (3) need not be satisfied for all stimuli in X but rather only those which represent "normal" conditions under which the system operates. Also, to realize a stable stratification or even a complete stratification under a restriction on the stimuli, it might be necessary to merge some adjacent strata into one stratum. It could very well turn out in some extreme cases that such mergers result in a single stratum, thereby eliminating stratification altogether.

Stratification implies a reduction in information sent up the hierarchy: many lower strata stimuli look alike to the higher strata units. This "upward reduction of information" has a variety of interesting consequence, one of them indicating a need for organizational-type multiechelon hierarchy.

Consider the case where the stimuli are entering the system only on the lowest level. Because of the reduction of information, the higher levels have to process a smaller amount of information. This has two important consequences:

(i) When the system is built out of building blocks which have a limited decision-making capacity, the higher strata will apparently be composed of a fewer number of units.

(ii) Reduction of information can be achieved in many ways, one of them being aggregation. Aggregation, as discussed before, leads to the partitioning of a family of variables into the subfamilies, such that each subfamily is represented by a single "aggregated" variable. This effectively represents a decomposition of a lower stratum into subsystems. Indeed, the information feedback can very well contain precisely the interaction variables between the subsystems, as in the case of the coordination by interaction prediction or balance. In short, reduction of information leads in a natural way to a horizontal decomposition of a stratum into subsystems. On each stratum, the decision units are concerned primarily with the operation of the subsystems, for the most part neglecting the interactions between the subsystems on the same stratum. The decision units on the higher strata are concerned then only with the interactions and interdependence between the subsystems, assuming each subsystem is functioning properly. This reasoning clearly leads to an organizational-type multiechelon hierarchy.

(b) *Layers - Levels of Decision Complexity*

Another concept of hierarchy appears in the context of a complex decision-making process. There are two quite obvious, very important, but conflicting, features of many real-life decision-making situations: on one hand, to avoid decision by default, there is a need to act without delay while, on another, there is an equally great need to take time to understand the situation better. In complex decision-making situations, the resolution of this dilemma is sought in a hierarchical approach. Essentially one defines a family of decision problems whose solution is attempted in a sequential manner. The solution of any problem in the sequence determines and fixes some parameters in the next problem so that the latter is completely specified and its solution can be attempted. In such a way the solution of a complex decision problem is substituted by the solution of a family of sequentially arranged simpler subproblems so that the solution of all the subproblems in the family implies the solution of the original problem. Such a hierarchy is termed a *hierarchy of decision layers*, and the entire decision-making system is referred to as a *multi-layer (decision) system.*

It is easy to give examples from everyday life of complex decision situations which are approached in a multi-layer fashion. Indeed, personal goals or objectives are as a rule vague and have to be translated into what might be called operational objectives which then provide a basis for the selection of a concrete course of action. For example, the goal of a young man might be to achieve happiness, but that vague goal has to be translated into objectives which lead to some specific actions: an objective has to be selected (say, to attend college, go into business, get married, etc.) which in turn leads to subgoals (say, to select a college, a major field of study, etc.). Very often only after a subgoal is achieved is one in the position to evaluate whether the original goal is approached.

As an example of multilayer hierarchies consider the functional hierarchy in decision problems under true uncertainties. This hierarchy emerges naturally in reference to three essential aspects of decision problems under uncertainty: (i) the selection of strategies to be used in the solution process; (ii) the reduction or elimination of uncertainties; and (iii) the search for a preferable or acceptable course of action under prespecified conditions. The functional hierarchy is shown in Figure 4. It contains three layers:

1) *Search layer:* The task of this layer is to select the course of action m.

2) *Learning or Adaptation Layer:* The task of this layer is to specify the uncertainty set U used by the search layer.

3) *Self-organizing layer:* This layer selects the structure, functions, and strategies which are used on the lower layers so that an overall goal (usually defined in terms which cannot easily be made operational) is being pursued as closely as possible.

The three layer hierarchy follows in a natural way from the basic formulation of a decision problem. In the general situation where the uncertainties are

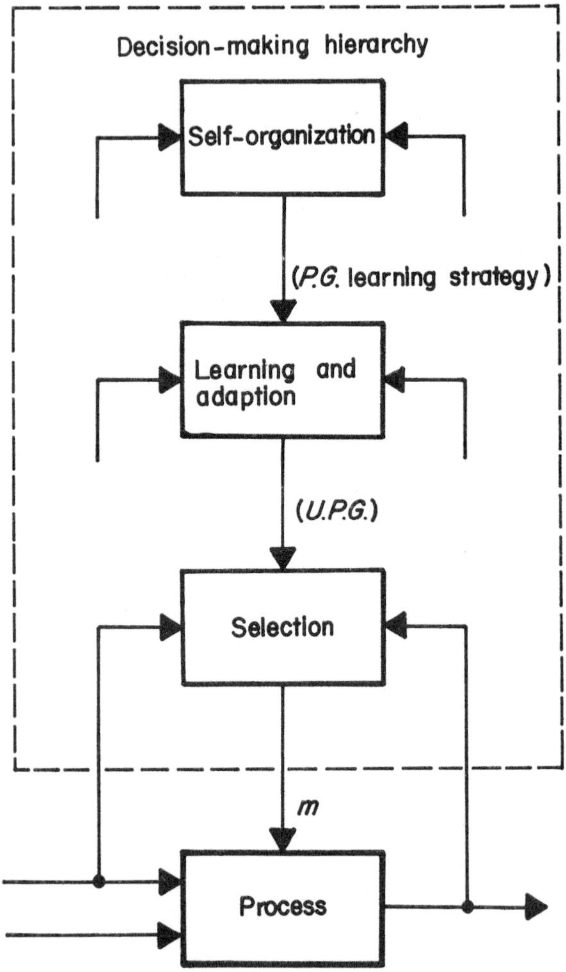

FIGURE 4

BASIC DECISION-MAKING LAYERS

present a decision problem is defined by a quadruple (g,τ,X^f,Ω) where g is a cost function, X^f is the set of feasible alternative decisions, Ω set of uncertainties (disturbance), g a cost function,

$$g: X^f \times \Omega \times V$$

where V, is the value set, and τ is the so-called tolerance function

$$\tau: \Omega \rightarrow V$$

The decision problem is then defined to be that of selecting a feasible decision $\tilde{x} \in X^f$ so that for every plausible uncertainty, $w \in \Omega$, the cost is below the tolerance limit as specified by τ, i.e.,

$$g(x,w) \leq \tau(w) \tag{4}$$

for all $w \in \Omega$. This formulation is termed the satisfaction approach to the (general) decision situation because the essential requirement is to satisfy the inequality (4).

We can now present the three layer hierarchy in a formal manner.

We represent the (first) *selection layer* by a mapping:

$$S_1: W_1 \times C_1 \times U \rightarrow M,$$

where W_1 is the set of elements representing feedback information from the controlled process (environment), C_1 is the set of inputs from the third layer which specify the structure of S_1, and U is the set of inputs from the second layer specifying the uncertainty sets for the first layer.

In the context of the satisfaction problem, an element from C_1 specifies the first three elements g, τ, and X^f of the problem; the elements of U specify the last element Ω of the problem.

The (second) *learning layer* is represented by a mapping:

$$S_2: W_2 \times C_2 \rightarrow U,$$

where the elements of the set W_2 represent information from the environment, while the elements of the set U specify the uncertainty sets for the first layer. C_2 is the set of parameters which determine the structure of the learning layer just as C_1 determines the structure of the selection layer.

Finally, the (third) *self-organizing layer* is represented by a mapping:

$$S_3: W_3 \rightarrow C_1 \times C_2,$$

where the elements of the set W_3 represent feedback information available to the self-organization layer.

(c) *Organizational Hierarchies: Multi-level Multi-goal Systems*

For this concept of hierarchy it is necessary that (i) the system consist of a family of interacting subsystems which are recognized explicitly, (ii) some of the subsystems are defined as decision (making) units and (iii) the decision units are arranged hierarchically in the sense that some of them are influenced or controlled by other decision units (Figure 5). This relationship between subsystems can be formalized and leads to a partial ordering relationship between subsystems.

These systems are denoted as multi-level multi-goal because, in general, various decision units comprising the system have conflicting goals. This conflict

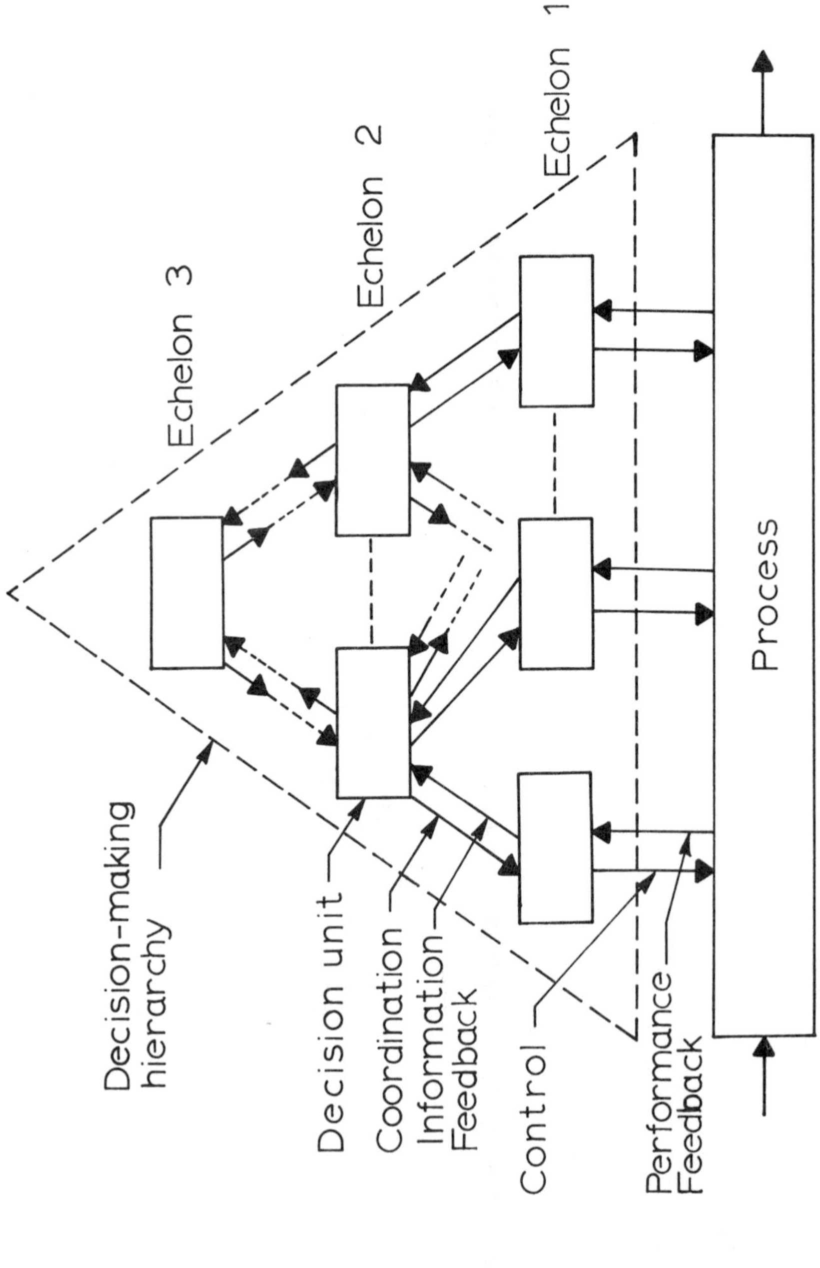

FIGURE 5

MULTILEVEL - MULTIGOAL (MULTIECHELON) SYSTEM

not only appears as the result of the evolution and composition of the system but also can be shown to be necessary (to a degree and in a given sense) for the efficient operation of the overall system.

4. APPLICATION OF THE MULTILEVEL APPROACH IN THE SYNTHESIS OF COMPLEX CONTROL SYSTEMS

Multilevel approach has been applied in the control of complex and large scale dynamic processes in two ways: (i) For the design of a total control system. Here the stratification and multilayer decision concepts are of primary importance: (ii) For the integration of the interacting control subsystems. Multiechelon concepts and coordination theory are of prime importance here. We shall give an example in the traffic control area†.

To speed the traffic from one side of a large city to another the urban freeways are built which have a limited access via controlled ramps. A major problem in operation as well as in the construction of new freeway systems of this kind is that they are in excessive use during short peak hours of heavy traffic and are used below capacity during the rest of time. There is obviously here a control problem: to control the access to the freeway so as to maximize the traffic flow during peak hours. This is done by appropriately changing the ramp signals. The measurement variables include: speed of the vehicles, gap between the consecutive vehicles, merging characteristics, etc. The control problem is a very complex one because of the unknown response of the "controlled system", need for a fast action and the effect of a large number of unpredictable events, such as the occurrence of an accident causing a congestion.

Using multilevel systems approach discussed in the preceding section a computer control system was built and implemented. The layer decomposition of the overall control function is shown in Figure 6. Specific control function on each of the layers is indicated in the diagram. Each of the control layer is in itself a rather complicated control system. Discussion of these is outside of the scope of this presentation but for the sake of illustration the diagram of the respective control system on each of the layers is given in Figures 7-9.

An additional practical advantage of the multilevel approach to complex control systems ought to be mentioned: Although the initial design includes the total control system the implementation can proceed in stages; i.e., starting from the lowest level the additional levels can be added when the technical, operational or economic conditions warrant.

---

† Drew, Donald, Application of Multilevel Systems Theory to the Design of a Freeway Control System, Proc. Fifth Systems Symposium "Systems Approach and The City", North-Holland Publishing Company, 1972.

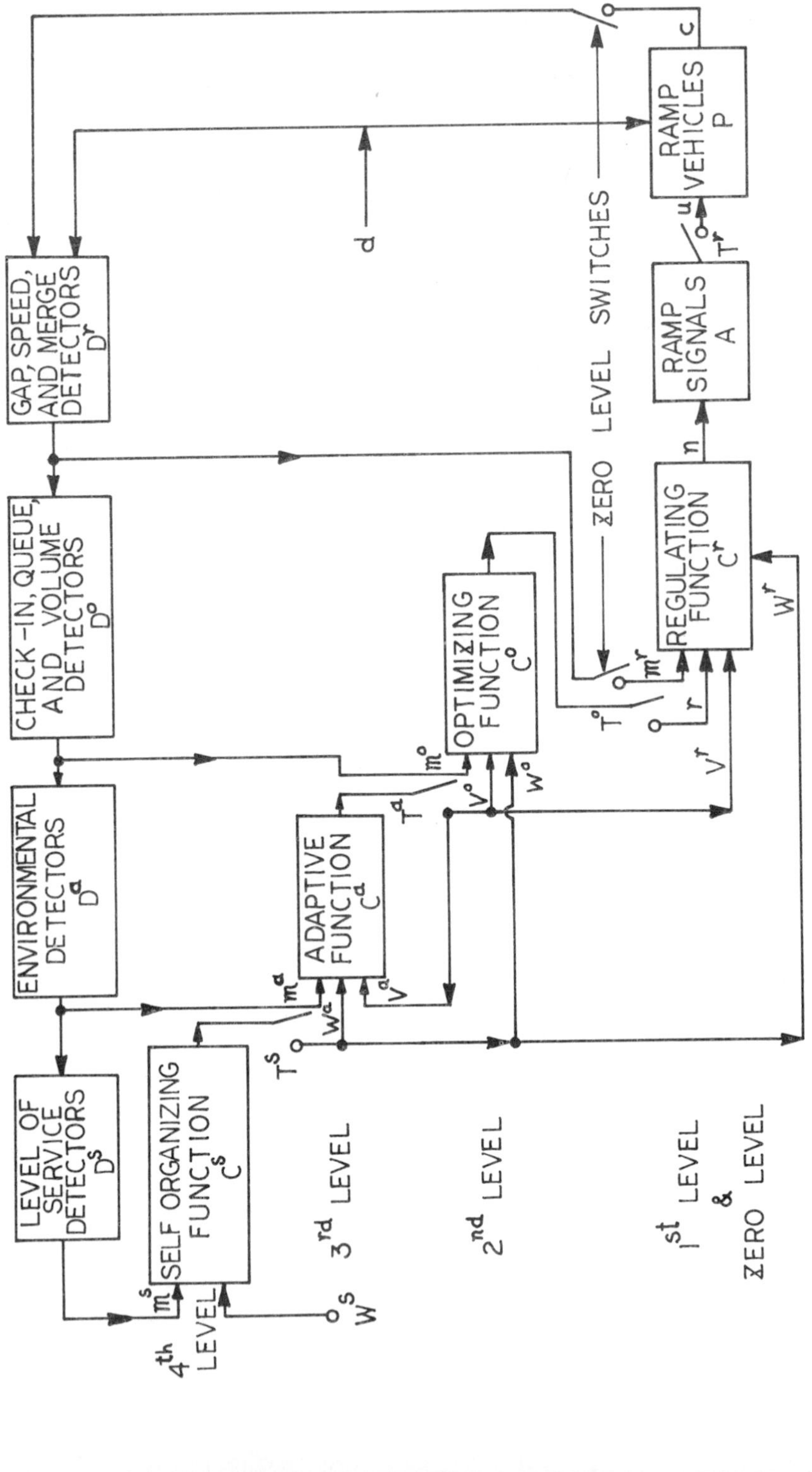

FIGURE 6
DECOMPOSITION OF FREEWAY CONTROL FUNCTION

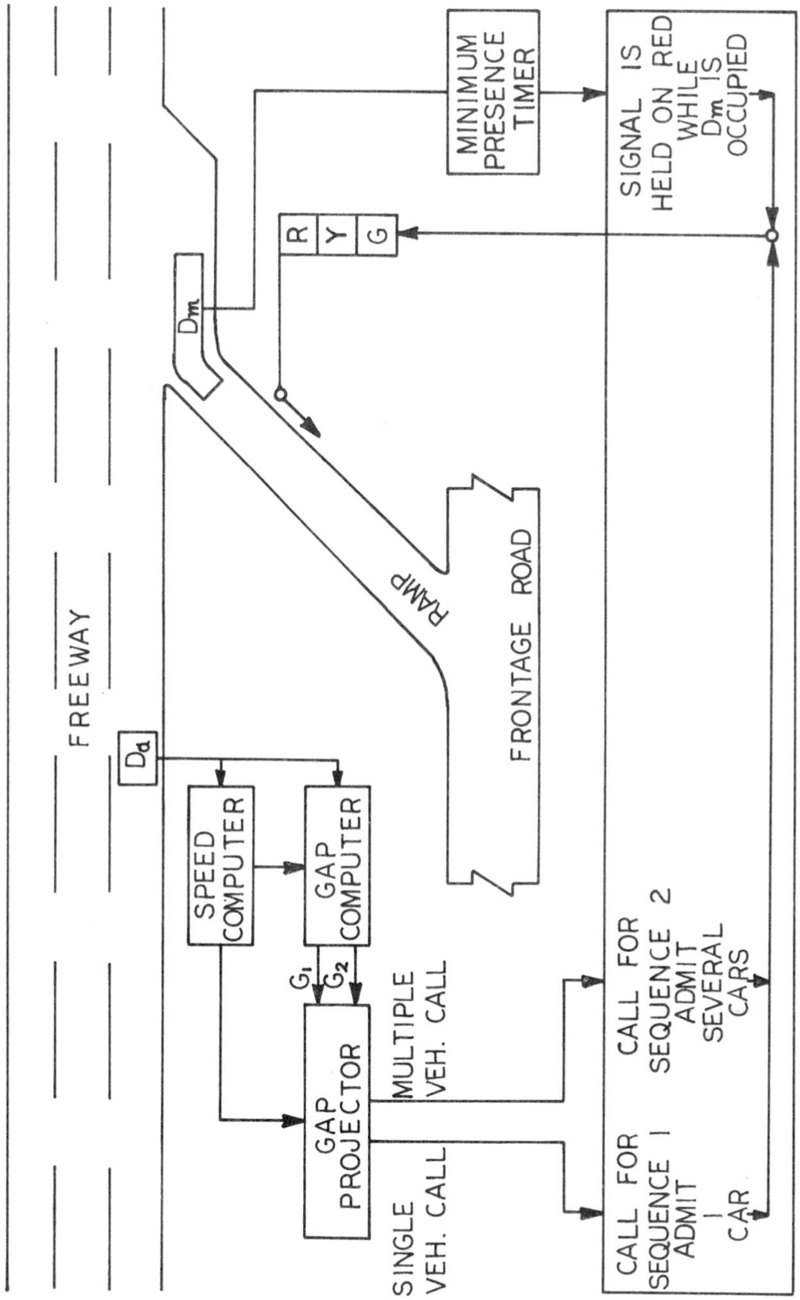

FIGURE 7

FIRST LEVEL FREEWAY CONTROL SHOWING REGULATING FUNCTION COMPONENTS

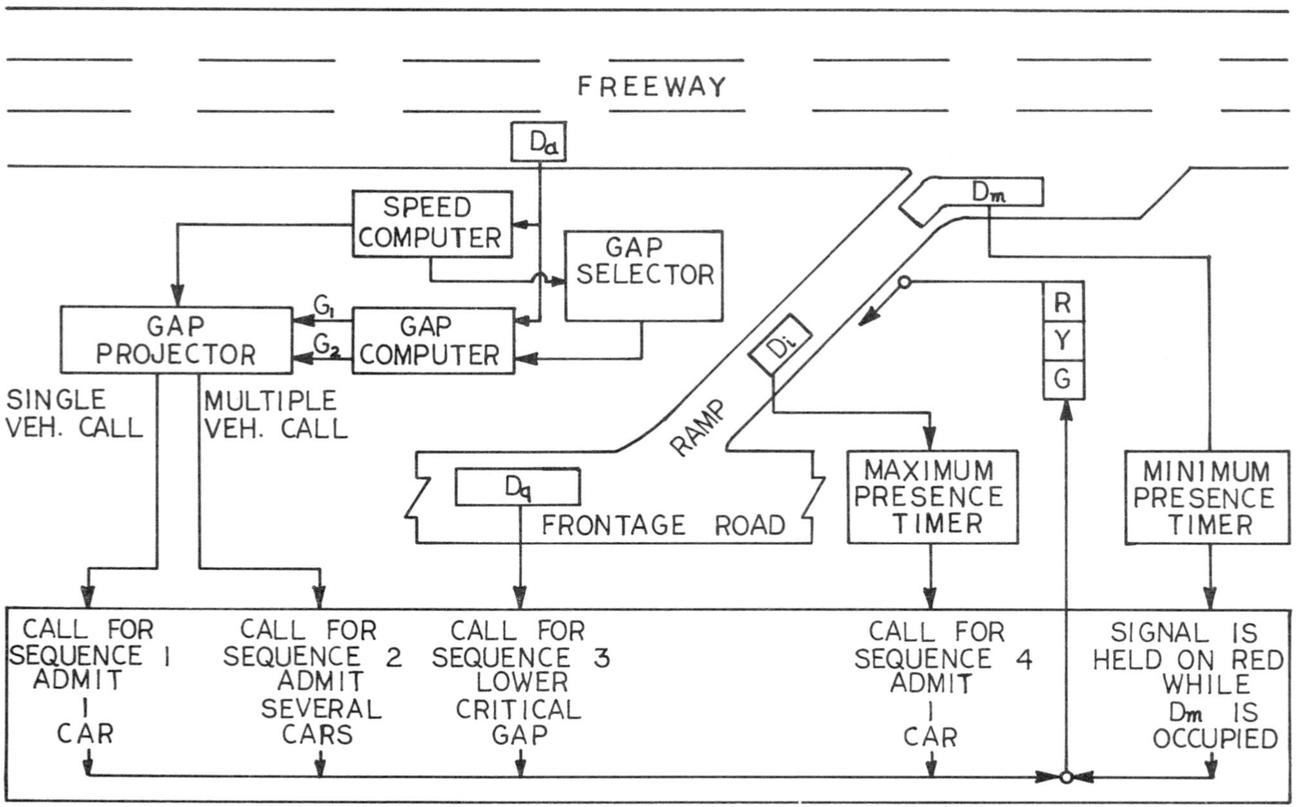

FIGURE 8

SECOND LEVEL FREEWAY CONTROL WITH OPTIMIZING FUNCTION COMPONENTS IDENTIFIED

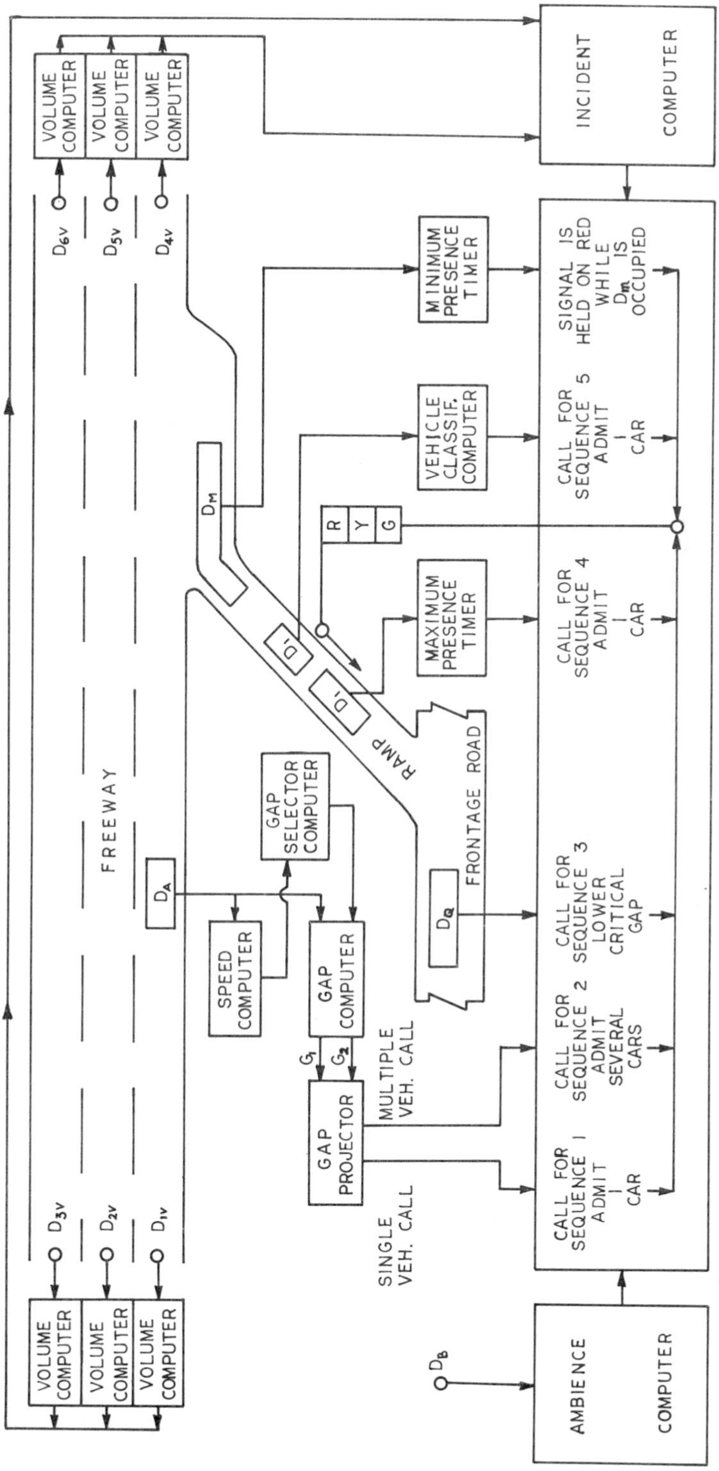

FIGURE 9

THIRD LEVEL FREEWAY CONTROL WITH ADAPTIVE FUNCNTION COMPONENTS IDENTIFIED

ROLE OF SEPARABILITY OF THE PERFORMANCE (COST) FUNCTION IN THE DECOMPOSITION OF DECISION PROBLEMS

MIHAJLO D. MESAROVIC
Systems Research Center
Case Western Reserve University
Cleveland, Ohio, USA

The decomposition of decision problems has been studied extensively in particular in the context of the linear and non-linear programming. Traditionally it is assumed that the overall performance is an additive function of the local performance, i.e.

$$g(m) = \sum_i g_i(m_i, u_i)$$

where g is the overall performance g_i are the local performance, m the overall decision, m_i a local decision while u_i is the interaction term and $m = (m_1, \ldots, m_n)$. In the non-linear programming g_i is in addition assumed to be convex. The proof of the validity of decomposition then usually proceeds by using the additive property of the Lagrangian associated with the overall decision problem and the constraints.

The objective of this note is to show that the additivity assumption is not necessary to decompose the overall problem and also that the use of the Lagrangian for the proof of the validity of decomposition is non-essential. Indeed, the decomposition of the Lagrangian into suitable local performances is only a special case of a more general property of separability. We shall start from a general framework which subsumed the programming formulation and will use the interaction balance coordination principle.†

Our paper has four parts: First, we shall review the basic concept of interaction balance coordination. Second, we shall introduce the concept of separability and related notions necessary for its application. Third, we shall present some of the decomposition theorems on which the application of separability is based; and fourth, we shall sketch a procedure of how to use separability in the decomposition of complex problems, (e.g. the non-linear case).

(1) <u>Interaction balance coordination principle</u>

Let $g: M \rightarrow V$ be a function, representing the cost function, where M is the decision set while V is the value set. We shall only require that V is linearly ordered while M will remain arbitrary. Assume that the overall decision problem is to minimize g over M. We shall consider here the case of the equality constraints given by the map $P: M \rightarrow Y$. To relate P and g it is assumed, as usual, that the cost function g is given in terms of a map $G: M \times Y \rightarrow V$ and is such that $g(m) = G(m, P(m))$. The decision problem can be then defined as:

Find $\hat{m} \in M$ such that

$$G(\hat{m}, P(\hat{m})) \leq G(m, P(m))$$

---

† M.D. Mesarovic, D. Macko, Y. Takahara, "Theory of Hierarchical, Multilevel Systems", Academic Press, 1970.

for all $m \in M$. In order to be able to decompose the above decision problem it will be assumed that the decision, m, is an n-tuple $m = (m_1,\ldots,m_n)$ and the decision set (for conveniency) is given by the cartesian product $M = {}^nM = M_1\times\ldots\times M_n$ where $m_i \in M_i$; Y is also represented in terms of n components, $Y = Y_1\times\ldots\times Y_n$ and the constraint map can be decomposed into n local constraint maps $P_i: M_i \times U_i \to U_i$ where U_i are the interacting terms. The relationship between P and P_i is established by the interconnection maps $K_i: M \to U_i$ which are such that

$$y = (y_i,\ldots,y_n) = (P_1(m_1,K_1(m),\ldots,P_n(m_n,K_n(m))$$

Notice that such a decomposition of the overall constraint map is always possible. To define local decision problems one ought to introduce local cost functions, defined most generally, on the respective local terms, $G_i: M_i \times Y_i \to V$. Notice however, that the output term y_i depends upon the interaction term u_i so that g_i has to be defined on U_i as well, i.e., $g_i: M_i \times U_i \to V$ such that $g_i(m_i,u_i) = G_i(m_i,P_i(m_i,u_i))$ for all $(m_i,u_i) \in M_i \times U_i$. It is appropriate to define the local decision problems as minimizations. The difficulty, as expected, is that the interaction term u_i depends upon other, non-local, decisions. To solve this dilemma the following procedure is used:

Local cost function g_i is modified by a parameter β; i.e., g_i is embedded in a family of functions $\{g_{i\beta}: \beta \in B\}$ such that $g_{i\beta}: M_i \times U_i \to V$. It is assumed then that a central, coordinating unit, selects the parameter β and in this way specifies the local cost function. The local decision unit then proceeds to solve a given decision problem by treating the interactions as additional decision variables. The local decision problem is then: Find $(m_i(\beta),\ u_i(\beta)) \in M_i \times U_i$ such that

$$G_{i\beta}(m_i(\beta),P_i(m_i(\beta),u_i(\beta))) \leq G_{i\beta}(m_i,u_i,P_i(m_i,u_i))$$

for all $(m_i,u_i) \in M_i \times U_i$, i.e., the cost function is minimum. $(m_i(\beta),u_i(\beta))$ are locally optimal decision and interaction variable. Notice that in general $m_i(\beta)$ and $u_i(\beta)$ are not necessarily compatible, i.e., do not satisfy the constraint

$$K_i(m(\beta)) = u_i \neq u_i(\beta)$$

where $m(\beta) = (m_i(\beta),\ldots,m_n(\beta))$.

This is so because $u_i(\beta)$ is selected to be optimal for $g_{i\beta}$ and as such cannot expect to be equal to u_i. This very fact is used in the decomposition procedure termed interaction balance. The procedure is the following: The central unit selects the coordination term β. The local units then minimize the local cost function with local constraint, which results in $m_i(\beta)$, $u_i(\beta)$, $i=1,\ldots,n$. If $u_i(\beta) \neq K_i(m(\beta))$, the central unit selects a new coordination term β until the equality $u_i(\beta) = K_i(m(\beta))$ is achieved. The interaction balance coordination principle (IBP) is given then by the following:

$$[(m,u) = (m(\beta),u(\beta)) \text{ and } u = K(m)] \to m = \hat{m} \tag{1}$$

which reads: If a decision and interaction variable are locally optimal and the equality $u = K(m)$, is satisfied the decision variable is optimal for the overall problem. The equality $u = K(m)$ is referred to as the interactions balance.

We have now two concepts:

(i) Interaction balance principle (IBP) is applicable if the statement (1) holds (for a given decision problem).

(ii) The local decision problems are coordinable by the interaction balance principle (IBP) if (1) is true and there exists the coordination parameters for which the balance is achieved.

The difference between (i) and (ii) is that the former does not require that the balance can indeed be achieved but only that if it is achieved the decision is the overall optimal. The second, stronger requirement, needs the existence of such a coordination term.

2. SOME CONDITIONS FOR THE APPLICATION OF THE INTERACTION BALANCE PRINCIPLE

A number of theorems have been proven about both applicability of IBP and the coordinability by IBP†. We shall mention here only some of them which are of particular interest for the separability problem considered in this note.

Th. 1: IBP is applicable whenever:

(i) there exists a map $\Psi: V^n \to V$ such that

$$G(m,y) = \Psi(G_1(m_1,P_1(m_1,K_1(m))),\ldots,G_n(m_n,P_n(m_n,K_n(m)))$$

$$v = \Psi(v_1,\ldots,v_n)$$

(ii) Ψ is monotone (order preserving) in each argument.

Th. 2: Let Th. 1 hold; the subproblems are coordinable by IBP whenever

$$\max_{B} \min_{M\times U} \Psi = \min G \tag{2}$$

Th. 3: Let Th. 1 hold and Ψ is strictly monotone; then, the subproblems are coordinable by IBP iff (2) holds.

Th. 4: Let Th. 1 hold and Ψ is strictly monotone and furthermore

$$G(m,P(m)) \leq \sup_{B} \Psi_B(\beta,m,u)$$

Then

$$\max_{B} \min_{M\times U} \Psi_B(\beta,m,u) = \min_{M\times U} \max_{B} \Psi_B(\beta,m,u)$$

3. SEPARABILITY AND BALANCED MODIFICATION

Let P and G define the overall problem, $P: M \to Y$, $G: M \times Y \to V$, and the local constraints are given by $P_i: M_i \times U_i \to Y_i$ with the interaction map $H_i: M\times Y \to U_i$. The local cost function will also have to reflect the interdependence between the subproblems and is given by $G_i: M_i \times U_i \times W_i \to V$ where W_i is the interconnection term. To select the local and overall cost functions it is always possible to find the mappings $\theta_i: M \times Y \to W_i$ and $\Psi: V^n \to V$ such that

$$G(m,y) = \Psi[G_1(m_1,y_1,\theta_1(m,y),\ldots,G_n(m_n,y_n)\ \theta_n(m,y))]$$

We shall say that the overall cost function is separable iff $\theta_i = H_i$, i.e., the local performance can be specified fully by the local variables and there exists an interlevel function Ψ. Let us give some examples of separability.

Let

$$G(m,y) = (m_1^2 + y_1^2)\ |\ m_2y_2\ |\ + \sin^2 y_3$$

Define the local cost functions by $G_1(m_1,y_1) = m_1^2 + y_1^2$, $G_2(m_2,y_2) = | m_2,y_2 |$ and $G_3(m_3,y_3) = \sin^2 y_3$. The overall cost function is then

$$G(m,y) = \Psi[G_1(m_1,u_1),G_2(m_2,u_2),G_3(m_3,y_3)] = G_1(m_1,y_1)G_2(m_2,y_2) + G_3(m_3,y_3)$$

Let the overall process P be given by the equations

$$y_1 = -3m_1 - m_2, \quad y_2 = -4m_1 - m_2$$

Suppose the subprocesses P_1 and P_2 are coupled according to the equations

$$u_1 = y_2 \equiv H_1(m,y), \quad u_2 = m_1 + y_1 \equiv H_2(m,y)$$

Because of the simplicity of this system, we can express the nonlocal variables of the ith subprocess as functions of its local variables. Therefore, G is separable let

$$G_1(m_1,y_1,u_1) = a_{11}m_1^2 + b_{11}y_1^2 - a_{12}m_1(3m_1 + y_1) + b_{12}y_1u_1,$$
$$G_2(m_2,y_2,u_2) = a_{22}m_2^2 + b_{22}y_2^2 - \frac{1}{4}a_{21}m_2(m_2 + y_2)$$
$$+ \frac{1}{4} b_{21}y_2(4u_2 + m_2 + y_2)$$

Hence

$$G(m,y) = G_1(m_1,y_1,H_1(m,y)) + G_2(m_2,y_2,H_2(m,y))$$

whenever $y = P(m)$. If B is diagonal, G_1 and G_2 are independent, respectively, of u_1 and u_2; if both A and B are diagonal, the decomposition is obvious.

In applications, when an overall cost function is given the first level cost function, as given or assumed, are not necessarily the best ones; e.g., the convergence of the coordination process is not satisfactory or even the coordinabilit conditions are not satisfied. It is of interest then to modify the local performances so as to improve the performance of the overall system. A convenient method to do that is by means of the balance modification which we shall define now.

Let G and $\overline{G} = \{G_1,\ldots,G_n\}$ be given and are such that there exists an inter-level function Ψ: $V^n \to V$, i.e.

$$G(m,y) = \Psi[G_1(m_1,v_1),\ldots,G_n(m_n,u_n)]$$

where the dependence of G_i on u is shown explicitly for the sake of conveniency.

Let $G_{i\beta}$ be the modified cost function. In general $G_{i\beta}(m_i,u_i) \neq G_i(m_i,u_i)$ even if $u_i = K_i(m)$, i.e., the interactions are balanced. However if the family of modified performances $\{G_{1\beta},\ldots,G_{n\beta}\}$ are such that

$$\Psi[G_1(m_1,u_1),\ldots,G_n(m_n,u_n)] = \Psi[G_1(m_1,u_1),\ldots,G_n(m_n,u_n)]$$

wherever $u_i = K(m)$, i.e., the interactions are balanced the modification is called balanced. Consider some examples:

Let the overall cost function be given by

$$G(m,y) = G_1(m_1,y_1)\ G_2(m_2,y_2)$$

and the constraint maps are:

$$y_1 = m_1 + u_1$$
$$y_2 = m_2 + 2u_2$$
$$u_1 = y_2$$
$$u_2 = m_1 + y_1$$

Let the modification be achieved by the multiplication functions

$$G_{i\beta}(m_i, u_i) = G_i(m_i, u_i)\,\mu_{i\beta}(m_i, u_i)$$

If the modification functions are given by

$$\mu_{1\beta}(m_1, u_1) = \frac{\beta u_1}{2m_1 + u_1} \quad ; \quad \mu_{2\beta}(m_2, u_2) = \frac{m_2 + 2u_2}{\beta u_2}$$

the modification is balanced.

As the second example let the overall cost be given by

$$G(m,y) = m_1^2 + m_2^2 + (y_1 - 1)^2 + (y_2 - 1)^2$$

with the constraint

$$y_1 = 2m_1 + u_1$$
$$y_2 = 2m_2 - u_2$$
$$u_1 = y_2$$
$$u_2 = y_1$$

A balanced modification is given then by

$$G_{1\beta} = m_1^2 + (y_1 - 1)^2 + \beta_1 u_1^2 - \beta_2 y_1^2$$
$$G_{2\beta} = m_2^2 + (y_2 - 1)^2 + \beta_2 u_2^2 - \beta_1 y_2^2$$

Indeed, $G(m,y) = G_{1\beta}(m_1, u_1) + G_{2\beta}(m_2, u_2)$ whenever $u = K(m)$ even if $G_{i\beta}(m_i, u_i) \neq G_i(m_i, u_i)$.

The importance of the balanced modification is primarily due to the following:

Th. 5: If the IBP is applicable for a given decomposition of the overall cost function it is applicable for any balanced modification of the local cost functions.

After a decomposition of the overall cost function is achieved such that IBP is applicable one can try a whole set of balanced modification in order to improve the performance, without fear that the applicability of IBP will be violated, i.e., that balancing of interactions will lead to an erroneous overall decision.

4. DECOMPOSITION BASED ON SEPARABILITY

We can now outline a decomposition procedure based on the interaction balance coordination principle. It involves the following steps:

(i) Decompose the overall cost function by using its separability property.

(ii) Check the monotonicity of the interlevel function. This assures the applicability of IBP.

(iii) Modify the local cost functions with a balanced modification to ensure the coordinability and improve the coordination process.

DECOMPOSITION AND MULTILEVEL APPROACH IN THE MODELING AND MANAGEMENT OF WATER RESOURCES SYSTEMS

YACOV Y. HAIMES
Associate Professor of Engineering
Systems Research Center
Case Western Reserve University
Cleveland, Ohio USA

Abstract: This is a survey paper which summarizes some of the author's work in the applications of decomposition and multilevel approach to water resources systems. This survey includes: (i) The modeling and identification of distributed aquifer parameters, (ii) Integrated system identification and optimization for conjunctive use of ground and surface water resources, (iii) Regional water quality control and management with a system of effluent charges, (iv) Regional planning for capacity expansion in water resources systems.

1.0 INTRODUCTION

The hierarchical-multilevel approach provides essential modeling and optimization methodologies for tackling large scale and complex problems. In particular, the applications of this approach to modeling for planning and management of water resources systems, which naturally lend themselves to hierarchical structures, have been successfully demonstrated.

Most problems in water resources planning, operation and management are regional and span over a wide spatial area and long time horizon (up to 30 to 50 years). Accordingly a meaningful decomposition of water resources systems into subsystems can be achieved in many ways depending on the specific objectives, constraints and the input-output relationships. The following are some characteristic decomposition schemes of water resources systems: (i) Hydrological base decomposition (e.g. subregions, riverbasins, watersheds, etc.), (ii) Geographical-political base decomposition (e.g. cities, counties, states, etc.), (iii) Time base decomposition (e.g. hours, days, weeks, months, years, etc.), (iv) Goal base decomposition (e.g. demand models, supply models, etc.). The above decomposition schemes are particularly essential for solving regional water resources problems where realistic considerations are taken into account. Such considerations may include the realization of a vector of objective functions, e.g. (i) minimization of cost for meeting water needs, (ii) enhancement of regional development, (iii) enhancement of water and environmental quality, (iv) enhancement of social well being, or the characterization of water quality with a vector of quality levels (e.g. (a) fresh water (b) brackish water (c) saline water, and (d) recycled and reclaimed water, etc.).

2.0 DECOMPOSITION AND MULTILEVEL APPROACH

The concept of the multilevel approach is based on the decomposition of large scale and complex systems and the subsequent modeling of the systems into independent subsystems. This decentralized approach, by utilizing the concepts of strata, layers and echelons, enables the system analyst to analyze and comprehend the behavior of the subsystems in a lower level and to transmit the information obtained to fewer subsystems of a higher level, Mesarovic, et al (1970). Whenever more decentralization is needed, the system is decomposed. This decomposition is accomplished by introducing new variables, which are called pseudo variables, into the system. Then each subsystem is separately and independently optimized where different optimization techniques are applied based on the nature of the subsystems' models. This is called a first level solution. The subsystems are joined together by coupling variables which are manipulated by second level controllers in order to arrive at the optimal solution of the whole system, which is called the second level solution. One way to achieve subsystem independence is by relaxing one or more of the necessary conditions for optimality and then satisfying this condition with the second level controller. Higher level controllers may be achieved in a similar way.

Decomposition and multilevel optimization approach have several significant advantages for the solution of large scale and complex optimization problems over the conventional optimization methods. This is due to the following characteristics of the approach. By decomposing the problem into several subproblems (subsystems), a conceptual simplification of the complex system is achieved. This is especially important for highly coupled systems where the outputs of one subsystem are the inputs to others. The decomposition yields a reduction in the dimensionality of the immediate problem at hand at the expense of having to solve several subproblems of lower dimensions. This in turn reduces the computational effort involved, such as problem formulation time, programming effort, debugging effort, and the number of cards to be punched. As far as the total computer time is concerned, there is no definite trend that can be pointed out. For example, in the water pollution taxation problem reported later in this paper, the decomposition and multilevel approach required less computer time than the Dantzig-Wolfe, or the quadratic programming method. In other works, however, different results are reported. A significant advantage of the approach is that none of the system model functions need be linear. Thus more flexible mathematical models can be constructed to represent the real system. Note that a major shortcoming and deficiency of systems engineering is in the often unbalanced relationship between system modeling and system optimization. This is reflected by the vast number of linearized models in the literature in order to take advantages of the Simplex method and its extensions. By applying the decomposition and multilevel optimization techniques, no such costly sacrifice of accuracy in modeling is needed and more representative and sophisticated nonlinear multivariable mathematical models can be constructed. Furthermore, interactions among subsystems are permissible, since at the lower levels the subsystems' independence are achieved via psuedo variables. The above tradeoff between system modeling and system optimization is minimized by the applicability of the approach to both static and dynamic systems. Thus the time domain which plays an important role in water resources systems, need not be discretized, imbedded or ignored in the analyses, as is the case in static models (e.g. linear programming). Therefore the water resource system can be modeled by both static algebraic equations and dynamic differential equations. Both centralized and decentralized decision making processes can be achieved via the hierarchical approach. This is particularly important for regional water resources management in general and for regional water quality control and pollution abatement in particular as is shown later in this paper.

In order to illustrate the decomposition and multilevel approach, consider a company with N departments. Each department has a manager who reports to the president of the company. The sole objective of the ith manager is to maximize the performance function of his own department by manipulating his own decision variables. The president's objective is to maximize the performance of the whole company. He does that by imposing internal prices for "sellings" and "buyings" between the departments to control the supply and demand for interdepartment products. The company's overall optimal policy is achieved through an interactive procedure. The separate optimization of each department does not necessarily imply the overall maximization of the company's performance, unless the performance of each department is coordinated by a higher level, i.e. by the president.

3.0 MODELING AND IDENTIFICATION OF AQUIFER SYSTEMS

The identification of underground reservoir (aquifer) parameters is an important problem in both groundwater hydrology and petroleum engineering. Considerable information concerning the reservoir may be derived from the transient behavior of existing wells. The resulting knowledge of the mechanism, structure, physics, and dynamics of the reservoir is of vital importance for further development of either water or petroleum resources. Once the properties of an aquifer are known, it is possible to apply appropriate physical laws and to predict the response of the aquifer system to demands placed on it. Without knowledge, prediction is impossible. Thus, knowledge of system properties is prerequisite to management of the system.

To find out system properties by direct measurement is not feasible, however. In the usual course of events, sampling during discovery and exploitation of an aquifer touches only an insignificant fraction of the system. The value of the resource being recovered, while large, is yet too small to warrant extensive and expensive sampling and testing. Thus the probable best method to determine system properties is by means of identification techniques. That is, the response to known changes imposed at the system boundaries is observed. Then this response is analyzed to see what are the most likely values of aquifer system parameters which would lead to the observed behavior.

The response of an aquifer primarily is determined by two parameters, transmissibility (T) and storage (s), in addition to boundary configuration (α). Transmissibility reflects the ease with which water moves through the porous medium comprising the system. Storage measures the quantity of recoverable water contained within a unit of the system. Well bore flow rate (q) and pressure, as functions of time (t), are the observables. Transmissibility and storage are distributed parameters, with values changing throughout the system. The differential equation describing the flow conditions in an unconfined aquifer is nonlinear, Dracup, Haimes and Perrine (1968). However, in a confined aquifer this equation is linear, Haimes (1967). Therefore, in the analysis of an unconfined aquifer the linear equation from the confined aquifer has been adapted using the assumptions of a small drawdown and an aquifer of infinite extent. For real problems we often work with a limited number of observations, errors, and variable parameters. Under these conditions the inverse relationship we seek is not necessarily unique, and the problem of finding the "best" description is not a simple one. The goal here is to find simple, efficient methods by which to select optimal estimates for the aquifer parameters. In turn, use of these values should permit the best predictions to be made regarding aquifer response to any demand.

3.1 Model Formulation

Several models for the identification of aquifer parameters via decomposition and multilevel approach have been constructed. Haimes (1967), Haimes, Perrine, and Wismer (1968), and Wismer, Perrine and Haimes (1970), developed two models applicable to confined aquifers. Dracup, Haimes and Perrine (1968) considered the identification of unconfined aquifers. Due to a space limitation only one model will be discussed here.

Consider a cluster of wells tapping a large aquifer. Within times considered no effect of boundaries is felt. Thus a two-dimensional system model is infinite in extent, comprising N producing wells with coordinates $Z_i = r_i \exp(J\theta_i)$, $i = 1, 2, \ldots, N$, in the complex domain. Fluid between two wells can be attracted to either. Thus, at any time, there is a line lying between wells across which no flow occurs. Given a cluster of wells in an infinite aquifer then, a reasonable and analytically tractable model is the decomposition of the aquifer into N wedge-shaped homogeneous regions, each enclosing a single well. The regions are separated by N straight, impermeable boundaries radiating from the origin at azimuths α_i. Hence $\theta_i<\alpha_i<\theta_{i+1}$, $i = 1, 2, \ldots, N$, where $\theta_{N+1} \equiv \theta_1$. The response of the system will be permitted to select the optimal location of each "no-flow" line between wells. Optimal parameter values also will be selected within each region.

The continuous line source solution, Ramey (1966), is applied to the system giving a finite number of well images, $Z_{i,k}$, $k = 1,2,\ldots,m_i$, where m_i, restricted to even integers, are the number of images corresponding to the ith well. Note that K = 1 indicates the original well. The following relation holds:

$$\alpha_i - \alpha_{i-1} + 2\pi\delta_{i,1} = \frac{2\pi}{m_i}, \quad i = 1,2,\ldots,N$$

where

$\delta_{i,1}$ is the Kronecker delta and $\alpha_o \equiv \alpha_N$.

The transient flow equation for the system considered leads to a diffusion equation as follows:

$$\tau\left[\frac{1}{r}\frac{\partial}{\partial r}\left(r\frac{\partial\phi}{\partial r}\right)\right] = S\frac{\partial\phi}{\partial t} \tag{3-1}$$

where ϕ denotes the model computed pressure to distinguish it from observed values, P, and r represents radial distance from the well. The initial and boundary conditions for the partial differential equation are the following, Ramey (1966), Haimes et al (1968).

$$\lim_{t\to 0}\phi(r,t) = \lim_{r\to\infty}\phi(r,t) = P_1, \text{ a constant}$$

$$\lim_{r\to 0} r\left(\frac{\partial\phi}{\partial r}\right) = \frac{q}{2\pi\tau}, \text{ a constant}$$

The positive constant production rate is given by q. These conditions define slow laminar flow of slightly compressible fluids in an infinite, horizonatal, radial isotropic system of uniform thickness.

Equation (3-1) is transformed into an ordinary differential equation using the Boltzmann transformation: $\mu = (Sr^2/4\tau t)$. The new equation is

$$\mu\frac{d^2\phi}{d\mu^2} + (1+\mu)\frac{d\phi}{d\mu} = 0$$

with boundary conditions:

$$\lim_{\mu\to\infty}\phi(\mu) = P_1; \quad \lim_{\mu\to 0} 2\mu\frac{d\phi}{d\mu} = \frac{q}{2\pi\tau}$$

The solution to this equation is obtained in terms of the exponential integral; Ei(u):

$$\phi = P_1 - \frac{q}{4\pi T}\int_{u=\mu}^{u=\infty}\frac{\exp(-u)}{u}\,du = P_1 + \frac{q}{4\pi\tau}\,\mathrm{Ei}\left(-\frac{Sr^2}{4\tau t}\right)$$

Since Ei(-u) is negative, the pressure will decrease as production occurs.

The use of images for extension of the solution to problems in a bounded, wedge-shaped, homogeneous region is straightforward. Suppose the ith actual wedge includes an angle of $2\pi/m_i$ radians, between azimuths α_i and α_{i-1}, where m_i is an even integer. The image system then contains m_i wedges, filling the whole plane. Each wedge has the same origin, and contains a well which is the mirror image of the actual well and/or image wells across each of its adjacent boundaries. The angle between the ith well with azimuth θ_i and its kth image is given by $2\zeta_{ik}$ where

$$\zeta_{ik} = [\tfrac{k}{2}](\alpha_i - \theta_i) + [\tfrac{k-1}{2}](\theta_i - \alpha_{i-1} + 2\pi\,\delta_{i1})$$

$$i = 1,2,\ldots,N \quad k = 1,2,\ldots,m_i$$

and δ_{i1} is the Kronecker delta. The notation [k/2] and [(k-1)/2] denotes that these quantities are truncated to integer values. The α_i define wedge boundaries.

Thus the pressure computed at the ith well and jth time is conveniently denoted by ϕ_{ij}, which in the bounded system results from production at a rate q_i. It is computed by superposition of the pressure effects caused by all m_i image wells, for each of N actual wells. The computed pressure response for this model for a given set of τ_i, S_i can be determined from simple geometric relationships as:

$$\phi_{ij} = P_1 + \frac{q_i}{4\pi\tau} \sum_{k=1}^{m_i} Ei\left(- \frac{S_i r_i^2 \sin^2 \zeta_{ik}}{\tau_i t_j} \right)$$

where r_i is the radial distance from the origin to the ith well and $r_i \sin \zeta_{i1}$ is defined to be r_{wi}, the bore hole radius for the ith well.

3.2 Model Decomposition

A nonlinear programming problem is formulated for any integer number of wells N, and number of discrete observations T. Knowing the production and pressure history of each well, the system input and output respectively, the identification problem for the aquifer system can be mathematically formulated as follows:

$$\min_{\tau_i, S_i, \alpha_i} \sum_{i=1}^{N} \sum_{j=1}^{T} (\phi_{i,j} - P_{i,j})^2 \qquad i = 1,2,\ldots,N$$

Subject to the following constraints:

$$\tau_i > 0; \quad S_i > 0; \quad \theta_i < \alpha_i < \theta_{i+1}; \quad \alpha_i - \alpha_{i-1} + 2\pi\delta_{i,1} = \frac{2\pi}{m_i};$$

$$i = 1,2,\ldots, N.$$

where, m_i are even integers and $P_{i,j}$ are observed pressure of the ith well at time t_j.

The identification problem posed above has 3N independent variables, N(T+1) equality constraints, 4N inequality constraints and in addition is highly nonlinear. Such problems become increasingly difficult to solve as N and T increase. For this reason, a multilevel optimization technique was employed. The optimal solution for the ith subsystem depends in general on the (i-1) subsystem, because of the relationship between α_i and α_{i-1}. To obtain the required subsystem independence, it is convenient to introduce pseudo-variables, σ_i, and require that:

$$\sigma_i = \alpha_{i-1}, \qquad i = 1,2,\ldots, N \tag{3.2}$$

By considering the σ_i as parameters in the ith subsystem, the overall aquifer is effectively decomposed into independent subsystems as required. The determination of the optimal τ_i, S_i, α_i in each subsystem for fixed σ_i will be referred to as the first-level optimization problem, and the task of determining the optimal σ_i, i = 1, 2,..., N will be termed the second-level optimization problem.

Equation (3-2) represents N additional constraints introduced by the decomposition. They are treated by defining the Lagrangian L:

$$L = \sum_{i=1}^{N} \sum_{j=1}^{T} (\phi_{i,j} - P_{i,j})^2 + \sum_{i=1}^{N} \lambda_i (\alpha_i - \sigma_{i+1})$$

where λ_i are Lagrange multipliers.

Decomposition of the Lagrangian into N sub-Lagrangians yields:

$$L(\underline{\tau}, \underline{S}, \underline{\alpha}, \underline{\sigma}, \underline{\lambda}) = \sum_{i=1}^{N} L_i(\tau_i, S_i, \alpha_i; \sigma_i, \lambda_i, \lambda_{i-1})$$

Note that the first three arguments are independent variables at the first level and the last three are considered as parameters.

The variable determination or first-level optimization is straight-forward and several schemes are applicable. On the other hand, the method for parameter determination, or second-level optimization, must be carefully chosen to assure that the iterative procedure converges to the overall system optimum. Several such schemes have been developed, each having its own merits and drawbacks. The first level optimization techniques employed are the Gradient method, and a direct search algorithm.

The second-level optimization technique employed is termed the Gauss-Seidel method, and has been used successfully for the optimization of distributed parameter systems, Wismer (1971). Schematic representation of the two-level coordination is presented by Figure (3-1).

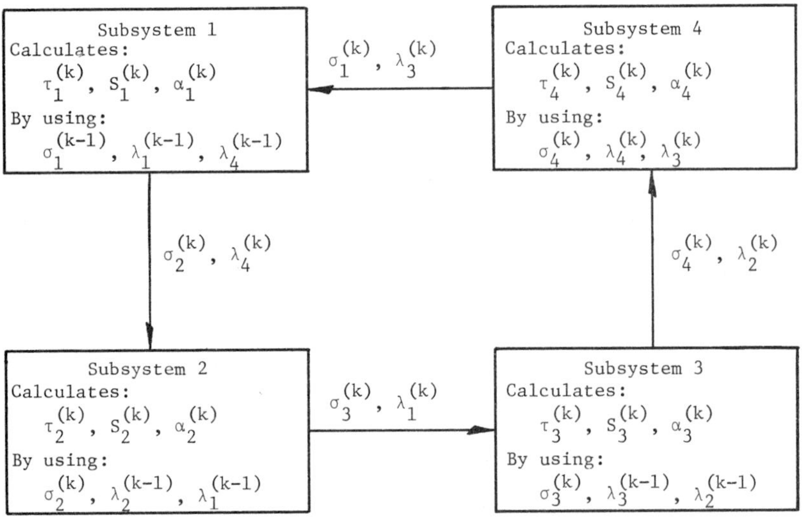

THE GAUSS-SEIDEL ALGORITHM FOR THE SECOND LEVEL FOR THE kth ITERATION

Figure (3-1)

3.3 Conclusions

Decomposition and multilevel optimization techniques have been applied to the identification of parameters characterizing flow behavior of an aquifer. The system is described by a nonlinear partial differential equation and the number of parameters may be large.

This complex identification problem has been successfully treated using a simple direct search for the subsystem optimization and Gauss-Seidel type algorithm for the second-level optimization. This combination was found: (a) to give fast convergence toward an optimum, (b) to require no computation either analytically or numerically of response surface derivatives, (c) to require little computation per iteration, and (d) to be capable of handling both inequality and equality constraints on the system. The latter is significant since additional inequality constraints reduce the feasible solution space and hence speed the convergence rate. While convergence of this two-level scheme has not been proven for this problem, example problems were run with several data sets and the convergence was good in each case.

4.0 INTEGRATED SYSTEM IDENTIFICATION AND OPTIMIZATION

4.1 Introduction

The necessity and the advantages of an integrated approach to the identification and optimization problems have recently been established in the literature both theoretically and computationally, Haimes (1970,1973), Haimes and Wismer (1971b, 1972a), Haimes, Lasdon and Wismer (1971c), McGrew (1971), McGrew and Haimes (1973), and Olagundoye (1971). The coupling relationship between these two concepts is inherent in the nature of the desired "optimal solution".

Any mathematical model consists of unknown variables and "known" parameters characterizing the system. In general, these parameters are not known exactly, but rather are estimated or determined under non-optimal conditions. Accordingly, the solution generated from such system models may be non-optimal. Clearly, the identification of the system's parameters, often referred to as system modeling, is essential in order to obtain an optimal control policy. Unfortunately, system modeling and system optimization are generally treated as two separate problems in the literature, and hence the optimal coordination between the above two concepts is often overlooked.

Several classes of integrated problems were studied in the above mentioned works and various solution procedures were introduced and applied. In this survey paper the discussion will be limited to the applications of the multilevel approach in solving the joint problem.

4.2 Problem Definition

The system optimization problem may be defined as follows:

$$\min_{\underline{u}} \; P'(\underline{y}, \underline{u}, \underline{\alpha})$$

subject to

$$\underline{y} = \underline{F}(\underline{u},\underline{\alpha}); \quad \underline{g}(\underline{y},\underline{u},\underline{\alpha}) < 0$$

where

$\underline{y}$ is an N-vector of system outputs; $\underline{u}$ is an M-vector of system inputs; $\underline{\alpha}$ is a p-vector of model parameters; $\underline{g}$ is a Q-vector of system constraints. (it is an elementary extension to also include equality constraints in the development.)

or, after substitution of the output relations,

$$\min_{\underline{u}} \; P(\underline{u},\underline{\alpha}) \qquad \text{s.t.} \quad g(\underline{u},\underline{\alpha}) \leq 0 \tag{4-1}$$

Here, $P(\underline{u},\underline{\alpha})$ is a scalar valued function whose value is an indication of system performance.

With this structure, the identification problem is to determine the system parameters, $\underline{\alpha}$, in such a manner that the deviation between the model and real system responses to a given class of inputs is as small as possible. The degree of "goodness of fit" is represented by the value of a pre-defined criterion function, $G(\underline{\alpha})$. For example, $G(\underline{\alpha})$ may represent the sum of the squared deviations between the model and real system outputs for various inputs. If J observations of the real system inputs, $\hat{\underline{u}}^j$, and outputs, $\hat{\underline{y}}^j$; $j = 1,\ldots, J$, are made and the least squares criterion is employed, then the identification problem may be presented as

$$\min_{\underline{\alpha}} \; G(\hat{\underline{y}},\hat{\underline{u}},\underline{\alpha})$$

$$\text{s.t.} \quad \underline{g}(\hat{\underline{y}},\hat{\underline{u}}^j,\underline{\alpha}) \leq 0; \qquad \hat{\underline{y}}^j = F(\hat{\underline{u}}^j,\underline{\alpha}); \quad j = 1,\ldots,J$$

or after substitution of the output relations

$$\min_{\underline{\alpha}} \; G(\underline{\alpha}) \qquad \text{s.t.} \quad \underline{g}(\hat{\underline{u}}^j,\underline{\alpha}) \leq 0 \quad j = 1,\ldots,J \tag{4-2}$$

where

$$G(\underline{\alpha}) = \sum_{j=1}^{J} [\hat{\underline{y}}^j - F(\hat{\underline{u}}^j,\underline{\alpha})]^T [\hat{\underline{y}}^j - F(\hat{\underline{u}}^j,\underline{\alpha})]$$

As suggested by Haimes (1970), an integrated solution to (4-1) and (4-2) can be viewed as finding a solution to the bicriterion minimization below

$$\min_{\underline{\alpha},\underline{u}} \begin{bmatrix} G(\underline{\alpha}) \\ P(\underline{u},\underline{\alpha}) \end{bmatrix} \qquad \text{s.t.} \quad \underline{g}(\underline{u},\underline{\alpha}) \leq 0 \tag{4-3}$$

Solutions to (4-3) are defined in terms of efficient points

Def. 1: $(\underline{u}°,\underline{\alpha}°)$ is an efficient point of (4-3) if and only if there does not exist another point $(\underline{u},\underline{\alpha})$ such that $G(\underline{\alpha}) \leq G(\underline{\alpha}°)$ and $P(\underline{u},\underline{\alpha}) \leq P(\underline{u}°,\underline{\alpha}°)$; with strict inequality holding for at least one expression.

It is quite clear that (4-3) may have many efficient points. However, it is reasoned that since optimization (4-1) must be based on the best possible system model, we may define the optimum solution to the joint identification and optimization problem as follows.

Def. 2: The point $(\underline{u}^*,\underline{\alpha}^*)$ is an optimum solution to the joint identification and optimization problem if
(1) $(\underline{u}^*,\underline{\alpha}^*)$ is an efficient point of (4-3)
(2) $\underline{\alpha}^*$ solves (4-2).

To find the above optimum solution, a parametric approach developed by McGrew (1971 will be employed. This approach reformulates (4-3) as the equivalent scalar minimization, (4-4), below.

$$\min_{\underline{u},\underline{\alpha}} \; \theta G(\underline{\alpha}) + (1-\theta)P(\underline{u},\underline{\alpha}) \qquad \text{s.t.} \quad \underline{g}(\underline{u},\underline{\alpha}) \leq 0; \quad 0 < \theta < 1 \tag{4-4}$$

It was proven by McGrew (1971) and McGrew and Haimes (1972) that the limiting solution to (4-4) satisfies the conditions for optimality of definition 2. That is the joint solution is

$$[\underline{\alpha}^*,\underline{u}^*] = \lim_{\theta \to 1^-} [\underline{\alpha}°(\theta),\underline{u}°(\theta)]$$

where $\alpha°(\theta)$ and $\underline{u}°(\theta)$ are optimum solutions to (4-4) for a fixed $\theta \varepsilon (0,1)$.

4.3 Multilevel Structure

Two multilevel hierarchies for solving the joint problem will be presented. The first is based on a feasible decomposition presented by Brosilow, et al (1965) and Lasdon (1970).

Forming the Lagrangian

$$\min_{\underline{u},\underline{\alpha},\underline{\mu}} \; \theta G(\underline{\alpha}) + (1-\theta)P(\underline{u},\underline{\alpha}) + \underline{\mu}^T \underline{g}(\underline{u},\underline{\alpha}) \qquad 0 < \theta < 1 \tag{4-5}$$

where $\underline{\mu}$ is a Q-vector of Kuhn-Tucker multipliers for appending the system constraints.

Decomposition of (4-5) is achieved by introducing the pseudo-variables $\underline{s}$ wherever the controls, $\underline{u}$, and model parameters, $\underline{\alpha}$, are coupled. The constraint, $\underline{s} = \underline{\alpha}$ is then forced. Form the new overall system Lagrangian L as follows

$$L(\underline{u},\underline{\alpha},\underline{s},\underline{\mu},\underline{\lambda},\theta) = \theta G(\underline{\alpha}) + (1-\theta)P(\underline{u},\underline{s}) + \underline{\mu}^T \underline{g}(\underline{u},\underline{s}) + \underline{\lambda}^T(\underline{s}-\underline{\alpha})$$

where $\underline{\lambda}$ is a P-vector of Lagrange multipliers for the additional equality constraints.

Assume the pseudo-variables, $\underline{s}$, and the parameter θ in the Lagrangian L are known at the first level. Then the system Lagrangian L may be decomposed into the following "independent" sub-Lagrangians.

$$L(\underline{u},\underline{\alpha},\underline{s},\underline{\mu},\underline{\lambda},\theta) = L_1(\underline{\alpha},\underline{\lambda};\underline{s},\theta) + L_2(\underline{u},\underline{\mu};\underline{s},\theta)$$

Thus the task of the first level is to:

subsystem 1: $\min\limits_{\underline{\alpha},\underline{\lambda}} \; L_1(\underline{\alpha},\underline{\lambda};\underline{s},\theta)$

subsystem 2: $\min\limits_{\underline{u},\underline{\mu}} \; L_2(\underline{u},\underline{\mu};\underline{s},\theta)$

where $\quad L_1(\underline{\alpha},\underline{\lambda};\underline{s},\theta) = \theta G(\underline{\alpha}) + \underline{\lambda}^T(\underline{s}-\underline{\alpha})$

and $\quad L_2(\underline{u},\underline{\mu};\underline{s},\theta) = (1-\theta)P(\underline{u},\underline{s}) + \underline{\mu}^T\underline{g}(\underline{u},\underline{s})$

$\underline{s}$ and θ are assumed known at the first level and to be determined by higher levels.

A second level problem can then be defined to determine the optimum value, $\underline{s}^*$, used in the two first level subproblems. The parameter θ will be determined at the third level. Thus the second level problem is

$$\min_{\underline{s}} \; \theta G[\underline{\alpha}^*(\underline{s})] + (1-\theta)P[\underline{u}^*(\underline{s})] + \underline{\lambda}^T[\underline{s}-\underline{\alpha}^*(\underline{s})] + \underline{\mu}^*(\underline{s})^T \, \underline{g}[\underline{u}^*(\underline{s})] \tag{4-6}$$

If analytic solutions to the first level problem are possible, i.e. $\underline{u}^*(s)$, $\underline{\alpha}^*(s)$, $\underline{\mu}^*(s)$ and $\underline{\lambda}^*(s)$; then the second level controller will easily yield to the integrated problem solution. This generally is not possible; the solution is then reached by iterating between the first and second levels.

The second level controller determines the values of the pseudo-variables, $\underline{s}$, used in the first level optimizations by the following algorithm.

$$\underline{s}^{I+1} = \underline{s}^I - K\underline{D} \tag{4-7}$$

where I is the iteration number and K is taken to be the step size which minimizes the joint problem down the search direction, $\underline{D}$. In this work, the search direction, $\underline{D}$, is determined by a modified Fletcher-Powell (1963) algorithm.

It is shown by Brosilow et al (1965) that convergence to the integrated problem solution when using (4-7) as a second level controller requires that each subsystem be minimized to at least a local optimum for any choice of the pseudo variables, $\underline{s}$. This should always be possible for subsystem 2. It is also true for subsystem 1, since it is not an optimization problem, but has the unique solution

$$\underline{\lambda}^* = \theta \nabla_{\underline{\alpha}} G(\underline{\alpha})$$

$$\underline{\alpha}^* = \underline{s}$$

θ and $\underline{s}$ are given by the higher levels.

5.0 INTEGRATED SYSTEM IDENTIFICATION AND OPTIMIZATION FOR CONJUNCTIVE GROUND AND SURFACE WATER MANAGEMENT

5.1 Introduction

Something more than half of the water supplies of the United States are extracted directly from groundwater systems. Groundwater also acts as an enormous regulating reservoir providing for the natural base flows. Increasingly man induced recharge is being used to augment natural replenishment of aquifers. In many locations groundwater reserves are utilized for interim economic development.

Groundwater is clearly one of the major elements of our water resources. Its wise management must be considered a necessity. Key to the optimal management of this resource for any of the multiple beneficial uses is an ability to predict with reasonable accuracy the response of the system to decisions affecting recharge and withdrawals. This in turn is dependent upon the values of the important parameters such as permeability, specific yield, compressibility, etc., as these are distributed over the aquifer formations in space and time.

Obtaining these values on a sufficiently dense space-time network by an extensive data observation system, would be prohibitively expensive. For this reason, most of the values used are deduced from the behavior of the system rather than by direct observation. Because of analytical limitation the values presently being obtained essentially represent an estimate of the parameters near the well being observed, i.e., a point observation. The aquifer as a system, however, responds to these values as distributed over the system, not just in the vicinity of the well. What is needed is an analytical procedure which will permit a determination of those parameters which will reflect, through an optimum weighting process, the effect of the distribution of their values as a function of time and space in a way which provides the most relevant information for the management optimization problems.

5.2 Model Formulation

The identification of aquifer system parameters, (e.g. transmissivity and storage functions) as part of system modeling, has traditionally been considered and treated separately from the optimization procedure. However, since the unknown aquifer parameters are used in determining the optimal decision variables an integrated system identification and optimization approach will be adopted for the optimal management of ground and surface water, Yu (1972) and Yu and Haimes (1972). The identification models discussed in section 3.0 will be utilized here.

Due to the competitive nature of groundwater pumping, the lack of appropriate legal provisions and policy instruments resulted in a nonoptimal exploitation of groundwater resources. As a remedy for this, a decentralized decision making structure is postulated, as in the following. The whole region is thought to be divided into several subregions in each of which there exists a local water agency who has an exclusive jurisdiction over development and management water of resources in that subregion. The existence of a regional authority who takes responsibility for the artificial and natural recharge of the aquifer is assumed. The goal of ground and surface water management posed here is the minimization of water supply costs for predetermined water demands. It is assumed that all engineering facilities have already been constructed and these are not considered. Each subregion satisfies its two water demands (the municipal/industrial demand and the agricultural demand) by pumping in its own subregion and by importing water from outside the region. The tasks of the regional authority are to determine: (a) the optimal recharge rates, (b) the corresponding tax rates, and (c) the optimal water heads and cross-boundary subsurface flows at the intersubregional boundaries.

The solution for the overall regional problem is iteratively implemented by the decomposition and multilevel approach (see Figure 5-1). The local agencies (the first level subsystems) optimize their own problems for given conditions (artificial recharge quantity, pumping tax, boundary water heads), and report their results (costs, pumpages, boundary subsurface flows) to the regional authority (the second level master controller). The regional authority then determines, based on the responses of the local agencies, the new set of conditions to be sent down to the local agencies, and so on. Here, the objective of the regional authority is to induce the overall solution to be physically feasible (i.e., the in-and-out subsurface flows along the inter-subregional aquifer boundaries should be coincident), and to be least costly (i.e. the artificial recharge operation and hence the tax rate should be optimal). The actual computation by the regional authority is conducted in three stages (see Figure 5-1). The first stage (the lowest) has direct communications with the first level subsystems and tries to get the physically feasible optimal solutions for a given set of artificial recharge rates and pumping tax by the upper stages. That is, it adjusts the boundary water heads until the algebraic sum of the cross-boundary subsurface flows reported by the first level subsystems becomes zero. Here, a Newton-Raphson type algorithm is used. The second stage is in between the third and the first stages and is to calculate the tax rate which provides the revenue for a nominal artificial recharge operation. A quadratic interpolation algorithm is used at this stage, iterating solutions down below. The third stage now minimizes the overall regional cost by adjusting the artificial recharge operation rate. A minimization algorithm without derivatives may be used at this stage.

The local agencies' problems lend themselves to nonlinear programming. The penalty function method is applied with the conjugate gradient algorithm for the maximization task.

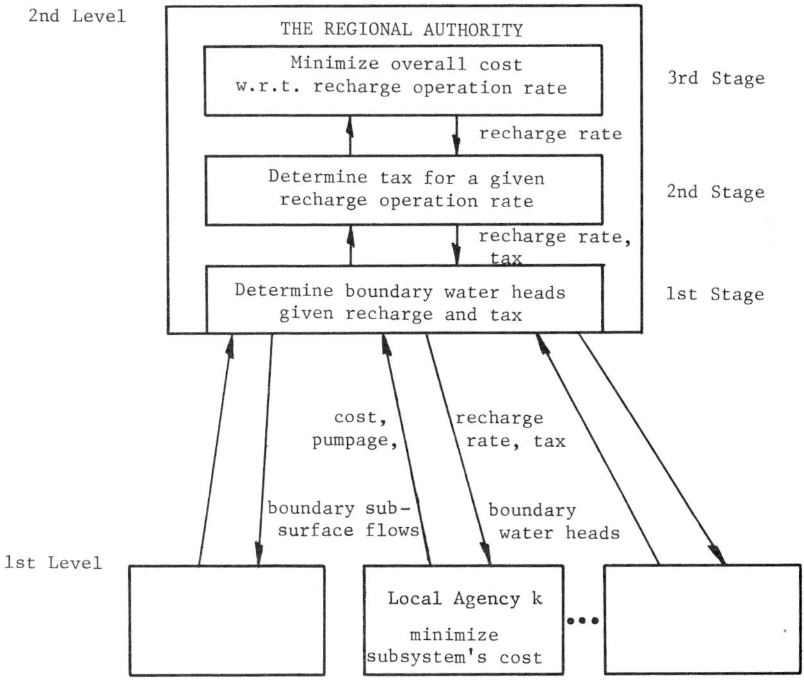

Figure 5-1: MULTILEVEL STRUCTURE

6.0 REGIONAL WATER QUALITY CONTROL AND MANAGEMENT

6.1 Introduction

The planning and operation of a regional water quality management system over a long period of time is considered. A regional water quality management system consists of a number of existing local wastewater treatment plants, both municipal and industrial, along with one or more regional wastewater treatment plants, and a network of pipelines connecting some or all of the local plants with the regional plants. In order to meet increased demands for wastewater treatment in the future, the expansion of local plants and the construction and/or expansion of regional plants are also considered. The capital and operating-and-maintenance costs of wastewater treatment plants and pipelines are expected to vary throughout the planning period, due to changes in the costs of construction, labor, chemicals, and electric power.

A Central Authority is postulated which is authorized to impose effluent charges on all dischargers in the basin. In addition, the cost allocation of regionally operated facilities via economic considerations is specifically dealt with. The hierarchical models allow many different pertinent economic-legal-political conditions to be reflected in the solution procedure, especially that of decentralized decision making. Finally the models yield a clear, conceptually simple, representation of the way in which the mathematical programming algorithms are used and allows ease in application from one river (or estuary) to another.

Operating policy will be determined by a multilevel optimization technique. Each polluter decides whether to treat his wastewater at his own local plant and discharge it to the river, to treat his wastewater at the local plant and to ship the effluent, via pipeline, to a regional plant for further treatment, or to ship his raw wastewater, without local treatment, to a regional plant for treatment. The polluter may use any combination of the above options and any or all of the regional plants, to realize the least cost combination. Effluent charges, based on the water quality constraints, will be used to induce the polluters to choose the configuration which will result in least cost to society, *while satisfying quality standards*.

The numerical value of the effluent charges is determined by including the actual average pollution load and the receiving water's hydrological characteristics. This is more equitable than the imposition of an arbitrary "x" cents per pound of waste as suggested by other proponents of effluent charges. Most importantly, the hierarchical approach proposed here has been demonstrated to apply in a situation where the important state variables are dissolved oxygen (DO), biochemical oxygen demand (BOD), temperature, and algae concentrations. In other words, the water quality was characterized by a vector of parameters. The important control variables were BOD and heat discharged.

6.2 Importance of Regional Water Resource Planning and Management

The goal of water resource planning and management is to allow for the most efficient use of existing water resource facilities and future expansion projects. At the local level resource management involves numerous activities. Water is to be supplied for industries and municipalities. The water quality level requirements usually differ among various users. Resource management must decide upon policies of usage of existing facilities such as the reservoir levels to be maintained or well discharge and recharge rates. The concept of water resources should be expanded to consider total "water services". By this is meant all activities required to provide water supplies of multiple quality level and the treatment activities required before and after water use.

A particular locality may have accurate estimates of their own water service over a planning horizon and may have a clear set of alternative policies available to meet these future service demands; however, in general, the situation is not so straight forward. There are numerous reasons why a local approach to resource management will more than likely fail:

1) Rarely can a relatively small locality be considered hydrologically independent from surrounding areas.

2) Often small localities do not possess the monetary resources available to make the necessary capital expenditures required by resource management policies.

3) The jurisdictional powers of a small locality will be inadequate to fully implement its management policies (this is one aspect of the great diffusion of authority that exists in water resources).

4) Small localities cannot be considered *economically* independent from surrounding areas.

6.3 Advantages of a System of Effluent Charges

For a successful implementation of the system of effluent charges, a freedom of choice should be offered to the polluter as to whether: (i) to treat his effluent by his own wastewater treatment plant, or (ii) to ship his effluent to a regional (governmental or a co-op type) wastewater treatment plant, and/or (iii) to pay the effluent charges imposed by the central authority and utilize whatever assimilative capacity of the water body is available.

The system of effluent charges has remarkable advantages and potential of success. The following are the major ones, Proxmire (1969):

(a) Economic incentives - There would be almost an immediate impact due to the natural desire of polluters to reduce their charges.

(b) Assign responsibility - It would assign responsibility for pollution control to those who are responsible for the pollution. The polluters, not the public at large, would pay for the damages.

(c) Source of revenue - It would provide revenue which could be utilized for further pollution control efforts.

(d) Regional water management - It would provide strong economic incentive for the creation of regional water management associations, and

(e) Ultimate solution - It would contribute to the ultimate solution of the water pollution problem.

6.4 Multilevel Structure

Several hierarchical models for water quality management have recently been developed at the Systems Research Center, Case Western Reserve University, Haimes (1971a, 1972b), Haimes et al (1972b, 1972c), Foley et al (1971), Rosen (1972), and Kaplan (1972). In the above mentioned works, hierarchical models with two, three and four levels have been introduced and applied to real data from various river basins in the United States.

Common to all hierarchical models is the decomposition of the water resource system into N reaches (subsystems). Water quality standards are satisfied for each reach via the many optional policies made available to each user (polluter). These options include local treatment by the user, shipping the effluent to a regional wastewater treatment plant, shipping the effluent to other reaches via a bypass piping, and/or paying the effluent charges and shipping partially treated effluent to the river. The last option can be exercised only when available, (e.g., there exists an excess of dissolved oxygen in the river).

6.5 Two-level Hierarchy

Consider a Water Resource System, WRS. Let $\underline{U}_i$ be the ith pollution effluent input to the WRS, $i = 1,2,\ldots,N$. $\underline{U}_i^T = [U_{i1}, U_{i2}, U_{im}]$, where U_{i1} represents the water quantity and $U_{i2},\ldots,U_{im}$ represents different water quality characteristics, e.g. BOD, DO, pH, conductivity, temperature, algae, phosphates, nitrates, etc. It is convenient to decompose the WRS into N subsystems (N reaches in the case of a river) each of which includes one pollution input $\underline{U}_i$. Let $\underline{W}_i$ be the input vector coming into the ith subsystem from other subsystems, $\underline{V}_i$ be the output vector of the ith subsystem going to other subsystems, and $\underline{S}_i$ be the output vector of the ith subsystem leaving the system, where $\underline{W}_i$, $\underline{V}_i$, and $\underline{S}_i$ are of the same dimensionality as $\underline{U}_i$, i.e. m-dimension (see Figure 6-1). The subsystem outputs $\underline{V}_i$ and $\underline{S}_i$ are assumed to be represented by the following functions:

$$\underline{V}_i = \underline{\Psi}_i(\underline{U}_i, \underline{W}_i) \qquad (6\text{-}1)$$

$$\underline{S}_i = \underline{\Phi}_i(\underline{U}_i, \underline{W}_i) \qquad (6\text{-}2)$$

where $\underline{\Psi}_i^T = [\Psi_{i1}, \Psi_{i2},\ldots, \Psi_{im}]$; $\underline{\Phi}_i^T = [\phi_{i1}, \phi_{i2},\ldots, \phi_{im}]$; $i = 1,2,\ldots,N$.

Let

$$\underline{G}_i(\underline{U}_i, \underline{W}_i, \underline{S}_i) \leq 0, \qquad i = 1,2,\ldots,N \qquad (6\text{-}3)$$

be the vector of constraints to be satisfied by the ith subsystem, $i = 1,2,\ldots,N$; where $\underline{G}_i(\underline{U}_i,\underline{W}_i,\underline{S}_i)$ are k-dimensional continuous functions. The inequality constraint (6-3) includes equality constraints and represents the physical, legal, economical and other system constraints. The overall WRS cost function, F, is the sum of wastewater treatment cost functions of each subsystem $F_i(\underline{U}_i,\underline{S}_i)$;

$$F = \sum_{i=1}^{N} F_i(\underline{U}_i, \underline{S}_i) \qquad (6\text{-}4)$$

It is assumed that the cost function F_i for the ith subsystem is an explicit function of the effluent input $\underline{U}_i$ and the outputs $\underline{S}_i$. It is assumed that the functions $F_i(\underline{U}_i,\underline{S}_i)$ are continuous. The optimization problem for the WRS can be summarized as follows:

$$\text{minimize} \quad F = \sum_{i=1}^{N} F_i(\underline{U}_i, \underline{S}_i)$$

Subject to the constraints:

$$\underline{V}_i = \underline{\Psi}_i(\underline{U}_i, \underline{W}_i); \quad \underline{S}_i = \underline{\Phi}_i(\underline{U}_i, \underline{W}_i);$$

$$\underline{G}_i(\underline{U}_i, \underline{W}_i, \underline{S}_i) \leq 0; \quad \underline{W}_{i+1} = \underline{V}_i; \qquad i = 1,2,\ldots,N$$

Where $\underline{W}_{N+1} = \underline{V}_N$ is the water leaving the WRS. Form the Lagrangian L:

$$L = \sum_{i=1}^{N} F_i(\underline{U}_i, \underline{S}_i) + \sum_{i=1}^{N-1} \underline{\lambda}_i^T[\underline{V}_i - \underline{W}_{i+1}] + \sum_{i=1}^{N} \underline{\delta}_i^T \underline{G}_i(\underline{U}_i, \underline{W}_i, \underline{S}_i)$$

where $\underline{\lambda}_i$ are m-dimensional Lagrange multipliers, $i = 1,2,\ldots,N-1$ and $\underline{\delta}_i$ are kth dimensional generalized Lagrange multipliers (Kuhn-Tucker multipliers) Kuhn and Tucker (1950), $i = 1,2,\ldots,N$.

Substituting the values for $\underline{S}_i$ and $\underline{V}_i$ from equations 6-1, 6-2 yields:

$$L = \sum_{i=1}^{N} F_i(\underline{U}_i,\underline{\Phi}_i) + \sum_{i=1}^{N-1} \underline{\lambda}_i^T[\underline{\Psi}_i(\underline{U}_i,\underline{W}_i)-\underline{W}_{i+1}] + \sum_{i=1}^{N} \underline{\delta}_i^T \underline{G}_i(\underline{U}_i,\underline{W}_i,\underline{\Phi}_i)$$

Note that

$$\sum_{i=1}^{N-1} \underline{\lambda}_i^T [\underline{\Psi}_i(\underline{U}_i, \underline{W}_i) - \underline{W}_{i+1}] = \sum_{i=1}^{N-1} \underline{\lambda}_i^T \underline{\Psi}_i(\underline{U}_i, \underline{W}_i) - \sum_{i=2}^{N} \underline{\lambda}_{i-1}^T \underline{W}_i$$

Then L is readily decomposed into N independent subsystems:

$$L = \sum_{i=1}^{N} L_i(\underline{U}_i, \underline{W}_i, \underline{\delta}, \underline{\lambda})$$

where

$$L_i = F_i(\underline{U}_i, \underline{\Phi}_i) + \underline{\lambda}_i^T \underline{\Psi}_i(\underline{U}_i, \underline{W}_i) - \underline{\lambda}_{i-1}^T \underline{W}_i + \underline{\delta}_i^T \underline{G}_i(\underline{U}_i, \underline{W}_i, \underline{\Phi}_i)$$

$$i = 1, 2, \ldots, N$$

and where by definition:

$\underline{\lambda}_N^T \underline{\Psi}_N(\underline{U}_N, \underline{W}_N) \equiv 0$; $\underline{\lambda}_1^T \underline{W}_1 \equiv 0$; $\underline{\lambda}$ and $\underline{\delta}$ are augmented vectors of $\underline{\lambda}_i$ and $\underline{\delta}_i$, $i = 1, 2, \ldots, N$.

At the first-level, $\underline{\lambda}$ and $\underline{\delta}$ will be assumed to be known, and thus the first-level optimization will consist of obtaining $\underline{U}_i(\underline{\lambda}, \underline{\delta})$ and $\underline{W}_i(\underline{\lambda}, \underline{\delta})$, which minimize the corresponding ith subsystem Lagrangian, $L_i(\underline{U}_i, \underline{W}_i, \underline{\delta}, \underline{\lambda})$.

At the second level, with known values of $\underline{U}_i^\circ(\lambda, \delta)$ and $\underline{W}_i^\circ(\lambda, \delta)$ from the first level, the Lagrangian L of the whole WRS is maximized with respect to $\underline{\lambda}$ and $\underline{\delta}$, assuming the existence of a saddle point $(\underline{U}_i^\circ, \underline{W}_i^\circ, \underline{\delta}^\circ, \underline{\lambda}^\circ)$ (see Figure 6-2). Different algorithms are available for carrying the minimizations of L_i at the first level and the optimization of L at the second level.

The Lagrange multipliers $\underline{\lambda}$ associated with the equality constraints are of special interest. The two-level optimization approach utilized here is analogous to the operation of a perfectly competitive economic system. The $\underline{\lambda}$ are prices imposed by the central authority for the pollution caused by the subsystems. These are prices for water improvements, for use of the water, and for wastewater discharges (or treatment). Each subsystem in turn is able to determine its own policy as to whether to pay the price specified by the central authority for causing the degradation of the quality of the WRS or to invest its money in developing a local treatment plant which in turn will improve the quality of its pollution effluent. An optimal policy for each subsystem can be reached while the total cost of improving the quality of the WRS by the central authority as a whole is minimized. Such an approach is both economically and politically desirable and feasible.

The Lagrange multipliers $\underline{\delta}$ associated with the inequality constraints $\underline{G}_i(\underline{U}_i, \underline{W}_i, \underline{S}_i) \leq 0$ are also of interest. The Kuhn-Tucker condition $\delta_i G_i = 0$, means that if G_i is not binding (i.e. $G_i < 0$), then the corresponding δ_i must be zero. In other words, when demand is strictly less than supply the shadow price is zero. If, for instance, the DO level is higher than the minimum requirement for the ith reach, then the corresponding shadow price is zero. The solution is then independent of the total supply of DO available, since there already is more than enough of DO relative to its use at the optimal point. On the other hand if the ith constraint is binding (i.e. $G_i = 0$) then the corresponding shadow price is $\delta_i \geq 0$.

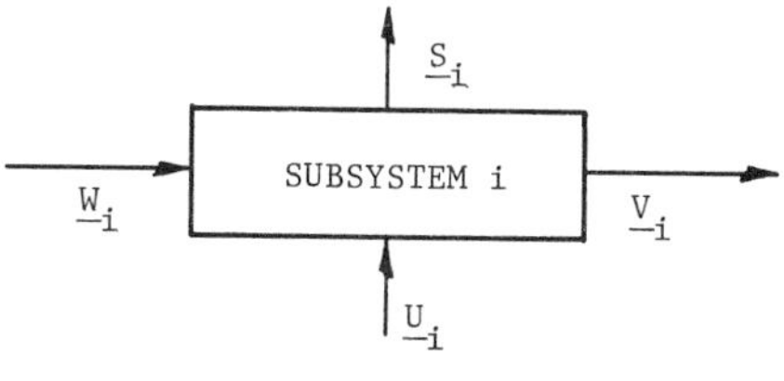

Figure 6-1: ith SUBSYSTEM REPRESENTATION

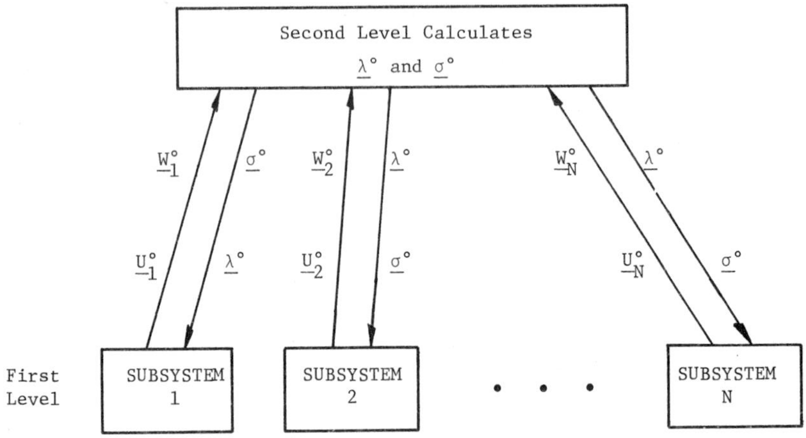

Figure 6-2: TWO LEVEL OPTIMIZATION

6.6 Three-Level Hierarchy

A three-level hierarchy takes advantages of the preceeding model by proposing the construction of regional treatment plants. The central authority determines the optimal treatment configuration for meeting water quality standards along a river basin, and simultaneously determines optimal pollution charges to realize this configuration. The optimal regional plant capacity and regional treatment level are also found, depending on which polluters decide to use the regional facilities. The central authority does not need to know the local treatment cost functions.

The central authority, the third level in the multilevel hierarchy, proposes a taxing structure to the regional plant and the individual polluters. The regional plant, the second level, assumes an optimal plant size and determines a regional treatment level to minimize the regional treatment cost. The individual polluters at the first level determine the optimal local treatment level, and decide whether or not to use the regional facilities. According to the first-level decisions, the second level adjusts the regional plant size, so that there is no excess capacity. The third level checks the water quality constraints and adjusts the effluent charges so that all constraints are satisfied, at the least cost to society.

Use of the technique was simulated with data for the Miami River Basin. A considerable savings over the cost of local treatment is realized when one regional plant is employed, Haimes et al (1972b). The technique has been extended to allow for several regional plants, possibly realizing greater savings. This extension of the technique involves a "transportation problem," that of determining which regional plant should be used by each polluter, Rosen (1972). The possibility of several polluters combining to build a co-op treatment plant is also allowed.

6.7 Summary

Decomposition and multilevel techniques have the following significant advantages for the solution of large scale optimization problems in water quality management: (a) Conceptual simplification of complex systems. (b) Reduction in dimensionability and therefore in the computational effort involved. (c) None of the system model functions need be linear. Thus more flexible models can be constructed to represent the real plant. (d) Simple programming and computational procedure. (e) Interactions among subsystems are permissible, subsystem's independence are achieved via psuedo variables. (f) Each subsystem can be handled with different optimization techniques depending on the nature of the subsystem. (g) The approach is applicable to both static and dynamic systems. (h) Central to the computational procedure is the duality theory of nonlinear programming. In particular, the economic interpretation of the Lagrange multipliers as shadow prices. (i) Excellent approach for regional management. (j) Handles parametric problems.

Clearly, the management of the water quality of a regional river basin is a complex task, which involves a high dimensional interacting decision variables. When we add the complexity induced by the vector of water characteristics and properties representing the water quality level, the nonlinearities in cost functions and physical relationships, the large number of polluters in the regional river basin, the various technical, economical, political, social and other constraints, the need for a capable methodology to handle this large scale system is evident. The multilevel approach has been proven to be an effective and successful methodology.

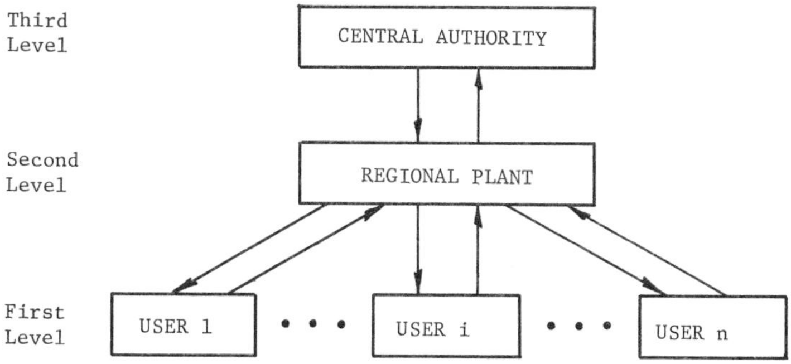

Figure 6-3: THREE LEVEL STRUCTURE

7.0 REGIONAL WATER RESOURCES PLANNING AND MANAGEMENT

7.1 Introduction

In the process of planning for a water resource management system, the planner is often confronted with a multitude of goals and objectives associated with a single project. This multigoal characteristic of a water resource project is often not only desirable, but rather essential, for the economic feasibility of the project's construction and operation. Furthermore, because of water resource projects' interdependence, an efficient coordination of the functions they serve is needed. For example, a reservoir with a high dam may generate hydroelectric power, supply water for municipal, industrial and agricultural use, provide ways and means for flood control, for proper flow of aquatic life, provide for water quality control, navigation, fishing and recreation to nearby streams, and provide a buffer for drought years and groundwater recharge, etc.

The consideration of the primary and secondary benefits of a single water resource project jointly with the optimal use of water, land, and other resources of a larger region, is the subject of this paper. The task of modeling to obtain an efficient solution strategy for such a comprehensive and large scale problem is obviously an intricate one. It is the availability of the hierarchical-multilevel approach however, which enables us to model and study small subsystems (local goals or objectives) at the lower levels and to coordinate all subsystems at a higher level to ultimately yield an optimal solution to the system as a whole, that makes this task both feasible and tractible. The basic principle in the modeling of a total water resource system is the utilization of and the coordination among many of the existing planning and management submodels reported in the literature. Investigations of hierarchical models that encompass the highly interactive economic, geographic and other aspects of regional water resources management have been conducted by Haimes (1971a, 1972e,f), Nainis and Haimes (1972), Butcher, Haimes, and Hall (1969), and Yu and Haimes (1972).

A regional area can be very complex within its network of water sources (streams, rivers, lakes, wells) and its network of water users and polluters. One faces the dilemma that if he attacks the problem at the local level there is often not enough jurisdictional power to control the disturbances from outside the local area and typically not enough local funds available to adequately provide for the resource needs of the local citizenry and industry. Also it appears obvious that when viewed from the larger, regional level that the resulting independent local development will not provide maximal regional benefits nor minimum expenditures to the region since no beneficial effects of inter-locality cooperation and thus incurred economies of scale will be utilized. On the other hand the total regional resource management problem represents too complex a problem to approach directly. The major proposed objective is to configure a regional area as a multilevel system with respect to its water resource planning and management policies. The regional area would be decomposed into smaller areas and their sub-systems would be investigated. The multilevel approach allows this sub-division to be made yet accounts for the interactions between the subsystems.

The satisfaction of economic goals and optimization of costs simultaneously is readily incorporated into the second and third levels of the multilevel system. No other mathematical approach can as easily include all the major interactions and constraints in a model for investigating all alternatives of regional management, and provide a structure which more realistically parallels the structure of the real-life problem itself.

The multilevel approach allows the system of equations to be subdivided into more easily managed optimization problems. Project construction and scheduling at the local level are readily optimized at the local level via dynamic programming yet inter-local resource transfer can be considered through multilevel techniques. Without the subdivisions of localities a regional dynamic program would become very large and unmanageable. The decomposition approach has the advantage of flexibility since any subsystem can be modified without directly affecting the other; this feature does not exist in an undecomposed approach.

The following aspects of the multilevel water resource system are investigated:

*The relative cost between local production of water (and water services) versus transportation of resources from locations within the region. The goal is to investigate the relative costs dynamically, i.e., over a planning horizon and to provide a minimal cost solution.

*The relationships between water consumption and regional economic activity predictions of gross regional product (GRP)/regional income and final demands for goods within the region.

*Rational methods for determining water and waste treatment prices and at the same time levels of government subsidy for water service projects. (e.g. by introducing a system of industrial and domestic effluent charges.)

*Techniques whereby new and influential data can be incorporated into the model in order to re-evaluate proposed decisions (the sensitivity of the model to inaccurate data for the initial stages of the planning horizon should be investigated).

*Better information to assist in the allocation of Federal, state and local government resources ($) for the development of water resources and related lands. In particular as to the effect of these investments to labor, future water needs, water quality, etc.

7.2 Hierarchical Structures

Two hierarchical models for regional water resources planning and management are presented here. Research activities are presently being undertaken on both models for their actual implementation.

7.2.1 Hierarchical Modeling of a Total Water Resource System

Consider the following hierarchical four level water resource model, Haimes (1972e). The overall objective at the fourth (highest) level may be the satisfaction of regional water requirements at a minimum cost, or the maximization of Gross Regional Product, etc. (See Figure 7-1) The third level is concerned with the water supply problem. Alternative ground and surface water projects will be analyzed for their optimal size and construction time. To be utilized for this purpose are submodels such as "dynamic Programming for the Optimal Sequencing of Water Supply Projects" by Butcher, Haimes & Hall, (1969); "Analysis of the Feasibility of Interim Water Supplies," by Hall, Haimes & Butcher, (1972); "A Multilevel Approach to Planning for Capacity Expansion in Water Resource Systems," by Nainis and Haimes, (1972); and Integrated System Identification and Optimization for Optimal Conjunctive Use of Ground and Surface Water," by Yu and Haimes, (1972). The second level is concerned with the water demand and the overall adaptation of the models' parameters. Projected water demand and optimal growth patterns will be obtained via Leontief Input-Output Model, while system identification submodels for the estimation of watershed and aquifer parameters such as: "Optimal Identification of Lumped Watershed Models," by Labadie & Dracup, (1969); "Identification of Aquifer Parameters by Decomposition and Multilevel Optimization," by Haimes, Perrine and Wismer, (1968), and "Modeling and Identification of Aquifer Systems of High Dimensions," by Wismer, Perrine and Haimes, (1970), can be utilized.

The first level is concerned with the water use submodels such as: "Optimum Operations for Planning of a Complex Water Resources System," by Hall & Shephard, (1967); "Modeling and Control of the Pollution of the Water Resources Systems via Multilevel Approach," by Haimes, (1971a); "A Multilevel Approach to Determining Optimal Taxation for the Abatement of Water Pollution," by Haimes, Kaplan, and Husar, (1972b); "Optimal Timing of Irrigation," by Hall & Butcher, (1968), etc.

The major task in the hierarchical modeling of a total water resource system is the submodels coordination with respect to time, space and goals. The fact that water resources problems are hierarchical in nature, and that technical and engineering solutions are obtained only after a long and tedious iterative procedure which satisfy economical, social, political, legal, institutional and other constraints, provide the basis for the success of the proposed hierarchical modeling.

7.2.2 Regional Planning for Capacity Expansion

Water resource projects, often of large magnitudes both physically and monetarily, are discrete in nature having been coined "one-shot". In order to decide which project to build out of a set of feasible candidates various techniques have been used such as cost-benefit ratio and rate-of-return analysis. These methods are not useful when taking into account the dynamic aspects of situations where projects must be planned over time to be available for use in the future.

A regional viewpoint is taken here, Nainis, Haimes (1972), where many projects located in various local sub-regions (river basins) must be considered as candidates for construction at particular times over a long planning horizon (50 years).

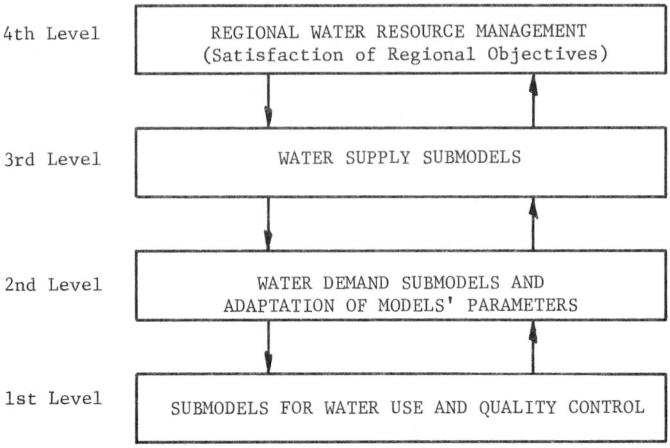

Figure 7-1: HIERARCHICAL STRUCTURE OF SUBMODEL COORDINATION

The North Atlantic Region (NAR) has been chosen as an example for study. It contains 21 separate river basins, and 25% and 30% respectively of the nations population and GNP. The large regional approach is motivated by the desire to place every feasible candidate project in perspective with all others so that a selection reflecting their relative benefits can be made. An area with sufficient economic activity (as a % of the national economy) was chosen in order to permit the use of an input-output multi-sectional model for relating the effects of water resource project decisions. Such modeling work for the NAR is being performed by the Corps of Engineers, Schaake, (1971).

The problem of capacity expansion on the regional level involves selecting a set of feasible alternative projects for each subregional area (basin). Once these are determined the selection and timing problem requires the determination of:

1) Water requirements over the planning horizon levels in each basin area.
2) Schedule (order and timing) of project construction in each basin area.
3) Amount of water transfers to make between basin areas when it is possible to do so.

The strategy used to model the above classes of decisions is to invoke a hierarchical structure. In general, the multilevel structure possesses three levels and can also be divided horizontally into a "supply" and a "demand" model.

The first level of the solution strategy utilizes dynamic programming algorithm to efficiently schedule projects to meet a *given* demand function at minimum present-worth cost for each basin area, Butcher, Haimes, Hall (1969). Within the "supply" model the 2nd level is designed to allocate interbasin area water transfers over time in order to present "effective" demand functions for the 1st level algorithms. The 2nd level attempts to satisfy all water demand functions at minimum total present worth cost.

The task of the third level is to coordinate the "supply" model with a simplified input-output econometric model of the region. In this way the basin area demand functions are determined. The input-output model is a balance, scaled down version of a 39 sector national model. Coefficients are included to relate the total regional economic activity by sector to each basin-area (and further).

The input-output model is dynamic; capital investment constraints and resource usage constraints (labor and water) are used to relate economic activity from one period to the next. The coordination between the "supply" and "demand" models include the water availability over time and the shadow prices from the linear programming solution to the input-output model where the objective function is the present worth of regional income over the *economic* planning horizon (typically less than the project scheduling horizon). The 3rd level coordinates the "supply" and "demand" components of the hierarchy until the total period per unit cost of water balances the shadow price for water. The total effect of the model is to produce an estimate of regional economic growth consistent with efficient investment in water resources projects.

References

Brosilow, C.B., L.S. Lasdon, & J.D. Pearson, (1965), "Feasible Optimization Method for Interconnected Systems," JACC, Rensselear Polytechnic Inst. June 22-25.

Butcher, W.S., Y.Y. Haimes & W.A. Hall, (1969), "Dynamic Programming for the Optimal Sequencing of Water Supply Projects," Water Resources Res., 5, p 1196.

Dracup, J.A., Y.Y. Haimes & R.L. Perrine, (1968), "Identification of Distributed Unconfined Aquifer Parameters," the 4th Am. Water Resources Conf., N.Y., Nov. 18-22, p 399.

Fletcher, R. & M.J.D. Powell, (1963), "A Rapidly Convergent Descent Method for Minimization," The Computer J., y, p 163.

Foley, J., & Y.Y. Haimes, (1972), "Decentralized System of Effluent Charges Implemented via Multilevel Approach," SRC, 27, Case Western Reserve Univ., Cleveland, Ohio.

Haimes, Y.Y., (1967), "Optimal Estimation of Reservoir Parameters," M.S. Thesis, Univ. of Calif., L.A., Calif.

__________, R.L. Perrine & D.A. Wismer, (1968), "Identification of Aquifer Parameters by Decomposition and Multilevel Optimization," Israel J. of Tech., 6(5), p 322.

__________, (1970a), "The Integration of System Identification and System Optimization," Univ. of Calif., L.A., Report No. UCLA-ENG-7029.

__________, (1971a), "Modeling and Control of the Pollution of Water Resources Systems via Multilevel Approach," Water Resources Bulletin, 7(1), p 104.

__________, & D.A. Wismer, (1971b), "Integrated System Identification and Optimization via Quasilinearization," J. of Opt. Theory & Appl., 8, p 100.

__________, L.S. Lasdon, & D.A. Wismer, (1971c), "On a Bicriterion Formulation of the Integrated System Identification and System Optimization," IEEE, Trans., SMC-1, p 296.

__________, & D.A. Wismer, (1972a), "A Computational Approach to the Combined Problem of Optimal Control and Parameter Estimation." To appear in Automatica, 7, May.

__________, M.A. Kaplan, & M.A. Husar, (1972b), "A Multilevel Approach to Determining Optimal Taxation for the Abatement of Water Pollution," Water Resources Res., 8(5), Oct.

__________, J. Foley & W. Yu, (1972c), "Computational Results for Water Pollution Taxation Using Multilevel Approach." To appear in Water Resources Bulletin.

__________, (1972d), "Decomposition and Multilevel Techniques for Water Quality Control." To appear in Water Resources Res., 8(2), June.

__________, (1972e), "Hierarchical Modeling of a Total Water Resource System." To be presented at the 8th Am. Water Resource Conf., St. Louis, Mo., Oct. 30.

Haimes, Y.Y., (1972f), "Multilevel Dynamic Programming Structure for Regional Water Resource Management." Invited paper to be presented at the NATO Inst. on Decom. as a Tool for Solving Large Scale Problems, Cambridge, England, July 17-26.

__________, (1973), "Integrated System Identification and Optimization." To appear in Advances in Cont. Sys., C.T. Leondes, Ed., Academic Press Inc.

Hall, W.A. & W.S. Butcher, (1968), "Optimal Timing of Irrigation," J. Irrigation Drainage Div., Am. Soc. Civil Engrs., 9y, p 267.

__________, Y.Y. Haimes & W.S. Butcher, (1972), "Analysis of the Feasibility of Interim Water Supplies," Water Resources Res., 8, p. 319.

__________, R.W. Shephard, (1967), "Optimum Operations for Planning of a Complex Water Resources System," Univ. of Calif. Water Res. Center, L.A., Contribution No. 122, p 75.

Kaplan, M., (1972), "The Planning and Operation of a Regional Water Quality Management System: A Multilevel Approach," M.S. Thesis, Sys. Engr. Div., CWRU, Ohio.

Kuhn, H.W. & A.W. Tucker, (1950), "Nonlinear Programming," Proc. 2nd Berkely Sym. on Math. Statistics & Probability, Univ. Calif. Press, Calif., 481.

Labadie, J. & J.A. Dracup, (1969), "Optimal Identification of Lumped Watershed Models," Water Resources Res., 5(3), p 583.

Lasdon, L.S., (1970), Optimization Theory for Large Systems, Macmillan Co., London.

McGrew, D.A., (1971), "A Parametric Approach to the Integrated System Identification and Optimization Problems," M.S. Thesis, Sys. Engr. Div., CWRU, Clev., Ohio.

__________, & Y.Y. Haimes, (1973), "A Parametric Solution to the Joint Systems Identification and Optimization Problem via Multilevel Approach." To appear in J. of Opt., Theory & Applications.

Mesarovic, M.D., D. Macko & Y. Takahara, (1970), Theory of Hierarchical Multilevel Systems, Academic Press, New York, N.Y.

Nainis, S., & Y.Y. Haimes, (1972), "A Multilevel Approach to Planning for Capacity Expansion in Water Resource Systems." Accepted for presentation to Internat'l. Conf. on Cybernetics & Soc., October 9-12, Washington, D.C.

Olagundoye, O., (1971), "Efficiency and the ε-Constraint Approach for Multi-Criteria Systems," M.S. Thesis, Sys. Engr. Div., CWRU, Cleveland, Ohio

Proxmire, W., (1969), "Introduction of the Regional Water Quality Act of 1970," Cong. Record-Senate, Proc. & Debates of the 91st Cong., 115, Washington, Nov.2

Ramey, H.J. Jr., (1966), "Application of the Line Source Solution to Flow in Porous Media--A Review," SPE 1361, AIChE Symposium, Dallas, Texas.

Rosen, S.O., (1972), "Regional Water Quality Management," M.S. Thesis, Sys. Engr. Div., CWRU, Cleveland, Ohio.

Schaake, J.C., Jr. (1971), "A Model for Regional Water Needs," presented at 52nd Annual Meeting, AGU Washington, D.C., April. To appear in Water Resources Res.

Wismer, D.A., R.L. Perrine & Y.Y. Haimes, (1970), "Modeling and Identification of Aquifer Systems of High Dimensions," Automatica, 6, p 77.

__________, Ed., (1971), "Optimization Methods for Large-Scale Systems, McGraw-Hill, New York, N.Y.

Yu, Wanyoung, (1972), "Modeling and Optimization for Conjunctive Use of Ground and Surface Water Resources," M.S. Thesis, Sys. Engr. Div., CWRU, Clev., Ohio.

__________, & Y.Y. Haimes, (1972), "Multilevel Optimization for Conjunctive use of Ground and Surface Water," SRC Tech Memo #28, CWRU, Cleveland, Ohio.

MULTILEVEL DYNAMIC PROGRAMMING STRUCTURE FOR REGIONAL WATER RESOURCE MANAGEMENT

YACOV Y. HAIMES
Associate Professor of Engineering
Systems Research Center
Case Western Reserve University
Cleveland, Ohio USA

Abstract: Multiregional planning for the development of water resources systems over a long time horizon is considered. In addition, the problems of optimal timing and sequencing of project development, the optimal allocation and scheduling of land, water and funds for crop growth along with the optimal timing for irrigation are analyzed. A three level hierarchical model is proposed in this paper for solving the above problems.

1.0 INTRODUCTION

The needs and importance of regional water resources planning and management have gained considerable attention in the literature as well as in all levels of government. A regional area can be very complex within its network of sources (streams, rivers, lakes, wells, dams, reservoirs, aqueducts, and others). The fact that the hydrology characterizing water resources systems spans local political boundaries makes river basin or regional water resource management a realistic and essential approach. It also allows for increased efficiency of operation through economies of scale, both in water production and treatment and for increased efficiency through use of technological advances made feasible by pooling of funds for research and development expenditures. Often small localities do not possess the monetary resources available to make the necessary capital expenditures required by water resource management policies, making a regional approach more practical. Furthermore, the jurisdictional powers of a small locality will be inadequate to fully implement its management policies. Small localities cannot be considered economically independent from the surrounding areas. A local area typically does not have the tax support to finance its own projects and therefore must request outside support in the form of bond issues or government subsidies.

Water demand is a function of water supply. The myth that water demand should *always* be met has no justification in view of limited resources. Social as well as economic considerations should be taken into account in the development of local and regional areas. In addition, a local water demand is often met by water transfer from outside. Thus the demand-supply problem is best solved through a regional consideration. Economic factors, of course, affect the demand for water within a region; however, the quality and cost of water resource service within an area may also affect the rates of growth (or decline) of water intensive industries within a local area. The flow of industry and population from one region to others can significantly change water resources needs.

In many parts of the country, ground water constitutes an important element of the water supply system. Over 50% of the U.S. water supply comes from ground water resources. Ground water acts as an enormous regulating reservoir, providing for the natural base flows in rivers. Thus a conjunctive management of groundwater and surface water resources, which is essential, has to cross local boundaries. A river basin often serves as a desired region.

Finally water quality control and management is best accomplished via regional management due to economic, jurisdictional and political factors.

2.0 PROBLEM FORMULATION

Let the projected water demand over a planning time horizon T years be D(t), $0 \leq t \leq T$. The function D(t) will be modified based on the region's water demand and supply balance. The region R is divided into I subregions such that an optimal crop growth policy can be implemented for the ith subregion, i = 1,2,...,I. One may identify a subregion by its soil and consequently may view the region as composed of I different soils. Let K be the number of seasons in the planning time horixon of T years, i.e. the planning horizon T is divided into K seasons of equal lengths. The subscript k will denote the kth season, k = 1,2,...,K. Let J denote the number of different types of crops under consideration in region R during the planning horizon T. The subscript j will denote the jth crop, j = 1,2,...,J. Let Q_k denote the number of units of water available for agricultural consumption at season k, k = 1,2,...,K. If storage facilities are available to store water from one season to another, i.e. through reservoirs or ground water aquifers, then Q_k can be viewed as the hydrological input of water to season k, and water can be stored from one season to another. Let S_k denote the amount of funds available for expenditure at season k. Note that it is assumed that in each subregion i one crop of type j will be grown. If a subregion is too large for single crop growth, it can be further divided into smaller subregions, etc.

Let M be the number of water resource projects available for construction in region R to meet the future water demand for the planning time horizon T. Assume that each such project, m,m = 1,2,...,M has a capacity C_m and its capital construction cost is g_m. Assume the existence of a regional authority whose jurisdictional and economic control will be defined in the following sections.

All system's constraints with respect to crops scheduling, production, funds and water availability, environmental quality, land, and other social political and legal constraints will be grouped and be represented by the vector $\underline{G} \leq 0$.

The regional authority has the following four objectives and goals: (1) Enhancement of the environmental quality of the region, (2) Maximization of net return from crop growth for the whole region over the total planning time horizon, (3) Minimization of total construction cost of water resource projects over the planning time horizon T while still meeting all projected water needs, & (4) Enhancement of regional development. Objective (4) will be maintained at the third (highest) level for different levels of regional development. Objectives (2) and (3) will be maintained at the second level of the hierarchy. Objective (1) will be maintained at a satisfactory level as inequality constraints throughout the hierarchy.

3.0 MODEL FORMULATION

A three level hierarchy is proposed for solving the problem defined in the previous sections. At the third (highest) level the degree of regional development will be determined via a balance coordination using Lagrange multipliers. At the second level a decision as to the optimal sequencing of construction of water supply projects for the whole region R and the entire planning time horizon T will be derived. For this purpose a dynamic programming model based on Butcher, Haimes, and Hall's model (1969) is utilized. This will be termed the 'supply model'. The objective function of the second-level supply model is to meet water needs at minimum cost. At the same level through a 'demand model', water, funds and crop type are allocated for each subregion via dynamic programming. At the first (lowest) level, the amount of water allocated from the second level demand model is optimally distributed over the growing season of each crop. This is accomplished through the use of Hall-Butcher's (1968) dynamic programming model, Hall and Dracup (1970).

The three level hierarchy is coordinated through the use of a system of prices based on Lagrange multipliers and the law of supply and demand, (see Figure 3.1).

3.1 Third Level Hierarchy

The enhancement of regional development is one of the objectives that the United States Water Resources Council (1971) has specified for consideration for any water resource project evaluation. From the summary of the needs and importance of regional planning and management presented in Section 1.0, it is evident that any decision with respect to this objective will be affected by legal, political, social, institutional, economic, financial, as well as technical factors. Therefore, since the purpose of the model and analyses presented in this paper is to aid the decision maker by providing him with a set of satisfactory (optimal) feasible alternatives, it is thus proper to consider objective (4) at the highest level of the model hierarchy. This hierarchical model is aimed at providing the decision maker with a trade-off analysis with respect to different levels of regional development.

The revised projected demand $D(t;\lambda)$, $t \leq T$ is, of course, dependent on the level of regional development specified by the price of water λ, which is determined at the third level, (see Sec. 4.0).

3.2 Second-level Hierarchy

The second level consists of two subsystem models: A demand model and a supply model.

3.2.1 Supply Model

Water supply projects are designed to meet multipurpose uses and needs. Often the construction of a dam, for example, will provide hydroelectric power along with water for industrial, municipal, agricultural, recreational and other needs. While the cost of the project is easily evaluated, the corresponding benefits, whether tangible or intangible, direct or indirect are much more difficult to assess.

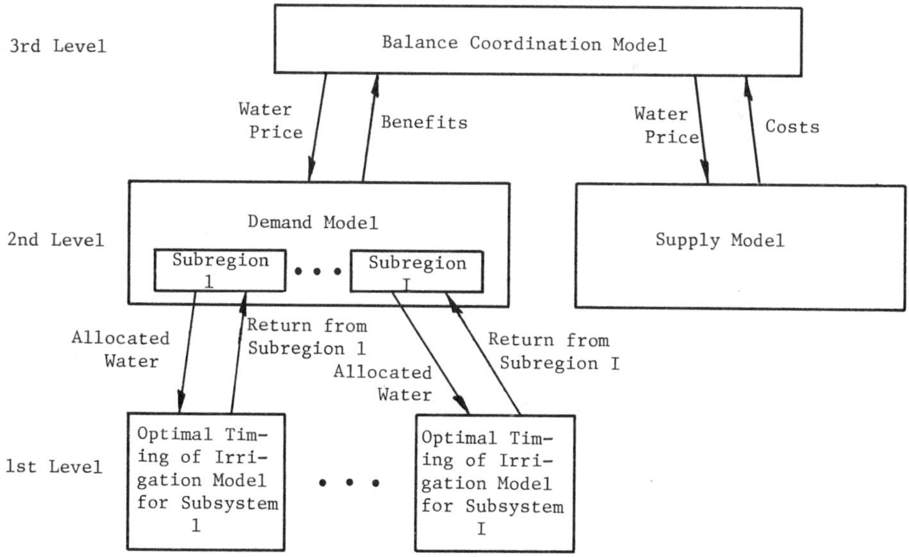

FIGURE 3.1 Multilevel Structure I

In the analysis of a total water resource system one should construct as many "demand models" at the second level as there are uses and corresponding benefits from the water supply projects, Haimes (1972). In the present paper a detailed analysis of the use and benefits of water for irrigation only is considered at the second-level demand model. It is proposed here that a similar analysis be conducted for the other benefits, such that the benefits from the total water use can be determined.

Given a specified projected water demand from the third level, $D(t;\lambda)$, one goal of the supply model of this level is to determine the optimal sequence of construction of water resource projects to meet that water demand $D(t;\lambda)$, $t \leq T$ at minimum cost, Hall, Haimes and Butcher (1972). Assume that M such projects are feasible for construction, each of which is of a capactiy C_m with a capital construction cost g_m, $m = 1,2,\ldots,M$. It is convenient at this stage to define a new variable cost function $g_m(C)$ as follows:

$$\begin{aligned} g_m(C) &= 0 \quad \text{when} \quad C = 0 \\ g_m(C) &= g_m \quad \text{when} \quad C \leq C_m \\ g_m(C) &= \infty \quad \text{when} \quad C > C_m \end{aligned} \tag{3-1}$$

Relations (3-1) provide that the cost of a project is zero if it is not built (serves no capacity requirement), is equal to its construction cost if it is built and serves some capacity within its capability, and is considered to have an infinite cost, i.e. infeasible, if it fails to meet the capacity required.

For the development of the dynamic programming recursive equation, Butcher, Haimes, and Hall (1969), define the function $F_1^{v_1}(C)$ as the minimum cost for meeting water demand C with only one project v_1. Thus,

$$F_1^{v_1}(C) = \min_{m=1,2,\ldots,M} [g_m(C)] \qquad 0 \leq C \leq \sum_{m=1}^{M} C_m \tag{3-2}$$

Define the function $F_2^{v_2}(C)$ as the minimum cost for meeting water demand C with no more than two projects to be constructed at sequence v_2. Based on Bellman's Principle of Optimality, Bellman and Dreyfus (1962), the following recursive relation can be developed:

$$F_2^{v_2}(C) = \min_{m=1,2,\ldots,M,\ \text{but}\ m \notin v_1} [g_m(C_2)(1+r)^{-t_2} + f_1^{v_1}(C-C_2)] \qquad 0 \leq C_2 \leq C \leq \sum_{m=1}^{M} C_m \tag{3-3}$$

where r is the interest rate and t_2 is the time that the second project is built. Relation (3-3) states that if a two-stage project is to be built to satisfy a demand capability C with the second stage to be built at a time $t_2 > 0$ with the intention of providing a portion of the demand capability C_2, $0 \leq C_2 \leq C$, then regardless of the decision that might be made at t_2, the combined system cannot be optimal unless the remaining demand capability $(C-C_2)$ is provided by an optimal one stage project. The optimal stage project for any value of demand capability is given by equation (3-2), hence the value for $(C-C_2)$ is known.

Equation (3-3) appears to have both C_2 and t_2 as decision variables, however since the actual demand must *always* be met at the second level and the function $t = t(C)$ is known, t_2 can be expressed as a function of C and C_2 as follows: $t_2 = t(C-C_2)$. Note that convention regarding interest calculations requires t_2 to be the next larger integral value of t(C). The general recursive relationship can be obtained from (3-3) and by substituting $t_n = t(C-C_n)$:

$$F_n^{v_n}(C) = \min_{\substack{m=1,2,\dots,M,\ \text{but } m \notin v_{n-1} \\ 0 \le C_n \le C \le \sum_{m=1}^{M} C_m}} \left[g_m(C_n)(1+r)^{-t(C-C_n)} + F_{n-1}^{v_{n-1}}(C-C_n)\right]$$

Note that

$$v_n = v_n(C) = m_{min}(C) + v_{n-1}(C).$$

(An extension of the dynamic programming model presented here, which takes into account operation, maintenance and replacement cost, in addition to a variable per unit cost of water, is currently under study.)

In summary, the supply model at the second level will determine the construction sequence of water resource projects with their associated cost. It will be shown later that the level of water demand, which has been determined so far at the third level, can be relaxed and that a feedback mechanism based on the economic law of supply and demand will be activated between the supply and demand models of the second level.

3.2.2 Demand Model

The objective of the demand model of this second level hierarchy is to maximize the overall return from crop yield for each season k, k = 1,2,...,K. This will be achieved by determining the optimal selection of water supply, funds, and crop type to each subregion i of region R, i = 1,2,...,I. For this subsystem model, the three decision variables for any season k and subregion i are: the water supply allocated to the ith subregion, q_i, the funds allocated for the ith subregion, s_i, and the crop type to be grown in the ith subregion ℓ_i. It is reasonable to assume that the above decisions for the kth season are not affected by the decision at the k+1st season. Hence the time domain at this level can be suppressed and the analysis will be repeated for each season.

The allocation problem in the demand model at the second level is solved via dynamic programming. There are two state variables: the quantity of water available for allocation at period k, Q_k, and the amount of funds available for allocation, S_k. The number of stages is equal to the number of subregions I. The three decision variables for each subregion were defined earlier. The subscript k will be eliminated hereafter for simplicity in notation.

Let $g_i(q_i,s_i,\ell_i;\lambda)$ represent the return function for the ith subregion where q units of water, s units of funds have been allocated for growing crop type ℓ_i, and λ is a water cost parameter determined at the third level (see sec. 4.0). For the development of the dynamic programming recursive equation, define the function $f_1^{j_1}(q,s)$ as the maximum return from allocating q units of water and s units of funds to one subregion for growing crop sequence j_1, where $j_1 = \ell_1$. Thus

$$f_1^{j_1}(q,s) = \max_{\substack{0 \le q_1 \le q \le Q \\ 0 \le s_1 \le s \le S \\ \ell_1 \in J}} g_1(q_1,s_1,\ell_1;\lambda) \qquad (3\text{-}5)$$

where J is the set of feasible selection of crops.

Define the function $f_2^{j_2}(q,s)$ as the maximum return from allocating water quantity q and funds s to two subregions, where j_2 is a vector representing the choice of crops for the two subregions.

Thus

$$f_2^{j_2}(q,s) = \max_{\substack{0 \le q_2 \le q \le Q \\ 0 \le s_2 \le s \le S \\ \ell_2 \in J}} [g_2(q_2,s_2,\ell_2;\lambda) + f_1^{j_1}(q-q_2,s-s_2)] \qquad j_2 = j_1 + \ell_2 \tag{3-6}$$

Finally, the general recursive relationship can be obtained from (3-6) for all I subregions:

$$f_I^{j_I}(q,s) = \max_{\substack{0 \le q_I \le q \le Q \\ 0 \le s_I \le s \le S \\ \ell_I \in J}} [g_I(q_I,s_I,\ell_I;\lambda) + f_{I-1}^{j_{I-1}}(q-q_{I-1},s-s_{I-1})] \qquad j_I = j_{I-1} + \ell_I \tag{3-7}$$

Note that the superscript j_I is an I dimensional vector where its hth component represents the crop type selected for the hth subregion.

In summary the demand model at the second level will determine the optimal allocation of water, funds and crops for each subregion at a given season. The coupling between water supply determined from the supply model and water demand determined at the demand level will be discussed in section 4.0.

3.3 First-level Hierarchy

Once the amount of water allocated for crop growth in a season has been determined, the seasonal distribution (in weeks) of that quantity is very important, Hall and Butcher (1968), Windsor (1970) and Smith (1970) studied the problem of optimal timing of irrigation. Hall and Butcher (1968), (p. 267) found that "When soil-moisture conditions are allowed to become less than optimum for plant growth, for any reason, a corresponding reduction in crop yield may be expected. Such reductions may exist locally in a field because of a lack of uniformity of application of the irrigation water. Probably the most important reason for a yiel reduction, from the point of view of water-use planning, is a lack of an adequate water supply for the entire season. Recent research results suggest that the magnitude of losses may depend almost as much on when the soil-moisture deficiency occurs as it does on the total magnitude of the seasonal shortage...".

In order to avoid the redevelopment of Hall, Butcher's (1968) model in this paper, only excerpt of the basic ideas of that model will be presented. As was mentioned earlier, this dynamic programming model for the optimal timing of crop irrigation will be utilized at the first level of the analysis.

The crop yield y, is assumed to be a function of the following form

$$y = \prod_{p=1}^{P} a_p(w_p)\, y_{max} \tag{3-8}$$

where the growing season is divided into P periods, not necessarily equal, w_p is th level of soil moisture at the pth period ($w_p \le w_f$, where w_f is the soil moisture capacity), y_{max} is the maximum yield of a crop which can be obtained under any given conditions of soil and $a_p(w_p)$ is defined by equation (3-8), and it represents the rate of yield reduction due to water defficiency w_p.

Let q_p^i represent the quantities of water to be applied to subregion i (with soil i and crop type ℓ_i) in the time period p. Let A represent the state of the crop at any time as a result of the possible difficiencies before the time period. In concept, A is the product of the coefficients from the beginning of the growing season up to the particular period in question.

The dynamic programming problem has two state variables, the state of the crop and the water available for irrigation. A general recursive equation is of the following form:

$$f_p^i(q^i,A^i) = \max_{\substack{0 \le q_p^i \le q^i \\ 0 \le A_p^i \le A^i}} [g_p^i(q_p^i,A_p^i) \cdot f_{p-1}^i(q^i-q_p^i,A^i-A_p^i)-C_p^i(q_p^i)]$$

where $f_p^i(q^i,A^i)$ is the maximum return at period p for subregion i with a state of crop A^i and water quantity q^i. The function $g_p^i(q_p^i,A_p^i)$ represents the return for subregion i at period p from allocating q_p^i water, where A_p^i represents the state of the crop at period p to subregion i. $C_p^i(q_p^i)$ is the irrigation cost (in addition to the water supply costs in the allocation model) to supply q_p^i units of irrigation water to subregion i at time period p (see Figure 3.1).

In summary, the above dynamic programming procedure is applied to each subregion i at the first level of the hierarchy, yielding the optimal timing of irrigation for a given quantity of water for the entire season.

4.0 COORDINATION OF THE SUBMODELS

In the preceding sections, the goals of the subsystems at each of the three levels in the hierarchy were introduced, along with a computational procedure for achieving those goals at an optimum performance level. Clearly an overall system coordination for the three-level hierarchy presented in this paper is essential and thus deserves a careful analysis, Haimes (1972), Nainis and Haimes (1972), Mesarovic et al (1970), Lasdon (1970), and Wismer (1971).

An optimal solution to the total system may be achieved through an iterative procedure whereby a new set of decisions at any level is determined only after convergence has been achieved at the lower levels. The balance coordination method, Mesarovic (1969), through the use of Lagrange multipliers will be utilized here for achieving an overall system optimum. "The principle of interaction balance states that the overall problem is solved when the iteractions which each of the first level units require (in order to optimize local performance) are in agreement with the interactions which actually occur after the respective controls are applied," Mesarovic (1969 , p 110).

The total cost, $F_n^{v_n}(Q)$, of supplying water to meet a demand $D(t;\lambda)$, $0 \le t \le T$, up to level of demand Q over a planning time horizon T is obtained by the second level supply model (see Figure 4.1). The total net benefits from crops grown in region R for one season k, with funds S_k and water Q_k is given by the recursive equation (3-7) to be

$$f_{I,k}^{j_I}(Q_k,S_k)$$

note that

$$f_{I,k}^{j_I}(Q_k,S_k) = \text{optimal } \{f_I^{j_I}(q,s)\}_k .$$

Thus the total net benefit B(Q,S) for the entire planning horizon T is:

$$B(Q,S) = \sum_{k=1}^{K} f_{I,k}^{j_I}(Q_k,S_k)(1+r)^{-k} \qquad (4-1)$$

where r is the interest rate and

$$Q = \sum_{k=1}^{K} Q_k, \quad S = \sum_{k=1}^{K} S_k$$

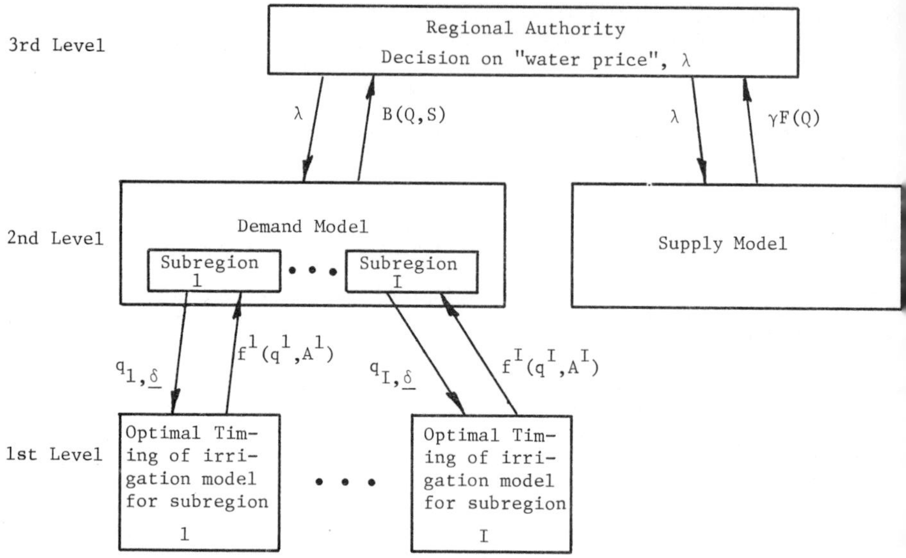

FIGURE 4.1 Multilevel Structure II

It is assumed here that values of $B(Q,S)$ for all other water uses can be determined in similar analysis at the second level whenever applicable. In this paper, water for irrigation will be the only purpose considered for the water supply projects. An extension of a multipurpose project can be added to the present model. Clearly, the total system benefits, $B(Q,S)$ determined at the second level demand model, and the total system cost, $F(Q)$, determined at the second level supply model have to be coordinated and balanced.

The goal of the third level is to adjust water demand and supply to maximize overall system benefits, i.e. to enhance regional development. This may be accomplished by considering the following overall third level Lagrangian:

$$L = \Phi(Q,S,R,E) + \underline{\mu}^T\underline{G}(Q,S,R,E) + \lambda[B(Q,S) - \gamma F(Q)] \tag{4-2}$$

where $\Phi(Q,S,R,E)$ reflects the third level's objectives, $Q,S,R,$ and E represents water supply, funds, land and environmental quality respectively. The vector $G(Q,S,R,E)$ represents all the system constraints, and $\underline{\mu}$ and λ are Lagrange multipliers, where $\underline{\mu}$ is a vector and λ and γ are scalars, (see Figure 4.1).

The necessary condition for stationarity is $\frac{\partial L}{\partial \lambda} = 0$. Thus the goal of this level is to minimize the error between the supply model adjusted cost, $\gamma F(Q)$, and the demand model $B(Q,S)$. The Lagrange multiplier, λ, can be viewed as a price which coordinates water supply and demand for the optimal use of regional resources

The parameter factor γ can be set = 1 if it is desirable to find the largest economically efficient level of development in an agricultural area. γ can be set > 1 if there exists a pre-determined cut-off benefit-cost ratio which should be satisfied in area of water resource development.

At the second level, the supply model will determine the minimum cost for meeting a given water demand, $D(t;\lambda)$ where λ is determined by the third level. The demand model will determine the maximum benefit from using a given water supply Q, along with utilizing funds and land for crop growth. The third level balances the above supply and demand for improving the overall system performance by proper choice of λ. Thus λ will directly affect water demand by raising the price of water if $\gamma F(Q) > B(Q,S)$ and lowering that price otherwise.

Similarly, the coordination between the demand model at the second level and the optimal timing model at the first level can be achieved via the balance coordination principle. At the first level the quantity of water allocated for the ith subregion, q_i, is fixed (it is determined by the second level demand model). Thus the first level can be viewed as a "demand" model in the sense that having more water allocated to subregion i for growing crop ℓ_i may improve yield. The second level demand model, on the other hand, can be viewed as "supply" model for the first level. Accordingly a similar Lagrange multiplier scheme with Lagrange multipliers $\underline{\delta}$ can be constructed to adjust water "supply" and "demand" between the first and second levels.

5.0 SUMMARY

The hierarchical multilevel structure has been shown to provide a tractible approach for the modeling and management of water resource systems. In particular, via proper model coordination several dynamic programming submodels have been utilized for the optimal timing and sequencing of project development, the optimal allocation and scheduling of land, water and funds for crop growth, along with the optimal timing of irrigation. It is hoped that this paper will encourage further study in this complex problem.

References

Bellman, R.E. and S.E. Dreyfus, (1962), Applied Dynamic Programming, Princeton Univ. Press, Princeton, N.J.

Butcher, W.S., Y.Y. Haimes, & W.A. Hall, (1969), "Dynamic Programming for the Optimal Sequencing of Water Supply Projects," Water Resources Res., 5, p 1196.

Haimes, Y.Y., (1972), "Hierarchical Modeling of a Total Water Resource System," to be presented at the 8th Am. Water Resource Conf., St. Louis, Mo., Oct. 30.

Hall, W.A. and W.S. Butcher, (1968), "Optimal Timing of Irrigation," J. Irrigation Drainage Div., Am. Soc. Civil Engrs., 9y, p 267.

_________ and J.A. Dracup, (1970), Water Resources Systems Engineering, McGraw Hill, New York, N.Y.

_________ , Y.Y. Haimes, & W.S. Butcher, (1972), "Analysis of the Feasibility of Interim Water Supplies," Water Resources Res., 8, p 319.

Lasdon, L.S., (1970), Optimization Theory for Large Systems, Macmillan Co., New York, N.Y.

Mesarovic, M.D., D. Macko, and Y. Takahara, (1970), Theory of Hierarchical Multilevel Systems, Academic Press, New York, N.Y.

Mesarovic, M.D., (1969), "Mathematical Theory of the General Systems and some Economic Problems," Mathematical Systems Theory and Economics I, Proc. of an International Summer School, Varenna, Italy, 1967, Kuhn & Szegö, Ed., Springer-Verlag, New York, N.Y.

Nainis, S. and Y.Y. Haimes, (1972), "A Multilevel Approach to Planning for Capacity Expansion in Water Resource Systems." Accepted for presentation at the International Conf. on Cybernetics and Soc., Oct. 9-12, Washington, D.C.

Smith, G., (1970), "Optimal Timing of Irrigation," unpublished paper, Civil Engr. Dept., Univ. of Illinois, Urbana.

Windsor, J.S., (1970), "Mathematical Model of a Farm Irrigation System," Ph.D. Dissertation, Univ. of Illinois, Urbana, Champaign.

Water Resources Council, (1971), "Proposed Principle and Standards for Planning Water and Related Land Resources," Federal Register, 36, No. 245, Dec., p 2414

Wismer, D.A., Ed., (1971), Optimization Method for Large Scale Systems, McGraw-Hill New York, N.Y.

Acknowledgments: The author wishes to express his thanks to F. Gembicki, S. Nainis and S. Rosen for reviewing this paper.

Decomposition Methods for Solving Steady State Process Design Problems

Arthur W. Westerberg
Department of Chemical Engineering
University of Florida
Gainesville, Florida 32601, U.S.A.

July 1972

Abstract

Steady state design for engineering systems requires decomposition strategies useful for optimizing system structure as well as for optimizing parameters for a fixed structure. This presentation discusses briefly the uses of heuristic and algorithmic methods of decomposition for this problem. The principal topics based on algorithmic methods include Lagrangian methods for decomposing, dual bounding and sensitivity studies; the projection method; and the policy of "restriction" which relies heavily on effective equation solving techniques while optimizing. The computer system, GENDER, under development by the author is discussed. GENDER is intended to be a system for deriving, modifying, and executing solution procedures for large sets of algebraic equations.

1. Introduction

This paper is largely tutorial and concerns itself with decomposition methods useful for solving steady state (nontime varying) process design problems. The nature of the process design problem considered will be made more evident in the paper, but the term decomposition as interpreted here requires some definition. We shall consider that a problem has been decomposed if its special structure and numbers have been accounted for in the development of its solution. Two commonly used decompositions are those in which the optimization has been decomposed and those in which the calculations in the analysis of a process system are decomposed. The former includes such methods as Linear Programming decomposition, Dynamic Programming, the two level method of Lasdon, and the use of projection. The latter emphasizes the use of tearing schemes or sparse matrix methods in conjunction with the Newton-Raphson procedure to develop solution procedures for solving sets of sparse equations.

The first topic covered in the paper is the synthesis or generation of a process flowsheet. The magnitude of this problem requires that heuristic methods will be required if any solution is to be found. Heuristic methods are rather like using rules of thumb, and they are developed to produce reasonable but usually not optimal solutions to a problem. The rules used in a heuristic method are specific to the problem, and they take advantage of the usual structure and numbers found in the type problem at hand. Major effects are usually accounted for first, then secondary effects, and so on. In this sense the problem is decomposed.

The second class of decompositions are algorithmic; that is, a proof can be developed that they can lead to an optimum in special cases. The ones considered are those appropriate to optimizing a given flowsheet by adjusting the decision variable values for the problem. Lagrangian based algorithms are considered first. The development shows that a Lagrange formulation to the problem has considerable structure, just like the original problem, and solution procedures suggested by this structure are presented. The two level or pricing algorithm of Lasdon is also discussed. The dual bounding theorem, on which this algorithm is based, follows, and its use to aid in the synthesis of flowsheets is presented briefly.

We next consider handling the original problem directly, looking at the commonly used strategy of projection to decompose the optimization problem. This method indicates the usefulness of information reversal in solving design problems. Having suggested information reversal, we consider an approach for solving inequality constrained optimization problems. In this approach the optimization problem can be automatically altered to include or drop constraints as the solution procedure proceeds, almost the same automatic capabilities as required for information reversal.

Finally we present a discussion of the GENDER system, a set of routines to aid in design analysis and optimization. These routines are being developed at Florida and are intended to permit the automatic capabilities described above for information reversal.

2. The Process Synthesis Problem--The Need for Heuristic Algorithms

Automatic process synthesis is a newly evolving area in computer aided design. Nine papers were presented on this topic at a recent symposium held at the 71st National Meeting of the AIChE in Dallas, Texas, this last February. They represent about one-third of the papers in the area in Chemical Engineering. A definition of synthesis might be as follows:

Using subsystems of known technology, construct a system of these subsystems which is capable of transforming given inputs into desired outputs. Ideally one wants to generate the best system possible.

We might consider a simple example by looking at Figure 1. The technology available is a distillation process "a" which can separate a multicomponent stream into two product streams, each with a subset of the components contained in it. Figure 1 illustrates how one might generate 5 different flowsheets to split a four component stream into four pure component streams using the single technology available, process "a". The possible number of flowsheets for this type of system is (Thompson (1972), Hendry (1972))

$$\text{Number of flowsheets} = \frac{(2(N-1))!}{N!(N-1)!} \cdot S^{N-1}$$

where N = Number of chemical components and
S = Number of separation methods (a, b, etc.)

If N=8 and S=5 (a modest engineering problem), the number of flowsheets is 33,575,625.

Similarly for heat exchanger networks, where one wishes to use the heat from a set of hot streams by using them to heat some cold streams, the combinatorial problem is unwieldly. For two hot and two cold streams, 47 simple flowsheets can be drawn.

To reduce the problem size one must use the special structure and the numbers. For example suppose in the separation problem one desired product is the mixture of A and B, not pure A and pure B. Adopting the heuristic, or intuitive rule, that first separating A from B and recombining will not generate the optimal flowsheet, we find only flowsheets 3 and 5, less the A-B separation step, are left as candidates. Or suppose A is a minor component and we conclude minor components should be separated last, then the same two flowsheets are again our only candidates. If all options can be enumerated after being reduced to a reasonable number by heuristic rules, the problem can probably be formulated as a nonlinear, integer programming problem. Certainly no scheme is likely which will guarantee the optimum structure. Of course, we might, in the face of this, resign ourselves to letting the experienced engineer choose the flowsheets to be investigated.

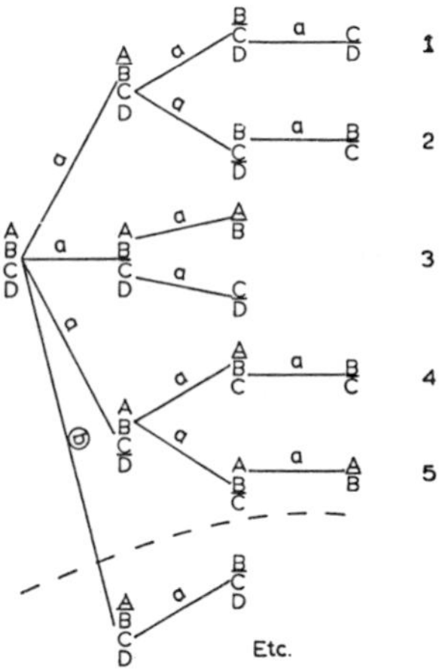

Figure 1. Generation of Separation Process Flowsheets

One approach used by Siirola (1971) for generating flowsheets automatically is to use ends-means concepts together with heuristics to pick the better ones. To generate a flowsheet, the raw materials and possible product streams are given. For each, the engineer specifies a set of required properties such as components present, total flow, composition, pressure, temperature, and phase. The different source and destination streams are tentatively matched and crude estimates are made of technical feasibility and the costs to remove differences in their given required properties. If reasonable the match is kept. To make the problem tractable, the order in which property differences are reduced is prespecified. First one reduces component differences by proposing chemical reactions. Once one (or more) reaction step is accepted, the reactor outputs and inputs are candidates for stream matching. More matches are proposed. Composition differences are reduced next using separators or mixers. The procedure continues by reducing flow differences, temperature differences and so forth. At each step when a match is proposed, its feasibility is checked, and, of those matches possible, the better ones are selected using heuristic rules to estimate costs for reducing the remaining differences. The final step consolidates tasks and chooses equipment types.

One should recognize the analogy here with programming a computer to play chess, only here the rules are more difficult. Experience with chess indicates only moderately good games can be played automatically. The use as an aid is not to be underplayed however.

A second approach to synthesis is to improve an existing flowsheet by systematically proposing modifications and investigating the economic consequences of each. If a modification represents an improvement, it is made, generating a new flowsheet to improve. Such an approach is rather like hill climbing in structure "space". We shall investigate one strategy for this approach using dual bounding techniques in the next section.

3. Algorithmic Methods for Decomposition

We shall next consider decomposition strategies for optimizing decision variable values for a fixed flowsheet. Two general approaches are presented: an approach based on Lagrange Multipliers and one based on information reversal methods. The intent is to present different strategies which can be evolved for both approaches, indicating the advantages and limitations of each as seen by the author.

Consider the simple flowsheet in Figure 2 and the following problem definition associated with it. (Nomenclature is given at the end of the paper.*)

$$\text{Min}\left\{\phi \middle| \phi = \sum_{i=1}^{n} \phi_i(x_{ik},\ k\varepsilon I(i);\ u_i)\right\} \tag{1}$$

subject to the following constraints.

Stage transformation and inequality constraints

$$y_{ij} = f_{ij}(x_{ik},\ k\varepsilon I(i);\ u_i) \tag{2}$$

$$g_i(x_{ik},\ k\varepsilon I(i);\ u_i) \le 0 \tag{3}$$

Connection constraints of the form

$$x_{\ell m} = y_{ij} \tag{4}$$

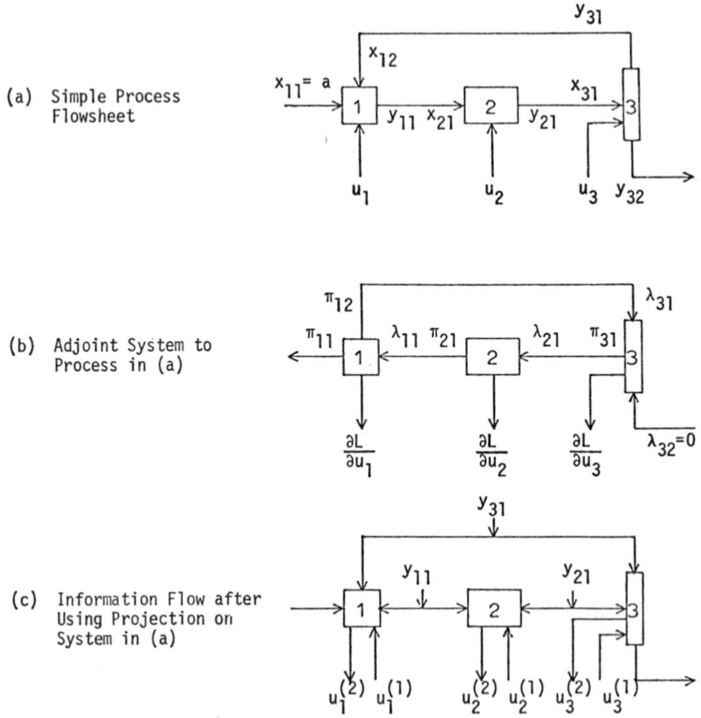

Figure 2. Example Process

---

*Often input and output subscripts (j and k above) and occasionally unit subscripts (i above) are dropped later in the paper. When dropped it should be understood we are referring to variables for all inputs or outputs or units.

For this particular problem, the equations are as follows.

$$\phi = \phi_1(x_{11}, x_{12}, u_1) + \phi_2(x_{21}, u_2) + \phi_3(x_{31}, u_3)$$

$$y_{11} = f_{11}(x_{11}, x_{12}, u_1) \tag{2a}$$

$$y_{21} = f_{21}(x_{21}, u_2) \tag{2b}$$

$$y_{31} = f_{31}(x_{31}, u_3) \tag{2c}$$

$$y_{32} = f_{32}(x_{31}, u_3) \tag{2d}$$

$$g_1(x_{11}, x_{12}, u_1) \leq 0 \tag{3a}$$

$$g_2(x_{21}, u_2) \leq 0 \tag{3b}$$

$$g_3(x_{31}, u_3) \leq 0 \tag{3c}$$

and

$$x_{11} = a \quad \text{(given)} \tag{4a}$$

$$x_{12} = y_{31} \tag{4b}$$

$$x_{21} = y_{11} \tag{4c}$$

$$x_{31} = y_{21} \tag{4d}$$

Equations (1) through (4) define what we shall refer to as our "original" problem throughout the rest of the paper. We shall consider several strategies possible for solving this problem.

3.1. Lagrangian Based Strategies: First we shall discuss strategies for solving the above problem based on Lagrangian methods. The Lagrange function for this problem is

$$\begin{aligned} L(x,y,u,\lambda,\mu,\pi) = \; & \phi_1 + \phi_2 + \phi_3 - \lambda_{11}^T(y_{11} - f_{11}) \\ & - \lambda_{21}^T(y_{21} - f_{21}) - \lambda_{31}^T(y_{31} - f_{31}) - \lambda_{32}^T(y_{32} - f_{32}) \\ & - \mu_1^T g_1 - \mu_2^T g_2 - \mu_3^T g_3 - \pi_{11}^T(x_{11} - a) - \pi_{12}^T(x_{12} - y_{31}) \\ & - \pi_{21}^T(x_{21} - y_{11}) - \pi_{31}^T(x_{31} - y_{21}) \; . \end{aligned} \tag{5}$$

We expand L in a Taylor Series to first order terms only to get

$$\Delta L = \left(\frac{\partial \phi_1}{\partial x_{11}^T} + \lambda_{11}^T \frac{\partial f_{11}}{\partial x_{11}^T} - \mu_1^T \frac{\partial g_1}{\partial x_{11}^T} - \pi_{11}^T\right) \Delta x_{11}$$

+ terms in the other Δx and in the Δy, Δu, $\Delta \lambda$, $\Delta \mu$, and $\Delta \pi$ variables.

For L to be stationary with respect to all these variables, it is necessary that the coefficients of each of the variations be zero. The resulting equations define the adjoint system defined by equations (6), (7) and (8).

Stage Transformation Equations

$$\pi_{11} = \frac{\partial \phi_1}{\partial x_{11}} + \frac{\partial f_{11}^T}{\partial x_{11}} \lambda_{11} - \frac{\partial g_1^T}{\partial x_{11}} \mu_1 \tag{6a}$$

$$\pi_{12} = \frac{\partial \phi_1}{\partial x_{12}} + \frac{\partial f_{11}^T}{\partial x_{12}} \lambda_{11} - \frac{\partial g_1^T}{\partial x_{12}} \mu_1 \tag{6b}$$

$$\pi_{21} = \frac{\partial \phi_2}{\partial x_{21}} + \frac{\partial f_{21}^T}{\partial x_{21}} \lambda_{21} - \frac{\partial g_2^T}{\partial x_{21}} \mu_2 \tag{6c}$$

$$\pi_{31} = \frac{\partial \phi_3}{\partial x_{31}} + \frac{\partial f_{31}^T}{\partial x_{31}} \lambda_{31} + \frac{\partial f_{32}^T}{\partial x_{31}} \lambda_{32} - \frac{\partial g_3^T}{\partial x_{31}} \mu_3 \tag{6d}$$

Connection Equations

$$\lambda_{11} = \pi_{21} \tag{7a}$$

$$\lambda_{21} = \pi_{31} \tag{7b}$$

$$\lambda_{31} = \pi_{12} \tag{7c}$$

$$\lambda_{32} = 0 \tag{7d}$$

In addition the coefficients for variations in u_i are:

$$\frac{\partial \phi_1}{\partial u_1} - \frac{\partial g_1^T}{\partial u_1} \mu_1 + \frac{\partial f_{11}^T}{\partial u_1} \lambda_{11} = 0 \tag{8a}$$

$$\frac{\partial \phi_2}{\partial u_2} - \frac{\partial g_2^T}{\partial u_2} \mu_2 + \frac{\partial f_{21}^T}{\partial u_2} \lambda_{21} = 0 \tag{8b}$$

$$\frac{\partial \phi_3}{\partial u_3} - \frac{\partial g_3^T}{\partial u_3} \mu_3 + \frac{\partial f_{31}^T}{\partial u_3} \lambda_{31} + \frac{\partial f_{32}^T}{\partial u_3} \lambda_{32} = 0 \tag{8c}$$

The set of equations obtained from the coefficients of $\Delta\lambda$ and $\Delta\pi$ are equations (2) and (4) again. If ϕ is to be a minimum, the Kuhn-Tucker conditions for inequalities (3) are also necessary:

$$\mu_i g_i = 0 \quad \text{and} \quad \mu_i \le 0 \quad \text{for} \quad i=1,2,3 . \tag{9}$$

Our minimization problem is now converted into one of solving the set of equations (2), (4), (6), (7) and (8) while satisfying inequalities (3) and the Kuhn-Tucker conditions (9). Several solution algorithms suggest themselves. The following one is commonly stated.

Forward/Backward Pass Algorithm:

1. Guess values for all decision variables u_i and all Kuhn-Tucker multipliers $\mu_i \leq 0$.

2. Solve equations (2) and (4) for the resulting values of the x and y variables. A possible solution sequence, based on "tearing", might be:

   A. Guess the recycle stream x_{12}.

   B. Using equations (4a) and (2a), calculate y_{11}.

   C. Then, in order, find x_{21}, y_{21}, x_{31}, and y_{31}.

   D. Compare x_{12} guessed with y_{31}. If not equal, adjust x_{12} and iterate steps A through D until they are equal.

   E. Evaluate y_{32}.

3. Using these values of x, y, and u variables, evaluate the Jacobian matrices needed for equations (6) and (8). (These can be evaluated only as needed in the next two steps.)

4. Solve the adjoint equations (6), (7) and (8). These solve most naturally in the reverse direction through the network. A possible solution algorithm based on tearing might be:

   A. Guess recycle variables λ_{31}.

   B. Use values for λ_{32} given by (7d) and guessed values for λ_{31} to evaluate π_{31} from equation (6d).

   C. Evaluate in order λ_{21}, π_{21}, λ_{11}, π_{12}.

   D. Compare λ_{31} guessed with π_{12}. If not equal, adjust λ_{31} and iterate steps A through D until they are equal.

   E. Evaluate π_{11}.

   Figure 2b indicates the direction of information flow in the above solution algorithm. It is clearly in the reverse direction to information flow in the original system. (We might note here that the adjoint system is defined by a set of linear equations if x, y, and u are fixed as is the case here.)

5. Solve equations (8) equal to zero for calculated u_i values. Adjust u_i using the guessed and calculated values of u_i.

6. Also evaluate g_i, the inequality constraints, and adjust the μ_i values to give

$$\mu_i = \begin{cases} 0 & \text{if } \mu_i - \varepsilon g_i > 0 \\ \mu_i - \varepsilon g_i & \text{Otherwise} \end{cases}$$

   where $\varepsilon > 0$ is some suitable number.

7. Iterate steps 2 through 6 until equations (2), (3), (4), (6), (7), (8) and (9) are satisfied.

Sensitivity Algorithm: If one changes the goal from finding the stationary point of L to minimizing L, the problem is solved in a quite similar manner. The only change exists in step (5). Rather than attempting to solve (8) for u_i, one can recognize that the left hand side of equations (8a), (8b) and (8c) are in fact the gradients of L with respect to u_1, u_2 and u_3 respectively. A gradient based optimization algorithm, such as the Fletcher-Powell (1963) algorithm can be used to minimize L, using an evaluation of the right hand sides of (8) to provide the needed gradients. It can be shown equations (8a), (8b), and (8c) are also the constrained gradients of ϕ with respect to the u_i variables (Jackson (1964), Westerberg and Director (1972)).

Discrete Minimum Principle: We define the Hamiltonian for each stage as (Brosilow (1965))

$$h_1 = \phi_1 + \lambda_{11}^T f_{11} - \mu_1^T g_1 ,$$

$$h_2 = \phi_2 + \lambda_{21}^T f_{21} - \mu_2^T g_2 , \tag{10}$$

and

$$h_3 = \phi_3 + \lambda_{31}^T f_{31} + \lambda_{32}^T f_{32} - \mu_3^T g_3 .$$

Substituting the connection equations (7) into (6) to eliminate the π variables gives the usual transition equations for λ:

$$\lambda_{31} = \frac{\partial h_1}{\partial x_{12}} , \qquad \lambda_{11} = \frac{\partial h_2}{\partial x_{21}} , \qquad \lambda_{21} = \frac{\partial h_3}{\partial x_{31}} , \qquad \lambda_{32} = 0 .$$

Also we can write, substituting the connection equations (4) into (2) to eliminate y:

$$x_{11} = a \quad , \qquad x_{12} = \frac{\partial h_3}{\partial \lambda_{31}} , \qquad x_{21} = \frac{\partial h_1}{\partial \lambda_{11}} , \qquad x_{31} = \frac{\partial h_2}{\partial \lambda_{21}} .$$

The method of choosing u_i is then

$$\min_{u_i} h_i$$

The above is the formulation typically encountered when using the discrete minimum principle.

The above, quite similar, approaches to solving our original optimization problem have a potentially serious disadvantage associated with them. The algorithms all require Jacobian matrices which, for a chemical process model, may be exceedingly difficult and/or time consuming to evaluate. Consider, for example, obtaining the Jacobian matrices

$$\left(\frac{\partial f_{31}}{\partial x_{31}^T}\right)_{x,u} , \qquad \left(\frac{\partial f_{32}}{\partial x_{32}^T}\right)_{x,u}$$

for the third unit, a separation unit. If the functions f_{31} and f_{32} are in the form of a large, complex subroutine in which several hundred algebraic equations are solved before inputs x_{31} and u_3 are converted to y_{31} and y_{32}, only numerical methods may prove feasible to evaluate these Jacobians. If x_{31} and x_{32} are n-dimensional vectors, n+1 evaluations of the subroutine will be needed to evaluate these matrices, a nontrivial task.

Up to now we have said little about using Newton-Raphson based iteration methods for solving the nonlinear equations defining our original process. For example several articles (Goldstein (1970), Naphtali (1971)) have appeared recently showing how the distillation process can be solved using the Newton-Raphson algorithm. In principle, the equations for the entire forward pass can be solved in this manner. Each iteration step for the Newton-Raphson procedure involves solving a set of sparse linear equations, and sparse matrix methods can be used to solve them efficiently. A bonus for solving the equations using this approach is that the required Jacobian matrices for the adjoint system can be readily produced from the Jacobians used during the final Newton-Raphson iteration for the forward pass (Director (1972)).

The principal advantage to the above Lagrangian based strategies is that one can obtain constrained gradients to the objective function and thus use gradient methods to adjust the u_i. For many simple problems on which these algorithms have been applied, generally excellent results occur (Lee (1969), Fan (1964)).

3.2. <u>Two Level Algorithm</u>: The two level algorithm (Brosilow (1965), Lasdon (1970)) is another approach based on a Lagrangian formulation. We rewrite our problem slightly by requiring the stage transformation constraints to hold, thus eliminating the y variables. In addition we require the inequality constraints be satisfied at each step in the search for an optimum. The Lagrange function can then be decomposed:

$$\begin{aligned} L(x,u,\pi) &= \phi_1 + \phi_2 + \phi_3 - \pi_{11}^T(x_{11} - a) \\ &\quad - \pi_{12}^T(x_{12} - f_{31}) - \pi_{21}^T(x_{21} - f_{11}) - \pi_{31}^T(x_{31} - f_{21}) \\ &= \{\phi_1 - \pi_{11}^T x_{11} - \pi_{12}^T x_{12} + \pi_{21}^T f_{11}\} \\ &\quad + \{\phi_2 - \pi_{21}^T x_{21} + \pi_{31}^T f_{21}\} \\ &\quad + \{\phi_3 - \pi_{31}^T x_{31} + \pi_{12}^T f_{31}\} + \pi_{11}^T a \\ &= \sum_{i=1}^{3} \ell_i(x_i,u_i,\pi) + \pi_{11}^T a \ . \end{aligned}$$

We define (note we dropped input subscripts on the x_i--see earlier footnote)

$$\begin{aligned} h(\pi) &= \min_{x,u} \{L(x,u,\pi) \,|\, g_i(x_i,u_i) \le 0 \ , \qquad i=1,2,3\} \\ &= \left[\sum_i \min_{x_i,u_i \varepsilon S_i} \ell_i(x_i,u_i,\pi)\right] + \pi_{11}^T a \end{aligned}$$

where regions S_i are those defined by $g_i(x_i,u_i) \le 0$. We then write the theorem (see Lasdon (1970) for a proof):

Theorem 1: $h(\pi) \le \phi^o = \text{Min}\,\{\phi(x,u)\,|\,g(x,u) \le 0\,,\ \text{Connection Constraints (4) hold}\}$

If equality holds, the x^* and u^* variable values found in evaluating $h^*(\pi^*)$ solve our original problem.

The two level algorithm suggested is

$$\underset{\pi}{\text{Max}}\ \{h(\pi) = \left[\sum_i \underset{x_i,u_i}{\text{Min}}\ \ell_i(x_i,u_i,\pi)\right] + \pi_{11}^T a\} \tag{11}$$

where the inner minimization requires us to solve three subproblem minimizations, one for each unit.

The two level algorithm has an apparent advantage. To use it, one no longer has to obtain the π variables by solving the adjoint system. Rather, one simply guesses them and adjusts them to maximize $h(\pi)$ as the solution proceeds.

Unfortunately the two level algorithm has some serious drawbacks also. Equality does not always hold in Theorem 1; solving our problem by the two level method may not lead to a solution of our original problem. If it does not solve the original problem, often it does not even provide an approximate answer except to the value of the objective function. A sufficient but not necessary condition for the two level procedure to lead to a solution of the initial problem is that the original problem be a convex problem. Unfortunately most chemical engineering problems are not convex. Also numerical experience with the procedure seems to indicate it is slower than solving the original problem directly (for example Jung (1972)).

A Strategy for Screening Flowsheet Modifications: We (McGalliard (1971) and (1972)) have proposed and are currently testing the usefulness of the "dual" bounding aspect of Theorem 1 in the area of flowsheet synthesis. Because $h(\pi)$ always provides a lower bound to the original problem, it can be used as a screening criterion for testing flowsheet modifications. Figure 3 illustrates the basic idea.

Suppose we have a proposed flowsheet, flowsheet 1, for which we have obtained a feasible primal solution with return ϕ_1. Also suppose we obtain a dual bound h_1. We now propose a modification to the flowsheet and wish to know if it represents an improvement. The strategy is first to calculate a dual bound, h_2, for the modified system. Suppose h_2 has the value shown for case A. Then clearly the modified flowsheet cannot give a lower, and thus better, return. If however h_2 is below h_1 as in case B, we can calculate a primal bound, ϕ_2. If ϕ_2 is less than h_1, the new flowsheet is guaranteed to be better. Case C indicates how we might have to refine our bounds if the initial intervals overlap. The first time h_{21} and ϕ_{21} are found but the interval they define overlaps that for our original flowsheet. After adjusting them, the values h_{22} and ϕ_{22} are found and indicate the modified flowsheet is better.

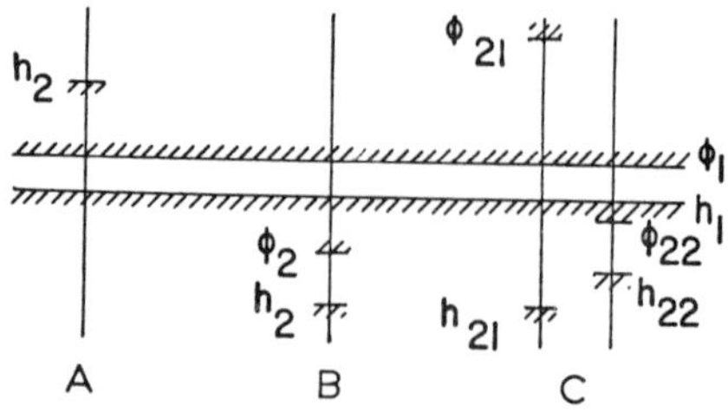

Figure 3. Use of Dual Bounding to Screen Flowsheet Modifications

One computational aspect of this procedure is the ease with which the first dual bound can be obtained for a modification; usually only the directly affected units have to be optimized. Thus it behooves one to propose several modifications at a time and screen to choose the best. Once the best is incorporated, the modified flowsheet becomes the new basic flowsheet.

Another computational aspect is that the flowsheets do not have to be fully optimized, either to get a primal or a dual bound. Thus one might hope the bounding intervals will quickly tighten, but the slow final steps to the actual optimum do not have to be performed except on the final flowsheet.

A difficulty not to be overlooked is the solving of the subproblems

$$\min_{x_i, u_i \varepsilon S_i} \ell_i(x_i, u_i, \pi)$$

Often these subproblems are unbounded unless S_i is carefully constructed and often they are nonconvex problems. Usually the unboundedness is discovered quickly and cured by adding a sufficient set of new constraints overlooked originally, but the nonconvexity, particularly if the problem is not unimodal, means the subproblems are just plain difficult to solve.

3.3. Information Reversal Strategy: We shall now consider an entirely different approach to solving our original problem; in some respects it will seem the more natural to the engineer solving the problem.

The strategy will use information reversal to decompose our optimization problem, and it corresponds to the technique labeled "projection" (Geoffrion (1970)). Consider the original problem again. Suppose we rewrite the solution routines for each of the units so all interconnecting stream data becomes input. A subset of the original decision variables in each unit will then have to become calculated or dependent variables. Equations (2) would have the new form

$$u_i^{(2)} = f_i'(x_i, y_i, u_i^{(1)}) , \qquad (12)$$

where the number of elements of $u_i^{(2)}$ is equal to the number of elements of y_i. Any additional u_i variables remain as decisions, $u_i^{(1)}$. (For the moment, we shall overlook the possibility that this reversal of information may not be possible as described.)

Figure 2c indicates the resulting information flow for our flowsheet. It is obvious that the y_i variables have become the decisions, and the optimization can be decomposed to read

$$\min_{y_i} \left\{ \sum_i \phi_i^*(y_i) \,\middle|\, \phi_i^*(y_i) = \min_{u_i^{(2)}} \phi_i(x_i, u_i) \text{ , equations (4) and (12) satisfied} \right\}$$

Note that the subproblems of finding the $\phi_i^*(y_i)$ are over each unit only.

If the inequalities (3) written for the system are explicit in only the original decision variables (u_i), this decomposition causes a problem not present before the decomposition in locating the feasible region for the y_i variables (Geoffrion (1970)). However for general constraints (3), written in terms of all the variables x_i, y_i, and u_i, the problem of handling them for the problem decomposed as above is no more difficult than for the original formulation in principle.

In practice for the typical unit subroutine, the required rewriting to reverse the information flow on the y_i and $u_i^{(2)}$ variables is not possible. For example, a distillation column has insufficient degrees of freedom to allow both its input and its output streams to be independently specified. However, one can and usually does break a large process flowsheet into a few subsystems of several units each, and for each subsystem reverses as needed the information flow on the interconnecting streams. Each subsystem is then isolated and can be analyzed and possibly optimized as just described. (For this reason, this approach was stated as being a more natural approach for the engineer.)

Often the information reversal cannot in theory be made completely for a stream and thus the subsystems must remain interconnected. Quite often, however, only one or two variables in the connecting stream have any significant effect on the calculation and/or the values of the other variables may be very easy to guess accurately. In this case reversing the direction of only the key variables, making them decisions, can have a profound effect on the optimization. Again this approach is commonly taken by an engineer but without being given a formal name. For example, consider our original problem. Suppose four chemical species A, B, C and D are involved in each stream, and further suppose the separation stage, unit 3, is separating for recycle the light component A from the mixture of less volatile components B, C, and D. Then although one cannot, in principle, specify the flows of all components out the top of the column, one can and would in practice do so by assuming components B, C, and D have essentially zero flows back to unit 1. Then the recycle flow of A can be made a decision, and the recycle loop is effectively broken from a computational standpoint.

Until automatic means become available to handle this type of decomposition, it seems likely the engineer will feel he can do the problem better by himself, using the computer to analyze one or two units at a time. Some work on automating the choosing of information flow through a process has been published (Lee (1967), Rudd (1968)).

4. A Strategy for Handling Inequality Constraints

Consider again our original problem which we will reformulate as

$$\text{Min } \{\phi(z) \mid f(z) = 0 , \quad g_j(z) \leq 0 , \quad j \in J\}$$

Our connection constraints (4) and stage transformation constraints (2) are in the set of equality constraints $f(z) = 0$, and our inequality constraints (3) form the inequality constraints $g_j \leq 0$ here.

Essentially two problems exist when solving this problem: finding a feasible point and, once feasible, finding the optimum. These two problems were the subject of a Ph.D. thesis by DeBrosse (1971). His project was formulated as:

"Suppose a person had the capability to generate, modify, and execute automatically the computer programs which solve a set of nonlinear equality constraints. How might one use this capability to aid in solving the above optimization problem?"

In essence then one capability anticipated is that the above problem is read into the computer essentially in the above form. The computer system automatically partitions the variables z into a set of calculated variables x and decision variables u, and it then produces the code (in some form) to calculate efficiently, using the equality constraints, values for the variables x given values for the decisions u. (The choice of which variables to use as decisions can be made to ease the calculations for the remaining x.)

The second, and equally important capability, also anticipated is that if a solution procedure exists for satisfying a given set of equality constraints, and one decides another equality constraint should also be held or released, the procedure can be quickly modified accordingly.

4.1. Finding a First Feasible Point: Several methods have been used for finding a feasible point; for example, guess a point or use penalty functions and solve as an optimization problem. Also, of course, phase 1 of the Simplex Algorithm can be used for problems with constraints linear in the decision variables.

The algorithm of DeBrosse (1971,1972) is rather a different approach and proceeds indirectly to find a feasible point. It relies on the following theorem.

Theorem 2: If we evaluate a point on every surface of "maximal intersection" for a set of constraints, then a point has been placed in every region defined by the constraints. If no such point is feasible with respect to the set of constraints, the problem is infeasible because of them.

A surface of "maximal intersection" occurs when the intersection of p constraints (where p is generally less than or equal to n, the dimension of the space) forms a surface which intersects no other constraint. For a simple case, p=n and such an intersection is simply a point.

The algorithm hypothesizes a set of constraints is infeasible, and it then proceeds to try to prove this hypothesis using Theorem 2. If the hypothesis is disproved, either a feasible point has been located or a new hypothesis is suggested. An example will help indicate the approach.

Figure 4 represents a set of four constraints in a two dimensional space. The algorithm generally starts by assuming n+1 constraints are infeasible. The following steps are taken for the initial hypothesis.

Hypothesis H_1: g_1, g_2, and g_3 cause our system to be infeasible.

Maximal Intersections Checked	Point Found	Constraints Violated
g_1,g_2	1	g_3
g_1,g_3	2	g_2
g_2,g_3	3	g_1
	4	g_4

Note both points 3 and 4 satisfy $g_2 = g_3 = 0$. Point 4 disproves our original hypothesis H_1 as that point satisfies g_1, g_2, and g_3. If constraint g_4 were not present, we would have found a feasible point. An alternate hypothesis suggested is the set of constraints in the last maximal intersection and one of the violated constraints.

H_2: g_2, g_3, and g_4

Maximal Intersections Checked	Point Found	Constraints Violated
g_2,g_3	3(again)	g_1

H_2 is rejected by point 3. No alternate hypothesis is suggested at any step involving only n+1 or 3 constraints that has not already been disproved. We therefore must create one involving 4 constraints.

H_3: g_1, g_2, g_3, g_4

Points 5, 6, and 7 are then found and represent all the maximal intersections possible. None is feasible and Theorem 1 proves that constraints g_1, g_2, g_3, and g_4 cause our system to be infeasibl

Note that the moving from one point of maximal intersection to another corresponds to removing one equality constraint and replacing it with another in our solution procedure. This is the automatic capability we claimed was to be available. To perform this task by hand corresponds to reprogramming the computer, which gets tiresome very quickly.

The complete feasible point algorithm contains provisions to handle nonintersecting constraints effectively and, provided the user always locates multiple intersections for any set of constraints, will guarantee finding a feasible point if one exists. If only a finite number of maximal intersections exist, the algorithm is finite.

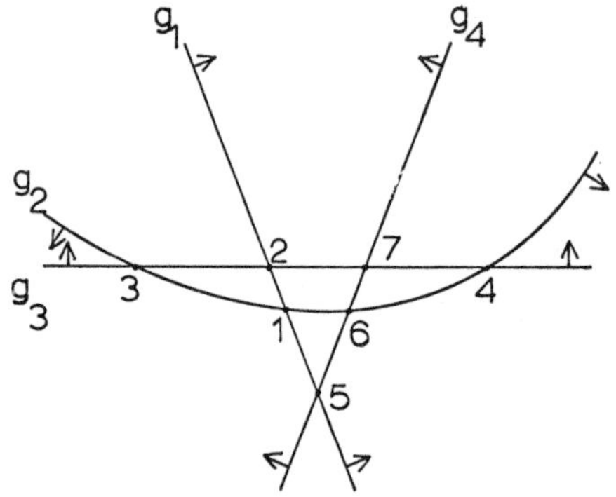

Figure 4. Feasible Point Example Four Inequality Constraints in a Two Dimensional Space

4.2. Optimizing Once Feasible: The optimization algorithm developed uses the general strategy of "restriction" (Geoffrion (1970)) to find the optimum. The general approach has two parts. First one must partition the inequality constraints g_j into three sets, based on the current point z_i:

$$R = \{g_j | g_j(z_i) < 0\}$$

$$S = \{g_j | g_j(z_i) = 0 \,, \quad g_j \text{ should be released}\}$$

$$T = \{g_j | g_j(z_i) = 0 \,, \quad g_j \text{ should be held}\}$$

Note sets S and T contain all the currently "active" constraints. Partitioning is based on evaluation and examination of the Kuhn-Tucker multipliers for the constraints at the current point z_i. The algorithm is as follows.

1. Put all active constraints ($g_j(z_i) = 0$) into set T.

2. Evaluate the Kuhn-Tucker multipliers for the constraints in set T.

3. Any multiplier, μ_k, greater than zero indicates the constraint g_k should be released. If one or more $\mu_k > 0$, release one constraint by putting it into set S.

4. Project the search direction (e.g. minus the gradient of the objective function, $-\nabla\phi$) onto the constraint surface in set T.

5. Check if this direction is into the infeasible region of any constraints in set S (it might be if set S has two or more constraints in it). If so move the last constraint put into S back to set T and either exit or try to

release a different constraint. If direction is feasible with respect to all constraints in S, repeat from step 2.

The solution procedure to move from point z_i, once the active constraints are partitioned is as follows.

1. Using an unconstrainted optimization algorithm, generate moves while satisfying the constraints

$$f(z) = 0$$

$$g_j(z) = 0 , \qquad g_j \varepsilon T$$

   Use, as coordinate directions for the search, the constraints in set S and a compatible set of decision variables.

2. If at any point a constraint g_j in either R or S is violated, add it to set T.

3. When the improvement in the objective becomes too slow, repartition the constraints. The repartitioning step in fact contains sufficient information to check the Kuhn-Tucker conditions for an optimum. If satisfied, exit. If not, continue the search.

Note that moving a constraint from sets S and T to R or from R to set T requires that one rewrite the programs which satisfy the equality constraints. In the former case, one is releasing an equality constraint whereas in the latter one is adding one.

Using constraints as coordinate directions is actually quite easy to do. If one writes the constraints with slack variables

$$g_j(z) + \varepsilon_j = 0$$

then choosing values for the slack variable ε_j is equivalent to using the constraint as a coordinate. For example choosing $\varepsilon_j = 0$ is equivalent to requiring the constraint to be tight. Choosing $\varepsilon_j > 0$ moves into the feasible region of g_j, and so forth.

Table 1 indicates the results of using this approach for solving several small optimization problems. Note that no initial points were needed here at which to start the optimizations. The feasible point algorithm supplied the point.

A final comment seems appropriate. The approach suggested here is a very natural one, except of course specific rules are given (so infeasibility can be proved if it exists for example). One commonly approaches the optimization of a process design by first trying to decide which inequality constraints are tight at the optimum. Once discovered, those inequality constraints are programmed into the problem as equality constraints and the system is optimized over the remaining degrees of freedom. This problem can be solved using an unconstrained optimization algorithm and is one a person has some confidence in solving.

5. The GENDER (General ENgineering DEsign Routines) System

Our group at Florida is currently pursuing research with a principal goal of producing the GENDER system, a system of subroutines which will aid in solving general design problems. Figure 5 is a rough sketch of the architecture of the initial GENDER system. It is written entirely in FØRTRAN using only

Problem	Number of Variables	Number of Constraints: Equality	Number of Constraints: Inequality	Number of Steps to Feasible Region	Additional Number of Steps to Optimum	Problem Description
1	3	0	6	5		Nonlinear constraints
2	3	0	6	7		Nonlinear constraints
3	4	0	12	13		Linear constraints
4	2	0	4	7*		Nonlinear constraints
5‡	10	3	28	8	15	Nonlinear alkylation process model
6	2	0	4	2	3	Linear constraints
7‡	4	0	7	?	4	Nonlinear constraints

*Number of steps to prove infeasibility.

‡Problems from Bracken and McCormick (1968).

Table 1. Results of Feasible Point and Optimization Algorithm

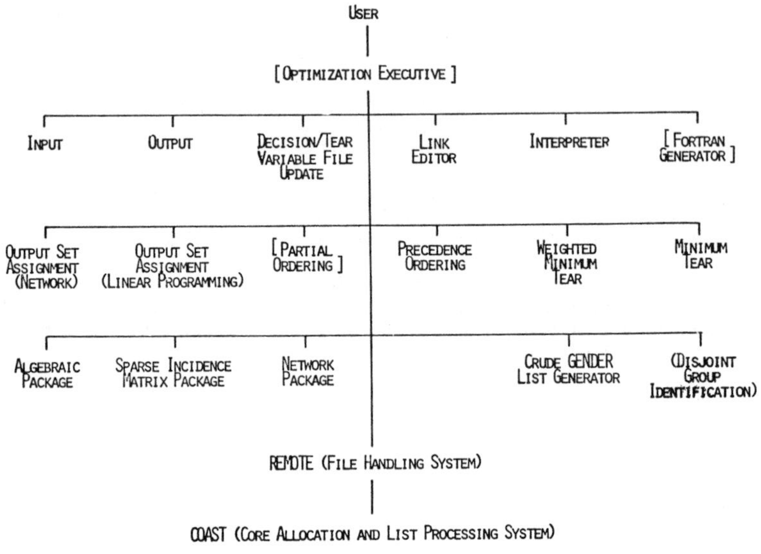

Figure 5. Architecture of GENDER. [ ] indicates routines not in Version 1. References for algorithms are Westerberg (1971), Gupta (1972), Sargent (1964), Christensen (1970), and Barkley (1970).

those instructions which will permit it to be usable on a variety of computer systems.

The user views GENDER as a set of powerful subroutines which can input the equations and data for a design problem, create a variety of solution procedures varying from very quick but inelegant to quite sophisticated, execute a derived solution procedure, display the results, and even modify the procedure as execution proceeds. The structure of GENDER is such that new algorithms useful in deriving solution procedures can be added directly to the system as they become known.

The aids, in the form of subroutine sets, are on five levels in Version 1. Underlying the system on levels 1 and 2 are the COAST and REMOTE systems which perform the basic operations involved in core allocation and retrieval and data structure manipulation. On level 3 are aids which manipulate useful data structures common to most of the rest of GENDER, an algebraic package, and some initialization routines for setting up an unstructured solution procedure.

Level 4 contains the simple to quite sophisticated algorithms that are used to derive equation set solution procedures. Level 5 in Version 1 has the service functions of input and output, and it contains the interpreter package to execute derived procedures. Above level 5 will reside the executive systems yet to be developed which will aid in optimization and synthesis studies. Currently the user must provide all the programs at this level when using GENDER.

Design Philosophy of GENDER: GENDER has been designed and implemented with several prime objectives in mind. First it is conceived of as being a system which is more than a simple prototype. Also the algorithms which prove effective for small equation sets may be unworkable for the very large sets (several hundred to a few thousand equations) encountered in real problems, thus we require the system to be an effective experimental apparatus for testing algorithms. For these reasons considerable effort was expended initially to develop a flexible and hopefully quite useful data handling capability. A rather extensive error detection and reporting package is also part GENDER.

From an ultimate user's (the design engineer) viewpoint, the input and output portion of any system is of prime importance. We have chosen however to slight this portion at present. Upgrading is planned during this next year.

The GENDER routines currently stress using tearing procedures to solve the algebraic equations involved in design. At a future date, it would be appropriate to add routines which use Newton-Raphson based methods. In this case Lagrangian based optimization algorithms, which require Jacobian matrices, will be more easily added. It is also interesting to note that a method for adding constraints to an existing Newton-Raphson based solution procedure is in the literature (Goldstein (1970)). Thus many of the ideas given here can be applied fairly directly even when using this approach.

6. List of Symbols

f_{ij} Vector of stage transformations converting inputs and controls to unit i into outputs y_{ij}

g_i Inequality functions for unit i

h_i Hamiltonian for unit i

$h(\pi)$ Dual function in two level optimization scheme

$I(i)$ Set of indices for input streams to unit i

J Index set for all inequality constraints in an optimization problem

L Lagrange function

S_i Feasible region for u_i and x_i variables as defined by the constraints $g_i(x_i,u_i) \le 0$ for unit i

u_i Vector of decision variables for unit i

x_{ij} Vector of input variables for the jth input stream to unit i

y_{ij} Vector of output variables for the jth output stream from unit i

Greek

ε Step size parameter for adjusting Kuhn-Tucker multipliers

λ_{ij} Vector of adjoint variables (Lagrange multipliers) entering unit i, stream j in adjoint system

μ_i Kuhn-Tucker multipliers for inequality constraints for unit i

π_{ij} Vector of adjoint variables (Lagrange multipliers) leaving unit i, stream j in adjoint system

ϕ Objective function for system

ϕ_i Objective function for unit i

7. References

Barkley, R. W. and Motard, R. L. (1970), Decomposition of Nets, University of Houston.

Bracken, J. and McCormick, G. P. (1968), Selected Applications of Nonlinear Programming, John Wiley, New York.

Brosilow, C. and Lasdon, L. (1965), A Two Level Optimization Technique for Recycle Processes, AIChE-IChemE Symp. Ser. No. 4, 75.

Christensen, J. H. (1970), The Structuring of Process Optimization, AIChE J., 16(2), 177.

DeBrosse, C. J. (1971), Feasible Point and Optimization Algorithms for Structured Design Systems in Chemical Engineering, Ph.D. Thesis, Univ. of Florida.

DeBrosse, C. J. and Westerberg, A. W. (1972), A Feasible Point Algorithm for Structured Design Systems in Chemical Engineering, accepted for publication in AIChE J.

Director, S. W. (1972), A Survey of Decomposition Techniques for Analysis and Design of Electrical Networks, Proceedings of this meeting.

Fan, L. T. and Wang, C. S. (1964), The Discrete Maximum Principle, John Wiley, New York.

Fletcher, R. and Powell, M.J.D. (1963), A Rapidly Convergent Descent Method for Minimization, Comput. J., 6, 163.

Geoffrion, A. M. (1970), Elements of Large-Scale Mathematical Programming, Management Science, 16(11), 652.

Goldstein, R. P. and Stanfield, R. B. (1970), Flexible Method for the Solution of Distillation Design Problems Using the Newton-Raphson Technique, 9(1), 78.

Gupta, P. K. (1972), Application of Analysis and Optimization Techniques for Structured Systems to the Design of a Double Effect Evaporator, M.Sc. Thesis, University of Florida.

Hendry, J. F. and Hughes, R. R. (1972), The Synthesis of Optimal Separation Sequence by List Processing Techniques, 71st National Meeting of the AIChE, Dallas, Texas.

Jackson, R. (1964), Some Algebraic Properties of Optimization Problems in Complex Chemical Plants, Chem. Engng. Sci., 19, 19.

Jung, B. S., Mirosh, W., and Ray, W. H. (1972), A Study of Large Scale Optimization Techniques, 71st National Meeting of the AIChE, Dallas, Texas.

Lasdon, L. S. (1970), Optimization Theory for Large Systems, The Macmillan Co., New York, Chapter 8.

Lee, E. S. (1969), Optimization of Complex Chemical Plant by a Gradient Technique, AIChE J., 15(3), 393.

Lee, W., Christensen, J. H., and Rudd, D. F. (1966), Design Variable Selection to Simplify Process Calculations, AIChE J., 12(6), 1104.

McGalliard, R. L. (1971), Structural Sensitivity Analysis in Design Synthesis, Ph.D. Thesis, University of Florida.

McGalliard, R. L. and Westerberg, A. W. (1972), Structural Sensitivity Analysis in Design Synthesis, 71st National Meeting of the AIChE, Dallas, Texas. Accepted for publication by Chem. Engng. J.

Naphtali, L. M. and Sandholm, D. P. (1971), Multicomponent Separation Calculations by Linearization, AIChE J., 17(1), 148.

Rudd, D. F. and Watson, C. C. (1968), Strategy of Process Engineering, John Wiley and Sons, Inc., New York, Chapter 3.

Sargent, R.W.H. and Westerberg, A. W. (1964), SPEED-UP in Chemical Engineering Design, Trans. Instn. Chem. Engrs., 42(5), T190.

Siirola, J. J. and Rudd, D. F. (1971), Computer-Aided Synthesis of Chemical Process Designs, Ind. Engng. Chem. Fundamentals, 10, 353.

Thompson, R. W. and King, C. J. (1972), Systematic Synthesis of Separation Schemes, 71st National Meeting of the AIChE, Dallas, Texas.

Westerberg, A. W. and Edie, F. C. (1971), Computer Aided Design Part 2, An Approach to Convergence and Tearing in the Solution of Sparse Equation Sets, Chem. Engng. J., 2(1), 17.

Westerberg, A. W. and Director, S. W. (1972), Parameter Sensitivity Analysis in Chemical Process Design--A Physical Interpretation, submitted for publication.

A Decomposition Method for Structured Nonlinear Programming Problems

Klaus Ritter
Department of Mathematics, Rutgers University
New Brunswick, New Jersey 08903

Abstract: This paper describes a decomposition method for a class of structured nonlinear programming problems with block diagonal constraints and coupling variables. The solution to the problem is approximated by solving a sequence of reduced nonlinear programs which is obtained by elimination of variables. The subproblems are solved by a method of conjugate directions which also provides an effective tool for the elimination of a maximal number of variables.

1. Introduction

In practice many large mathematical programming problems have specially structured objective functions and constraints. The most important class is represented by problems exhibiting block diagonal structure with coupling constraints or variables or both. Beginning with the decomposition algorithm for linear problems by Dantzig and Wolfe (1961) many methods for the solution of large structured linear and nonlinear programming problems have been proposed. Detailed references may be found in Geoffrion (1969) and Grigoriadis (1971).

In this paper a method is described for problems with a nonlinear objective function and linear inequality constraints with block diagonal structure and coupling variables. The algorithm is based on the observation that for fixed values of the coupling variables the problem decomposes into small nonlinear subproblems. Using the constraints which are active for an approximate solution of these problems for the elimination of variables an auxiliary problem is defined. An approximate solution of this problem provides improved values of the coupling variables.

A major difficulty in this procedure arises from the fact that for nonlinear objective functions the number of active constraints at the optimal solution is normally much smaller than the number of variables. In unfavorable cases this may result in an auxiliary problem which has almost as many variables as the original one. To decrease the number of variables in the auxiliary problem Grigoriadis (1971) has suggested the use of the pseudoinverse of the matrix corresponding to the active constraints. In the present paper the above mentioned difficulty is overcome by the application of a method of conjugate directions to the solution of the subproblems. This method associates with each point a square matrix whose dimension is equal to the number of variables in the problem under

---

This work was supported by the National Science Foundation under Research Grant GP - 33033.

consideration. Using this matrix for the elimination of variables it is possible to obtain an auxiliary problem which contains the coupling variables only.

The next section contains a precise statement of the structured nonlinear problem. Section 3 gives an outline of the method of conjugate directions used for the solution of the subproblems. In Section 4 the decomposition algorithm is described in detail and convergence results are stated.

2. Formulation of the problem, definitions and notation

For $i = 1, \cdots, p$ let $x_i \in E^{n_i}$ be a n_i-dimensional column vector and $y \in E^{n_o}$ a n_o-dimensional column vector. Assume that $F_1(x_1,y)$, $\cdots$, $F_p(x_p,y)$, $F_o(y)$ are real valued functions. For $i = 1, \cdots, p$ let A_i be a (m_i, n_i)-matrix and, for $i = o, \cdots, p$, let B_i be a (m_i, n_o)-matrix and b_i a m_i-dimensional column vector. A prime denotes the transpose.

We consider the problem of minimizing the function

$$\sum_{i=1}^{p} F_i(x_i,y) + F_o(y)$$

subject to the constraints

$$A_i x_i + B_i y \leq b_i, \quad i = 1, \cdots, p$$

$$B_o y \leq b_o.$$

In the following this problem is denoted by P and abbreviated in the form

$$\min \{F(x,y) \mid (x,y) \in R\}$$

where

$$x = (x_1, \cdots, x_p), \quad F(x,y) = \sum_{i=1}^{p} F_i(x_i,y) + F_o(y)$$

and

$$R = \{(x,y) \mid A_i x_1 + B_i y \leq b_i, B_o y \leq b_o, \quad i = 1, \cdots, p\}.$$

Any $y \in E^{n_o}$ for which there exist x with $(x,y) \in R$ is said to be feasible. For any fixed feasible y the given problem reduces to p smaller problems of the form

$$\min_{x_i} \{F_i(x_i,y) \mid x_i \in R_i(y)\}, \quad i = 1, \cdots, p$$

where

$$R_i(y) = \{x_i \in E^{n_i} \mid A_i x_i \leq b_i - B_i y\}.$$

We denote these subproblems by $P_i(y)$ in order to emphasize that their solution depends on the chosen y.

If $F_i(x_i,y)$ is differentiable at a point (x_i^j,y^j) we denote its gradient with respect to x_i or y at (x_i^j,y^j) by $\nabla_x F_{ij}$ or $\nabla_y F_{ij}$, respectively. Similarly, ∇F_{oj} denotes the gradient of $F_o(y)$ at y^j.

Throughout this paper we make the following

Assumption 1:

Let $(x^o,y^o) \in R$ and $S_o = \{(x,y) \in R \mid F(x,y) \leq F(x^o,y^o)\}$.
There exists a bounded convex subset $S \subset E^{n_1} \times \cdots \times E^{n_p} \times E^{n_o}$ such that $S_o \subset S$ and $F(y), F_1(x_1,y), \cdots, F_p(x_p,y)$ are continuously differentiable on S.

This assumption guarantees that P has an optimal solution. Furthermore, let $\bar{y}$ be feasible such that there is $\bar{x}$ with $(\bar{x},\bar{y}) \in S_o$. Then it follows from Assumption 1 that

$$S_i(\bar{y}) = \{x_i \in R_i(\bar{y}) \mid F_i(x_i,\bar{y}) \leq F_i(\bar{x}_i,\bar{y})\}, \quad i = 1, \cdots, p$$

is bounded and that $F_i(x_i,\bar{y})$ is continuously differentiable on some convex set containing $S_i(\bar{y})$. Therefore, each problem $P_i(\bar{y})$ has an optimal solution.

For later reference we add the following

Definition 1:

A point $(x^j,y^j) \in R$ is said to be a stationary point if there are vectors $u_i \in E^{mi}$, $i = 0, \cdots, p$, such that

$$\nabla_x F_{ij} = A'_i u_i$$

$$\sum_{i=1}^{p} \nabla_y F_{ij} + \nabla F_{oj} = \sum_{i=1}^{p} B'_i u_i + B'_o u_o$$

$$u'_i(A_i x_i^j + B_i y^j - b_i) = o, \quad u_i \leq o, \quad i = 1, \cdots, p$$

$$u'_o(B_o y^j - b_o) = o, \quad u_o \leq o.$$

It is well-known that every optimal solution (x,y) of P is a stationary point. Conversely, if $F(x,y)$ is pseudo-convex, then every stationary point is an optimal solution of P (see e.g. Mangasarian 1969). It should be observed that a stationary point is not necessarily an optimal solution of P if it is only required that each function $F_1(x_1,y), \cdots, F_p(x_p,y), F_o(y)$ is pseudo-convex since the sum of pseudo-convex functions need not be pseudo-convex.

The decomposition method proposed in this paper is based on a method of conjugate directions for minimization problems with linear inequality constraints. This method has been developed in Ritter (1972). In order to facilitate the description of the decomposition method an outline of the method of conjugate directions is given in the next section.

3. A method of conjugate directions

The method outlined in this section applies to problems of the following type:

$$\min \{F(x) \mid Ax \leq b\}.$$

Here $x \in E^n$, A is a (m,n)-matrix and b is a m-dimensional column vector.

It is assumed that the method starts with an x_o such that $Ax_o \leq b$, the set $S = \{x \in E^n \mid Ax \leq b$ and $F(x) \leq F(x_o)\}$ is bounded and $F(x)$ is continuously differentiable on some convex set containing S.

Suppose

$$a'_i x_o = b_i, \qquad i = 1, \cdots, q$$

$$a'_i x_o < b_i, \qquad i = q+1, \cdots, m,$$

where the a'_i are the rows of A and the b_i are the components of b. Assume that the vectors $a_1, \cdots, a_q$ are linearly independent, set

$$d_{io} = a_i, \quad i = 1, \cdots, q,$$

and let

$$d_{q+1,o}, \cdots, d_{no}$$

be any set of column vectors such that with

$$D'_o = (d_{1o}, \cdots, d_{no})$$

the inverse

$$D_o^{-1} = (c_{1o}, \cdots, c_{no})$$

exists.

Furthermore, set $\beta_o = 1$ and let

$$J(x_o) = \{\alpha_{1o}, \cdots, \alpha_{no}\}$$

be an ordered set of n elements with

$$\alpha_{io} = i \qquad \text{for} \qquad i = 1, \cdots, q$$

$$\alpha_{io} = o \qquad \text{for} \qquad i = q+1, \cdots, n.$$

A general cycle of the algorithm consists of three steps which are described below. At the beginning of the jth cycle the following data are available: x_j with $Ax_j \leq b$, $\nabla F(x_j) = g_j$, β_j,

$$D_j^{-1} = (c_{ij}, \cdots, c_{nj}) \text{ and } J(x_j) = \{\alpha_{ij}, \cdots, \alpha_{nj}\}.$$

Here $\beta_j = 1$ if a constraint is active for x_j which was inactive for x_{j-1} and $\beta_j = o$ otherwise. A positive α_{ij} gives the index of a constraint $a'_k x \leq b_k$ which is active for x_j and corresponds to the ith row of D_j. A nonpositive α_{ij} indicates that the ith row of D_j does not correspond to a constraint and was changed for the last time during the $-\alpha_{ij}{}^{th}$ cycle.

In each cycle the direction of descent is proportional to a column c_{ij} of D_j^{-1}. An active constraint is dropped if and only if the corresponding α_{ij} is positive. The step size is chosen in such a way that, under appropriate assumptions, it approaches the global minimizer of $F(x)$ in $\{x \mid Ax \leq b\}$ along the chosen direction of descent. Finally, the matrix D_j^{-1} is updated by means of an exchange of one row of D_j'.

Step I: Computation of the direction of descent s_j

Let

$$u_j' = g_j' D_j^{-1}, \qquad u_j' = ((u_j)_1, \cdots, (u_j)_n)$$

and define $l = l_j$ and $k = k_j$ such that

$$(u_j)_l \geq (u_j)_i \quad \text{for all } i \text{ with } \alpha_{ij} > o,$$

$$|(u_j)_k| \geq |(u_j)_i| \quad \text{for all } i \text{ with } \alpha_{ij} \leq o,$$

where α_{ij} are the elements of $J(x_j)$. If all $\alpha_{ij} \leq o$ set $(u_j)_l = o$ and if all $\alpha_{ij} > o$ set $(u_j)_k = o$.
Let $\beta > o$ be any constant and set

$$s_j = \begin{cases} c_{kj}(u_j)_k & \text{if } \beta|(u_j)_k| \geq (u_j)_l \text{ or } (\beta_j = 1 \text{ and} \\ & |(u_j)_k| > o) \text{ or } (|(u_j)_k| = o \text{ and} \\ & (u_j)_l \leq o) \\ c_{lj}(u_j)_l & \text{if } (\beta|(u_j)_k| < (u_j)_l \text{ and } \beta_j = o) \text{ or} \\ & ((u_j)_l > o \text{ and } |(u_j)_k| = o) \end{cases}$$

If $s_j = o$ stop; otherwise go to Step II.

Step II: Computation of the step size σ_j

Choose constants $o < \delta < \frac{1}{2}$, $o < \alpha_1$ and $o < \alpha_2$.
Let

$$\delta_j = \max \{|(u_j)_k|, \quad (u_j)_l\}$$

and

$$\sigma_j^* = \min \{\frac{a'_i x_j - b_i}{a'_i s_j} \text{ for all } i \text{ with } a'_i s_j < o\}.$$

If, for $i = 1, \cdots, m$, $a'_i s_j \geq o$ set $\sigma_j^* = 1$.

Set

$$\hat{\sigma}_j = \begin{cases} \min \{\sigma_j^*, \dfrac{g'_j s_j}{2(F(x_j - s_j) - F(x_j) + g_j' s_j)}\} & \text{if } F(x_j - s_j) - F(x_j) + g_j' s_j > o \\ \min \{\sigma_j^*, 1\} & \text{if } F(x_j - s_j) - F(x_j) + g_j' s_j \leq o \end{cases}$$

and

$$\tilde{\sigma}_j = \begin{cases} \hat{\sigma}_j & \text{if } \alpha_1 \delta_j \leq \hat{\sigma}_j \leq \alpha_2 \delta_j^{-\frac{1}{2}} \\ \sigma_j^* & \text{otherwise} \end{cases}$$

Let ν_j be the smallest nonnegative integer for which

$$F(x_j) - F(x_j - (\tfrac{1}{2})^{\nu_j} \tilde{\sigma}_j s_j) \geq \delta g'_j s_j (\tfrac{1}{2})^{\nu_j} \tilde{\sigma}_j.$$

Choose

$$\sigma_j = (\tfrac{1}{2})^{\nu_j} \tilde{\sigma}_j$$

and set

$$x_{j+1} = x_j - \sigma_j s_j.$$

Compute g_{j+1}. If $g_{j+1} = o$ stop, otherwise go to Step III.

Step III: Computation of D_{j+1}^{-1}, $J(x_{j+1})$ and β_{j+1}

Choose numbers $\gamma > o$ and $c > o$.

Case 1: $\sigma_j = \sigma_j^*$, i.e., a new constraint becomes active for x_{j+1}. Set $\beta_{j+1} = 1$. Let $\xi = \xi_j$ be the index for which the minimum σ_j^* is attained.

1.1: $s_j = c_{1j}(u_j)_1$, i.e., an active constraint is dropped. Suppose $|c'_{1j} a_\xi| \geq \gamma \, || c_{1j} ||$ (1). Replace the 1th row of D_j by a'_ξ, denote the new matrix by D_{j+1}, and compute D_{j+1}^{-1}. Set

(1) The case that this assumption is not satisfied is discussed in Ritter (1972).

$$J(x_{j+1}) = \{\alpha_{1,j+1}, \cdots, \alpha_{n,j+1}\}$$

where $\alpha_{i,j+1} = \alpha_{ij}$ for $i = 1, \cdots, n, \ i \neq 1$

$$\alpha_{1,j+1} = \xi_j .$$

1.2: $s_j = c_{kj}(u_j)_k$; i.e., no active constraint is dropped. Proceed as in Case 1.1 with 1 replaced by k.

Case 2: $\sigma_j < \sigma_j^*$, i.e., no new active constraint occurs. Set $\beta_{j+1} = o$.

2.1: $s_j = c_{1j}(u_j)_1$, i.e., an active constraint is dropped. Let

$$d_j = \frac{g_j - g_{j+1}}{\|\sigma_j s_j\|} .$$

Set

$$d_{1,j+1} = \begin{cases} d_j & \text{if } |c'_{1j} d_j| \geq \gamma \|c_{1j}\| \text{ and } \|d_j\| \leq c \\ \dfrac{c_{1j}}{\|c_{1j}\|} & \text{otherwise} \end{cases}$$

Replace the 1st row of D_j by $d'_{1,j+1}$, denote the new matrix by D_{j+1} and compute D_{j+1}^{-1} from D_j^{-1}. Set

$$J(x_{j+1}) = \{\alpha_{1,j+1}, \cdots, \alpha_{n,j+1}\}$$

where

$$\alpha_{i,j+1} = \alpha_{ij} \quad \text{for} \quad i = 1, \cdots, n, \quad i \neq 1$$

$$\alpha_{1,j+1} = -j .$$

2.2: $s_j = c_{kj}(u_j)_k$, i.e., no active constraint is dropped. Proceed as in Case 2.1 with 1 replaced by k.

In Ritter (1972) it is shown that this algorithm has the following properties:

Theorem 1

Let the sequence $\{x_j\}$ be generated by the above algorithm. Then:

1) $F(x_{j+1}) < F(x_j)$.

2) The sequence $\{x_j\}$ terminates with some element x_q if and only if x_q is a stationary point.

3) If $\{x_j\}$ is an infinite sequence then every cluster point of $\{x_j\}$ is a stationary point.

Theorem 2

Let z be a cluster point of $\{x_j\}$ with

$$a'_i z = b_i, \qquad i = 1, \cdots, q$$

$$a'_i z < b_i, \qquad i = q+1, \cdots, m$$

$$\nabla F(z) = \sum_{i=1}^{q} \lambda_i a_i, \qquad \lambda_i < o, \quad i = 1, \cdots, q.$$

Suppose that $F(x)$ is twice continuously differentiable in some neighborhood of z and that the Hessian matrix $G(z)$ of $F(x)$ at z has the property

$$x'G(z)x > o \quad \text{for every} \quad x \neq o \quad \text{with} \quad a'_1 x = o, \quad i = 1, \cdots, q.$$

Then we have

$$\frac{\| x_{j+n-q} - z \|}{\| x_j - z \|} \longrightarrow o \quad \text{as} \quad j \longrightarrow \infty.$$

For any x_j we can write g_j as a linear combination of the columns d_{ij} of D_j',

$$g_j = \sum_{i=1}^{n} \lambda_{ij} d_{ij}.$$

Thus

$$(u_j)_i = g'_j c_{ij} = \lambda_{ij}, \quad i = 1, \cdots, n.$$

Since $\{\| D_j^{-1} \|\}$ is bounded and every cluster point of $\{x_j\}$ is a stationary point it follows that for every $\varepsilon > o$ there is

$j(\varepsilon)$ such that for every $j \geq j(\varepsilon)$

$$|(u_j)_k| \leq \varepsilon \quad \text{and} \quad (u_j)_1 \leq \varepsilon.$$

This leads to the following:

Definition 2

Let $\varepsilon > o$. An $x_j \in \{x_j\}$ is said to be a ε-stationary point if

$$|(u_j)_k| \leq \varepsilon \quad \text{and} \quad (u_j)_1 \leq \varepsilon.$$

4. The Algorithm

Before going into the details of the algorithm we first describe the general idea of the decomposition method.

Suppose a $\bar{y} \in E^{n_o}$ is given with $R_i(\bar{y}) \neq \phi$ for $i = 1, \cdots, p$. If such a feasible $\bar{y}$ is not known it can be constructed by solving a linear problem. For this fixed $\bar{y}$, P reduces to the smaller problems $P_1(\bar{y}), \cdots, P_p(\bar{y})$. Usually it is assumed that each of the subproblems $P_i(\bar{y})$ is solved by a suitable method. Denote the solution by $\bar{x}_i$ and partition the constraints of $P_i(\bar{y})$ as follows into active and inactive constraints:

(1)
$$A_{i1}\bar{x}_i + B_{i1}\bar{y} = b_{i1}$$

$$A_{i2}\bar{x}_i + B_{i2}\bar{y} < b_{i2}.$$

The equations (1) are now used to eliminate from P as many of the x-variables as possible. The elimination results in an auxiliary problem of the following form:

AP
$$\min \{G(v_1, \cdots, v_p, y) \mid M\binom{v}{y} \leq d\}.$$

Here v_i is a vector containing all components of x_i which were not eliminated by means of the equations (1), $v = (v_1, \cdots, v_p)$, and $M\binom{v}{y} \leq d$ denotes all constraints of P which were not used in the elimination of variables. Now the auxiliary problem is solved. Denote the solution by $(\hat{v}_1, \cdots, \hat{v}_p, \hat{y})$. Using this solution and the equations (1) a new value $\hat{x}_i$ for the complete vector x_i is obtained.

It is then shown that $(\hat{x}_1, \cdots, \hat{x}_p, \hat{y})$ is a solution of the problem

$$(2) \quad \min \left\{ F(x,y) \;\middle|\; \begin{matrix} A_{i1}x_i+B_{i1}y = b_{i1} \\ A_{i2}x_i+B_{i2}y \le b_{i2} \end{matrix} \;,\; B_o y \le b_o, \quad i = 1, \cdots, p \right\}$$

and that after a finite number of the above cycles the solution of (2) coincides with the solution of P.

This procedure encounters the following three difficulties:

1) If $F_i(x_i,y)$ is not linear or quadratic in x_i it is in general not possible to obtain the solution of $P_i(\bar{y})$ in a finite number of steps.

2) If $F_i(x_i,y)$ is not linear in x_i it is well-known in practice that the number of active constraints in $P_i(\bar{y})$ is usually much smaller than the number of variables in this problem. Since the elimination of variables is based on the active constraints (1), the elimination procedure can result in an auxiliary problem AP which has almost as many variables as P, thus rendering the decomposition method computationally ineffective.

3) If $G(v,y)$ is not linear or quadratic in general only an approximation $(\bar{v},\bar{y})$ to the solution (v,y) of AP can be obtained. The resulting $(\bar{x}_1, \cdots, \bar{x}_p,\bar{y})$ is then an approximation to the solution of (2). In this case, however, it need not be true that the number of cycles is finite. Thus the need for a general convergence proof arises.

The first difficulty can easily be overcome since it suffices to obtain an approximation to the solution of $P_i(\bar{y})$. In our method we require only an ε-stationary point. The method described in Section 3 yields such a point in a finite number of steps. Since Theorem 2 guarantees superlinear convergence under rather weak assumptions, the method can be expected to be efficient.

Grigoriadis (1971) discusses an interesting way to deal with the second difficulty. He suggests to use the equations (1) and the pseudoinverse of A_{i1} to eliminate all components of x_i, so that AP contains only y. Even if $\hat{y}$ is the exact solution of AP the resulting $(\hat{x}_1, \cdots, \hat{x}_p,\hat{y})$ is in general not a solution of (2). To preserve the concept of a finite number of cycles, at least under the assumption that AP can be solved exactly, an acceleration procedure is used which appends extra variables to AP. In unfavorable cases this results in an auxiliary problem AP with as many variables as the one obtained by means of (1).
In the present method the matrix D_j^{-1}, associated with each point constructed by the algorithm of Section 3, is used as a convenient tool to eliminate all x-variables. Therefore, we deal exclusively with auxiliary problems containing y only.

Since in general AP cannot be solved exactly it cannot be shown that the number of cycles is finite. We require that an ε-stationary point $\bar{y}$ for AP is obtained. Using the method of Section 3 again, we can compute $\bar{y}$ in a finite number of steps. Inserting $\bar{v}$ into the linear equations used for the elimination of the x_i we obtain a new point $(\bar{x}_1, \cdots, \bar{x}_p, \bar{y})$ at the end of each cycle.[1] Theorem 3 below states the convergence properties of the sequence of these points.

We now describe a general cycle of the algorithm in detail. In the algorithm a sequence of numbers ε_j with

$$\varepsilon_j > o \quad \text{and} \quad \lim_{j\to\infty} \varepsilon_j = o$$

is used to determine ε_j-stationary points. A possible choice for ε_j is $\varepsilon_j = j^{-1}$. At the beginning of the jth cycle it is assumed that a point $(\bar{x}_1{}^j, \cdots, \bar{x}_p{}^j, \bar{y}^j) \in R$ is available.

<u>Step I</u>: Computation of an ε_j-stationary point for

$$P_i(\bar{y}^j), \quad i = 1, \cdots, p.$$

Using $\bar{x}_i{}^j \in R_i(\bar{y}^j)$ as starting point for the method described in Section 3 an ε_j-stationary point $z_i{}^j$ of $P_i(\bar{y}^j)$, $i = 1, \cdots, p$, is computed in a finite number of steps. Let D_{ij}^{-1} and $J_i(z_i{}^j)$ be the matrix and ordered set, respectively, associated with $z_i{}^j$.

<u>Step II</u>: Definition of the auxiliary problem AP.

For $i = 1, \cdots, p$ partition A_i, B_i and b_i into A_{i1}, A_{i2}, B_{i1}, B_{i2} and b_{i1}, b_{i2}, respectively, in such a way that

$$A_{i1}\, z_i{}^j = b_{i1} - B_{i1}\, \bar{y}^j \tag{3}$$

and

$$A_{i2}\, z_i{}^j < b_{i2} - B_{i2}\, \bar{y}^j .$$

Let

$$J_i(z_i{}^j) = \{\alpha^i_{1j}, \cdots, \alpha^i_{n_i j}\}$$

and define the νth component of the vectors $b_i{}^j$ and $c_i{}^j$ as follows:

$$(b_i{}^j) = \begin{cases} o & \text{if } \alpha^i_{\nu j} \leq o \\ (b_i)_\nu & \text{if } \alpha^i_{\nu j} > o \end{cases}$$

$$(c_i^j) = \begin{cases} (z_i^j)_\nu & \text{if } \alpha_{\nu j}^i \leq o \\ (D_{ij}^{-1} b_i^j)_\nu & \text{if } \alpha_{\nu j}^i > o. \end{cases}$$

Similarly, depending on the elements of $J_i(z_i^j)$ let the rows of the (n_i, n_i)-matrix C_{ij} be zero or equal to the appropriate rows of B_i.

Substitute

(4) $$x_i = c_i^j - D_{ij}^{-1} C_{ij} y$$

into $F_i(x_i,y)$ and $A_{i2}x_i \leq b_{i2} - B_{i2} y$ to obtain

(5) $$G_i(y) = F_i(c_i^j - D_{ij}^{-1} C_{ij} y,y)$$

and

(6) $$(B_{i2} - A_{i2}D_{ij}^{-1}C_{ij})y \leq b_{i2} - A_{i2}c_i^j.$$

Define the auxiliary problem AP as follows:

$$\min \sum_{i=1}^{p} G_i(y)+F_o(y) \mid M_i y \leq d_i,\ B_o y \leq b_o,\quad i = 1, \cdots, p\},$$

where, for $i = 1, \cdots, p$,

$$M_i = B_{i2} - A_{i2}D_{ij}^{-1}C_{ij} \quad \text{and} \quad d_i = b_{i2} - A_{i2} c_i^j.$$

Remark:

It follows from the definition of D_{ij}, C_{ij} and c_i^j that

(7) $$A_{i1}x_i = b_{i1} - B_{i1} y$$

is a subsystem of

(8) $$D_{ij}x_i = D_{ij}c_i^j - C_{ij} y.$$

Hence, every solution of (4) satisfies (3). There are two special cases. If all elements of $J_i(z_i^j)$ are positive, i.e., if z_i^j is an extreme point of $R_i(\bar{y}^j)$ then $D_{ij}^{-1} = A_{i1}^{-1}$ and $C_{ij} = B_i$.

In this case (7) and (8) coincide since $D_{ij} = A_{i1}$ and $C_{ij} = B_i$. If all elements of $J_i(z_i^j)$ are nonpositive, i.e., if z_i^j is an interior point of $R_i(\bar{y}^j)$ then $C_{ij} = 0$ and $c_i^j = z_i^j$. Therefore, (5) and (6) reduce to $G_i(y) = F_i(z_i^j, y)$ and $B_i\, y \leq b_i - A_i z_i^j$, respectively.

Step III: Computation of an ε_j-stationary point of AP.

Using $\bar{y}^j$ as a starting point for the method described in Section 3 an ε_j-stationary point $\bar{y}^{j+1}$ is computed in a finite number of steps.

Step IV: Computation of $(\bar{x}_1^{j+1}, \cdots, \bar{x}_p^{j+1}, \bar{y}^{j+1})$

Use $\bar{y}^{j+1}$ and equations (4) in Step II to compute

$$\bar{x}_i^{j+1} = c_i^j - D_{ij}^{-1} C_{ij} \bar{y}^{j+1}.$$

We have the following

Theorem 4

Let Assumption 1 be satisfied and suppose that the sequence $\{\bar{x}_1^j, \cdots, \bar{x}_p^j, \bar{y}^j\}$ is generated by the above algorithm. Then, for $j = 0, 1, \cdots,$

1) $$\{\bar{x}_1^j, \cdots, \bar{x}_p^j, \bar{y}^j\} \in R$$

2) $$\sum_{i=1}^{p} F_i(\bar{x}_i^{j+1}, \bar{y}^{j+1}) + F_0(\bar{y}^{j+1}) < \sum_{i=1}^{p} F_i(\bar{x}_i^j, \bar{y}^j) + F_0(\bar{y}^j)$$

3) Every cluster point of the sequence $\{\bar{x}_1^j, \cdots, \bar{x}_p^j, \bar{y}^j\}$ is a stationary point of P.

The proof of this theorem is omitted. It is similar to the proof of Theorem 1 given in Ritter (1972). As an immediate consequence of Theorem 4 we have the following

Corollary

Let the assumptions of Theorem 4 be satisfied. If $F(x,y)$ is pseudoconvex, then every cluster point of $\{\bar{x}_1^j, \cdots, \bar{x}_p^j, \bar{y}^j\}$ is an optimal solution of P. If $F(x,y)$ is strictly convex, then $\{\bar{x}_1^j, \cdots, \bar{x}_p^j, \bar{y}^j\}$ converges to the optimal solution of P.

REFERENCES

Dantzig, G. B. and P. Wolfe, "The Decomposition Algorithm for Linear Programs," Econometrica, Vol. 29, 1961, pp. 767-778.

Geoffrion, A. M., "Elements of Large Scale Mathematical Programming," Western Management Science Institute, Working Paper No. 144, University of California, Los Angeles, 1969.

Grigoriadis, M. D., "A Projective Method for Structured Nonlinear Programs," Mathematical Programming, Vol. 1, 1971, pp. 321-358.

Mangasarian, O. L., Nonlinear Programming, McGraw-Hill, New York, 1969.

Ritter, K., "A Method of Conjugate Directions for Nonlinear Programming Problems," to appear.

A COMPUTATIONAL INVESTIGATION INTO NONLINEAR DECOMPOSITION

CARL-LOUIS SANDBLOM

University of Birmingham, England

Abstract: A brief description is given of a nonlinear decomposition algorithm due to T.O.M. Kronsjö. Some computational results, obtained by K.P. Wong, are also reviewed. A test problem of a highly decomposable structure is then solved several times numerically, using the above decomposition algorithm, and breaking it down in different ways every time. Using as an efficiency criterion the computer time required to solve the test problem, an attempt is made to find the best decomposition policy.

1. INTRODUCTION

The purpose of this paper is to examine the numerical efficiency of a nonlinear decomposition algorithm by T.O.M. Kronsjö (1969). We shall first describe the decomposition algorithm itself.

Suppose that the convex problem:

$$\min f(x) \quad \text{subject to} \quad g(x) \leqq 0 \qquad (A)$$

($f: R^n \to R$, $g: R^n \to R^m$) is separable in the following sense:

$f(x) = f_1(x_1) + f_2(x_2) \qquad ((x_1, x_2) = x)$,

$\{x | g(x) \leqq 0\} = \{x \mid g_1(x_1) + g_2(x_2) \leqq 0, x_1 \in C_1\}$

where

a) f_1, f_2, g_1, g_2 are convex, differentiable functions,

b) $\emptyset \neq C_1$ is a convex, compact set,

c) the set $C_2 = \{x_2 \mid g_1(x_1) + g_2(x_2) \leqq 0 \text{ for some } x_1 \in C_1\}$ is non-empty and compact,

d) $x_1 \in C_1 \Rightarrow \exists x_2 \in C_2 : g_1(x_1) + g_2(x_2) < 0$.

We conclude immediately:

a), b), c) $\Rightarrow C_2$ is convex.

Thus A is feasible and has an optimal solution.

Using the Lagrange functions

$L = f + ug$, $L_1 = f_1 + ug_1$, $L_2 = f_2 + ug_2$

where u is the row vector of dual variables (Lagrange multipliers) we propose the scheme:

1) Take $x_1^0 \in C_1$. Put $q = 1$.

2) Solve the problem:

$$\min_{x_2} f_1(x_1^{q-1}) + f_2(x_2) \quad \text{subject to } g_1(x_1^{q-1}) + g_2(x_2) \leqq 0 \tag{SUB}$$

An optimal solution of SUB is denoted by x_2^q, u^q (u^q are dual variables in the Wolfe sense). L^q is the corresponding f- value.

3) Solve the problem:

$$\min_{h,\, x_1} h \quad \text{subject to } L(x_1, x_2^j, u^j) \leqq h,\ j = 1, \ldots, q, \quad x_1 \in C_1 \tag{MASTER}$$

An optimal solution of MASTER is denoted by x_1^q; the corresponding h- value is l^q. Go back to 2) with $q+1$ replacing q.

In this scheme, SUB is always feasible as, by construction, $x_1^{q-1} \in C_1$, $q = 1, 2, \ldots$. MASTER is always feasible as $L(x_1, x_2^j, u^j)$, being continuous on the compact C_1, is bounded for each j, $j = 1, 2, \ldots$.

The quantities l^q and L^q, defined by this scheme, have an important property:

<u>Theorem 1</u>: $l^q \leqq \bar{f} \leqq L^q$, $q = 1, 2, \ldots,$

where $\bar{f}$ is the optimum value of (A).

<u>Proof</u>: $\bar{f} \leqq L^q$ because (x_1^{q-1}, x_2^q) is feasible to (A) and yields the value L^q.

Turning to the first inequality of the theorem, we first observe that, by definition,

$$l^q = \min_{h,\, x_1} h \quad \text{s.t. } L(x_1, x_2^j, u^j) \leqq h,\ j = 1, \ldots, q, \quad x_1 \in C_1.$$

If K is the index set for which the constraints are binding, we thus have

$$l^q = L(x_1^q, x_2^k, u^k), \quad k \in K \subset \{1, \ldots, q\}.$$

If $(\bar{x}_1, \bar{x}_2)$ is optimal to (A), we must have for at least one binding constraint k:

$$L(x_1^q, x_2^k, u^k) \leqq L(\bar{x}_1, x_2^k, u^k).$$

Then:
$$L(\bar{x}_1, x_2^k, u^k) = f_1(\bar{x}_1) + u^k g_1(\bar{x}_1) + f_2(x_2^k) + u^k g_2(x_2^k) = f_1(\bar{x}_1) + f_2(\bar{x}_2) +$$
$$+ u^k[g_1(\bar{x}_1) + g_2(\bar{x}_2)] + L_2(x_2^k, u^k) - L_2(\bar{x}_2, u^k)$$

But $u^k \geqq 0$ and $g_1(\bar{x}_1) + g_2(\bar{x}_2) = g(\bar{x}) \leqq 0$ (being feasible to (A)), hence:

$$L(\bar{x}_1, x_2^k, u^k) \leqq \bar{f} + L_2(x_2^k, u^k) - L_2(\bar{x}_2, u^k)$$

But $L_2(x_2^k, u^k) - L_2(\bar{x}_2, u^k) \leqq (x_2^k - \bar{x}_2)^T \nabla_{x_2} L_2(x_2^k, u^k) = 0$ (by convexity of L_2 in x_2 and by the optimality of (x_2^k, u^k) to SUB)

and so we have $L(\bar{x}_1, x_2^k, u^k) \leqq \bar{f}$.

Remembering that we have already proved $l^q \leqq L(\bar{x}_1, x_2^k, u^k)$ we conclude:

$l^q \leqq \bar{f}$. □

With some additional assumptions one can prove:

<u>Theorem 2</u>: $L^q - l^q \to 0, \; q \to \infty$.

<u>Proof</u>: See T.O.M. Kronsjö (1969). □

We see that theorems 1 and 2 provide the theoretical justification of the following method: Work through our scheme above iteratively till some stop criterion $L^q - l^q < \varepsilon$ is satisfied. Then $\frac{l^q + L^q}{2}$ approximates $\bar{f}$ with an error of less than ε.

This decomposition procedure is quite advantageous when used on problems of the following kind:

$$\begin{aligned}
\min\ & f_0(x_o) + f_1(x_1) + f_2(x_2) + \dots + f_s(x_s) \\
\text{s.t. } & g_{00}(x_o) \leqq 0 \\
& g_{10}(x_o) + g_{11}(x_1) \leqq 0 \\
& g_{20}(x_o) + g_{22}(x_2) \leqq 0 \\
& \vdots \qquad\qquad \ddots \\
& g_{s0}(x_o) + g_{ss}(x_s) \leqq 0
\end{aligned}$$

Iteration k will then be:

a) Subproblems:

$$\min_{x_i} f_i(x_i)$$

$$\text{s.t. } g_{io}(x_o^{k-1}) + g_{ii}(x_i) \leqq 0$$

$i = 1, \ldots, s$. Denoting solutions to these subproblems by (x_i^k, u_i^k), we define

$$L^k = f_o(x_o^{k-1}) + \sum_{i=1}^{s} f_i(x_i^k).$$

b) Master:

$$\min_{h, x_o} h$$

$$\text{s.t.} \quad f_o(x_o) + \sum_{i=1}^{s} [f_i(x_i^j) + u_i^j(g_{io}(x_o) + g_{ii}(x_i^j))] \leqq h, \; j = 1, \ldots, k,$$

$$g_{oo}(x_o) \leqq 0$$

Denoting the optimum value of the master by l^k, we get as before: $l^k \leqq \bar{f} \leqq L^k$.

The computational efficiency of our decomposition algorithm has been tested on the ICL KDF9 Computer at the University of Birmingham. The Algol programme used for solving the test problems was originally written by K.P. Wong (1970, 1971) and uses the sequential unconstrained minimization technique (SUMT) by A.V. Fiacco and G.P. McCormick (1964). This version of SUMT incorporates the variable metric method by R. Fletcher and M.J.D. Powell (1963) and the Golden Section search method (see e.g. D.J. Wilde (1964)). To impose uniformity the stop criterion $\frac{L^k - l^k}{L^k} \leqq 0.01$ has been used in all cases.

2. PREVIOUS TESTS

K.P. Wong (1970, 1971) has compared solution times for 38 different test problems when decomposed and when directly solved. The computational effort required to solve the problems was measured in terms of computer time needed. The comparison between decomposition and direct solution was carried out with regard to run time as well as to execution time. (Roughly speaking, execution time equals run time minus time for compilation.) The problems were of different sizes ranging from (N, M) = (6, 2) to (N, M) = (41, 32), where N stands for the number of variables in the problem and M for the number of constraints. The number of variables in the master problem (n_o) ranged between 2 and 6, and the number of variables in the subproblems was 2-3, except for four of the test problems which contained subproblems of sizes 4, 5, 6 and 8, respectively.

As was expected, K.P. Wong found that the larger the N, the more favourable the decomposition. For N > 20 (which was the case for 21 of the test problems) decomposition was always better than direct solution. In the extreme case N = 41, he found T_{dir}, direct solution time, was 192.13 minutes and T_{dec}, decomposition solution time, was 25.48 minutes. This was for run time; for execution time the difference was even more striking: $T_{dir} = 189.60$, $T_{dec} = 9.65$.

Performing loglinear regression of T_{dir} on N, K.P. Wong showed that, for run time, $T_{dir} = 0.0094 . N^{2.56}$ ($R^2 = 0.93$) where R^2, the coefficient of determination, is defined as the explained variance divided by the sample variance. For execution time the corresponding equation is $T_{dir} = 0.00060 . N^{3.38}$ ($R^2 = 0.97$).

If, on the basis of K.P. Wong's data, we try to find some dependence of

T_{dir} on M, n_o and m, the number of subproblems, we find little significance, e.g. a t- test on a loglinear regression of T_{dir} (run time) on N, M, n_o and m showed that, apart from N, none of the other variables were significantly non-zero at a level of 0.05. Even at the 0.15- level, only m was significant in addition to N. Our conclusion is that information about N is sufficient to explain the behaviour of T_{dir}.

Turning to decomposition the picture is less clear. If we try a t- test on a loglinear regression of T_{dec} (run time) on N, M, n_o and m, we find that except for the highly significant N only n_o was significant at the level 0.05. In this case we got $R^2 = 0.82$. Using this information a loglinear regression was performed of T_{dec} (run time) on N and n_o. To eliminate dependence on m, the number of subproblems was kept less than 4. This meant 34 observations instead of 38 as previously. We got:

$$T_{dec} = 0.253\, N^{1.13}\, n_o^{0.33} \quad (R^2 = 0.80).$$

Although these figures put beyond doubt the superiority of decomposition over straightforward solution for large N (solution time being a nearly linear function of N), the comparatively low R^2- value in the last regression equation is not quite satisfactory. Allowing for the different computational complexity of the different test problems by repeating the regression with T_{dec}/T_{dir} instead of T_{dec} did not improve the figures.

To analyse more closely the dependence of T_{dec} on n_o and m we have used an approach different from that of K.P. Wong. This approach is described in Section 3.

3. NEW TESTS

To simplify matters, let us look at problems where all the subproblems are about equally large, i.e. have about the same number of variables. Denoting the average subproblem size by $\bar{n}$, we would then expect the solution time under decomposition to be a function of N, n_o, m and $\bar{n}$:

$$T = f(N, n_o, m, \bar{n})$$

We have left out M as K.P. Wong's previous results (as well as theoretical considerations) make us believe that T does not really depend on M (for moderate variations of M). Observing that $\bar{n} = \dfrac{N - n_o}{m}$ we find that it is sufficient to look at N, n_o and m:

$$T = f(N, n_o, m).$$

From the previous results we found a well documented dependence on N, less so for n_o, and m was not significant at all. To gain more knowledge about the dependence of n_o and m we ran a new series of tests as follows:

A problem with 18 variables, 13 constraints and a highly decomposable structure (see Fig. 1) is solved several times at different levels of decomposition and with different sizes of the master problem. Although this approach deviates from that of Section 2 in that we do not always break the problem down as far as its structure will allow us, it is closely connected to a problem that will arise in a practical case.

$$f(x)=(x_1-7)^2-x_2+e^{-x_3}+x_3+x_4^2+e^{x_5}+(x_6+3)^4+(x_7-4)^2+e^{-x_8}+12x_8+(x_9+x_{10})^2+e^{x_9}+(x_{11}-x_{12})^4+e^{x_{13}}+e^{-x_{14}}+(x_{15}+x_{16})^2+e^{x_{17}}-2x_{17}+x_{18}^2$$

$$\begin{array}{llr}
\text{s.t.} & x_1^2 & \le \frac{3}{2} \\
& x_1^2+e^{x_2} & \le 2 \\
& -2x_1 \quad +(x_3-1)^2 & \le 4 \\
& e^{-x_1}+x_1 \quad +x_4 & \le 5 \\
& (x_1+2)^2 \quad e^{-x_5}-x_5 & \le 11 \\
& 7x_1 \quad -2x_6 & \le 10 \\
& e^{-x_1}+x_1^2 \quad +x_7^2 & \le 15 \\
& (x_1-3)^2 \quad -x_8 & \le 12 \\
& -3x_1 \quad +(x_9-x_{10})^2 & \le 2 \\
& x_1^4+x_1 \quad +x_{11}+e^{-x_{12}} & \le 12 \\
& e^{x_1}-10x_1 \quad +x_{13}^2-5x_{14} & \le 20 \\
& x_1^4-5x_1 \quad +e^{-x_{15}}+(x_{15}-x_{16})^2 & \le 35 \\
& (x_1+4)^4 \quad +e^{-x_{17}}-10x_{18} & \le 300
\end{array}$$

Fig. 1

Given a structured problem that is to be solved, should the structure of it be exploited fully (i.e. "total" decomposition be carried out), partly ("partial" decomposition) or not at all (direct solution)? Strangely enough, this question, so obvious and challenging, seems to have been overlooked by most people in the field. To the author's knowledge, only a few attempts to answer this question have been made: C. Labro (1964), N.A. Rasmussen (1971) and O.B.G. Madsen (1972). However, these attempts only deal with linear programming, and from the results of these investigations really nothing can be concluded about the nonlinear case.

A related question is the one of almost-structured problems. If a large problem contains variables and/or constraints that from other considerations are known to affect the total solution very little (e.g. some variable might be restricted to very small magnitudes), it is likely that a more structured problem would be obtained if we omitted these non-essential variables and/or constraints. A quite natural decomposition approach would then be the following. Using "optimal" decomposition on the structurised problem, run through a few decomposition iterations till an approximate solution has been found. Then introduce the non-essential parts of the original problem and continue iterating with some different level of decomposition, using the previously obtained approximate solution as a starting point (ascribing suitable values to the new variables which have just been introduced).

4. COMPUTATIONAL RESULTS

When solved by direct application of our SUMT algorithm, our test problem required 4.25 minutes of execution time. For run time the figure was 5.78 (time will always be given in minutes). The solution times with decomposition are displayed in Table 1 (execution time) and Table 2 (run time).

Table 1. Decomposition execution times (minutes)

n_o \ m	1	2	3	4	8	12
1	26.77	12.33		8.27		8.08
5	24.46	19.64	23.75		17.40	
10	32.01	36.45		28.93		

Table 2. Decomposition run times (minutes)

n_o \ m	1	2	3	4	8	12
1	29.05	14.72		10.88		11.22
5	26.70	21.97	26.20		20.15	
10	34.17	38.67		31.55		

Although these figures confirm our anticipations that the more you decompose, the better, and that the size of the master is more crucial than the size of the subproblems, the results for the case $(n_o, m) = (5, 3)$ might look puzzling. To show how the decomposition procedure converges we give the values of l^k and L^k of the iterations in tabular form below:

Table 3. Values of l^k and L^k for $(n_o, m) = (1, 12)$

iteration no. k	l^k	L^k
1	14.19	31.84
2	20.08	2032
3	24.26	542
4	26.81	156
5	28.25	59.86
6	29.03	36.50
7	29.46	30.97
8	29.75	29.78

Table 4. Values of l^k and L^k for $(n_o, m) = (1, 4)$

iteration no. k	l^k	L^k
1	17.85	32.85
2	22.54	2032
3	25.96	550
4	28.07	159
5	29.27	61.35
6	29.91	37.60
7	30.25	32.05
8	30.42	30.79
9	30.43	30.58

Table 5. Values of l^k and L^k for $(n_o, m) = (1, 2)$

iteration no. k	l^k	L^k
1	17.84	32.86
2	22.56	2031
3	25.97	543
4	28.08	157
5	29.27	60.87
6	29.91	37.52
7	30.24	31.99
8	30.41	30.79
9	30.49	30.59

Table 6. Values of l^k and L^k for $(n_o, m) = (1, 1)$

iteration no. k	l^k	L^k
1	11.32	32.93
2	16.78	2031
3	20.88	540
4	23.50	154
5	25.09	57.81
6	26.23	34.88
7	30.47	30.59

Table 7. Values of l^k and L^k for $(n_o, m) = (5, 8)$

iteration no. k	l^k	L^k
1	15.69	33.35
2	20.31	2031
3	24.30	541
4	26.83	156
5	28.25	59.85
6	29.03	36.51
7	29.45	30.98
8	29.74	29.78

Table 8. Values of l^k and L^k for $(n_o, m) = (5, 3)$

iteration no. k	l^k	L^k
1	19.35	34.36
2	306	2031
3	22.80	307
4	26.05	546
5	28.13	158
6	29.31	61.09
7	29.94	37.57
8	30.25	32.03
9	30.42	30.83
10	30.51	30.59

Table 9. Values of l^k and L^k for $(n_o, m) = (5, 2)$

iteration no. k	l^k	L^k
1	4.25	34.36
2	24.99	2033
3	13.86	196
4	15.81	60.56
5	18.58	32.60
6	30.20	32.30
7	30.39	30.93
8	30.39	30.61

Table 10. Values of l^k and L^k for $(n_o, m) = (5, 1)$

iteration no. k	l^k	L^k
1	19.35	34.36
2	111	2030
3	27.23	367
4	28.80	91.21
5	29.65	44.81
6	30.11	33.74
7	30.34	31.16
8	30.46	30.64

Table 11. Values of l^k and L^k for $(n_o, m) = (10, 4)$

iteration no. k	execution time(mins)	l^k	L^k
1	4.28	23.12	151
2	9.51	25.79	627
3	13.95	27.96	177
4	19.17	29.22	65.77
5	21.57	29.89	38.67
6	23.97	30.24	32.26
7	26.15	30.42	30.84
8	28.93	30.52	30.58

Table 12. Values of l^k and L^k for $(n_o, m) = (10, 2)$

iteration no. k	execution time(mins)	l^k	L^k
1	3.87	23.14	151
2	14.14	25.99	621
3	18.05	28.00	175
4	21.80	29.24	65.45
5	26.77	29.92	38.60
6	31.28	30.26	32.28
7	33.98	30.44	30.85
8	36.45	30.52	30.58

Table 13. Values of l^k and L^k for $(n_o, m) = (10, 1)$

iteration no. k	execution time(mins)	l^k	L^k
1	2.82	8.11	151
2	10.19	11.28	637
3	16.04	14.01	172
4	19.07	15.96	58.09
5	21.59	19.02	32.06
6	24.62	30.26	32.63
7	27.02	30.44	30.81
8	32.01	30.52	30.58

Looking at table 8 we see that the case $(n_o, m) = (5, 3)$ required 10 iterations and that this case is computationally less well-behaved than the others with $n_o = 5$. From the tables we also see that the computation times displayed do not always give a fair comparison between the cases. For example, table 5 shows that after 8 iterations we were just short of satisfying our stop criterion (and we had by then used 10.92 minutes of execution time). Thus a 9th iteration (using about 1.4 minutes of execution time) had to be performed in order to reach the required accuracy. Table 6, on the other hand, shows that before the last iteration was performed, we were still a long way from the required accuracy. Tables 11 - 13 include execution time to show another kind of computational irregularity.

5. CONCLUSIONS

Although decomposition of our test problem (Fig. 1) could not compete with direct solution using our SUMT programme, this did not worry us as the superiority of decomposition over direct solution has already been firmly established, (c.f. section 2). Our purpose was instead to compare against each other different ways of decomposition (N.B. always using the same Kronsjö decomposition algorithm). Unlike O.B.G. Madsen (1972), who found that some cases will favour "partial" decomposition, our limited results suggest that decomposition should always be carried out as far as possible and that bringing down the master problem size is the most important thing to do.

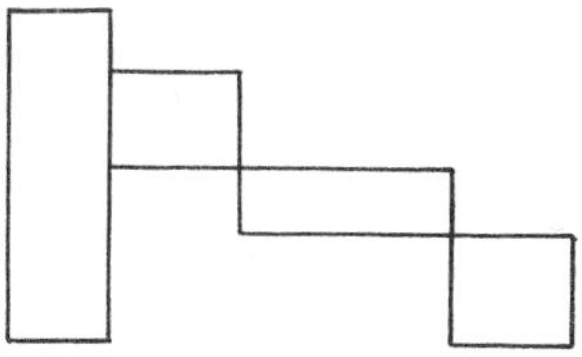

Fig. 2

Problems suitable for decomposition with the Kronsjö algorithm will often occur in centralised planning (Fig. 2). The decomposition procedure can then be interpreted as a dialogue between the central planning office (corresponding to the master problem) and the regional planning offices (corresponding to the subproblems). The costs incurred in carrying out the information flow represented by this dialogue correspond in our computational tests to computer time. This time used for "administration" can be quite substantial, and in the forthcoming tests that are currently under preparation we are therefore separating "calculating time" from "administration time", hoping that this will enable us to gain a deeper insight into the computational nature of decomposition.

References

Fiacco, A.V. and McCormick, G.P.: The sequential unconstrained minimisation technique for nonlinear programming: A primal-dual method. Management Science, 10, 1964, pp. 360-366.

Fletcher, R. and Powell, M.J.D.: A rapidly convergent descent method for minimisation. The Computer Journal, 6, 1963, pp. 163-168.

Kronsjö, T.O.M.: Decomposition of a large nonlinear convex separable economic system in the dual direction. Economics of Planning, 9, 1969, pp. 71-94.

Labro, C.: Efficiency and degrees of decomposition. Working paper no. 98, Center for Research in Management Science, University of California, Berkeley, 1964.

Madsen, O.B.G.: The connection between decomposition algorithms and optimal degree of decomposition. Page 239 in this volume.

Rasmussen, N.A.: Decomposition in large-scale systems. Thesis, The Institute of Mathematical Statistics and Operations Research, The Technical University of Denmark (in Danish), 1971.

Wilde, D.J.: Optimum seeking methods, Prentice-Hall, Englewood Cliffs, N.J., 1964, pp. 32-35.

Wong, K.P.: An Algol programme for SUMT. N.E.P. Research Papers, No. 41, University of Birmingham, 1970.

Wong, K.P.: Decentralised planning by vertical decomposition of an economic system: A nonlinear programming approach. Ph.D. thesis, University of Birmingham, 1970.

Wong, K.P.: Computer implementation of the decomposition of nonlinear convex separable programmes. N.E.P. Research Papers, No. 40 (Rev.), University of Birmingham, 1971.

Wong, K.P.: Decomposition of nonlinear convex separable programmes: An illustrative computer programme. N.E.P. Research Papers, No. 52, University of Birmingham, 1971.

A DECOMPOSITION PRINCIPLE FOR MINIMAX PROBLEMS

LYNN McLINDEN

Mathematics Research Center
University of Wisconsin
Madison, Wisconsin 53706

Abstract: This paper treats minimax problems in which there is an additively separable saddle function subject to coupling constraints in each argument. Relationships with a dual minimax problem and a Lagrangian minimax problem are explored. Methods of solution are discussed, including a decomposition principle whereby one solves a dual problem followed by some modified subproblems, each having much lower dimensionality. An application to multilevel resource allocation is suggested.

1. INTRODUCTION

Decomposition in linear and nonlinear programming has been the subject of many articles since 1960, when Dantzig and Wolfe published their decomposition principle for linear programs in Dantzig (1960). Extensive references to the literature can be found in Dantzig (1970), Geoffrion (1970) and Lasdon (1970). In this paper we show that a decomposition principle holds for another class of optimization problems--minimax problems.

To indicate one possible application of this result, as well as to introduce the problem it concerns, consider a multilevel resource allocation problem of the following type. Imagine there are two "central controllers" who compete against each other in each of r separate "enterprises." Suppose each controller's aim is to maximize the sum of the "benefits" he derives from each of the r enterprises, and suppose also that each controller has only a limited amount of "resources" to allocate among the enterprises. An economic interpretation of this situation might be duopoly in which two conglomerates play the roles of the central controllers. Other practical situations can also be interpreted in this abstract resource allocation framework.

One can model this situation mathematically by means of coupled minimax problems. To do this, let each enterprise correspond to a two-person zero-sum game having payoff $(x_i, y_i) \to K_i(x_i, y_i)$, where, without loss of generality, each K_i is maximized in x_i and minimized in y_i. Then let the central controllers' problem correspond to another two-person zero-sum game having payoff

$$(x_1, \ldots, x_r, y_1, \ldots, y_r) \to K_1(x_1, y_1) + \ldots + K_r(x_r, y_r)$$

and constrained by the "resource constraints"

$$x_1 + \ldots + x_r = a, \quad y_1 + \ldots + y_r = b \ .$$

Here one controller determines the x_i's and the other controller determines the y_i's. If it were not for the two coupling constraints, this minimax problem could be solved by solving the r minimax subproblems determined by the K_i's individually.

We present a decomposition principle whereby such a coupled minimax problem can in fact be broken down into $r+1$ minimax subproblems, each normally having much lower dimensionality. This is based on formulating a problem dual to the given one and then using the solutions of the dual problem together with each of the K_i's to form r modified subproblems. The solutions of the r subproblems then combine to furnish a solution of the coupled problem. A method is described by which the dual problem itself can be solved indirectly by working entirely within the original, "primal" space. To implement this method, in many cases it suffices to be able to evaluate a saddle function and find its gradient (or merely a subgradient) at any given point. As a sidelight, we describe also another approach to solving the above coupled problem. This involves working with an equivalent, Lagrangian problem.

To carry out the dual approach described above, we assume each K_i is a closed proper concave-convex function on $R^{m_i} \times R^{n_i}$. On the other hand, the K_i's are allowed to take the values $+\infty$ and $-\infty$, which effectively permits each K_i itself to represent a constrained minimax problem on $R^{m_i} \times R^{n_i}$. However, since our concern here is the troublesome coupling constraints, we use the $\pm\infty$ possibilities for K_i to suppress all explicit constraints except the coupling constraints. Thus, nonnegativity constraints such as $x_i \geq 0$ or $y_i \geq 0$ would be suppressed. Also, in order to treat a variety of different affine equality and/or inequality constraints simultaneously, we actually deal with model coupling constraints of the form

$$A_1x_1 + \ldots + A_rx_r \leq_P a, \quad B_1y_1 + \ldots + B_ry_r \leq_Q b \ ,$$

where $\leq_P$ and $\leq_Q$ are any partial orders induced by nonempty closed convex cones and the A_i's and B_i's are linear transformations.

While the dual approach to decomposition taken here has been used by a number of authors in treating coupled nonlinear or convex programming problems, the present work has perhaps been influenced most by Rockafellar (1970) (pp. 285-290). The proofs depend on new results for minimax problems due to McLinden (1972a, 1972b). For an iterative approach to the solution of coupled minimax problems via Everett's generalized Lagrange multipliers, see Penn (1971).

The paper is divided into sections as follows: Notation and Problem Statements (§2), Duality Theorems (§3), Methods of solution (§4), and Remarks (§5).

2. NOTATION AND PROBLEM STATEMENTS

The definitions used in this paper can be found in Rockafellar (1970).

For each $i = 1, \ldots, r$ let K_i be a closed proper concave-convex function from $R^{m_i} \times R^{n_i}$ to $[-\infty, +\infty]$, and let K_i^* be any conjugate of K_i. For convenience, write $\text{dom}\, K_i = C_i \times D_i$, $\text{dom}\, K_i^* = C_i^* \times D_i^*$, $C = C_1 \times \ldots \times C_r$ and $D = D_1 \times \ldots \times D_r$. Also for each $i = 1, \ldots, r$ let $A_i\colon R^{m_i} \to R^p$ and $B_i\colon R^{n_i} \to R^q$ be linear transformations, and let A_i^* and B_i^* denote the corresponding adjoint linear transformations. Let $(a, b) \in R^p \times R^q$.

Suppose P and Q are given nonempty closed convex cones in R^p and R^q, respectively, and define partial orderings as follows:

$$u_1 \leq_P u_2 \iff u_2 - u_1 \in P, \quad v_1 \leq_Q v_2 \iff v_2 - v_1 \in Q \ .$$

Let $P°$ and $Q°$ denote the polar cones, that is,

$$P° = \{z \in R^p \mid \langle z, u\rangle \leq 0, \ \forall u \in P\}, \quad Q° = \{w \in R^q \mid \langle w, v\rangle \leq 0, \ \forall v \in Q\} \ .$$

Observe that the choice $P = \{0\}$ yields $u_1 \leq_P u_2$ if and only if $u_1 = u_2$, and $P° = R^p$. Also, $P = R_+^p$ yields the usual coordinatewise partial ordering and $P° = -R_+^p$.

We are concerned with the saddle point problem

$$(\mathcal{P}) \qquad \underset{X \quad Y}{\text{maximin}}\{K_1(x_1, y_1) + \ldots + K_r(x_r, y_r)\} \ ,$$

where

$$X = \{(x_1, \ldots, x_r) \in C \mid A_1x_1 + \ldots + A_rx_r \leq_P a\} \ ,$$

and

$$Y = \{(y_1, \ldots, y_r) \in D \mid B_1y_1 + \ldots + B_ry_r \leq_Q b\} \ .$$

To treat $(\mathcal{P})$ we introduce several other saddle point problems. The first is the following <u>dual</u> of $(\mathcal{P})$:

$$(\mathcal{D}) \qquad \underset{Z \quad W}{\text{minimax}}\{\langle z, a\rangle + \langle w, b\rangle - K_1^*(A_1^*z, B_1^*w) - \ldots - K_r^*(A_r^*z, B_r^*w)\} \ ,$$

where

$$Z = \{z \varepsilon -P^\circ \mid A_i^* z \varepsilon C_i^*, \ \forall i\}$$

and

$$W = \{w \varepsilon Q^\circ \mid B_i^* w \varepsilon D_i^*, \ \forall i\} .$$

Another is the following Lagrangian problem:

$$(\mathfrak{L}) \qquad \underset{C\times Q^\circ D\times P^\circ}{\text{maximin}} \ \{K_1(x_1, y_1) + \ldots + K_r(x_r, y_r) - \langle z, a - \Sigma A_i x_i \rangle + \langle w, b - \Sigma B_i y_i \rangle\}$$

Finally, for any $(z, w) \varepsilon R^p \times R^q$ and any $i = 1, \ldots, r$ we consider the ith modified subproblem determined by (z, w):

$$(\mathcal{P}_i) \qquad \underset{C_i \ D_i}{\text{maximin}} \{K_i(x_i, y_i) - \langle x_i, A_i^* z \rangle - \langle y_i, B_i^* w \rangle\} .$$

Corresponding to each of these saddle point problems are lower and upper saddle values. These quantities will be denoted by self-explanatory notations such as inf sup($\mathcal{P}$) or sup inf($\mathcal{P}$). We shall use notations such as maximin($\mathcal{P}$) to indicate that a saddle value actually exists and is finite.

In what follows, the hypotheses of "strong consistency" will play an important role. We say that ($\mathcal{P}$) is strongly consistent if and only if there exist pairs $(x_i, y_i) \varepsilon \operatorname{ri}(\operatorname{dom} K_i)$ such that $a - \Sigma A_i x_i \varepsilon \operatorname{ri} P$ and $b - \Sigma B_i y_i \varepsilon \operatorname{ri} Q$. (The notation "ri" signifies the relative interior operation.) This condition is analogous to the familiar Slater constraint qualification in nonlinear programming. We say that ($\mathcal{D}$) is strongly consistent if and only if there exist $z \varepsilon -\operatorname{ri}(P^\circ)$ and $w \varepsilon \operatorname{ri}(Q^\circ)$ such that $(A_i^* z, B_i^* w) \varepsilon \operatorname{ri}(\operatorname{dom} K_i^*)$ for $i = 1, \ldots, r$. It can be shown, for example, that ($\mathcal{D}$) is strongly consistent if each set $\operatorname{dom} K_i$ is bounded.

3. DUALITY THEOREMS

As a theoretical basis for the actual solution methods to be discussed in §4, we need some theorems establishing the relationships between problems ($\mathcal{P}$), ($\mathcal{D}$) and ($\mathfrak{L}$). These theorems are given in this section.

Our strategy is essentially to obtain these results as corollaries to the theorems of McLinden (1972b). To do this, we introduce a linear transformation and saddle functions as follows. Let $m = \Sigma m_i$ and $n = \Sigma n_i$, and define a "product" linear transformation A from $R^m \times R^n$ to $R^p \times R^q$ by

$$A(x_1, \ldots, x_r, y_1, \ldots, y_r) = (A_1 x_1 + \ldots + A_r x_r, \ B_1 y_1 + \ldots + B_r y_r) .$$

Also, define a convex-concave "indicator" saddle function L on $R^p \times R^q$ by

$$L(u, v) = \begin{cases} 0 & \text{if } u \varepsilon a - P \text{ and } v \varepsilon b - Q \\ +\infty & \text{if } u \not\varepsilon a - P \text{ and } v \varepsilon b - Q \\ -\infty & \text{if } \phantom{u \varepsilon a - P \text{ and }} v \not\varepsilon b - Q , \end{cases}$$

and a concave-convex "separable" saddle function K on $R^m \times R^n$ by

$$K(x, y) = \begin{cases} K_1(x_1, y_1) + \ldots + K_r(x_r, y_r) & \text{if } x \varepsilon C \text{ and } y \varepsilon D \\ +\infty & \text{if } x \varepsilon C \text{ and } y \not\varepsilon D \\ -\infty & \text{if } x \not\varepsilon C \end{cases} ,$$

where $x = (x_1, \ldots, x_r)$ and $y = (y_1, \ldots, y_r)$.

With these elements K, L and A, it can be shown that problems (I), (II) and (III) of McLinden (1972b) specialize to yield ($\mathcal{P}$), ($\mathcal{D}$) and ($\mathfrak{L}$), respectively.

Accordingly, we present the following results with only a brief indication of proof. The few verifications necessary for complete proofs are left for the most part to the reader.

THEOREM 1. <u>The following inequalities always hold</u>:

$$\begin{array}{ccccc} & & \sup\inf(\mathcal{P}) \le \inf\sup(\mathcal{P}) & & \\ & \rotatebox{45}{$\le$} & & \rotatebox{-45}{$\le$} & \\ \sup\inf(\mathcal{L}) & & & & \inf\sup(\mathcal{L}) \; . \\ & \rotatebox{-45}{$\le$} & & \rotatebox{45}{$\le$} & \\ & & \sup\inf(\mathcal{D}) \le \inf\sup(\mathcal{D}) & & \end{array}$$

PROOF. By Theorem 1 of McLinden (1972b).

THEOREM 2. <u>If $(\mathcal{D})$ is strongly consistent, then</u>

$$\sup\inf(\mathcal{D}) \le \sup\inf(\mathcal{P}) \le \inf\sup(\mathcal{P}) \le \inf\sup(\mathcal{D}) \; .$$

<u>If $(\mathcal{P})$ is strongly consistent, then</u>

$$\sup\inf(\mathcal{P}) \le \sup\inf(\mathcal{D}) \le \inf\sup(\mathcal{D}) \le \inf\sup(\mathcal{P}) \; .$$

PROOF. This follows from Theorem 2 of McLinden (1972b), making use of the particular K, L and A defined above. One uses Theorem 7(iv) of McLinden (1972a) to see that a conjugate of K is given by

$$K^*(s,t) = \begin{cases} K_1^*(s_1,t_1) + \ldots + K_r^*(s_r,t_r) & \text{if } s \in C^* \text{ and } t \in D^* \\ +\infty & \text{if } s \in C^* \text{ and } t \notin D^* \\ -\infty & \text{if } s \notin C^* \end{cases} ,$$

where $s = (s_1, \ldots, s_r)$, $t = (t_1, \ldots, t_r)$, $C^* = C_1^* \times \ldots \times C_r^*$, and $D^* = D_1^* \times \ldots \times D_r^*$. A conjugate of L can be seen by direct computation to be

$$L^*(z,w) = \begin{cases} \langle z, a\rangle + \langle w, b\rangle & \text{if } z \in -P^\circ \text{ and } w \in Q^\circ \\ +\infty & \text{if } z \notin -P^\circ \text{ and } w \in Q^\circ \\ -\infty & \text{if } \qquad\qquad\qquad\; w \notin Q^\circ \end{cases} ,$$

while the adjoint of A is given by

$$A^*(z,w) = (A_1^*z, \ldots, A_r^*z, B_1^*w, \ldots, B_r^*w) \; .$$

Given these facts, the result follows.

We turn now to theorems involving the solutions of $(\mathcal{P})$, $(\mathcal{D})$ and $(\mathcal{L})$. Surprisingly, under the mild assumption that $(\mathcal{P})$ is strongly consistent, its solutions are "stable" with respect to perturbations in the right-hand-sides of the coupling constraints, that is, with respect to replacing the "feasible" sets X and Y by

$$X_u = \{(x_1, \ldots, x_r) \in C \mid A_1x_1 + \ldots + A_rx_r \le_P a - u\}$$

and

$$Y_v = \{(y_1, \ldots, y_r) \in D \mid B_1y_1 + \ldots + B_ry_r \le_Q b - v\} \; ,$$

respectively. In order to be precise, it is helpful to introduce some auxilliary functions.

For any pair $(x,y) = (x_1, \ldots, x_r, y_1, \ldots, y_r) \in X \times Y$, define a function g_y on R^p and a function f_x on R^q by setting

$$g_y(u) = \sup\{K_1(x_1', y_1) + \ldots + K_r(x_r', y_r) \mid x' \in X_u\}$$

and

$$f_x(v) = \inf\{K_1(x_1, y_1') + \ldots + K_r(x_r, y_r') \mid y' \in Y_v\} .$$

(We use the usual conventions that $\sup \emptyset = -\infty$ and $\inf \emptyset = +\infty$.) It follows from Lemma 3 of McLinden (1972b) that g_y is concave and f_x is convex. Moreover, (x, y) is a solution of $(\mathcal{P})$ and $\text{maximin}(\mathcal{P}) = \alpha$ if and only if $g_y(0) = f_x(0) = \alpha \in R$. We define a solution (x, y) of $(\mathcal{P})$ to be <u>stable</u> if and only if

$$\lim_{\lambda \downarrow 0} \frac{1}{\lambda}[g_y(\lambda u) - \alpha] < +\infty, \ \forall u, \quad \text{and} \quad \lim_{\lambda \downarrow 0} \frac{1}{\lambda}[f_x(\lambda v) - \alpha] > -\infty, \ \forall v ,$$

where $\alpha = \text{maximin}(\mathcal{P})$.

THEOREM 3. *If $(\mathcal{P})$ is strongly consistent, then every solution of $(\mathcal{P})$ is stable.*

PROOF. By Theorem 5 of McLinden (1972b).

It is of interest to know conditions which (stable) solutions of $(\mathcal{P})$ must necessarily satisfy. These are given in the next theorem; because of their importance, we give them a name. For any two pairs $(x, y) \in R^m \times R^n$ and $(z, w) \in R^p \times R^q$, we say that (x, y) *and* (z, w) *satisfy the Kuhn-Tucker conditions* if and only if

$$a - \Sigma A_i x_i \text{ is normal to } P^\circ \text{ at } -z, \quad b - \Sigma B_i y_i \text{ is normal to } Q^\circ \text{ at } w ,$$

and

$$(A_i^* z, B_i^* w) \in \partial K_i(x_i, y_i) \text{ for } i = 1, \ldots, r .$$

THEOREM 4. *If (x, y) is a stable solution of $(\mathcal{P})$, then there exists a pair (z, w) such that (x, y) and (z, w) together satisfy the Kuhn-Tucker conditions.*

PROOF. By Corollary 2 to Theorem 4 of McLinden (1972b).

It is immediate from Theorems 37.4 and 36.3 of Rockafellar (1970) that, of the conditions given above, the last r say precisely that (x_i, y_i) is a solution of $(\mathcal{P}_i)$ for $i = 1, \ldots, r$, where $(\mathcal{P}_1), \ldots, (\mathcal{P}_r)$ are the modified subproblems determined by (z, w) . Also, notice that for the case of equality constraints (i.e., $P = \{0\}$ and $Q = \{0\}$), the normality conditions reduce to $A_1x_1 + \ldots + A_rx_r = a$ and $B_1y_1 + \ldots + B_ry_r = b$, as expected.

According to the following theorem, the necessary conditions of Theorem 4 are also sufficient.

THEOREM 5. *Pairs (x, y) and (z, w) together satisfy the Kuhn-Tucker conditions if and only if (x, w, y, z) is a solution of $(\mathcal{L})$, and in this event (x, y) is a stable solution of $(\mathcal{P})$, (z, w) is a solution of $(\mathcal{D})$, and*

$$\text{maximin}(\mathcal{P}) = \text{minimax}(\mathcal{D}) = \text{maximin}(\mathcal{L}) .$$

Proof. By Theorem 4 and Lemma 3 of McLinden (1972b).

There is an "equilibrium price" interpretation for those pairs (z, w) for which the Kuhn-Tucker conditions can be satisfied. Suppose that each central controller can "buy" perturbations of the right-hand-side of his coupling resource constraint at a cost of either $\langle z, u \rangle$ or $\langle w, v \rangle$, as appropriate. If (z, w) and (x, y) together satisfy the Kuhn-Tucker conditions, it can be shown that

$$\langle z, u \rangle + g_y(u) \leq \alpha \leq f_x(v) + \langle w, v \rangle ,$$

or equivalently,

$$\sup\{\Sigma K_i(x_i', y_i) + \langle z, u \rangle \mid x' \in X_u\} \leq \alpha \leq \inf\{\Sigma K_i(x_i, y_i') + \langle w, v \rangle \mid y' \in Y_v\} ,$$

for any (u, v), where $\alpha = \text{maximin}(\mathcal{P})$. The vectors z and w can thus be regarded as "prices per unit of perturbation" such that neither central controller can gain an advantage by buying any perturbation of his resource constraint at that price.

By virtue of Theorems 1, 3, 4 and 5, solving $(\mathcal{P})$ is essentially equivalent to solving $(\mathcal{L})$. On the other hand, Theorems 2, 3, 4, 5 and 6 (below) show that solving $(\mathcal{P})$ is essentially equivalent to solving $(\mathcal{D})$.

THEOREM 6. *If $(\mathcal{D})$ is strongly consistent and (z, w) is a solution of $(\mathcal{D})$, then there exists a pair (x, y) such that (z, w) and (x, y) together satisfy the Kuhn-Tucker conditions.*

PROOF. By Theorem 3 of McLinden (1972b) and Theorem 37.5 of Rockafellar (1970).

The theorems so far have characterized the solutions of $(\mathcal{P})$, $(\mathcal{D})$ and $(\mathcal{L})$ fairly completely, but they make no existence assertion. For this, we give the following result.

THEOREM 7. *Assume that $(\mathcal{P})$ is strongly consistent and that there exists a pair $(x, y) \in X \times Y$ such that the sets*

$$X \cap \{x' \mid K_1(x_1', y_1) + \ldots + K_r(x_r', y_r) \geq \alpha\}$$

and

$$Y \cap \{y' \mid K_1(x_1, y_1') + \ldots + K_r(x_r, y_r') \leq \alpha\}$$

are bounded for every $\alpha \in R$. Then the Kuhn-Tucker conditions are satisfiable. Moreover, $(\mathcal{D})$ is strongly consistent and the set of solutions of $(\mathcal{P})$ is bounded.

PROOF. By Theorem 6 of McLinden (1972b).

4. METHODS OF SOLUTION

For simplicity we confine our attention in this section to the case of equality constraints, that is, we assume $P = \{0\}$ and $Q = \{0\}$. (Hence $P^\circ = R^p$ and $Q^\circ = R^q$.) The reader can modify the following remarks as necessary for other choices of P and Q.

First, observe that an especially nice situation for computation is that in which each K_i is finite. In this case $C = R^m$ and $D = R^n$, which means that the Lagrangian problem $(\mathcal{L})$ consists of finding the saddle points of an everywhere-finite, completely unconstrained saddle function which is differentiable if each K_i is. We shall not discuss this case further.

More common, perhaps, is the case in which not all of the K_i's are finite. As noted in the Introduction, some or all of the K_i's can themselves represent constrained saddle point problems. In this case, the situation computationally is still quite good as long as each K_i is *cofinite*, that is, has a finite conjugate. A saddle function K_i is cofinite if, for example, $\text{dom}\, K_i$ is bounded. More generally, K_i is cofinite provided it satisfies a certain "growth" condition to the effect that it curves upwards sufficiently steeply in its convex argument and it curves downwards sufficiently steeply in its concave argument.

Assume now that each K_i is cofinite. Then $Z = R^p$ and $W = R^q$, so that the dual problem $(\mathcal{D})$ consists of locating the saddle points of an everywhere-finite, completely unconstrained saddle function. Moreover, under this hypothesis $(\mathcal{D})$ is strongly consistent, which by Theorem 6 implies that once a solution (z, w) of $(\mathcal{D})$ has been found the Kuhn-Tucker conditions must be solvable with this (z, w) for some pair (x, y). That is, if $(\mathcal{P}_1), \ldots, (\mathcal{P}_r)$ denote the modified subproblems determined by (z, w), then there must exist a pair (x, y) such that

$$A_1x_1 + \ldots + A_rx_r = a, \quad B_1y_1 + \ldots + B_ry_r = b \ ,$$

and

$$(x_i, y_i) \text{ solves } (\mathcal{P}_i) \text{ for } i = 1, \ldots, r \ .$$

This gives us, then, a way of decomposing $(\mathcal{P})$ into $r + 1$ problems $(\mathcal{D})$, $(\mathcal{P}_1)$, $\ldots$, $(\mathcal{P}_r)$, where $(\mathcal{D})$ is completely unconstrained and each $(\mathcal{P}_i)$ has dimensionality much less than $(\mathcal{P})$.

Use of this decomposition principle depends upon two things. First we must know that a solution of $(\mathcal{D})$ in fact exists, and second, we must be able to compute one.

The first point is handled relatively easily. By Theorems 3 through 6, it is immediate that $(\mathcal{D})$ has a solution if and only if either $(\mathcal{P})$ has a stable solution or else $(\mathcal{P})$ is strongly consistent and has a solution. Hence Theorem 7 or other existence theorems for $(\mathcal{P})$ can be invoked. Observe that when $(\mathcal{P})$ is strongly consistent and each $\operatorname{dom} K_i$ is bounded, each K_i is cofinite and the hypotheses of Theorem 7 are met. Thus under this simple hypothesis the present approach to solving $(\mathcal{P})$ can be used, provided $(\mathcal{D})$ can be solved.

Solving $(\mathcal{D})$ is not necessarily easy, but we shall now show that it might be tractable even when the expressions $K_i^*(A_i^*z,\ B_i^*w)$ which it involves are not available in explicit form (cf. Rockafellar (1970), pp. 289-290).

Let H denote the saddle function appearing in $(\mathcal{D})$, that is,

$$H(z, w) = \langle z, a \rangle + \langle w, b \rangle - \Sigma K_i^*(A_i^*z,\ B_i^*w) \ .$$

Suppose any (z, w) is given. Since $(\mathcal{D})$ is strongly consistent, it follows from Theorems 1 and 4 of McLinden (1972a) that

$$\partial H(z, w) = \{(a, b)\} - \Sigma(A_i \times B_i)\, \partial K_i^*(A_i^*z,\ B_i^*w) \ .$$

Also, from Theorems 37.5, 37.4 and 36.3 of Rockafellar (1970) it follows that

$$(x_i,\ y_i) \ \varepsilon\ \partial K_i^*(A_i^*z,\ B_i^*w)$$

if and only if (x_i, y_i) solves

$$\operatorname*{maximin}_{C_i \ \ D_i} \{K_i - \langle \cdot,\ A_i^*z \rangle - \langle \cdot,\ B_i^*w \rangle\} \ ,$$

in which case the saddle value itself is $K_i^*(A_i^*z,\ B_i^*w)$. Combining these facts and letting $(\mathcal{P}_1), \ldots, (\mathcal{P}_r)$ denote the modified subproblems determined by (z, w), we have that

$$(s, t) \ \varepsilon\ \partial H(z, w)$$

if and only if

$$(s, t) = (a, b) - \Sigma(A_ix_i,\ B_iy_i) \ ,$$

where (x_i, y_i) solves $(\mathcal{P}_i)$ for $i = 1, \ldots, r$, and in this case

$$H(z, w) = \langle z, a \rangle + \langle w, b \rangle - \Sigma \operatorname{maximin}(\mathcal{P}_i) \ .$$

Solving the r modified subproblems determined by (z, w) thus amounts to "dualizing" the expressions $K_i^*(A_i^*z,\ B_i^*w)$ and $\partial K_i^*(A_i^*z,\ B_i^*w)$ and hence furnishes us with a means of calculating $H(z, w)$ and an element of $\partial H(z, w)$. Therefore, as long as it is relatively easy to solve the modified subproblems determined by any given (z, w), we could solve $(\mathcal{D})$ by any method which demanded only the ability to calculate $H(z, w)$ and an element of $\partial H(z, w)$.

5. REMARKS

The context treated in this paper includes, of course, the case in which the K_i's are differentiable saddle functions. In this case the subdifferentials ∂K_i coincide with the ordinary gradient mappings ∇K_i.

As a general rule, results involving saddle functions can be made to subsume corresponding results for purely concave and purely convex functions. One simply introduces "degenerate" saddle functions of the form $K(x, y) = g(x)$ and $K(x, y) = f(y)$, for g concave and f convex. Thus, any of the r coupled saddle point problems represented by the K_i's might in fact be trivial in one or the other argument, as just indicated. This would correspond to an enterprise in which only one of the two central controllers was involved. Of course, if each of the r enterprises involves only one of the two controllers, then the overall problem reduces to two completely independent coupled nonlinear programming problems, a coupled concave problem for the maximizing controller and a coupled convex problem for the minimizing controller.

Although we have dealt with only one coupling constraint in the concave argument and one in the convex argument, this actually entails no loss of generality. Indeed, suppose $(\mathcal{P})$ were of the form

$$\underset{X\quad Y}{\text{maximin}}\,\{K_1(x_1, y_1) + \ldots + K_r(x_r, y_r)\} ,$$

where

$$X = \{x \in C \mid A_{j1}x_1 + \ldots + A_{jr}x_r \leq_{P_j} a_j \text{ for } j = 1, \ldots, s\} ,$$

$$Y = \{y \in D \mid B_{k1}y_1 + \ldots + B_{kr}y_r \leq_{Q_k} b_k \text{ for } k = 1, \ldots, t\} ,$$

and

$$A_{ji}: R^{m_i} \to R^{p_j}, \quad j = 1, \ldots, s \text{ and } i = 1, \ldots, r ,$$

$$B_{ki}: R^{n_i} \to R^{q_k}, \quad k = 1, \ldots, t \text{ and } i = 1, \ldots, r .$$

Then simply put $p = \Sigma p_j$, $q = \Sigma q_k$,

$$a = (a_1, \ldots, a_s) \in R^p, \qquad b = (b_1, \ldots, b_t) \in R^q ,$$

$$P = P_1 \times \ldots \times P_s \subset R^p, \quad Q = Q_1 \times \ldots \times Q_t \subset R^q ,$$

and define $A_i: R^{m_i} \to R^p$ and $B_i: R^{n_i} \to R^q$ by

$$A_i(x_i) = (A_{1i}x_i, \ldots, A_{si}x_i), \quad B_i(y_i) = (B_{1i}y_i, \ldots, B_{ti}y_i) .$$

This converts that problem into the form treated in this paper.

The theorems of §3 by no means exhaust the possible conclusions for $(\mathcal{P})$, $(\mathcal{D})$ and $(\mathcal{L})$ which follow from McLinden (1972b). For example, one may define a notion of stability for the solutions of $(\mathcal{D})$ and then conclude that $(\mathcal{P})$ has a stable solution if and only if $(\mathcal{D})$ does, in which case $\text{maximin}(\mathcal{P}) = \text{minimax}(\mathcal{D})$. We shall not pursue this further, though, except to point out that the present results are "invariant under equivalence." That is, if new problems $(\tilde{\mathcal{P}})$, $(\tilde{\mathcal{D}})$, $(\tilde{\mathcal{L}})$ and $(\tilde{\mathcal{P}}_i)$ are defined by replacing each saddle function K_i (resp. K_i^*) by any saddle function $\tilde{K}_i$

(resp. $\tilde{K}_i^*$) which is equivalent to K_i (resp. K_i^*) in the sense of Rockafeller, then $(\tilde{\mathcal{P}})$, $(\tilde{\mathcal{D}})$, $(\tilde{\mathcal{L}})$ and $(\tilde{\mathcal{P}}_i)$ are exactly the same saddle point problems as $(\mathcal{P})$, $(\mathcal{D})$, $(\mathcal{L})$ and $(\mathcal{P}_i)$, respectively.

Finally, we note that many important choices of P and Q are polyhedral convex cones. In this event the hypotheses of strong consistency for $(\mathcal{P})$ and $(\mathcal{D})$ can be relaxed by deleting the relative interior operation "ri" from in front of P, Q, P° and Q°. All the theorems of §3 are then still true using these weaker hypotheses. The proof of this fact rests on establishing "polyhedral refinements" of the results of McLinden (1972a, 1972b).

REFERENCES

G. B. Dantzig (1970), "Large scale systems and the computer revolution," in Proceedings of the Princeton Symposium on Mathematical Programming, H. W. Kuhn, ed., Princeton University Press, Princeton.

G. B. Dantzig and P. Wolfe (1960), "Decomposition principle for linear programs," Operations Res. 8, 101-111.

A. M. Geoffrion (1970), "Elements of large-scale mathematical programming," Management Sci. 16, 652-691.

L. S. Lasdon (1970), Optimization Theory for Large Systems, Macmillan, New York.

L. McLinden (1972a), "Dual operations on saddle functions," TSR #1229, Mathematics Research Center, University of Wisconsin, Madison. (To appear in Trans. Amer. Math. Soc.)

L. McLinden (1972b), "An extension of Fenchel's Duality Theorem to saddle functions and dual minimax problems," TSR #1242, Mathematics Research Center, University of Wisconsin, Madison.

A. I. Penn (1971), "A generalized Lagrange-multiplier method for constrained matrix games," Operations Res. 19, 933-945.

R. T. Rockafellar (1970), Convex Analysis, Princeton University Press, Princeton.

LEAST SQUARES SOLUTION OF SPARSE SYSTEMS OF NON-LINEAR EQUATIONS BY A MODIFIED MARQUARDT ALGORITHM

J.K. REID

Theoretical Physics Division,
A.E.R.E., Harwell, Didcot,
Berks, England.

Abstract: Marquardt's algorithm is adapted to exploit sparsity in a non-linear least squares problem that may or may not be overdetermined. Consideration is given to cases when derivatives are available analytically and where they have to be estimated. The results of numerical experiments are presented.

1. INTRODUCTION

We consider the solution of a system of m non-linear algebraic equations

$$r(x) = 0 \tag{1.1}$$

in n unknowns x_i in the case where a large number of derivatives $\partial r_i/\partial x_j$ are known to be zero everywhere so that the m x n Jacobian matrix

$$J = \{\partial r_i \,/\, \partial x_j\} \tag{1.2}$$

is sparse. We solve equation (1.1) in the sense of least squares, that is look for a minimum of the sum of squares

$$S(x) = \sum_{i=1}^{m} [r_i(x)]^2. \tag{1.3}$$

In the important case when the number n of unknowns x_j is equal to the number m of equations we of course hope to find a zero of S(x) corresponding to a solution of equation (1.1).

Systems of equations of this kind arise, for example, in the solution of the algebraic equations produced by finite-difference or finite-element discretization of boundary-value problems for ordinary and partial differential equations and they also arise in various network flow problems.

When solving a set of linear equations a choice has to be made between the use of an iterative method and a direct one. In general an iterative method is preferable if the matrix of the equations is large, very sparse and well-conditioned. For non-linear problems an iterative method is necessarily required but the same basic choice remains for we may either linearize the problem as a whole and use a direct method for solving the resulting linear equations or we may avoid ever solving a linear system by using such a method as non-linear successive over-relaxation (see Concus (1969), for an example with m=n). As in the linear case, we expect that each will show advantages over the other for certain classes of problems. Here we consider a method which linearizes the problem as a whole and uses sparse matrix techniques to solve the linearized system directly.

We began by trying to adapt the 'dog-leg' algorithm of Powell (1970) for the case m=n but found that this was sometimes unsuccessful, giving very slow convergence and we have therefore abandoned this approach. The reaons for this failure are interesting in themselves, so we describe this algorithm briefly in section 2 and explain why it failed. It should be emphasised that it was the adaptation that failed and that we do not wish to imply any criticism of the original method, designed for full systems.

It is our second attempt at this problem, using the algorithm of Marquardt (1963), that is the principle subject of this paper. Apart from considerations of sparsity we follow the recommendations of Fletcher (1971), and describe the algorithm including Fletcher's recommendations in section 3.

Our purpose at Harwell is to provide the computer user with a reliable general-purpose subroutine. We allow both for the case where the user calculates the Jacobian matrix (1.2) analytically and where he wishes it to be approximated by differencing. For efficiency in the case where many of the derivatives $\partial r_i/\partial x_j$ are known constants, we take r(x) to be of the form

$$r(x) = f(x) + Ax \tag{1.4}$$

where A is the matrix of known constants, so that the Jacobian of f has many zeros. Details of how this Jacobian is first approximated by differences and how the approximation is updated during the iteration are given in section 4.

In section 5 we explain those details of our subroutine which are a direct consequence of the fact that the Jacobian is sparse. Finally in section 6 we report on the performance of the subroutine on some test problems.

2. THE DOG-LEG ALGORITHM

Powell's (1970) algorithm is designed for the important case when there are as many variables as unknowns (i.e. m = n). It is a compromise between the method of steepest descent which is convergent but is often very slow and Newton's method which usually gives very rapid convergence when the iterates are near the solution but often gives unreliable results in the early stages of an iteration.

Powell keeps and updates a parameter Δ which is intended to indicate that the linear approximation

$$r(x + \delta) \approx r(x) + J\delta \tag{2.1}$$

is satisfactorily accurate in a sphere of radius Δ about the current iterate x (i.e. if $\sum_i \delta_i^2 < \Delta^2$) and he limits the correction made to x so that the new iterate lies in this sphere, taking it to be on the 'dog-leg' formed by joining x to the steepest descent point g and g to the Newton point v. If the Newton point is within the sphere of radius Δ then he tries this for the new iterate and otherwise takes the point where the dog-log intersects the surface of the sphere. The new point is used as the next iterate only if the sum of squares (1.3) shows an improvement and in any case the residuals there are used to check the accuracy of the approximation (2.1) and Δ is adjusted accordingly. Thus if the function is very non-linear then Δ will be very small and a step in the direction of steepest descent is taken; if it is nearly linear then the Newton iterate is used; in an intermediate case a compromise betwen the two is taken.

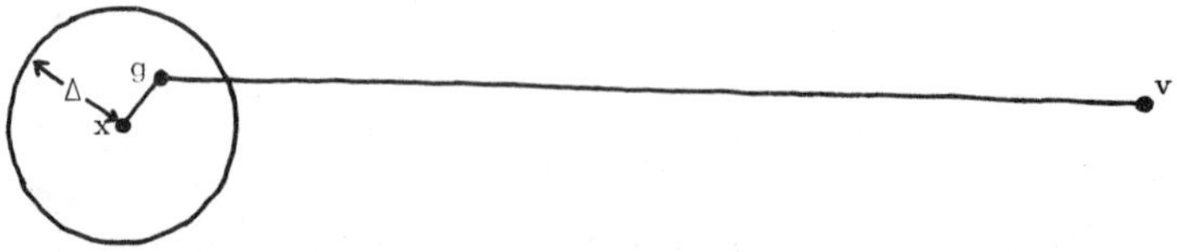

Figure 1.

The difficulty that we found with the method was associated with J being moderately ill-conditioned. In this case the length of the steepest-descent correction is likely to be much less than the length of the Newton correction. Now suppose the non-linearities are such that the radius Δ is significantly greater than the length of the steepest descent step and significantly less than the length of the Newton step (see Figure 1). In such a case we take a point on the sphere which is a long way from the Newton point v and so may not give a value for the sum of squares much better than that at the steepest descent point g. The Marquardt algorithm, on the other hand, produces a correction δ which is such that the sum of squares of the approximate residuals (2.1) is minimized on the sphere centre x passing through $x + \delta$. In two examples we found difficulties of this kind with our adaptation of the dog-leg algorithm and in each case the Marquardt algorithm performed perfectly satisfactorily. In Figure 2 we show the progress of the two algorithms on one of our cases. We plot the sum of squares at each of the points at which it was evaluated, although the iteration is continued from the previous point whenever the sum of squares shows an increase. It may be seen that although initially the dog-leg does rather better and must eventually show quadratic convergence its behaviour between iterations 10 and 60 is very poor. The Marquardt results here were obtained with Fletcher's subroutine for full systems.

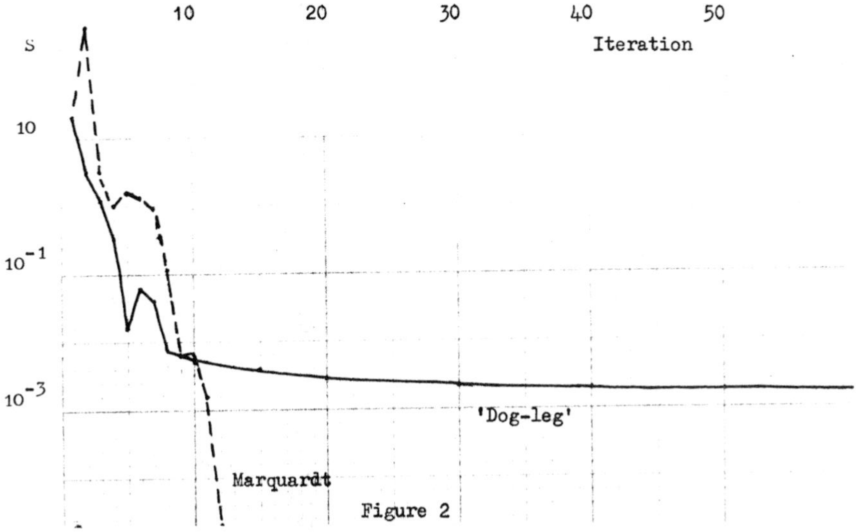

Figure 2

3. THE MARQUARDT ALGORITHM

In this section we describe the Marquardt algorithm and the detailed suggestions of Fletcher (1971). As in the case of the dog-leg algorithm the aim is to compromise between the methods of steepest descent and Newton. It has the additional advantage that is is applicable to overdetermined equations $(m > n)$.

A typical step involves solving the linear least squares problem

$$\begin{bmatrix} J \\ \lambda^{\frac{1}{2}} I \end{bmatrix} \delta = \begin{bmatrix} -r \\ 0 \end{bmatrix} \tag{3.1}$$

where J and r (see (1.1) and (1.2)) are evaluated at the current estimate of the solution, λ is a scalar known as the Marquardt parameter and the solution δ

gives a correction to the current estimated solution. This problem may be solved either by the formation and solution of the normal equations

$$(J^TJ + \lambda I) \quad \delta = - J^Tr \tag{3.2}$$

or by the premultiplication of equation (3.1) by an orthogonal matrix P to yield an equivalent system

$$\begin{bmatrix} U \\ O \end{bmatrix} \delta = \begin{bmatrix} s \\ t \end{bmatrix} \tag{3.3}$$

where U is upper triangular. Equation (3.2) is normally solved by Cholesky decomposition

$$(J^TJ + \lambda I) = LL^T \tag{3.4}$$

followed by forward and backward substitution and equation (3.3) is solved by a straightforward back-substitution. It may be shown (Wilkinson (1965) for example) that L^T and U are essentially identical in the sense that $L^T = DU$ where D is a diagonal matrix with diagonal elements ± 1. In the sparse case it is desirable to order the variables for sparsity in U. In our subroutine we work with the normal equations and choose the ordering by analysing the sparsity pattern of these equations just once before we commence interating.

After finding δ we check whether the inequality

$$S (x + \delta) < S (x) \tag{3.5}$$

holds, that is whether the sum of squares of residuals (1.3) would decrease if we replaced x by x + δ. If it holds then we make this replacement. In either case we adjust λ in the light of the size of the reduction in the sum of squares.

If λ is very large then it is clear from equation (3.2) that the correction is approximately $-\lambda^{-1} J^Tr$ which is an incremental step along the direction of steepest descent of the sum of squares (1.3). On the other hand if $\lambda = 0$ then equation (3.1) becomes that for the correction in the Gauss - Newton method. We adjust λ by comparing the actual reduction in the sum of squares, $S(x) - S(x+\delta)$, and the reduction

$$-(2r^TJ\delta + \delta^TJ^TJ\delta) \tag{3.6}$$

that would have taken place in the sum of squares if the linear model

$$r (x + \delta) = r + J \delta \tag{3.7}$$

had been exactly correct. If the ratio R of actual reduction divided by predicted reduction is near to 1 then the linear model is good and λ should be reduced to permit longer steps to be taken, whereas a negative or near-zero value of R suggest on the contrary that smaller steps should be taken. Fletcher makes no change to λ if R is in the range $[\rho,\sigma]$ choosing the values $\rho = 0.25$, $\sigma = 0.75$ on the basis of numerical experimentation. If $R < \rho$ then he increases λ by the factor

$$2 + \frac{S(x) - S(x + \delta)}{r^TJ\delta} \tag{3.8}$$

unless this lies outside the range $[2,10]$ in which case either 2 or 10 is used according as the factor is less then 2 or greater than 10.

Formula (3.8) is based on the fact that the sum of squares and its derivative is available at x and the sum of squares is available at x + δ, so that a quadratic approximation to the sum of squares on the line joining x to x + δ may be found and minimized. If $R > \sigma$ then Fletcher simply halves λ unless this would make it less than a cut-off value λ_c in which case he sets λ to zero. If later on an increase in λ is required and its value is zero then a new λ_c is calculated, λ is set to $\lambda_c/2$ and the same procedure for increasing λ is used, based on expression (3.8). On practical computing grounds we do not calculate λ_c in the way Fletcher does but follow the spirit of his recommendation and find a value that approximates the smallest eigenvalue of J^TJ. The intention is that λ_c be such that an iteration with $\lambda=\lambda_c$ will not give substantially different results from an iteration with $\lambda = 0$. The iteration is commenced with $\lambda = 0$.

4. THE APPROXIMATION OF THE JACOBIAN

We describe in this section the procedure adopted when the user does not supply code to evaluate $\{\partial f_i/\partial x_j\}$ analytically. We begin by approximating it by differences. For our Harwell subroutine we require the user to write a subroutine that enables f(x) to be found from x and use central differences with the procedure of Curtis and Reid (1972) for choosing steps and the procedure of Curtis, Powell and Reid (1972) for economizing the number of evaluations of f(x). It would have been more straightforward to calculate the non-zero derivatives $\partial f_i/\partial x_j$ if we had required the user to supply code to find each component $f_i(x)$ individually, but this approach has the disadvantages that such code is less straightforward to write, no advantage is taken of any subexpressions common to several components and the subroutine-call overheads will be substantially increased.

We correct the approximation whenever x + δ is taken as the next iterate. We evaluate the vector

$$\gamma = f(x + \delta) - f(x) \tag{4.1}$$

of differences and require the new approximate Jacobian J* to satisfy the equation

$$J^* \delta = \gamma. \tag{4.2}$$

Powell uses the updating formula

$$J^* = J + (\gamma - J\delta)\,\delta^T \;/\; \|\delta\|^2 \tag{4.3}$$

which unfortunately is not suitable in our case since it does not preserve the sparsity pattern. We therefore follow the suggestion of Schubert (1970) and use the formula

$$J_i^* = J_i + (\gamma_i - J_i\,\delta^{(i)})\,\delta^{(i)T} \;/\; \|\delta^{(i)}\|^2, \quad i = 1,2,\ldots m, \tag{4.4}$$

where J_i^* and J_i denote the ith rows of J* and J respectively, and $\delta^{(i)}$ is obtained from δ by replacing by zero each component which corresponds to a known zero in the ith row of the Jacobian. It may be readily verified that J* has the correct sparsity pattern and satisfies equation (4.2).

An interesting property of formula (4.3) is that it produces the matrix J* that minimizes $\|J^* - J\|$ among all those matrices satisfying equation (4.2), where the norm is the Frobenius norm

$$\|A\| = \left\{ \sum_{ij} a_{ij}^2 \right\}^{\frac{1}{2}} \tag{4.5}$$

To prove this we first remark that for the ith row of J we wish to minimize $\|J_i^* - J_i\|^2$ subject to the condition $J_i^*\delta=\gamma_i$ and by using a Lagrange multiplier we may immediately verify that formula (4.3) achieves this. Adding over all rows we find that $||J^* - J||^2$ has been minimized. In the present case an almost identical argument shows that formula (4.4) minimizes $||J^* - J||^2$ over all matrices J^* that have the correct sparsity pattern and satisfy equation (4.2).

A further important property of formula (4.3) is that if f is linear with exact Jacobian $\bar{J}$ then $\gamma = \bar{J}\,\delta$ so that equation (4.3) may be written in the form

$$J^* - \bar{J} = (J - \bar{J})\ \ (I - \delta\,\delta^T / ||\delta||^2). \tag{4.6}$$

We may deduce the inequality

$$||J^* - \bar{J}|| \leqslant ||J - \bar{J}||. \tag{4.7}$$

In the sparse case we may replace equation (4.6) by the equations

$$J_i^* - \bar{J}_i = (J_i - \bar{J}_i)\ \ (I - \delta^{(i)}\delta^{(i)T} / ||\delta^{(i)}||^2) \tag{4.8}$$

from which the truth of inequality (4.7) may be deduced.

Formula (4.4) breaks down if $\delta^{(i)} = 0$ (which may, of course, happen to non-zero δ) and in this case the only sensible strategy is to make no correction to the ith row of the Jacobian because we have no new information about this row. We next consider the case when $\delta^{(i)}$ is small. It might be thought that in the presence of roundoff the corrections of equation (4.4) might become unbounded as $\delta^{(i)} \to 0$, but if the computed values of f(x) satisfy the Lipschitz condition

$$|\bar{f}_i\,(x + \delta) - \bar{f}_i\,(x)| \leqslant L_i\ \ ||\delta^{(i)}|| \tag{4.9}$$

then the correction is bounded by the inequality

$$||J_i^* - J_i|| \leqslant L_i + ||J_i||. \tag{4.10}$$

Also $||J^*||$ can increase at most linearly for we may also deduce the inequality

$$||J_i^*|| < L_i + ||J_i\left(I - \delta^{(i)}\delta^{(i)T} / ||\delta^{(i)}||^2\right)||$$

$$\leqslant L_i + ||J_i|| \tag{4.11}$$

from equation (4.4) and the Lipschitz condition (4.9). Good code is likely to result in such a Lipschitz condition holding for all but a few exceptional values of x (e.g., perhaps, x = 0) so there is little danger of catastrophically large corrections being made. We therefore make no test on the smallness of $\delta^{(i)}$. This means that there is a possibility of the elements of one row becoming inaccurate, but later iterations are likely to correct it and in any case, for reasons to be explained in the next paragraph, we use finite differences to evaluate an entirely fresh approximate Jacobian at regular intervals.

Powell (1970) includes safeguards against the directions δ generated being almost linearly dependent, since in this case little information will be fed into the approximation to J about the variation of f in some subspace of directions. Powell's precautions are not practicable in our case because they require too much storage. We therefore guard against this possibility by calculating a fresh approximation by differences at regular intervals.

5. FURTHER DETAILS

The solution of equation (3.1) is normally quite an expensive operation since it involves either the orthogonal decomposition of $\begin{bmatrix} J \\ \lambda^{\frac{1}{2}} I \end{bmatrix}$ or the formation and Cholesky decomposition of $(J^TJ + \lambda I)$. We had hoped that if λ remained unchanged from one iteration to another then use of the old decomposition, coupled with iterative refinement, would enable an acceptable approximation to δ to be found without performing a new decomposition. Unfortunately, numerical experiments showed that comparatively few decompositions can be saved in this way and these savings are mostly near the end of the sequence of iterations which is just when it is worthwhile to calculate δ accurately. We therefore decompose the matrix at each iteration.

For the parameter λ_c (see end of section 3) we require an estimate of the smallest eigenvalue of J^TJ. Fletcher uses $1/\|(J^TJ)^{-1}\|$, but in our application this would involve too much work since $(J^TJ)^{-1}$ is, in general, full. Instead we form a vector α with pseudo-random components, solve the system

$$J^TJ \quad \beta = \alpha \tag{5.1}$$

and use the estimate

$$\lambda_c = \|\alpha\| \,/\, \|\beta\|. \tag{5.2}$$

The Cholesky decomposition of J^TJ is available (see section 3) so system (5.1) is easy to solve. Normally expression (5.2) gives a sufficiently accurate estimate of the smallest eigenvalue of J^TJ, but since it is always an overestimate and an unfortunate choice of α could make it a severe overestimate, we use the algorithm itself to check it. If λ is reduced to zero and stays there for several iterations then λ_c has performed its function satisfactorily. On the other hand if it is reduced to zero for one iteration only, this indicates that λ_c was too large; in such a case we calculate a fresh estimate (5.2) and take for λ_c the smaller of this and half the old value.

6. NUMERICAL RESULTS

As a test we have used an adaptation to the sparse case of the trigonometric equations of Fletcher and Powell (1963). The equations we used take the form

$$\sum_{j=1}^{n} (a_{ij} \sin x_j + b_{ij} \cos x_j) + \sum_{j=1}^{n} c_{ij} x_j = e_i \quad , \; i=1,2,\ldots,n \tag{6.1}$$

where $\{a_{ij}\}$ and $\{b_{ij}\}$ have the same sparsity pattern, including non-zeros on the diagonal and $\{c_{ij}\}$ has a separate sparsity pattern. Each is arranged to be a band matrix of band-width (2b-1) and the square of a pseudo-random integer, uniformly distributed in $[1,\mu]$, is used for the spacing between non-zeros in the columns, the spacing being interpreted as running on from the last element of a column that is within the band structure to the first element of the next column that is with the band structure. We generated the non-zero coefficients a_{ij}, b_{ij} by using a pseudo-random number generator which gives a uniform distribution on $[-1,1]$ and the solution coefficients x_j by a similar generator on $[-\pi,\pi]$; the coefficients e_i were evaluated to give this solution. For a first approximate solution we perturbed the components of the true solution by pseudo-random numbers in the range $[-\pi/10, \pi/10]$.

By varying n,b and μ we generated a range of problems, some of which were pathologically ill-conditioned, and ran each, in double-length arithmetic on the IBM 370/165 computer, with and without the use of analytic derivatives. The results of a few of these are summarized in Table 1. Most cases showed the satisfactory convergence of the cases (50,2,2), (90,2,7) and (100,10,2). When using analytic derivatives the new subroutine showed slight differences from Fletcher's subroutine for full matrices because of the different way λ_c is calculated, but neither was consistently better than the other; the case illustrated in Figure 2 is (100,10,2) and here the new subroutine happened to perform a little better. The two cases in which the soubroutine had difficulty are included in the table; both have extremely ill-conditioned Jacobians. In the (50,10,10) case good accuracy was obtained eventually, although convergence was very slow, requiring a non-zero λ almost all the time. In the (100,10,7) case the subroutine was left with $S=3.5\times10^{-8}$ because the steps being taken were pathologically small.

The results of Table 1 bring into question the wisdom of using the procedure of Curtis and Reid (1972) to approximate the Jacobians by differences for this computation often dominates the total number of function evaluations needed. A cruder differencing algorithm would perhaps suffice, but our algorithm does have the advantage that the user is not asked to specify a suitable step-length.

Tests have also been performed on the finite-difference approximations to the minimal-surface equation considered by Concus (1969). Our provisional conclusion from these tests is that the new algorithm is likely to be more expensive computationally than the block successive overrelaxation used by Concus. On the other hand it does have the advantage that no estimation of a relaxation parameter is necessary.

				Number of function evaluations to reduce S below				approximate	
n	b	μ	anal. derivs	10^{-4}	10^{-7}	10^{-10}	10^{-20}	first J	later Js
50	2	2	no	18	20	24	26	13	0
			yes	6	8	10	12		
50	10	10	no	21	62	105	131	17	18
			yes	8	10	50	81		
90	2	7	no	18	22	24	26	9	0
			yes	9	11	12	13		
100	10	2	no	71	79	82	90	61	0
			yes	7	8	9	10		
100	10	7	no	28	40	129*	129*	21	24
			yes	10	15	72*	72*		

* Accuracy not achieved when pathologically small steps being taken.

TABLE 1.

REFERENCES

P. Concus (1969) Numerical solution of the minimal surface equation by block non-linear successive over-relaxation. Information Processing 68, 153 - 158, North Holland.

A.R. Curtis, M.J.D. Powell and J.K. Reid (1972) On the estimation of sparse Jacobian matrices. A.E.R.E. report TP 476.

A.R. Curtis and J.K. Reid (1972) The choice of step lengths when using differences to approximate Jacobian matrices. A.E.R.E. report TP 477.

R. Fletcher (1971) A modified Marquardt subroutine for non-linear least squares. A.E.R.E. report R 6799. HMSO.

R. Fletcher and M.J.D. Powell (1963) A rapidly convergent descent method for minimization. Comp. J. 6, 163 - 168.

D.W. Marquardt (1963) An algorithm for least squares estimation of non-linear parameters. J. SIAM, 11 , 431 - 441.

M.J.D. Powell (1970) Chapters 6 and 7 of "Numerical methods for non-linear algebraic equations", edited by P. Rabinowitz, Gordon and Breach, London.

L.K. Schubert (1970) Modification of a quasi-Newton method for non-linear equations with a sparse Jacobian. Math. Comp. 24, 27 - 30.

J.H. Wilkinson (1965) The algebraic eigenvalue problem. Clarendon Press, Oxford.

UNIFIED PIVOTING PROCEDURES FOR LARGE STRUCTURED LINEAR SYSTEMS AND PROGRAMS

Michael D. Grigoriadis

IBM Corporation
1350 Avenue of the Americas
New York, N.Y., U.S.A.

Abstract: Pivoting procedures preserving the structure of the basis inverse for a system of linear algebraic equations and their efficient implementation are reviewed. It is suggested that existing linear programming codes may be used to accomplish this task. Partitioning methods in linear programming are discussed in the context of these pivoting procedures. It is argued that a number of partitioning methods may be synthesized by appropriately defining the "pivot selection" criteria (e.g. a particular pricing strategy of the primal or dual simplex method) and then applying the appropriate pivoting procedure. Partitioning algorithms derived from the primal, dual or primal-dual simplex pivot selection criteria require no further theoretical justification. Most known partitioning methods belong to this category. One such example is discussed in detail. Separating the pivoting effort (which can be measured or bounded) from the question of convergence allows realistic experimental testing of a proposed partitioning method via an existing large scale LP code and a minimum amount of programming effort. Partitioning approaches to some specially structured network flow problems are briefly reviewed.

1. LARGE STRUCTURED LINEAR SYSTEMS

Large block diagonal systems of linear algebraic equations of the form (LS):

(1.1) $$\Sigma_{k=1}^{L} A_k x_k = h_0$$

(1.2) $$B_k x_k = h_k; k=1,\ldots,L$$

frequently arise from large engineering and operations research applications. The A_k and B_k are given (m_0,n_k) and (m_k,n_k)-matrices respectively with $n_k \geqq m_k$, h_k are given m_k-vectors and x_k are n_k-vectors of variables. With no loss in generality, we assume LS to be of full row rank and we consider systems with $m \leqq n$ where $m = \Sigma_{k=0}^{L} m_k$ and $n = \Sigma_{k=1}^{L} n_k$. A double subscript on vectors and a triple subscript on the coefficients of LS will be used. Thus a_{kjr} will refer to the r-th element of column a_{kj} of A_k. A prime will indicate the transpose.

We are interested in large systems LS whose overall size far surpasses the capabilities of existing computational tools for handling LS directly. An underlying assumption, borne out in practice, is that the A_k and B_k are "sparse", i.e., they have very few nonzero elements. Sparse matrix methods for solving linear equations by direct methods have been used successfully in linear programming since the late fifties. In most LP codes the product form of the inverse, a form of the Gauss-Jordan elimination method, is used to solve linear equations. Elaborate facilities exist in these codes for input output, error detection, pivoting for accuracy and for preserving sparseness of the inverse, scaling, etc.

This paper will discuss computational methods for solving LS via implementations of the product form of the inverse method, taking full advantage of the structure exhibited by LS and the services offered by existing large LP codes. Although when m=n elementary knowledge of numerical analysis dictates avoiding the computation of the inverse as a procedure for solving LS, we assume that there is adequate interest in computing this inverse, e.g. when LS is to be solved against many right hand sides. The power and sophistication of commercial LP codes provides an ideal tool for solving large sparse linear systems of equations. Curiously, this capability has not been adequately exploited in practice.

Equally interesting, but not discussed here, is the implementation of the elimination form of the inverse, a form of Gaussian elimination, the LU decomposition, or other factorizations (e.g. see Brayton(70), Saunders(72), Willoughby(69)), adapted to take advantage of the structure of LS.

The case m=n is straightforward since pivoting procedures developed for m<n are directly applicable. When LS is underdetermined (m<n), we can segregate the variables into a *basic* or *dependent* set and a *nonbasic* or *independent* set of variables. The basic variables are uniquely determined for arbitrary values assigned to the nonbasic variables. There are, of course, finitely many bases of LS and each corresponding *basis matrix* G necessarily exhibits the structure of LS. It is of interest then to devise computationally efficient methods for constructing the inverse of a given basis matrix of LS and most importantly, for determining the representation of the inverse of a new basis matrix differing from the previous one (for which the inverse matrix is at hand) by exactly one column. The computational efficiency sought is derived not only from the reduction of primitive operations to be performed but also from the reduction of core storage requirements. Thus, it is essential for such methods to also preserve the structure of the basis inverse matrix.

2. THE BASIS MATRIX AND ITS INVERSE

A basis G for LS will necessarily contain m_k columns chosen from each block k=1,...,L. The additional m_0 columns of G will be chosen from at least one block, say from the first p<L blocks. The remaining L-p blocks contribute only m_k columns to G. After Dantzig and Van Slyke(67) we refer to k=1,...,p as "essential" and to k=p+1,...,L as "inessential" blocks. We denote them by superscripts E and I respectively. The B_k^I are square and nonsingular and the B_k^E are rectangular ($m_k<n_k$) guaranteed to contain at least one set of exactly m_k linearly independent columns. We let $B_k^E = (B_k^{BE}|B_k^{KE})$ where B_k^{KE} is an *arbitrarily chosen* nonsingular (m_k,m_k)-submatrix of B_k^E. The columns of G containing columns of B_k^{KE} and B_k^{BE} are referred to as "*key essential*" and "*basic essential*" respectively. The resulting partitioning of the coefficient matrix, along with some additional notation is shown in Figure 1.

The basis matrix is characterized by nonsingular blocks along the diagonal and by m_0 linking rows and m_0 linking columns. The latter also exhibits a block diagonal structure generally with p($\leq$L) blocks. From basis considerations it is obvious that when L$>m_0$,

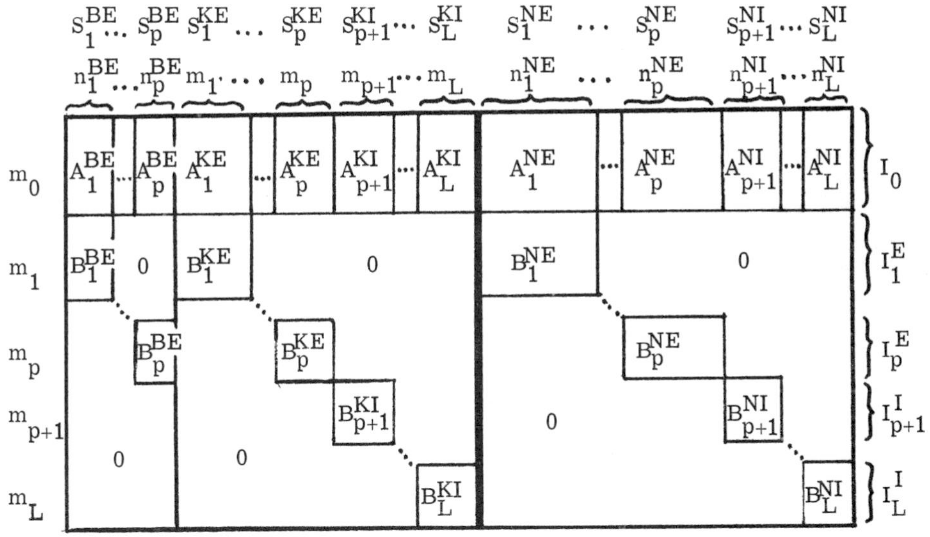

Figure 1. Partitioning induced by basis: $(G \mid \bar{G})$.

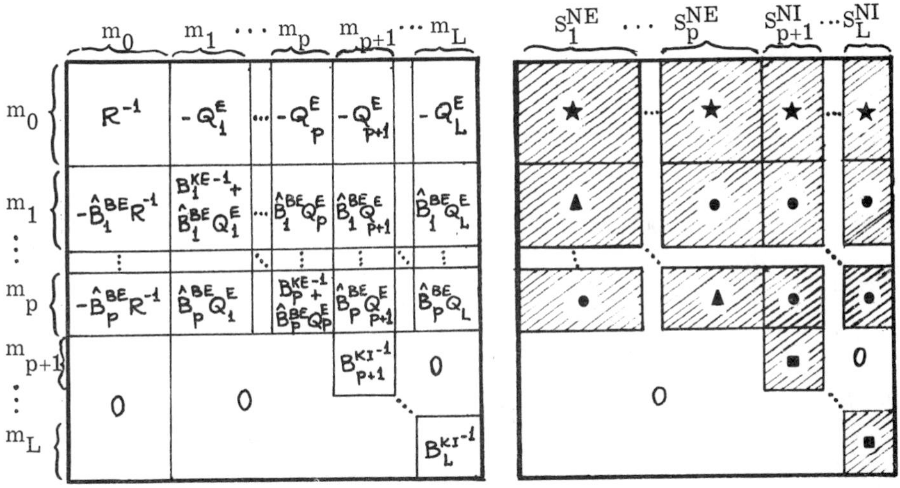

Figure 2. Basis inverse (G^{-1}) and tableau $(G^{-1}\bar{G})$ structure.

$p < m_0$. In practice, $p \ll m_0$ depending on the particular application. Both the basis and the nonbasic part of the original matrix retains the original structure. The superscript N denotes "nonbasic".

Figure 2 depicts the structure of G^{-1} where

(2.1) $$R = (D_1^{BE} | \ldots | D_p^{BE})$$

(2.2) $$D_k^{BE} = A_k^{BE} - A_k^{KE}\bar{B}_k^{BE}$$

$$(2.3) \qquad \bar{B}_k^{BE} = B_k^{KE^{-1}} B_k^{BE} \; ; \qquad \hat{B}_k^{BE} = B_k^{KE^{-1}} (B_k^{BE} \mid 0)$$

$$(2.4) \qquad Q_k^E = R^{-1} A_k^{KE} B_k^{KE^{-1}} \; ; \; k=1,\dots,p$$

are (m_0,m_0), (m_0,n_k^{BE}), (m_k,m_0) and (m_0,m_k) matrices respectively. We immediately observe that R^{-1}, the block basis inverses $B_k^{K^{-1}}$;k=1,...,L and knowledge of the current partitioning (i.e. the sets $S_1^{BE},\dots,S_p^{BE}$, $S_1^{KE},\dots,S_p^{KE}$, $S_{p+1}^{KI},\dots,S_L^{KI}$) are sufficient to determine any element of G^{-1}. This figure also indicates the possible nonzero (tableau) coefficients of $G^{-1}\bar{G}$ as hatched areas.

Since R^{-1} plays a central role in computing G^{-1} by partitioning, we are naturally led to examine means of constructing and revising R^{-1}. The following three step procedure will produce at least one R^{-1} and will define the partitioning depicted in Figure 1.

1) Obtain a basic solution to (1.2) and the basis inverses $B_k^{K^{-1}}$; k=1,...,L by <u>any</u> method, e.g., by a suitable pivoting strategy, by using the simplex method, etc.

2) Compute

$$(2.5) \qquad D_k^{\bar{K}} = A_k^{\bar{K}} - A_k^K B_k^{K^{-1}} B_k^{\bar{K}} \; ; \; k=1,\dots,L,$$

$$(2.6) \qquad h^D = h_0 - \Sigma_{k=1}^L A_k^K B_k^{K^{-1}} h_k$$

where $B_k = (B_k^K, B_k^{\bar{K}})$; $A_k = (A_k^K, A_k^{\bar{K}})$

3) Obtain a basic solution of the "Derived System" (DS):

$$(2.7) \qquad \Sigma_{k=1}^L D_k^{\bar{K}} x_k^{\bar{K}} = h^D$$

to define the partitioning

$$(2.8) \qquad (R \mid R^N) = (D_1^{BE},\dots,D_p^{BE} \mid D_1^{NE},\dots,D_p^{NE}, D_{p+1}^{NI},\dots,D_L^{NI})$$

and obtain R^{-1} by any suitable method.

If an LP code were used to accomplish steps 1 and 3, the $B_k^{K^{-1}}$, $\bar{h}_k = B_k^{K^{-1}} h_k$; k=1,...,L and the R^{-1}, $\bar{h}^D = R^{-1} h^D$ would be available as part of the solution. Furthermore, all facilities offered by such a code for updating and performing operations using these inverses would be available. For instance, nonbasic columns of DS expressed in terms of R:

$$(2.9) \qquad \bar{d}_{kj} = R^{-1} d_{kj} \; ; \quad j \in S_k^N \; ; \quad k=1,\dots,L$$

may be easily obtained.

Most modern LP codes offer facilities for handling "free" variables $(-\infty \leq x_j \leq \infty)$, but in the absence of this capability, each free variable in DS may be represented as the sum of two nonnegative variables in the usual manner (e.g. see Hadley(62)).

Computationally, the use of G^{-1} is limited to premultiplying or postmultiplying a given column or row vector. Accordingly, we derive formulae for performing these operations involving only the R^{-1}, $B_k^{K^{-1}}$ and the original problem data.

1) Premultiplication: Given the m-vector $q = (q_0, q_1, \ldots, q_L)'$,

$$(2.10)\qquad G^{-1}q = (\bar{q}^D, \bar{q}_1 - \hat{B}_1^{BE}\bar{q}^D, \ldots, \bar{q}_p - \hat{B}_p^{BE}\bar{q}^D, \bar{q}_{p+1}, \ldots, \bar{q}_L)'$$

where $\bar{q}_k = B_k^{K-1} q_k$, $\bar{q}^D = R^{-1}q^D$, $q^D = q_0 - \Sigma_{k=1}^{L} A_k^{KE}\bar{q}_k$ and $\hat{B}_k^{BE}$ is (2.3).

Three special cases of this operation are of interest. First, when $q=h=(h_0, h_1, \ldots, h_L)'$, $\bar{q}_k; k=1,\ldots,L$ and $\bar{q}^D$ are readily available. Second, when $q = \bar{g}_{kj}^N$; $j \in S_k^{NE}$; $k \in \{1,\ldots,p\}$, is a nonbasic essential column,

$$(2.11)\qquad \gamma_{kj}^{NE} = G^{-1}\bar{g}_{kj}^N = (\bar{d}_{kj}^{NE}, -\hat{B}_1^{BE}\bar{d}_{kj}^{NE}, \ldots, -\hat{B}_{k-1}^{BE}\bar{d}_{kj}^{NE}, b_{kj}^{NE} - \hat{B}_{kj}^{BE}\bar{d}_{kj}^{NE}, -\hat{B}_{k+1}^{BE}\bar{d}_{kj}^{NE}, \ldots, -\hat{B}_p^{BE}\bar{d}_{kj}^{NE}, 0, \ldots, 0)'$$

Third, when $q = \bar{g}_j$; $j \in S_k^{NI}$; $k \in \{p+1,\ldots,L\}$, is a nonbasic inessential column,

$$(2.12)\qquad \gamma_{kj}^{NI} = G^{-1}\bar{g}_{kj}^N = (\bar{d}_{kj}^{NI}, -\hat{B}_1^{BE}\bar{d}_{kj}^{NI}, \ldots, -\hat{B}_p^{BE}\bar{d}_{kj}^{NI}, 0, \ldots, -\bar{b}_{kj}^{NI}, 0, \ldots, 0)'$$

2) Postmultiplication:

$$(2.13)\qquad \hat{q}' = q'G^{-1}$$

$$(2.14)\qquad \hat{q}_0' = \bar{q}_0' - \Sigma_{k=1}^{p} \bar{q}_k'\hat{B}_k^{BE}R^{-1}$$

$$(2.15)\qquad \hat{q}_k' = -\bar{q}_0'A_k^K B_k^{K-1} + \bar{q}_k' + \Sigma_{i=1}^{L} \bar{q}_i'\hat{B}_i^{BE}Q_k^E\ ; \quad k=1,\ldots,L$$

$$(2.16)\qquad \bar{q}_k' = q_k'B_k^{K-1}\ ; \quad k=1,\ldots,L \text{ and } \bar{q}_0' = q_0'R^{-1}.$$

In applying the simplex method to linear programs with constraints of the form LS the following two cases with the obvious resulting simplifications of (2.14)-2.16), are also important.

a) $q_0' = (1,0,\ldots,0)'$; $q_k = 0$ for all k.

b) $q_k' = 0$, for all k except $q_\sigma = (0,\ldots,0,1,0,\ldots,0)'$ for some block $\sigma \in \{0,1,\ldots,L\}$ with the 1 in the s-th position.

3. PIVOTING OPERATIONS

The main reasons for maintaining the inverse of the basis for LS is to efficiently examine a large number of bases of LS and/or to solve for many right hand sides. The selection of bases could be dictated by completely heuristic search procedures, by more sophisticated criteria such as those of the simplex method or by a combination of the two. Customarily, search procedures examine a sequence of "adjacent" basic solutions of LS i.e. differing from each other only by one column. To compute the "new" basic solution and basis inverse from the "current" basic solution and inverse, one performs a "pivot operation" as in the Gauss-Jordan method of elimination. This implements a "variable exchange" step required to remove a variable from the basic set replacing it with a variable from the nonbasic set. It also updates the current basic solution and inverse to reflect these changes. The question of selecting the "pivot" will be examined later. Efficient procedures taking advantage of the problem structure to implement a given variable exchange are now developed by elementary transformations on R^{-1}, B_k^{K-1}, updating $\bar{h}_k$, $\bar{h}^D$ and by occasionally revising D_k in (2.7).

We first define a Procedure for the Rearrangement of Essential Columns (PREC) for some k, assuming that a rearrangement of the current partition S_k^{BE}, S_k^{KE}, resulting in a new nonsingular B_k^{KE}, exists. We have at hand B_k^{KE-1}, R^{-1}, $\bar{h}^D$ and seek their updated form reflecting the rearranged partition in which column b_{kr}^{KE}, perhaps along with some other columns of B_k^{KE} has been replaced by one or more columns from B_k^{BE} with the obvious corresponding effect on A_k^{KE} and A_k^{BE}.

If $\bar{b}_{ksr}^{BE} \neq 0$ for at least one $s \in S_k^{BE}$ then a direct exchange between b_{kr}^{KE} and b_{ks}^{BE} is possible and is implemented by letting

(3.1) $$B_k^{KE^{-1}} \leftarrow E^{-1} B_k^{KE^{-1}} \quad ; \quad \bar{h}_k^* \leftarrow E^{-1} \bar{h}_k$$

(3.2) $$R^{-1} \leftarrow T^{-1} R^{-1}$$

(3.3) $$\bar{h}^D \leftarrow T^{-1} \bar{h}^D + T^{-1} R^{-1} (A_\rho^K \bar{h}_\rho - A_\rho^{K*} \bar{h}_\rho^*)$$

where E^{-1} is an elementary column matrix which consists of an (m_k, m_k) -unit matrix with its r-th column replaced by the vector

$$(1/\bar{b}_{\rho rs})(-\bar{b}_{\rho 1s}, \ldots, -\bar{b}_{\rho, r-1, s}, 1, -\bar{b}_{\rho, r+1, s}, \ldots, -\bar{b}_{\rho m_k s})' .$$

T^{-1} is the elementary row matrix which consists of an (m_o, m_o)-unit matrix with its s'-th row replaced by the vector δ' with components

$$\delta_i = \begin{cases} -\bar{b}_{\rho jr}^{BE} & \text{if } i \Leftrightarrow j \in S_\rho^{BE} \\ 0 & \text{otherwise} \end{cases}$$

where s' is the row index of R^{-1} corresponding to column s. The submatrix A_k^{K*} is the A_k^K with its r-th column replaced by $a_{\rho s}^{BE}$. Alternately, we may express (3.3) as:

(3.4) $$\bar{h}^D \leftarrow T^{-1} \bar{h}^D + T^{-1} R^{-1} (\hat{A}_\rho^K (\bar{h}_\rho - \bar{h}_\rho^*) + (a_r - a_s/\bar{b}_{\rho rs}) \bar{h}_{\rho r})$$

where $\hat{A}_k^K$ is A_k^K with its r-th column replaced by zeroes.

If $\bar{b}_{kjr}^{BE} = 0$ for all $j \in S_k^{BE}$, the implementation of the rearrangement requires a "block pivot" e.g. a sequence of variable exchanges such that a subset of variables $\bar{S}_\rho^{KE} \subseteq S_\rho^{KE}$ leaves, in exchange for a subset $\bar{S}_\rho^{BE} \subseteq S_\rho^{BE}$ which enters the subproblem basic set. In practice, this requires a small number of pivots, usually two or three, provided that their sequence is known a priori. Although many tedious schemes may be devised, the easiest and most reliable way to accomplish this task is to solve the linear program

(3.5) $$\min \{ |x_{\rho r}^{KE}| \; \big| \; B_\rho^{BE} x_\rho^{BE} + B_\rho^{KE} x_\rho^{KE} = h_\rho \}$$

starting from the existing $B_\rho^{KE^{-1}}$. As indicated earlier, this linear program may readily be put into the standard form. The optimal basic solution, necessarily has $x_{\rho r}^{KE}=0$ and provides the new $B_\rho^{KE^{-1}}$ (in the form of one or more E^{-1} applied to the current inverse). For each iteration an elementary matrix T^{-1}, as defined above, is generated and (3.1)-(3.3) are applied.

We now examine in detail all possible basic/nonbasic variable exchanges for LS listed in Table 3.1. The possible pivot positions are also indicated in Figure 2 by the symbols listed in the third column of this table.

	Selected pivot position			Pivot	Corresp. variable exchange	
	row	column	Loc.	element	x_r leaves	x_s enters
Case 1	$r\in I_0$	$s\in S_\sigma^N$	★	$d_{\sigma sr}\neq 0$	$r\in S_\rho^{BE}$	$s\in S_\sigma^N$
Case 2	$r\in I_\rho^I$	$s\in S_\rho^{NI}$	■	$b_{\rho sr}^{NI}\neq 0$	$r\in S_\rho^{KI}$	$s\in S_\rho^{NI}$
Case 3a ($\rho\neq\sigma$)	$r\in I_\rho^E$	$s\in S_\sigma^N$	●	$\gamma_{\sigma sr}^N\neq 0$	$r\in S_\rho^{KE}$	$s\in S_\sigma^N$
Case 3b	$r\in I_\rho^E$	$s\in S_\rho^{NE}$	▲	$\gamma_{\sigma sr}^{NE}\neq 0$	$r\in S_\rho^{KE}$	$s\in S_\rho^{NE}$

Table 3.1 - Cases of pivoting

Case 1: The basis exchange and the new basic solution is obtained by:

(3.6) $$R^{-1} \leftarrow P^{-1}R^{-1} \;; \qquad \bar{h}^D \leftarrow P^{-1}\bar{h}^D$$

where P^{-1} is an elementary matrix which is a unit matrix with its $\bar{r}$-th column replaced by

(3.7) $$(1/\bar{d}_{\sigma sr})(-\bar{d}_{\sigma s1},\ldots,-\bar{d}_{\sigma,s,r-1},1,-\bar{d}_{\sigma,s,r+1},\ldots,-\bar{d}_{\sigma sm_0})'$$

Case 2: The basis exchange and the new basic solution is obtained by:

(3.8) $$B_\rho^{KI^{-1}} \leftarrow E^{-1}B_\rho^{KI^{-1}} \;; \quad \bar{h}_\rho^* \leftarrow E^{-1}\bar{h}_\rho$$

(3.9) $$\bar{h}^D \leftarrow \bar{h}^D + R^{-1}(A_\rho^{KI}\bar{h}_\rho - A_\rho^{KI*}\bar{h}_\rho^*)$$

where A_ρ^{KI*} is A_ρ^{KI} with its r-th column replaced by $a_{\rho s}^{NI}$. Alternately, we may update $\bar{h}^D$ via (3.4).

Case 3: Two possibilities may arise.
Case 3a. $\rho\neq\sigma$ which implies that $\bar{b}_{\sigma sr}^N=0$ and thus no exchange is possible at the subproblem level. However, since $\gamma_{\sigma sr}^N\neq 0$, it is easily shown that there exists at least one alternate subproblem basis B_ρ^{KE} formed from the columns of B_ρ^{BE} and the current B_ρ^{KE} but not containing $b_{\rho r}^{KE}$. This is accomplished by PREC. Then, we apply Case 1

to exchange $x^{BE}_{\rho r}$ (formerly $x^{KE}_{\rho r}$) and $x^{N}_{\sigma s}$.

Case 3b. $\rho=\sigma$. If $\bar{b}^{NE}_{\rho s r}=0$, apply case 3a. Otherwise, a direct exchange is possible at the subproblem level. We distinguish two possible implementations:

Implementation 1:

$$(3.10) \qquad B_{\rho}^{KE^{-1}} \longleftarrow E^{-1}B_{\rho}^{KE^{-1}} \quad ; \quad \bar{h}_{\rho} \longleftarrow E^{-1}\bar{h}_{\rho}$$

Unfortunately, one cannot devise an elementary transformation similar to T^{-1} which will reflect this change into R^{-1}. If this exchange were carried out directly, it would be necessary to recompute $D^{\bar{K}}_{\rho}$ via (2.5) and to re-solve DS. Some measure of efficiency may be achieved by trying to salvage as much of R^{-1} as possible. This may be accomplished by going through a "starting phase" in DS:

$$(3.11) \qquad \min\{ \Sigma_{j\in S_{\rho}^{BE}} |x^{BE}_{\rho j}| \;\Big|\; \Sigma^{p}_{k=1}D^{BE}_{k}x^{BE}_{k} + D^{BE*}_{\rho}x^{BE*}_{\rho} = h^{D*}\}$$

where D^{BE*}_{ρ} is the "BE" part of the recomputed $D^{\bar{K}}_{\rho}$ and $h^{D*}=h^{D}+A^{K}_{\rho}\cdot\bar{h}_{\rho} - A^{K*}_{\rho}\bar{h}^{*}_{\rho}$. This phase is started with the existing R^{-1}. At its conclusion, $x^{BE}_{\rho}=0$ and the new R^{-1} is available. Then, the columns of D^{BE}_{ρ} and the variables x^{BE}_{ρ} are discarded.

Implementation 2: There are several special classes of problems, e.g. network problems, for which the redefinition of D^{K}_{ρ} does not require its recomputation and is therefore clearly advantageous ,e.g. Grigoriadis (68-72) . In general, however, computational intuition suggests exploring possibilities of avoiding recomputation of D^{K}_{ρ}. One such possibility is to first try PREC to remove $x^{KE}_{\rho r}$ from S^{KE}_{ρ} in exchange for $x^{BE}_{\rho s}$, for some $s_1\in S^{BE}_{\rho}$. The E^{-1} and T^{-1} derived in this fashion are regarded as "provisional". If PREC is successfully completed, Case 1 is attempted in order to exchange $x^{BE}_{\rho r_1}$ and $x^{NE}_{\rho s}$ where r_1 is the new index given to r. If Case 1 is also successfully completed, the provisional transformations E^{-1},T^{-1} and P^{-1} are retained and (3.1)-(3.3), (3.6) are applied. Otherwise, if either PREC or Case 1 fails to be completed, all provisional elementary transformations are discarded and Implementation 1 is performed instead. The effort wasted in such a case may well be outweighed by the savings accrued for all successful completions of Implementation 2. This question must be settled experimentally.

4. APPLICATION TO LARGE LINEAR PROGRAMS

The most obvious application of the pivoting tools developed in the earlier paragraphs is the synthesis of partitioning methods for large, sparse linear programs of the form P:

$$(4.1) \qquad \max \left\{ x_{01} \;\middle|\; \begin{array}{l} \Sigma^{L}_{k=0} A_k x_k = h_0 \\ B_k x_k = h_k \;;\; k=1,\dots,L \\ x_k \geqq 0 \end{array} \right\}$$

where x_{01} is a free basic variable equal to the linear objective function, the "cost row" occupies the first row of A_k; $k=0,1,\dots,L$ and $a_{01}=(1,0,\dots,0)'$; $h_{01}=0$. For clarity A_0 may be thought to correspond to $B_0=0$. We assign Lagrange multipliers u_k,$k=0,1,\dots,L$ corresponding to the constraints of P. The constraints of P are precisely the same as LS except for the nonnegativity constraints on the variables.

In solving P by the simplex method, the choice of the pivot will remain the central problem, with its resulting profound effect on convergence, and will require operations (2.10),(2.13) involving the partitioned basis inverse in standardized form.

In the primal simplex method and its sophisticated computer implementations, this choice is primarily effected through the use of elaborate "pricing strategies" or options for selecting the entering variable. Once this is selected, the user cannot exercise any control to guide the pivot position so that it triggers the case of pivot operation most amenable to his problem data. Nevertheless, as we have seen, any choice will be handled by one of the three cases of pivoting. The computational effort attributed to each case is isolated and may be estimated,measured or bounded. The overall effort expended in implementing pivot operations also depends on the particular pricing strategy used since it determines the frequency of occurence of each case.

Any primal partitioning algorithm devised to handle P, upon close examination will correspond to a particular pricing strategy. For instance, Bennett(66) and Kaul(65) apply the usual pivot selection criteria of the simplex method with no attempt to control the ensuing pivoting effort. Bennett employs cases of variable exchanges similar to the above but with the aid of modification methods for updating the inverse (see e.g. Householder(53), Goldfarb(71)) which generally require an excessive amount of core storage thus limiting the size of P. Kaul uses similar exchanges for the revised simplex method although he does not discuss PREC and Case 3b. Clearly, these two approaches will exhibit the same convergence characteristics with those of the simplex method, but will differ on running times due to the manner in which pivoting is implemented. Similarly, methods for special structures of P as in Dantzig(67), Hartman(70), Cord(71) have used the same pivot selection criteria with specialized cases for pivoting. Of these, the Generalized Upper Bounding problem (GUB) is a common structure for P, in which B consist of one row vector each with elements +1 or -1.

Based on the above, it is not difficult to see that studying the convergence characteristics of a given class of problems P via a standard LP code (to which this strategy has been specified) provides an experimental tool for studying existing and future proposals for primal partitioning methods. The expense of such experimentation is dwarfed by the immense effort and specialized talent required to build experimental or production partitioning codes.

Furthermore, small unsophisticated programs are not generally believed to be of any great experimental value in the field. The difficulty stems from our inability to extrapolate claims of computational efficacy based on results obtained from a small experimental partitioning program to those expected from large LP systems solving large problems. The lack of appreciation for the need of a professional "team effort" to accomplish such tasks in the field of linear programming has been argued repeatedly by Orchard-Hays(68), Dantzig(70) and others. However, where it was "unrealistic to expect any single investigator to make major advances" in this field, the proposed approach makes this possible and allows the design and implementation of a versatile LP partitioning system encompassing most promising pivot selection methods.

In the dual simplex method, the pivot choice is more flexible. Elaborate options for selecting the existing variable may be used to guide the pivot to a more amenable position for the structure of P. As before, it may be argued that dual partitioning methods correspond to particular <u>dual</u> pricing strategies. This will now be examined in more detail to indicate this correspondence and the use of the tools for pivoting developed earlier. The very same procedure will yield corresponding primal or primal-dual partitioning methods from the primal or the primal-dual simplex method.

We first review the fundamental steps of the dual simplex method. (The notation in the next three paragraphs should not be confused with that in the rest of the paper).

Let problem P be: $\max_x\{cx \mid Ax = b;\ x \geq 0\}$ where A is an (m,n)-matrix of full rank and $b \geqq 0$. Let u be the vector of dual variables associated with $Ax = b$ and $I = \{1,\ldots,m\}$; $J = \{1,\ldots,n\}$. The dual simplex method may then be stated as follows (see e.g., pp. 119 in Hadley(62)):

0. Let a basis B (i.e. the partition: $J=(J^B,J^N)$; $A=(B,N)$; $c=(c^B,c^N)$; $x=(x^B,x^N)$) such that $\bar{c}^N=N'u-c^N \geq 0$ (dual feasibility) be given.

1. Let $\bar{b}=B^{-1}b$. If $\bar{b}_i \geq 0\ \forall i$, then $x^B=\bar{b}$; $x^N=0$ solves P. Terminate. Otherwise choose a row by $\bar{b}_r=\min\{\bar{b}_i \mid \bar{b}_i<0\ i\in I\}$.

2. Let $\bar{n}_r=n_rB^{-1}$. If $\bar{n}_r \geq 0$, then P has no feasible solution. Otherwise choose a column by:

$$\bar{c}^N_s/\bar{n}_{rs} = \max_j\{\ \bar{c}_j/\bar{n}_{rj} \mid \bar{n}_{rj} < 0\ ;\ j\in J^N\ \} \tag{4.2}$$

3. Perform a pivot step with $\bar{n}_{rs}$ as the pivot element. Return to 1.

Two remarks are relevant. First, in practical situations, an optimal u vector corresponding to a slightly modified problem may be available. Such a vector, even if it is dual infeasible, may for the problem to be solved, provide an excellent approximation to the optimal u. Using this as an initial solution will, in general, substantially reduce the overall computational effort. In order to allow the use of such an initial solution, P will be augmented by an artificial constraint of the form:

$$p'x + x_{n+1} = \Gamma \tag{4.3}$$

Where x_{n+1} is a slack variable, Γ is a sufficiently large positive number and p is the appropriate sum vector. Let λ be the dual variable associated with this constraint so that $w = (u,\lambda)$ is the vector of dual variables associated with the augmented problem. If for any given u^o the scalar λ is chosen so that $\lambda^o= \max\{(c-Au^o)_j\ ,0\}$, then $w^o = (u^o,\lambda^o)$ is clearly dual feasible and may be used to obtain a dual feasible basis. If in the optimal solution $x_{n+1}=0$ then P has an infinite solution.

Second, the minimization in step 1 is a simplification of a more complicated operation. We recall that for linear programs one can guarantee a steepest descent behavior of the simplex method if and only if steps 1 and 2 require that indices r and s are chosen by:

$$\min_i\{\ \bar{b}_i\ \ \max_j\{\ \bar{c}^N_j/\bar{n}_{ij} \mid \bar{n}_{ij}<0;\ j\in J^N \mid \bar{b}_i < 0;\ \ i\in I\ \} \tag{4.4}$$

Thus, the minimization in step 1, although computationally much more convenient, is nevertheless a gross simplification of (4.4) and perhaps not the best. Recent primal LP codes allow various "pricing" and "suboptimization" strategies which have reportedly led to better convergence than the corresponding step of the primal simplex method.

The reason similar dual pricing and suboptimization strategies have not been popular for the dual simplex method is related to programming efficiency. The potential benefits in convergence rates seem to be offset by the complicated programming and excessive core requirements, in particular when $m \ll n$ and when the matrix A is stored in column order. This objection, however, is removed when the constraint matrix has the structure of P. Dual pricing strategies can be effectively used to guide the chosen pivot into "favorable" positions. The general procedure outlined in the next paragraph provides this capability. We now return to our earlier notation.

To apply the dual simplex method to P, we assume that the m_o-th row of the coupling constraints is the bounding relation (4.3):

$$(4.5) \qquad \Sigma_{j=2}^{n} a_{0jm_0} x_{0j} + \Sigma_{k=1}^{L} \Sigma_{j=1}^{m} a_{kjm_0} x_{kj} = \Gamma$$

where $a_{kjm_o}=1$ for all j,k and $a_{02}=(0,\dots,0,1)'$. We further assume that u_o is dual feasible (e.g. $u_o=(1,0,\lambda)$ with λ chosen as above).

<u>Step 0</u>. Construct an initial feasible basis G in accordance with the three step procedure of Section 3. In particular, we first determine the subproblem basic solutions so that u_k is also dual feasible, i.e. $u_k'B_k \geqq u_o'A_k$, for each $k=1,\dots,L$. Although primal feasibility for the key variables is not necessary, it is advisable to enforce it so that infeasibilities in rows $i\in\{I_1,\dots,I_L\}$ of P can be detected at this early stage. Thus we solve the L subproblems:

$$(4.6) \qquad \min_{x_k} \{ u_0'A_k x_k \mid B_k x_k = h_k; \quad x_k \geq 0 \} \; ; \quad k=1,\dots,L$$

which give the set of key variables S_k^K and $\bar{h}_k \geqq 0$, B_k^K, B_k^{K-1} , u_k for each k.

Second, compute (in some special cases only implicitly) the $D_k^{\bar{K}}$ and h^D by (2.5) and third, obtain a basic solution to DS by solving the Derived Linear Program (DLP):

$$(4.7) \qquad \max \left\{ x_{01} \;\middle|\; \begin{array}{l} A_0 x_0 + \Sigma_{k=1}^{L} D_k^{\bar{K}} x_k^{\bar{K}} = h^D \\ x_0,\; x_k^{\bar{K}} \geq 0;\; k=1,\dots,L \end{array} \right\}$$

which defines the partitioning (R,R^N), $R=(D_1^{BE},\dots,D_p^{BE})$; $p \leqq L$ and provides R^{-1} and $\bar{h}^D$.

Now apply the three iterative steps of the dual simplex method:

<u>Step 1</u>. Select a variable to leave the overall basis by:

$$\beta_r = \text{operator} \{ \beta_i \mid \beta_i < 0 \; ; \; i\in I \}$$

where $\beta = (\bar{h}^D, \bar{h}_1 - \hat{B}_1^{BE}\bar{h}^D, \dots, \bar{h}_p - \hat{B}_p^{BE}\bar{h}^D, \bar{h}_{p+1}, \dots, \bar{h}_L)'$, $I=\{I_o,I_1,\dots,I_L\}$ and the "operator" is chosen with wide latitude, e.g. it could be the minimum over $i\in I$; the minimum over $i\in\bar{I}\subset I$, say $\bar{I}=\{I_1,\dots,I_L\}$; the minimum over $i\in\bar{I}\subset I$ and if $\bar{I}$ contains no i with $\beta_i<0$ then $i\in\hat{I}\subset I$, etc. in a hierarchical fashion; or it could be a random choice over all $i\in I$ or over subsets of I, etc. Such strategic choices of "dual pricing" are numerous and each may lead to a slightly different partitioning algorithm.

If $\beta_i \geq 0$ for all $i \in I$, then β solves P. Terminate. Otherwise, variable $x_{kr}; r \in I_k \subset I$ <u>leaves</u> the overall basis G.

<u>Step 2</u>. Select the variable to <u>enter</u> the overall basis by

$$(4.8) \qquad \gamma^N_{\sigma s1}/\gamma^N_{\sigma sr} = \max_{k,j} \{ \gamma^N_{kj1}/\gamma^N_{kjr} \mid \gamma^N_{kjr} < 0;\ j \in S^N_k;\ k=1,\ldots,L \}$$

where $\gamma^N_{kj1} = \bar{d}^N_{kj1}; j \in S^N_k; k=1,\ldots,L$, are the "reduced costs" in DLP and γ^N_{kjr} are the "tableau" coefficients given by (2.11) or (2.12).

If $\gamma^N_{kjr} \geq 0$ for all $j \in S^N_k; k=1,\ldots,L$, no feasible solution to P exists. Terminate. Otherwise, variable $x_{\sigma s}; s \in S^N_\sigma$ <u>enters</u> the overall basis G.

<u>Step 3</u>. Perform a pivot step, with $\gamma^N_{\sigma sr}$ as the "pivot" element, via the procedure applicable to the case defined by ρ, r, σ and s. Optionally, go to step 1, provided that the new $\beta_i < 0$ for some $i \in I_o$. Otherwise, go to step 4.

<u>Step 4</u>. Optionally, select some row $r \in S^{KE}_k$ guaranteed to result in a successful rearrangement of the partition S^{BE}_k, S^{KE}_k for some k and apply PREC (nonnegativity in (5.1) required). Return to step 1.

It is easy to see how one would proceed to show that a given dual partitioning method is a special case of these four steps. Let us examine Rosen's(64) Primal Partition Programming (PPP), a dual method in spite of its name.

<u>Step 0</u>. Exactly as before. Rosen refers to (4.4) as "Problem I" and to DLP as "Problem II".

<u>Step 1</u>. Essentially the same as above, but $r \in I_o$.

<u>Steps 2-3</u>. Same as above, confined to DLP coefficients.

In PPP steps 1-3 are repeated until $\beta_i \geq 0$ for all $i \in I_o$. Actually, the primal simplex method is used to reoptimize DLP in lieu of repetitive applications of steps 1-3. In either approach only Case 1 of pivoting is performed. The effectiveness of using the (primal) simplex method to reoptimize DLP has been found questionable at least for a special class of problems P in Grigoriadis(70,72).

<u>Step 4</u>. Performed for at least one essential block k such that $\beta_i < 0$ for at least one $i \in I^E_k$ and it amounts to performing PREC. The aim is to reduce the sum of all $\beta_i < 0$; $i \in I^E_k$ to zero by removing that variable $x^{KE}_{kr} > 0$ from the basis which is first reduced to zero when the x^{BE}_r variables are increased up from zero in (4.4). This gives the row index $\bar{r} \in I^E_k$ (corresponding to $x^{KE}_{k\bar{r}}$) from:

$$(4.9) \qquad \theta_{k\bar{r}} = \min_i \{\theta_{ki} \mid \beta_i < 0; i \in I^E_k\};\ \theta_{ki} = \bar{h}_{ki}/(\bar{h}_{ki} - \beta_i) = \bar{h}_{ki}/(\hat{B}^{BE}\bar{h}^D)_i;\ i \in I^E_k$$

with which PREC is executed. Go to 1.

This simple exercise emphasizes two aspects of PPP. First, no theoretical arguments are necessary to demonstrate the finiteness and monotonocity of the method since it has been derived as a straightforward specialization of the dual simplex method with a particular pricing strategy. The same argument holds for any partitioning method derived from the primal, dual or primal-dual simplex methods, including "parametric" methods. Second, an existing dual simplex code with minor modifications may be used as a vehicle for testing the convergence of PPP in the large scale environment provided by commercially available LP codes. This is easily implemented by using the "free" variables option , available in most of these codes, in conjunction with a specialized pricing routine as follows:

a) Fix all key variables in their current position in the basis by specifying them as "free" variables to the LP code. Price out only rows $i \in I_0$ until all $\beta_i \geqq 0$, $i \in I_0$ and count the pivots performed for this task. Go to (b).

b) For each block k with $\beta_i < 0; i \in I_k^E$ select the pivot row $\bar{r}$ via (4.9). Enter a simplified PREC subroutine which will only identify the new key variables, i.e. the new partition $\{\bar{S}_k^{BE}, \bar{S}_k^{KE}\}$ such that $\bar{r} \notin \bar{S}_k^{KE}$; $\bar{r} \in \bar{S}_k^{BE}$, but will not implement the transformations (3.1)-(3.3). Go to (a).

The "simplified" PREC routine could use another in-core LP subroutine with large enough capacity to handle the linear program (3.5) in main storage, e.g. Clasen's(61) MSUB. If one wishes to be more sophisticated, the commercial LP code used for this experiment can also be used for PREC. In any case, the method of handling (3.5) will not influence the desired large scale computational environment of the production LP code. Clearly, the specialized PREC routine is not required for testing the convergence of other partitioning methods for which a one pivot subproblem variable exchange is guaranteed.

The pivot steps performed in (3.5) are not implemented but simply counted and recorded for later reference. At the conclusion of the run, the pivots performed in (a) represent Case 1 exchanges (Table 3.1) while those counted by PREC in (b) (but not performed) represent subproblem pivots. Disregarding effects of degeneracy, the sequence of pivots obtained from the test coincides with the sequence of pivots which would be obtained from a large scale implementation of a PPP code.

The question to be answered next is the amount of computer resources (number of instructions, file accesses, etc.) required by each one of these pivot steps. The "cost per pivot" may be estimated, possibly by analyzing performance data for the same code on a set of test problems with appropriate solution parameter settings to resemble DLP and (3.5), the amount of main storage available, various pricing strategies, etc. This cost may also be estimated using other tools, e.g. discrete simulation, or it may be bounded via analytical arguments.

The number of times (a) is executed reflects the number of times DLP is revised and solved, while the number of infeasible blocks handled in (b) indicates the amount of subsequent revision of the DLP coefficients. The cost of revising and "setting-up" of DLP may also be estimated or bounded.

In conclusion, the convergence characteristics of a partitioning method are first obtained experimentally by solving large problems by some existing large LP code. A minimal amount of programming effort is required for this task. The, the results of the experiments are used to estimate or to find an upper bound of the total computer resources which would have been required by an implementation of a large scale PPP code for the method under investigation.

Beale's(63) method of Pseudo-Basic Variables is easily shown to correspond to the above pricing strategy with minor modifications. Similar modifications of the same strategy and the application of the parametric right hand side algorithm for linear programming results in Orchard-Hays'(68) PARRHS Block Pivot algorithm. The dual GUB algorithm in Grigoriadis(71) and the transportation algorithm in (68) are merely special cases of the above four step procedure. Other specializations with impressive computational advantages include network flow problems arising from various applications such as goods distribution, logistics and more recently communication problems (White(72)). These are briefly discussed in section 6.

The method for accomplishing a "subproblem exchange" in PPP has caused some confusion. An attempt is made here to clarify this point.

In applying the straightforward dual simplex method to problem P, it is necessary to perform only one pivot step in the subproblem to reduce the primal infeasibility in a chosen row in I_k^E by causing the corresponding subproblem variable to leave the subproblem basis and by introducing a nonbasic variable in its place such that the dual feasibility and the overall basis structure are maintained. In this case, PREC would terminate in exactly one pivot step thus rendering the linear program (3.5) unnecessary. However, the use of (3.5) may still be desirable on an optional basis as indicated in step 4 of the dual simplex method outlined above.

As we have seen, PPP is a particular implementation of the dual simplex method which stipulates that primal feasibility must be restored in step 4 in at least one subproblem which has become infeasible due to steps 1-3. At this point a basis to the overall problem has been defined. An "essential" subproblem, say the k-th, contributes m_k columns to the overall basis and DLP (or PII) in steps 1-3 contributes another n_k^{BE} columns. Now, due to this latter choice, it is possible to obtain $\beta_i < 0$ for at least one $i \in I_k^E$. Rosen(64) proposes (4.9) to select one among the negative β_i as the subproblem pivot row (exiting variable). The pivot column is, of course, to be chosen from the n_k^{BE} columns dictated by DLP, so that primal feasibility, i.e. optimality of the subproblem, is restored.

Although it has been suggested by other workers in this field that this can be accomplished in exactly one pivot step (the point is correctly stated in the published literature, e.g. Lasdon(70), Grigoriadis(69,69b,70,72)), the following example, constructed by W.W. White(71), shows that, generally, more than one pivot operation is necessary to restore primal feasibility in the subproblem. It also illustrates the need for (3.5) in PREC.

Let

$$\begin{matrix} & x_1 & x_2 & & x_3 & x_4 \end{matrix}$$

$$(B_k^{KE} \mid B_k^{BE}) = \left(\begin{array}{cc|cc} 1 & 0 & 4 & -1 \\ 0 & 1 & 4^* & 0 \end{array}\right); \quad \bar{h}_k = \begin{pmatrix} 4 \\ 5 \end{pmatrix}$$

and suppose that DLP gives $x_3 = x_4 = 4$, i.e. $\bar{h}^D = (4,4)'$. Then,

$$\beta = \bar{h}_k - B_k^{BE}\bar{h}^D = \begin{pmatrix} 4 \\ 5 \end{pmatrix} - \begin{pmatrix} 4 & -1 \\ 4 & 0 \end{pmatrix}\begin{pmatrix} 4 \\ 4 \end{pmatrix} = \begin{pmatrix} 4 \\ 5 \end{pmatrix} - \begin{pmatrix} 12 \\ 16 \end{pmatrix} = \begin{pmatrix} -8 \\ -11 \end{pmatrix}$$

Computing θ_1, θ_2 via (4.9): $\theta_1 = 4/4-(-8) = 16/48$; $\theta_2 = 5/5-(-11) = 15/48$; $\min(\theta_1, \theta_2) = \theta_2$. Choose the second row, i.e. x_2 leaves the subproblem basis, and since the only pivot available is 4 (indicated by *), x_3 must enter. Performing one pivot operation we obtain:

$$\begin{matrix} x_1 & x_2 & & x_3 & x_4 \end{matrix}$$

$$\left(\begin{array}{cc|cc} 1 & -1 & 0 & -1^{\oplus} \\ 0 & 1/4 & 1 & 0 \end{array}\right); \quad \beta = \begin{pmatrix} -1 \\ 5/4 \end{pmatrix}$$

which results in an infeasibility ($x_1 = -1$). Restoration of feasibility requires an <u>additional</u> pivot step with -1 (indicated by ⊕) as the pivot.

5. A GENERALIZATION OF PROBLEM P

A more general structure of P, which is frequently encountered in practice, is the block diagonal structure with both coupling constraints and coupling variables (GP), i.e.

$$\max \left\{ x_{01} \;\middle|\; \begin{array}{c} \Sigma_{k=1}^{L} A_k x_k + C_0 y = h_0 \\ B_k x_k + C_k y = h_k ; \\ y \geq 0,\ x_k \geq 0;\ k=1,\dots,L \end{array} \right\} \tag{5.1}$$

where the C_k are given (m_k, n_o)-matrices.

The basis matrix for GP will have the same structure as G with two important exceptions: First, the basic essential part of G will contain coupling columns, thus generally eliminating the possibility of key inessential blocks. Second, there might be more than m_o basic essential columns resulting in some blocks B_k^{KE} with $m_k > n_k$. In such a case, the structure of G^{-1} would be destroyed along with the effectiveness of any partitioning method devised for GP. Fortunately, however, if we assume that the number of basic essential columns exceeds m_o only by a small amount, the difficulty may be circumvented. This assumption has been found to hold for randomly generated problems reported in Grigoriadis(69). The rationale for this approach was first proposed by Ritter(67) with a subsequent generalization to nonlinear problems and experimental evidence reported in Grigoriadis(69), and further studied by Webber(68).

In our context, Ritter's approach may be summarized as follows. In order to retain the structure of the inverse, one must retain the block diagonal nature of G with the nonsingular blocks B_k^{KE}. As long as the number of basic essential columns does not exceed m_o, the usual three cases of pivoting obtained in section 3 apply. Whenever an "excess" column is slated to enter G, the size of G is increased by one in the coupling part , i.e. $m_o \leftarrow m_o+1$, by appending an "additional constraint" to the coupling rows and pivoting via case 1. These additional inequality constraints simply express the nonnegativity of a variable x_{kr}^{KE} (as detected by checking the corresponding $\beta_{ir} < 0$) as a coupling rather that as a subproblem constraint. The method follows the dual simplex method and the general model of flow in section 4, with the exception of step 0 which does not require primal feasible solutions to (4.6). Various strategies are also used to "delete" additional constraints which become inactive in DLP.

A direct pivoting approach, with side calculations, for solving GP has been proposed by Heesterman(65) and a method using modification methods for the staircase structure, a special case of GP, has been presented in Bennett(62b). Kaul's(65) method of generalized upper bounding for solving P has been extended to handle GP in Hartman(70) with attractive experimental results.

Excellent discussion and review of most decomposition and partitioning methods in mathematical programming and a large compilation of references may be found in Geoffrion (70) and Lasdon (70). Recently, Rutten (72) has presented a brief survey of primal and dual partitioning methods.

6. SPECIALLY STRUCTURED NETWORK FLOW PROBLEMS

The above concepts of partitioning may be directly applied to various classes of single or multicommodity network flow problems. Of interest is the general LP formulation of the minimum cost multistage multicommodity network flow problem, though at this time, no efficient methods exist for its solution.

(6.1) $\min \Sigma_t \Sigma_k c_{tk} x_{tk}$

subject to

(6.2) $\Sigma_k E_{tk} x_{tk} \leqq h_{t0}$; $t=1,\ldots,T$

(6.3) $B_{tk} x_{tk} - y_{t-1,k} + y_{tk} = h_{tk}$; $t=1,\ldots,T$; $k=1,\ldots,K$

(6.4) $h^{\ell}_{tk} \leqq x_{tk} \leqq h^{u}_{tk}$; $g^{\ell}_{tk} \leqq y_{tk} \leqq g^{u}_{tk}$; $t=1,\ldots,T$; $k=1,\ldots,K$

$y_{0k}=0$; $k=1,\ldots,K$

where the x_{tk} variables are the arc flows of commodity k in time period t and y_{tk} are the amounts of commodity k stored at the nodes for the time interval $[t,t+1]$. The B_{tk} are the incidence matrices for the network for all t,k and E_{tk} is a diagonal matrix with elements e_{ij} which are used in (6.2) to bound the "weighted" sum of commodity flows on each arc. Usually, $e_{ij}=1$. The $h^{\ell}_{tk}, h^{u}_{tk}, g^{\ell}_{tk}, g^{u}_{tk}$ are lower and upper bounds on the flow in each arc and the h_{tk} are exogenous demands (loads) imposed on the network, for each k and t. This problem is illustrated in Figure 6.1, where interaction between Commodity 1 and 2 is present due to the arc capacities h_{t0}. Clearly, (6.1)-(6.4) is a special structure of (5.1) and may be (painfully) treated by applying Ritter's partitioning algorithm as discussed by Grigoriadis (71b).

Important and common cases of (6.1)-(6.4) are the minimum cost multistage single commodity (MSC) and the single stage multicommodity (SMC) network flow problems. The first of these corresponds to the left column (k=1), while the second corresponds to the first row (t=1) of Figure 6.1 with the obvious simplifications of (6.1)-(6.4). It is well known that problem MSC has a completely unimodular constraint matrix and may be solved by any of the efficient minimum cost network flow computer codes. Thus any partitioning method directed to the solution of this problem must compete with the best available such code both in reducing the storage requirements and in improving overall solution efficiency. One such proposal was made in (71b).

The constraint matrix for SMC is not unimodular and thus no basic integer solutions are guaranteed. However, as it has been observed by many practitioners in the field, the number of fractional answers obtained by an LP solution to SMC is usually much smaller than expected. The reason for this pleasant characteristic may be partially explained by considering the special case of multicommodity

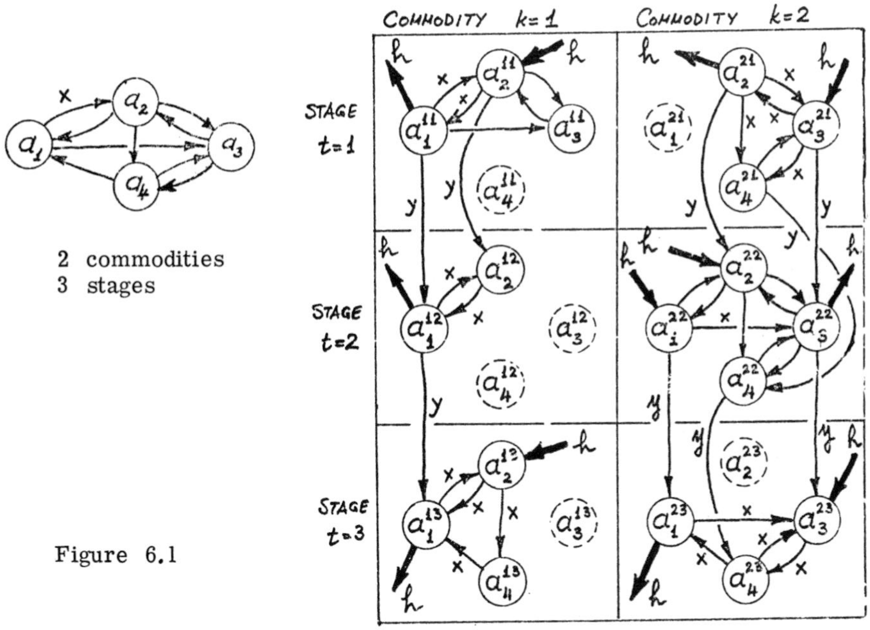

Figure 6.1

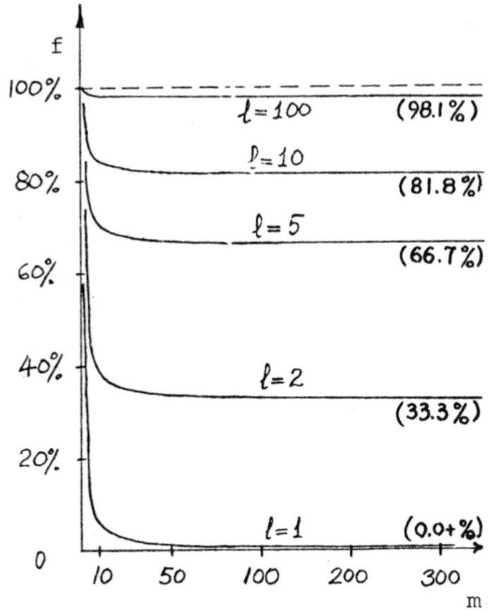

Figure 6.2 Percent integers in solution

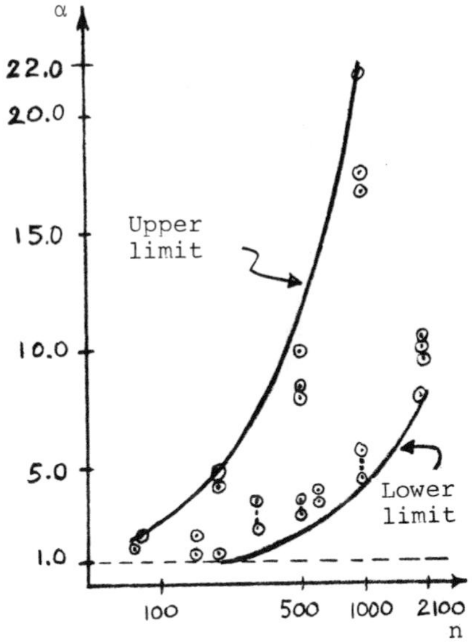

Figure 6.3 Comparison of test results

transportation problems. Using simple basis arguments it can be shown that (71b) for most transportation problems arising in practice (i.e. with $m \ll n$, m=number of sources, n=number of destinations), only a small percentage of the components in any basic solution may be fractional as indicated in Figure 6.2 where $\ell = n/m$.

Since the size of an explicitly stated practical problem SMC may be very large, use of standard LP codes is not satisfactory, though LP is the only tool available for general use. Therefore, it is sufficiently realistic to construct partitioning methods which will compare favorably with the size and efficiency obtained by LP codes operating on SMC. A dual method modeled after the general procedure of sections 1-4 with some additional computational devices such as the use of "secondary constraints" is given in Grigoriadis (70,72).

Early experimentation with this algorithm indicates a definite advantage over linear programming, shown in Figure 6.3 where α is the ratio of computing time for this method to the time for an efficient commercial LP code under the same testing conditions and n is the number of arcs in the problem. The dots indicate experiments performed.

A primal method also using the devices of sections 1-4 has been proposed by Hartman (70) with no computational results reported

REFERENCES

Beale, E.M.L.,(1963):"The Simplex Method Using Pseudo-Basic Variables for Structured Linear Programming Problems" in Recent Advances in Mathematical Programming (P.Wolfe and R.L.Graves, Eds.), p.133, McGraw-Hill,New York.

Bennett, J.M.,(1966):"An Approach to Some Structured Linear Programming Problems, Operations Research, 14,1,p.636.

Bennett, J.M. and D.R. Green,(1966b):"A Method for Solving Partially Linked Linear Programming Problems", Technical Report 38, Basser Computing Laboratory, The University of Sydney.

Brayton, R. and F.G. Gustavson, R.A.Willoughby, (1970): "Some Results on Sparse Matrices", Mathematics of Computation,24,112.

Clasen, R.J., (1961): "RS MSUB- Linear Programming Subroutine", The RAND Corporation, Santa Monica, California.

Cord,T., (1971): "Generalized Upper Bounds with Upper Bounded Variables", New York Scientific Center report, 320-2962, IBM Corporation, New York, N.Y.

Dantzig, G.B., (1970): "Large Scale Linear Programs- A Survey", presented by the 7th Mathematical Programming Symposium, The Hague, The Netherlands.

Dantzig,G.B. and R.M. Van Slyke, (1967): "Generalized Upper Bounding Techniques", Journal of Computer and System Sciences, 1,p.213.

Geoffrion,A.M., (1970): "Elements of Large-Scale Mathematical Programming", Management Science,16,p.652.

Glover,F., et al (1971): "A Comparison of Computation Times for Various Starting Procedures, Basis Change Criteria and Solution Algorithms for Distribution Problems", CS 44, Center for Cybernetics Studies, The University of Texas, Austin Texas.

Grigoriadis, M.D.,(1969): "Partitioning Methods in Mathematical Programming", 10th American Meeting of TIMS, Atlanta, Georgia.

Grigoriadis, M.D., (1971): "A Dual Generalized Upper Bounding Technique", Management Science, 17,5,p. 269.

Grigoriadis,M.D.,(1971b):"Minimum Cost Multistage Multicommodity Network Flows", NATO Conference on Large-Scale Resouce Allocation Problems, Elsinore, Denmark.

Grigoriadis,M.D. and K. Ritter,(1969): " A Decomposition Method for Structured Linear and Nonlinear Programs", Journal of Computer and System Sciences,3,p. 335.

Grigoriadis, M.D. and W.F. Walker, (1968): " A Treatment of Transportation Problems by Primal Partition Programming", Management Science ,14,9,p.565.

Grigoriadis, M.D. and W.W. White,(1970): " A Partitioning Algorithm for the Multicommodity Network Flow Problem", To Appear in Mathematical Programming, 1972, Also Philadelphia Scientific Center Report 320-2996, IBM Corporation, Philadelphia, Pa.

Grigoriadis,M.D. and W.W. White,(1972): " Computational Experience with a Multicommodity Network Flow Algorithm", To Appear in Optimization Methods for Resource Allocation, (R. Cottle and J. Krarup, eds.), English Universities Press,1972. Also Philadelphia Scientific Center Report 320-3011, IBM Corporation, Philadelphia, Pa.

Goldfarb,D.(1971):"Modification Methods for Inverting Matrices and Solving Systems of Linear Algebraic Equations", Philadelphia Scientific Center Report 320-2998, IBM Corporation, Philadelphia, Pa.

Hadley, G., (1962): Linear Programming, Addison-Wesley, Reading,Mass.

Hartman, J.K. and L.S. Lasdon, (1970): "A Generalized Upper Bounding Algorithm for Multicommodity Flow Networks", Technical Memorandum No.193, Operations Research Department, Case Western Reserve University, Cleveland, Ohio.

Hartman, J.K. and L.S. Lasdon,(1970b):"A Generalized Upper Bounding Method for Doubly Coupled Linear Programs", NLRQ,p.411.

Heesterman,A.R.G.(1968):"Special Simplex Algorithm for Multi-Sector Problems", Numerische Matematik, 12, p.288.

Householder, A.S., (1953): Principles of Numerical Analysis, p. 79, McGraw Hill, New York.

Kaul, R.N.,(1965): "An Extension of Generalized Upper Bounding Techniques for Linear Programming", ORC-62-27, Operations Research Center, University of California ,Berkeley.

Kronjo,T.O.M., (1968): "The Decompostiion of Any Linear Programme in Primal and Dual Directions", Report RC/A-12, University of Birmingham.

Lasdon, L.S. (1970): Optimization Theory of Large Systems, Macmillan.

Orchard-Hays, W., (1968): Advanced Linear Programming Computing Techniques, McGraw-Hill.

Ritter, K.(1967):"A Decomposition Method for Linear Programming Problems with Coupling Constraints and Variables", Mathematics Research Center Report 739, University of Wisconsin.

Rosen, J.B. (1964): Primal Partition Programming for Block Diagonal Matrices, Numerische Mathematik, 6, p. 250.

Rutten,D.P. (1972): "A Theoretical Comparison of Some Partitioning Methods for Solving Structured Linear Programs", Graduate School of Business, Indiana University. Presented at the ORSA Meeting, New Orleans, La.

Saunders, M.A., (1972): " Large Scale Linear Programming Using the Cholesky Factorization", Stan-CS-72-252, Computer Science Department, Stanford University.

Webber, W. and W.W. White, (1968): "A Partitioning Algorithm for Structured Linear Programming Problems", New York Scientific Center Report 320-2946,IBM Corporation.

White, W.W. (1971): Private communication.

White, W.W. , (1972): "Mathematical Programming Multicommodity Flows and Communications Nets",Philadelphia Scientific Center Report 320-3012, IBM Corporation, Philadelphia, Pa.

Willoughby, R.A.,(ed.), (1969): " Sparse Matrix Proceedings", RA-1, 11707, T.J. Watson Research Center, IBM Corporation .

DECOMPOSITION IN LARGE SCALE SYSTEMS

THEORY AND APPLICATIONS OF STRUCTURAL ANALYSIS IN PARTITIONING, DISJOINTING AND CONSTRUCTING HIERARCHICAL SYSTEMS

by

A.K. KEVORKIAN and J. SNOEK

KONINKLIJKE/SHELL-LABORATORIUM, AMSTERDAM
(Shell Research N.V.)

Abstract: A large scale integrated system such as an oil refinery, a chemical plant or a managerial organization, is decomposed by structural analysis into its maximum number of possible disjoint systems. The order of complexity in each disjoint system is reduced by reformulating these systems into minimally levelled hierarchical structures. Each hierarchical structure, here, may comprise a number of irreducible subsystems which are, within a hierarchical level, independent of each other and influenced only by the irreducible subsystems belonging to the higher levels in this system hierarchy. In this manner the study of a large scale system such as process design, control and optimization is replaced by the separate study of the irreducible subsystems.

1. INTRODUCTION

The control of a large scale integrated system - such as a complete oil refinery, a whole chemical plant or study of a managerial organization - as one complex process, may be considered possible only when the inherent structure of such a system is thoroughly analysed. In this manner a large scale system S may be broken up into the maximum number of subsystems, ordered into a given sequence, such that each of these subsystems $S_i \varepsilon S$, where $i \varepsilon I_k$, is studied independently of the remaining part of the system. This procedure is denoted in the literature as Partitioning. Moreover, structure analysis identifies the possible groups of subsystems $V_j \subset S$, $j \varepsilon I_p$, such that the intersection $V_i \cap V_j$ for $i \neq j; i,j \varepsilon I_p$ results in an empty set. The latter procedure is indicated in the literature as Disjointing. Furthermore, structural analysis allows the identification of the minimally levelled hierarchical structures associated with each of the disjoint systems V_i, $i \varepsilon I_p$, such that a hierarchical level comprises a number of subsystems which are independent of each other and are influenced only by the subsystems belonging to the higher levels in the system hierarchy.

A unified structural analysis procedure which partitions, disjoints and constructs the hierarchical structure associated with a large scale system may have wide applications in the fields of engineering and mathematics. For example, the representation of a process flow sheet, which comprises a large number of units, in the defined hierarchical form may facilitate the control scheme design of the corresponding system. It may also yield valuable information regarding the consequences of a failure in an individual process unit. Another application, by KEVORKIAN (1972), which is successfully being used, is in the solving of large sets of non-linear simultaneous equations such as required in the steady-state design of large scale systems.

Several methods of partitioning and disjointing large scale systems have been proposed. The main concept in partitioning, as discussed by LEDET (1970), is to find a set of vertices connected by a closed path, namely the maximal loop $\mathscr{L}$, such that every other loop in the flow graph is either contained in $\mathscr{L}$ or has no vertex in common with it. These loops are identified either by tracing paths of information flow in the adjacency matrix or by using powers of this matrix.

The main concept in identifying disjoint systems, unlike that in partitioning, has not been clearly defined. However, there exist procedures, such as that proposed by HIMMELBLAU (1967), for identifying disjoint systems. It should be noted that there has not been any attempt in the literature to define a uniform concept which combines partitioning and disjointing such that the information obtained from solvin one problem could be efficiently used to solve the other.

The partitioning concept presented here can be stated as follows: The arranging of an occurrence matrix into the kth order quasi-triangular matrix $\mathcal{C}$ breaks up a set of non-linear simultaneous equations into k subsets of equations $g_i \varepsilon R^{\theta i}$, where $i \varepsilon I_k$, which may be solved serially for the subsets of variables $z_i \varepsilon R^{\theta i}$. A more detailed description of this formulation is given in Section 3. The paired subsets of variables and equations (z_i, g_i), $i \varepsilon I_k$, are associated in Section 4 to a subsystem S_i which is defined as irreducible if and only if k corresponds to the maximum number of diagonal cells in $\mathcal{C}$.

DULMAGE (1962) and HARARY (1960 and 1962) have shown that the quasi or block-triangularization of matrices facilitates the inversion, and eigenvalue calculation, of sparse matrices. STEWARD (1965) has also shown that the quasi-triangularization of the occurrence matrix, corresponding to a non-linear (linear) set of equations, is equivalent to the partitioning of the set into smaller subsets and thus justifyin the application of DULMAGE's and HARARY's results to non-linear systems. The presented algorithms, however, are not suitable and perhaps prohibitive for large scale systems.

For each variable x_i, $i \varepsilon I_n$, a "Characteristic set" $\hat{\Lambda}_i$ is defined. In Section 5 it is shown that the determination of this characteristic set is sufficient for the quasi-triangularization of an occurrence matrix and, hence, the identification of the irreducible subsets of equations g_i and variables z_i where $i \varepsilon I_k$ constituting the set $f \varepsilon R^n$.

In Section 6 it is stated that the rearranging of a kth order quasi-triangular matrix $\mathcal{C}$ into a pth order quasi-diagonal matrix D disjoints the irreducible subsets of variables and equations (z_j, g_j), $j \varepsilon I_k$, into the groups of subsets of variables and equations (y_i, h_i) where $h_i \supset g_j^i$, $y_i \supset z_j^i$, $i \varepsilon I_p$, $j \varepsilon I_{\rho i}$ and $\Sigma_{i=1}^{p} \rho_i = k$. The pair (y_i, h_i) is associated in Section 7 to a disjoint system V_i which comprises the irreducible subsystems S_1^i , S_2^i , ..., $S_{\rho_i}^i$. The system V_i may not be decomposed further if and only if p corresponds to the maximum number of possible diagonal cells in D. The transformation of a disjoint system into its equivalent hierarchical struc ture corresponds to the quasi-triangularization of the D_{ii} cells in the D matrix suc that the resulting diagonal cells in D_{ii}, namely D_{rr}^i where $r \varepsilon I_1$, are diagonal in themselves. Such a matrix is defined in Section 8 as the "Multi-level matrix". In the mathematical procedure developed the disjoint systems are identified once the overall hierarchical structure corresponding to the system S is constructed.

Section 9 proves the following important relationship

$$\hat{\Lambda}_{j,i} = \{z_1 , z_2 , \ldots, z_i\} , j \varepsilon I_{\theta_i} \quad (1.1)$$

in which the characteristic sets are associated with the θ_i non-linear functions and variables describing the irreducible subset g_i. This relationship clearly indicates that one characteristic set, such as $\hat{\Lambda}_{r,i}$, comprising i elements, is sufficient to identify the irreducible subsystem g_i. Furthermore, equation (1.1), with the use of the quasi-triangular matrix $\mathcal{C}$, implies that the set which contains the subsets of variables which may influence the output of the irreducible subset g_i, namely z_i, is given by

$$\hat{\Omega}_i = \hat{\Lambda}_{1,i} \setminus \{z_i\} = \ldots = \hat{\Lambda}_{\theta_i,i} \setminus \{z_i\} , i \varepsilon I_k \quad (1.2)$$

The set $\hat{\Omega}_i$ is defined in this article as the "Condensed characteristic set".

The determination of the k-condensed characteristic sets is sufficient for the construction of the hierarchical structure corresponding to the system S (Lemma V) and, hence, for the disjointing of the resulting hierarchical structure into p non-intersecting hierarchical systems (Lemmas VI and VII). The last part here is presented in Section 10.

The application of the user-oriented digital program DECOMP, which implements the theoretical developments presented in this article, is illustrated in Section 11 by an analysis of the structure of a process flow sheet comprising sixty-three process units. It should be noted that the new tearing concept developed by KEVORKIAN (1972) is also implemented in the program DECOMP. Finally, the computer storage required by the program DECOMP and its execution time for analysing several problems is presented in tableform.

2. SYSTEM REPRESENTATION

Structural analysis may be applied to large scale systems which are formulated as a set of non-linear and/or linear simultaneous equations

$$f_i(x_1, x_2, \ldots, x_n, u_1, u_2, \ldots, u_m) = 0 \quad i\varepsilon I_n \tag{2.1}$$

in which x is a set of n-unknown variables and u is a set of m-known variables. Moreover, structural analysis can be used for identifying the inherent structure associated with a large process flow sheet or an organizational chart comprising a set σ of individual units or blocks.

In the case of (2.1) analysis is initiated by first forming a Boolean matrix, termed by STEWARD (1962) as the occurrence matrix, which has its rows and columns corresponding to the elements in f and x, respectively, such that:

$$C_i = \{c_{ij} \mid c_{ij} = 1 \text{ when } x_j \varepsilon f_i \text{ otherwise } c_{ij} = 0 \quad j\varepsilon I_n\} \; i\varepsilon I_n \tag{2.2}$$

The formulation, here, is completed by establishing the direction of information flow in the system equations. This is achieved by assigning to each equation f_i an output variable x_j, $i,j\varepsilon I_n$ (STEWARD, 1962) for which the equation f_i is solved. This is done in such a way that each equation has a single output variable and each variable is assigned to only one equation. It should be noted that x_j may be an output variable in f_i only if x_j is contained in f_i. Thus information flows from equation f_i, through the variable x_j, directly to any other equation in the set f which contains the variable x_j.

A pair such as (x_j,f_i) is defined as the "Output pair" and the set of output pairs is denoted as the "Output pairs set". The choice of (x,f) in partitioning as illustrated in the following section is arbitrary.

The direction of information flow in a given process flow sheet or organizational chart is inherently defined. As a result structural analysis is directly initiated by forming a Boolean matrix termed (LEDET, 1970) as the adjacency matrix, which has its rows and columns corresponding to the elements in σ such that:

$$A_i = \{a_{ij} \mid a_{ij} = 1 \text{ when } \sigma_j \rightarrow \sigma_i \text{ otherwise } a_{ij} = 0 \quad j\varepsilon I_n\} \; i\varepsilon I_n \tag{2.3}$$

The procedures developed in this article make use of an occurrence matrix in which the chosen output pairs set is arranged such that it occupies the main diagonal in C. Hence, in the analysis of a process flow sheet, or a given network, the obtained adjacency matrix is transformed to the desired C form by replacing its zero diagonal elements with Boolean '1's.

The structural analysis procedures developed here are performed on the system occurrence matrix. However, for the sake of convenience the system to be considered in the following sections will be of the type given by equation (2.1).

3. THE BREAKING UP OF A LARGE SET OF SIMULTANEOUS EQUATIONS INTO SMALLER SUBSETS OF EQUATIONS

Let it be considered that the C matrix is arranged, by row and column exchanging, to the following quasi-triangular form:

$$\mathcal{C} = \begin{bmatrix} \mathcal{C}_{11} & & \mathbf{0} & \\ \mathcal{C}_{21} & \mathcal{C}_{22} & & \\ & \cdot & \cdot \;\; \cdot \;\; \cdot & \\ \mathcal{C}_{k1} & \mathcal{C}_{k2} & \cdots & \mathcal{C}_{kk} \end{bmatrix} \quad (3.1)$$

in which the ith row and jth column of Boolean cells are associated with the subset of equations g_i and variables z_j, respectively, where

$$\bigcup_{i=1}^{k} g_i = f \quad \text{and} \quad \bigcup_{j=1}^{k} z_j = x \quad (3.2)$$

$$g_i \bigcap_{i \neq j} g_j = \phi \quad \text{and} \quad z_i \bigcap_{i \neq j} z_j = \phi \;, \; i,j \varepsilon I_k \quad (3.3)$$

where ϕ is an empty set, and such that

$$g_i ,\; z_i \varepsilon R^{\theta i} \qquad i \varepsilon I_k \quad (3.4)$$

in which

$$\sum_{i=1}^{k} \theta_i = n \quad (3.5)$$

The subsets of equations associated with $\mathcal{C}$ can thus be written as:

$$g_i(z_1, z_2, \ldots, z_i) = 0 \qquad i \varepsilon I_k \quad (3.6)$$

The formulation given above indicates that g_1 contains only z_1 and, hence, g_1 must be solved for z_1. Similarly, as g_2 contains z_1 and z_2, then g_2 can be solved only for z_2 and so forth. Thus it can be concluded that g_i may be solved only for z_i. As a result equation (3.6) may be formulated as:

$$g_i(\overline{z}_1, \overline{z}_2, \ldots, \overline{z}_{i-1}, z_i) = 0 \qquad i \varepsilon I_k \quad (3.7)$$

in which $\overline{z}_j$, j=1, 2, ..., i-1, denotes the subsets of known variables which must be available in order to solve g_i for z_i. It is important to note that, which ever output set is assigned to f, the output variables associated with the equations belonging to a subset g_i must belong to the subset of variables z_i in which case the form of equation (3.6) or (3.7) remains unchanged. Thus the choice of an output set does not alter the elements of the subsets of equations and variables associated with the quasi-triangular matrix given in equation (3.1).

Hence, the arranging of an occurrence matrix into the quasi-triangular form given by (3.1) breaks up a large set of non-linear simultaneous equations into k subsets of equations $g_1, g_2, \ldots, g_k$ which can be solved serially for the subsets of unknown variables $z_1, z_2, \ldots, z_k$, respectively. Obviously, if an occurrence matrix cannot be arranged in the form given by equation (3.1), then the corresponding set of equations f cannot be broken into smaller subsets of equations.

4. THE IRREDUCIBLE SUBSYSTEMS WHICH MAY BE ASSOCIATED WITH A LARGE SCALE SYSTEM

Let S_1, S_2, ..., S_k denote the subsystems which are described by the paired subsets of variables and equations (z_1, g_1), (z_2, g_2), ..., (z_k, g_k), respectively. Here S_i, $i \varepsilon I_k$, may be broken down to two or more smaller subsystems if and only if the rows and columns associated with the diagonal cell $\mathcal{C}_{ii}$ in the $\mathcal{C}$ matrix can be exchanged so that the resulting cell becomes quasi-triangular.

The procedure outlined in the following section results in a quasi-triangular matrix with the maximum number of possible diagonal cells. As a result S_i, $i \varepsilon I_k$, will be denoted, henceforth, as an irreducible subsystem.

In this manner a large scale system S is replaced by k irreducible subsystems where

(a) Each subsystem comprises either a single equation (unit) or a number of equations (units) connected to each other in closed paths of information flow.

(b) The information flow between the subsystems, as shown in Fig. 1, is in one direction only.

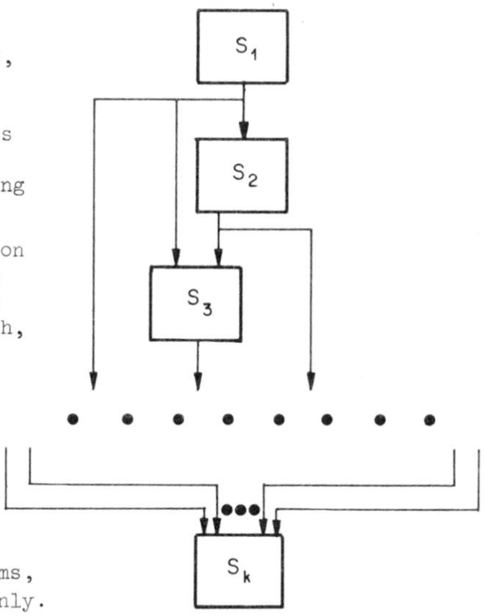

Figure 1 - The irreducible subsystems S_1, S_2, ..., S_k connected in the precedence order given by equation (3.1)

5. A NEW PROCEDURE FOR THE QUASI-TRIANGULARIZATION OF BOOLEAN MATRICES-PARTITIONING

Let a non-linear algebraic equation

$$f_i(x_1, x_2, \ldots, x_n) = 0 \quad , x_j \varepsilon f_i \tag{5.1}$$

be written in the following form:

$$x_j = f_i'(x_1, \ldots, x_{j-1}, x_{j+1}, \ldots, x_n) \tag{5.2}$$

which indicates that x_j is chosen as an output variable (STEWARD, 1962) in f_i and subsequently (x_j, f_i) is the resulting output pair.

Let us now denote the set of variables contained in f_i' by Λ_j, where

$$x_i \varepsilon \Lambda_j \quad \text{if} \quad x_i \varepsilon f_i' \tag{5.3}$$

From equations (5.2) and (5.3) it is clear that x_i is required in solving equation f_i for the variable x_j. This implies that x_i has a direct influence on the variable x_j. Furthermore if

$$x_q \varepsilon \Lambda_i \quad \text{and} \quad x_q \notin \Lambda_j \tag{5.4}$$

then x_q would have a direct influence on the variable x_i and subsequently an indirect influence on x_j. Let us denote the set $\hat{\Lambda}_j$ which contains x_j and the variables which have a direct or indirect influence on the variable x_j as the "Characteristic set".

The definition of the characteristic set gives rise to the following observations:

Lemma I: The relationship $\quad x_i \varepsilon \hat{\Lambda}_j \quad$ (5.5)

would imply $\quad x_i, x_j \varepsilon z_r \quad$ (5.6)

iff $\quad x_j \varepsilon \hat{\Lambda}_i \quad$ (5.7)

Proof - The relationship $x_i \varepsilon \hat{\Lambda}_j$ indicates that the variable x_i influences the variabl x_j. Similarly the relationship $x_j \varepsilon \hat{\Lambda}_i$ indicates that the variable x_j influences the variable x_i. As a result the variables x_i and x_j are related to each other in a closed path of information flow and, hence, the equations which have x_i and x_j as the output variables must be solved simultaneously.

Lemma II: The relationships

$$x_i \varepsilon \hat{\Lambda}_j \tag{5.8}$$

$$x_j \notin \hat{\Lambda}_i \tag{5.9}$$

imply that the equation with the output variable x_i must be solved prior to the equation with the output variable x_j.

The proof here can be easily derived from Lemma I.

In the computer program developed Lemma I is applied for identifying the subsets z_i, $i \varepsilon I_k$, after which Lemma II is used for ordering the irreducible subsets g_i, $i \varepsilon I_k$, such that the information flow between them is in one direction only.

5.1. Illustrative example

Consider the set of 13 non-linear simultaneous equations given in the first column of Table I.

TABLE I

System equations	Output pairs set	Form in which the eqs. have to be solved
$f_1(x_4, x_9) = 0$	(x_4, f_1)	$x_4 = f'_1(x_9)$
$f_2(x_1, x_5, x_{10}) = 0$	(x_5, f_2)	$x_5 = f'_2(x_1, x_{10})$
$f_3(x_{11}, x_{12}) = 0$	(x_{11}, f_3)	$x_{11} = f'_3(x_{12})$
$f_4(x_3, x_7) = 0$	(x_7, f_4)	$x_7 = f'_4(x_3)$
$f_5(x_2, x_{12}, x_{13}) = 0$	(x_2, f_5)	$x_2 = f'_5(x_{12}, x_{13})$
$f_6(x_4, x_9) = 0$	(x_9, f_6)	$x_9 = f'_6(x_4)$
$f_7(x_1, x_5, x_7) = 0$	(x_1, f_7)	$x_1 = f'_7(x_5, x_7)$
$f_8(x_2, x_5, x_6) = 0$	(x_6, f_8)	$x_6 = f'_8(x_2, x_5)$
$f_9(x_7, x_{10}) = 0$	(x_{10}, f_9)	$x_{10} = f'_9(x_7)$
$f_{10}(x_8, x_{13}) = 0$	(x_8, f_{10})	$x_8 = f'_{10}(x_{13})$
$f_{11}(x_4, x_{12}) = 0$	(x_{12}, f_{11})	$x_{12} = f'_{11}(x_4)$
$f_{12}(x_2, x_4, x_8, x_{13}) = 0$	(x_{13}, f_{12})	$x_{13} = f'_{12}(x_2, x_4, x_8)$
$f_{13}(x_3, x_{10}) = 0$	(x_3, f_{13})	$x_3 = f'_{13}(x_{10})$

If the output pairs set listed in the second column is adopted, the 13 equations can be written in the form of (5.2) as shown in the third column of the same Table. The corresponding occurrence matrix C is shown in Fig. 2 where the output pairs correspond to the main diagonal.

The elements of the set Λ_j, $j \varepsilon I_{13}$, as defined by equation (5.3) are indicated in Fig. 2 by the non-zero off-diagonal elements occurring in the row where x_j occurs on the main diagonal.

The characteristic sets $\hat{\Lambda}_i$ are determined by replacing the elements c_{ij}, $j \epsilon I_{13}$, occupying the ith row of the C matrix by $\bar{c}_{ij}$, where

$$\bar{c}_{ij} = c_{ij} \dotplus c_{qj} \quad \text{if} \quad c_{iq} = 1 \tag{5.10}$$

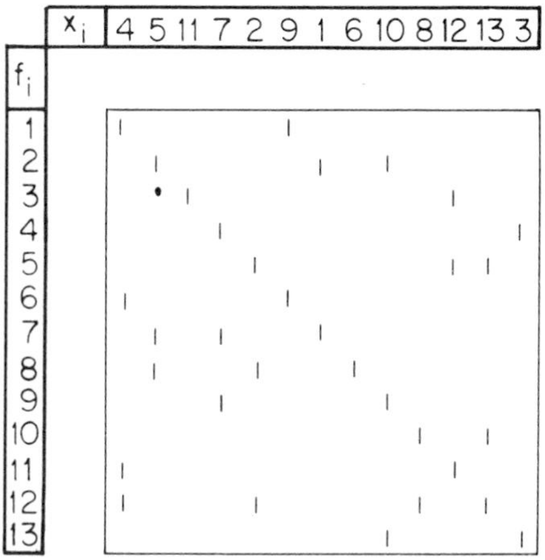

Fig. 2 - Occurrence matrix C with the output variables appearing on the main diagonal

For the given example the resulting Boolean matrix $\bar{C}$, denoted as the "Characteristic occurrence matrix", is shown in Fig. 3. Here a characteristic set $\hat{\Lambda}_j$ comprises the variables corresponding to the non-zero elements occupying the row where x_j occurs on the main diagonal.

Hence, the application of Lemmas I and II in the matrix shown in Fig. 3 results, by row and column permutation, in the quasi-triangular matrix shown in Fig. 4.

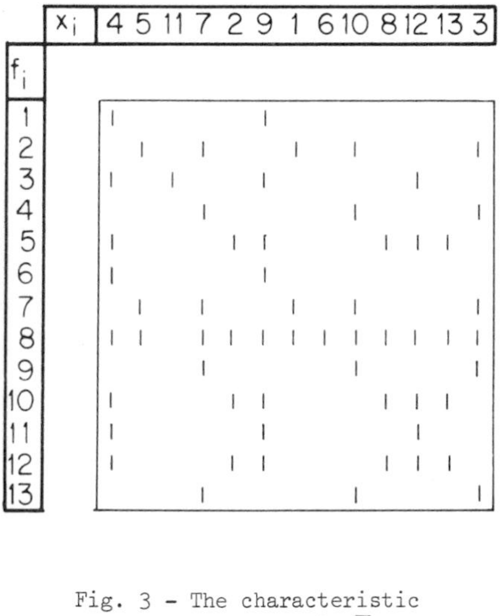

Fig. 3 - The characteristic occurrence matrix $\bar{C}$

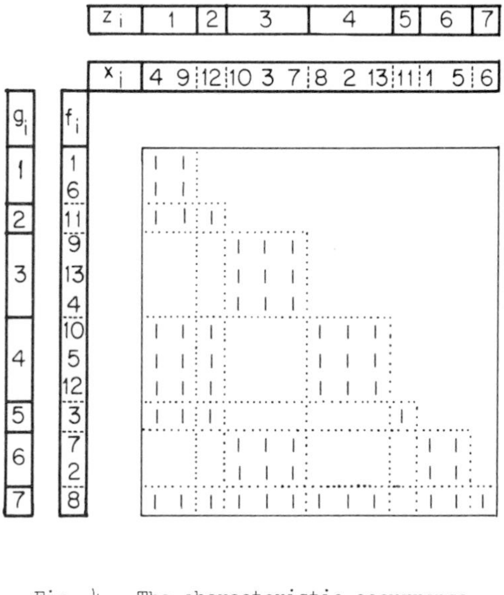

Fig. 4 - The characteristic occurrence matrix in quasi-triangular form

6. THE TRANSFORMATION OF A SET OF IRREDUCIBLE SUBSETS OF EQUATIONS INTO A SET COMPRISING DISJOINTED GROUPS OF EQUATIONS

Let it be considered that the quasi-triangular matrix $\mathcal{C}$ is arranged, by making use of its zero off-diagonal cells, into the quasi-diagonal form:

$$D = \begin{bmatrix} D_{11} & & & \mathbf{0} \\ & D_{22} & & \\ & & \ddots & \\ \mathbf{0} & & & D_{pp} \end{bmatrix} \qquad p \leqslant k \tag{6.1}$$

in which the ith row and column correspond to the sets of equations h_i and variables y_i where

$$\bigcup_{j=1}^{\rho_i} g_j^i = h_i \quad \text{and} \quad \bigcup_{j=1}^{\rho_i} z_j^i = y_i \ , \ i \varepsilon I_p \tag{6.2}$$

in which

$$\sum_{i=1}^{p} \rho_i = k \tag{6.3}$$

$$\bigcup_{i=1}^{p} h_i = f \quad \text{and} \quad \bigcup_{i=1}^{p} y_i = x \tag{6.4}$$

and

$$h_i \bigcap_{i \neq j} h_j = \emptyset \quad \text{and} \quad y_i \bigcap_{i \neq j} y_j = \emptyset \tag{6.5}$$

The p sets of equations associated with D can thus be written as:

$$h_i(y_i) = 0 \quad i \varepsilon I_p \tag{6.6}$$

which clearly indicates that the group of irreducible subsets $g_1^i, g_2^i, \ldots, g_{\rho_i}^i$ belonging to h_i form one disjoint set. Hence, the arranging of the quasi-triangular matrix $\mathcal{C}$ into the quasi-diagonal form D transforms the irreducible subsets of equations $g_1, g_2, \ldots, g_k$ into the disjointed sets of equations $h_1, h_2, \ldots, h_p$. Obviously, if the $\mathcal{C}$ matrix cannot be transformed into the diagonal form D given by equation (6.1), then the irreducible subsets of equations g_i, $i \varepsilon I_k$, belong to the same undisjointed set of equations f.

7. A LARGE SCALE SYSTEM COMPRISING DISJOINT SYSTEMS

Let $V_1, V_2, \ldots, V_p$ denote the disjoint systems, existing in the large scale system S, such that

$$V_i = \bigcup_{j=1}^{\rho_i} S_j^i \quad i \varepsilon I_p \tag{7.1}$$

in which $S_1^i, S_2^i, \ldots$, denote the irreducible subsystems which are associated with the subsets of equations $g_1^i, g_2^i, \ldots$, belonging to the set h_i. In other words:

$$S = \bigcup_{i=1}^{p} V_i \tag{7.2}$$

and

$$V_i \bigcap_{i \neq j} V_j = \emptyset \quad i, j \varepsilon I_p \ . \tag{7.3}$$

The system V_i, $i \varepsilon I_p$, may be disjointed further to two or more disjoint systems if and only if:

$$p < p_{max} \tag{7.4}$$

in which p_{max} represents the maximum number of possible diagonal cells in D. The procedure presented in Section 9 insures that $V_1, V_2, \ldots, V_p$ represent the maximum number of possible disjoint systems in S.

8. THE IRREDUCIBLE SUBSYSTEMS, BELONGING TO A DISJOINT SYSTEM, ARRANGED IN A HIERARCHICAL FORM

Let it be assumed that each of the cells in the quasi-diagonal matrix D is arranged in the form:

$$D_{ii} = \begin{bmatrix} D^i_{11} & & \mathbf{0} & \\ D^i_{21} & D^i_{22} & & \\ & \cdot\ \cdot & \cdot\ \cdot & \\ D^i_{l1} & D^i_{l2} & \cdots & D^i_{ll} \end{bmatrix} \quad i\varepsilon I_p \tag{8.1}$$

in which the diagonal cells are given by:

$$D^i_{rr} = \begin{bmatrix} d^{i,r}_{11} & & \mathbf{0} \\ & d^{i,r}_{22} & \\ & & \ddots & \\ \mathbf{0} & & & d^{i,r}_{\tau_r\tau_r} \end{bmatrix}, \quad r\varepsilon I_l \tag{8.2}$$

where

$$\sum_{j=1}^{l} \tau_j = \rho_i , \tag{8.3}$$

while each row of the first subdiagonal cells in D_{ii}, denoted by $(\mathbf{d}^i_{r\ r-1})_q$ where $q\varepsilon I_{\tau_r}$ contains at least one non-zero element. In other words:

$$(\mathbf{d}^i_{r\ r-1})_q \neq 0 \quad q\varepsilon I_{\tau_r} \tag{8.4}$$

The jth row and column of the cell D^i_{rr} correspond to the subset of equations $g^{i,r}_j$ and subset of variables $z^{i,r}_j$, respectively, where

$$\bigcup_{r=1}^{l} \bigcup_{j=1}^{\tau_r} g^{i,r}_j = h_i \quad \text{and} \quad \bigcup_{r=1}^{l} \bigcup_{j=1}^{\tau_r} z^{i,r}_j = y_i \ , \ i\varepsilon I_p \tag{8.5}$$

Let $S^{i,r}_1, S^{i,r}_2, \ldots, S^{i,r}_{\tau_r}$ where $r\varepsilon I_l$, denote the irreducible subsystems associated with the subsets of equations $g^{i,r}_1, g^{i,r}_2, \ldots, g^{i,r}_{\tau_r}$, respectively.

Now we are in a position to state the following preposition:

Preposition I: A quasi-triangular matrix D_{ii}, satisfying the relationships given by (8.2) and (8.4), identifies the minimum number of hierarchical levels associated with the disjoint system V_i such that:

(a) The irreducible subsystems belonging to a given hierarchical level are independent of each other.

(b) At a given level the irreducible subsystems are influenced by at least one subsystem at one level higher in the system hierarchy.

The validity of this preposition can be proved as follows: The off-diagonal elements in the cells D^i_{rr}, $r \varepsilon I_l$, are all zero. As a result the subsets of equations $g_1^{i,r}$, $g_2^{i,r}$, ..., $g_{\tau_r}^{i,r}$, corresponding to the rows of the cells D^i_{rr}, are disjointed from each other within the rth row of the quasi-triangular matrix D_{ii}. Hence, the irreducible subsystems $S_1^{i,r}$, $S_2^{i,r}$, ..., $S_{\tau_r}^{i,r}$, henceforth denoted by the set $W^{i,r}$, where

$$V_i = \bigcup_{r=1}^{l} W^{i,r} \tag{8.6}$$

and
$$W^{i,r} = \bigcup_{j=1}^{\tau_r} S_j^{i,r}, \quad r \varepsilon I_l \tag{8.7}$$

are independent of each other and belong to the same level of the defined hierarchical structure. In this manner the l rows of the matrix D_{ii}, henceforth defined as the "Multi-level matrix", identify the hierarchical levels associated with the disjoint system V_i as illustrated in Fig. 5.

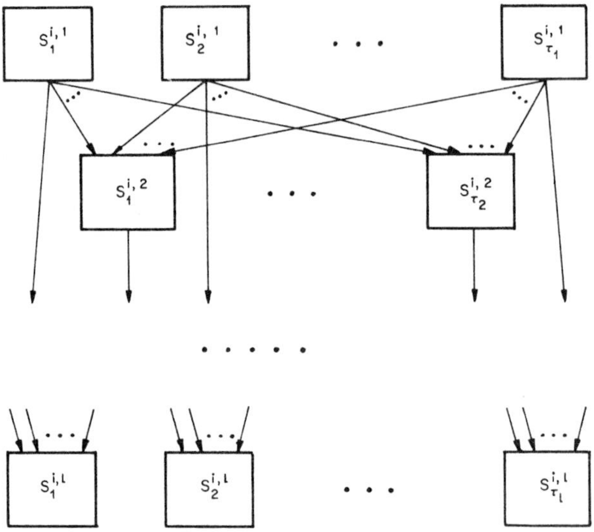

Fig. 5 - The hierarchical structure associated with the disjoint system V_i

The number of rows in the multi-level matrix D_{ii} can be reduced if and only if:

$$D^i_{rr-1} = 0 \quad r = 2, 3, \ldots, l \tag{8.8}$$

The relationship given by (8.4), however, implies that none of the τ_r rows in the cell D^i_{rr-1} can be zero. Therefore, the rows in the D_{ii} correspond to the minimum number of hierarchical levels associated with V_i.

Let us now consider the situation in Fig. 6a where the irreducible subsystems belonging to $W^{i,r+1}$ are not influenced by the subsystem $S_j^{i,r}$ occurring in the rth level of the hierarchical structure. As a result condition (a), given above, implies the situation shown in Fig. 6b where

$$S_j^{i,r} \varepsilon W^{i,r+1} \tag{8.9}$$

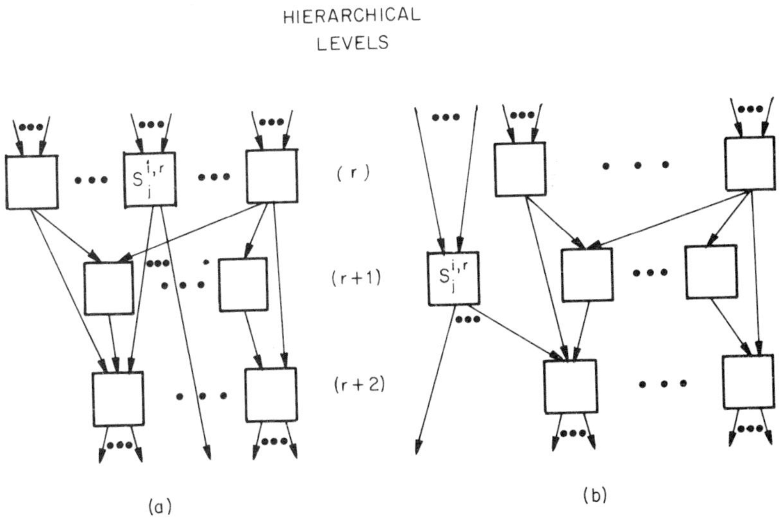

Fig. 6 - Two possible configurations of a hierarchical structure

Thus restricting preposition I with condition (a) only may result in the two possible configurations of a hierarchical structure which in fact convey the same information regarding the partitioned subsystems and disjointed systems in S. Therefore, from a consistency point of view condition (b) is also introduced in preposition I and thus the situation corresponding to that shown in Fig. 6b is eliminated.

9. A PROCEDURE FOR THE CONSTRUCTION OF A HIERARCHICAL STRUCTURE

Let the irreducible subsystem

$$g_i(z_1, z_2, \ldots, z_i) = 0 \quad i\varepsilon I_k \tag{9.1}$$

be written in the following form:

$$z_i = g_i'(z_1, z_2, \ldots, z_{i-1}) \tag{9.2}$$

Let us denote the set of irreducible subsets of variables contained in g_i' by Ω_i, where

$$z_j \varepsilon \Omega_i \quad \text{if} \quad z_j \varepsilon g_i' \quad i\varepsilon I_k \tag{9.3}$$

Moreover let us denote the set $\hat{\Omega}_i$ which contains the subsets of variables influencing z_i, directly or indirectly, as the "Condensed characteristic set".

The relationship between $\hat{\Omega}_i$ and the $\hat{\Lambda}$'s can be easily established through the following two observations:

Lemma III: The relationship $\quad x_i \varepsilon \hat{\Lambda}_s$ (9.4)

where $\quad x_i, x_j \varepsilon z_r$ (9.5)

implies $\quad z_r \subset \hat{\Lambda}_s$ (9.6)

Proof: The variables belonging to z_r are related to each other in closed paths of information flow. Thus the relationship (9.4) immediately implies that $x_j \varepsilon \hat{\Lambda}_s$. As a result we have $z_r \subset \hat{\Lambda}_s$.

Lemma IV: The relationship $x_i,\ x_j \varepsilon z_r$ (9.7)

$$\text{implies} \qquad \hat{\Lambda}_i = \hat{\Lambda}_j \tag{9.8}$$

and vice versa.

Proof: Using Lemma I, we have

$$x_i \varepsilon \hat{\Lambda}_j \tag{9.9}$$

and

$$x_j \varepsilon \hat{\Lambda}_i \tag{9.10}$$

Equation (9.9) implies by definition that all the variables which may influence x_i, namely $\hat{\Lambda}_i$, are also contained in $\hat{\Lambda}_j$, namely

$$\hat{\Lambda}_i \subset \hat{\Lambda}_j \tag{9.11}$$

Similar interpretation of equation (9.10) yields

$$\hat{\Lambda}_j \subset \hat{\Lambda}_i \tag{9.12}$$

and hence

$$\hat{\Lambda}_i = \hat{\Lambda}_j \tag{9.13}$$

Thus we could see that

$$\hat{\Omega}_i = \hat{\Lambda}_{1,i} \setminus \{z_i\} = \ldots = \hat{\Lambda}_{\theta_i,i} \setminus \{z_i\}\ ,\quad i \varepsilon I_k \tag{9.14}$$

where

$$\hat{\Lambda}_{j,i} = \{z_1,\ z_2,\ \ldots,\ z_i\} \qquad j \varepsilon I_{\theta_i} \tag{9.15}$$

in which $\hat{\Lambda}_{1,i},\ \ldots,\ \hat{\Lambda}_{\theta_i,i}$ denote the characteristic sets which are associated with the θ_i non-linear functions and variables describing g_i. From equation (9.14) it should be noted that

$$z_j \notin \hat{\Omega}_i \quad \text{if} \quad j = i, i+1,\ \ldots,\ k \tag{9.16}$$

which could be verified by the repeated use of equation (9.2).

The construction of the defined hierarchical structure, corresponding to the system S, is now accomplished by making use of the following observation.

Lemma V: An irreducible subsystem S_i will belong to the rth level of the hierarchical structure if and only if

$$\hat{\Omega}_i \setminus \{\bigcup_{j=1}^{i-1} z_j\} = \emptyset \qquad i \neq j \tag{9.17}$$

in which $\{\bigcup_{j=1}^{i-1} z_j\}$ is a set comprising the outputs of the irreducible subsystems belonging to the first (r-1) levels of the hierarchical structure.

Proof: The case $\hat{\Omega}_j = \emptyset$ implies that $g_j(z_j) = 0$, which indicates that the irreducible subset g_j can be solved independently of the other subsets in f. Hence, the corresponding subsystem S_j belongs to the first hierarchical level. In a similar manner equation (9.17) indicates that the condensed characteristic set $\hat{\Omega}_i$ is an empty set with respect to the outputs of the subsystems belonging to the remaining levels (r+1), (r+2), ..., of the hierarchical structure. Therefore it follows that S_i belongs to the rth hierarchical level.

10. A NEW PROCEDURE FOR DISJOINTING A LARGE SCALE SYSTEM

The hierarchical system constructed in Section 9 may comprise two or more disjoint systems. The identification of these disjoint systems is accomplished by performing the following observations:

Lemma VI: The maximum number of disjoint systems p which may exist in a given system S must satisfy the following two inequalities:

$$1 \leqslant p \leqslant \nu_{in} \tag{10.1}$$

$$1 \leqslant p \leqslant \nu_{out} \tag{10.2}$$

in which ν_{in} denotes the number of sets $\hat{\Omega}_i$, $i \varepsilon I_k$, which are empty while ν_{out} refers to the number of subsets z_j which are not contained in any of the sets $\hat{\Omega}_i$, $i \varepsilon I_k$.

Proof: In Lemma VI ν_{in} represents the maximum number of irreducible subsystems which belong to the first level of the hierarchical structure associated with the overall system S. Subsequently none of these subsystems can belong to two different disjoint systems. As a result the maximum number of disjoint systems cannot exceed the number of irreducible subsystems belonging to the first hierarchical level.

In a similar way ν_{out} represents the total number of irreducible subsystems which may belong to the last hierarchical levels of the disjoint systems in S. As a result the proof here reduces to that given above.

Lemma VI becomes a useful observation in disjointing only when ν_{in} or ν_{out} equals one. For more complicated cases, however, where ν_{in}, $\nu_{out} > 1$ the following observation is necessary for identifying the possible disjoint systems in S.

Lemma VII: The relationship $\hat{\Omega}_i \cap \hat{\Omega}_j \neq \emptyset$ (10.3)

implies that the irreducible subsystems S_i, S_j, together with those whose outputs comprise elements of the sets $\hat{\Omega}_i$ and $\hat{\Omega}_j$, belong to one disjoint system.

The proof here can be derived from Lemma VI.

In the computer program developed Lemma VII is applied successively in separating a set of irreducible subsystems into disjoint systems. It should be noted that it is sufficient to consider only the condensed characteristic sets associated with the ν_{out} irreducible subsystems which might belong to the last hierarchical levels.

10.1 The illustrative example continued

The application of Lemmas III and IV, which were introduced in the previous sections reduces the quasi-triangular matrix shown in Fig. 4 to the condensed form shown in Fig. 7. Here, each element represents a cell of the matrix in Fig. 4. Moreover, the non-zero off-diagonal elements occurring in the ith row of the matrix in Fig. 7 correspond to the subsets of variables belonging to the set $\hat{\Omega}_i$. For example, $\hat{\Omega}_1 = \emptyset$, $\hat{\Omega}_2 = \{z_1\}$ etc.

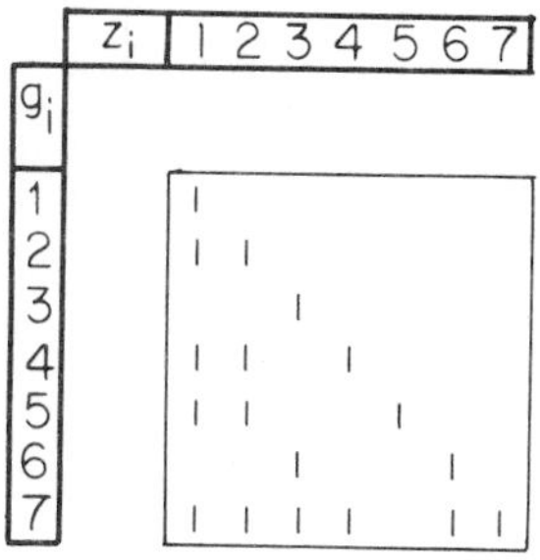

Fig. 7 - The condensed characteristic occurrence matrix

The application of Lemma V reduces the modified characteristic occurrence matrix to the hierarchical form shown in Fig. 8. Here, $\nu_{in} = 2$ because g_1 and g_3 are the only irreducible subsets belonging to the first level of the overall hierarchical structure, while $\nu_{out} = 2$ because z_5 and z_7 are the only subsets of variables which do not occur in any of the sets $\hat{\Omega}_i$ where $i \varepsilon I_7$. Subsequently, Lemma VI is not sufficient for concluding whether all the seven irreducible subsets belong to one disjoint set.

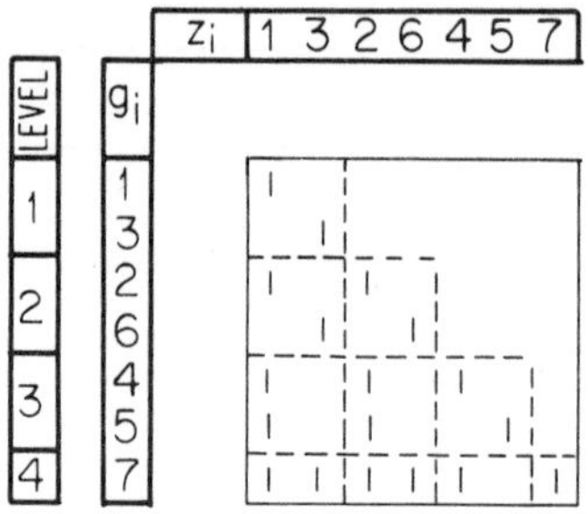

Fig. 8 - The condensed characteristic occurrence matrix arranged in a hierarchical form

The condensed characteristic sets associated with the irreducible subsets which might belong to the last hierarchical level are given by:

$$\hat{\Omega}_7 = \{z_1, z_2, z_3, z_4, z_6\} \quad \text{and} \quad \hat{\Omega}_5 = \{z_1, z_2\} . \tag{10.4}$$

As a result we have

$$\hat{\Omega}_7 \cap \hat{\Omega}_5 = \{z_1, z_2\} \neq \emptyset \tag{10.5}$$

which implies, by the use of Lemma VII, that the subsets of variables z_i, $i\varepsilon I_7$, belong to one disjoint system.

The resulting multi-level matrix is shown in Fig. 9 and the corresponding hierarchically structured network of equations is shown in Fig. 10.

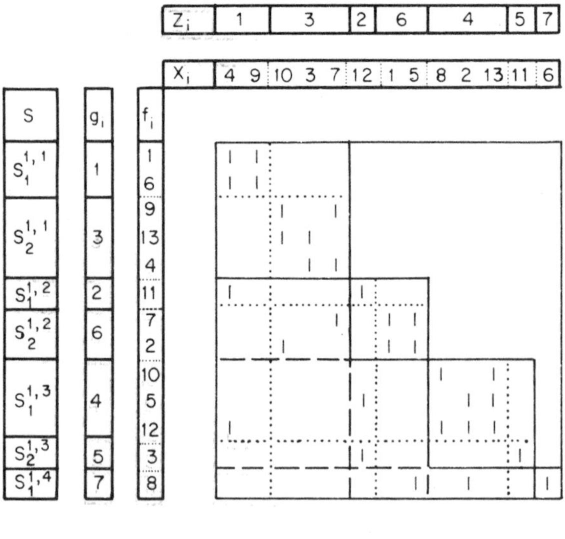

Fig. 9 - The multi-level matrix

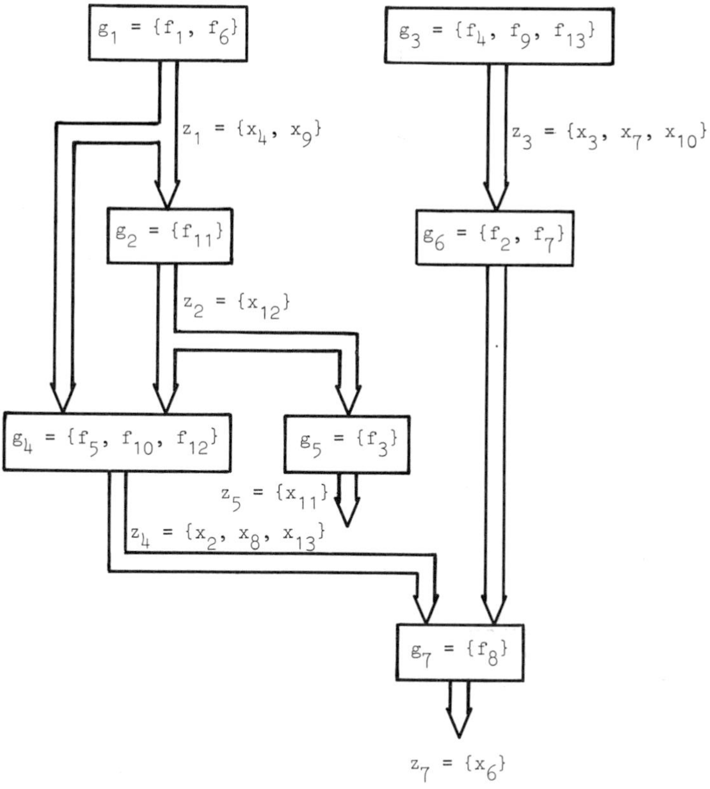

Fig. 10 - The set of equations given in the first column of Table I arranged in hierarchical form

11. APPLICATIONS

The integrated scheme shown in Fig. 11 is part of the block diagram of a typically large scale process. The numbered boxes, here, represent individual process units such as a complex distillation column or a simple pump. These units are connected to one another by one or more streams as indicated by the numbered lines.

The structure shown in Fig. 11 may in fact resemble an information flow network such as associated with an organizational chart.

The occurrence* matrix corresponding to the block diagram in Fig. 11 was inserted in the digital program DECOMP and the resulting computer output is illustrated in Figs. 12, 13 and 14.

----

* namely the adjacency matrix in which the zero diagonal elements are replaced by Boolean '1's.

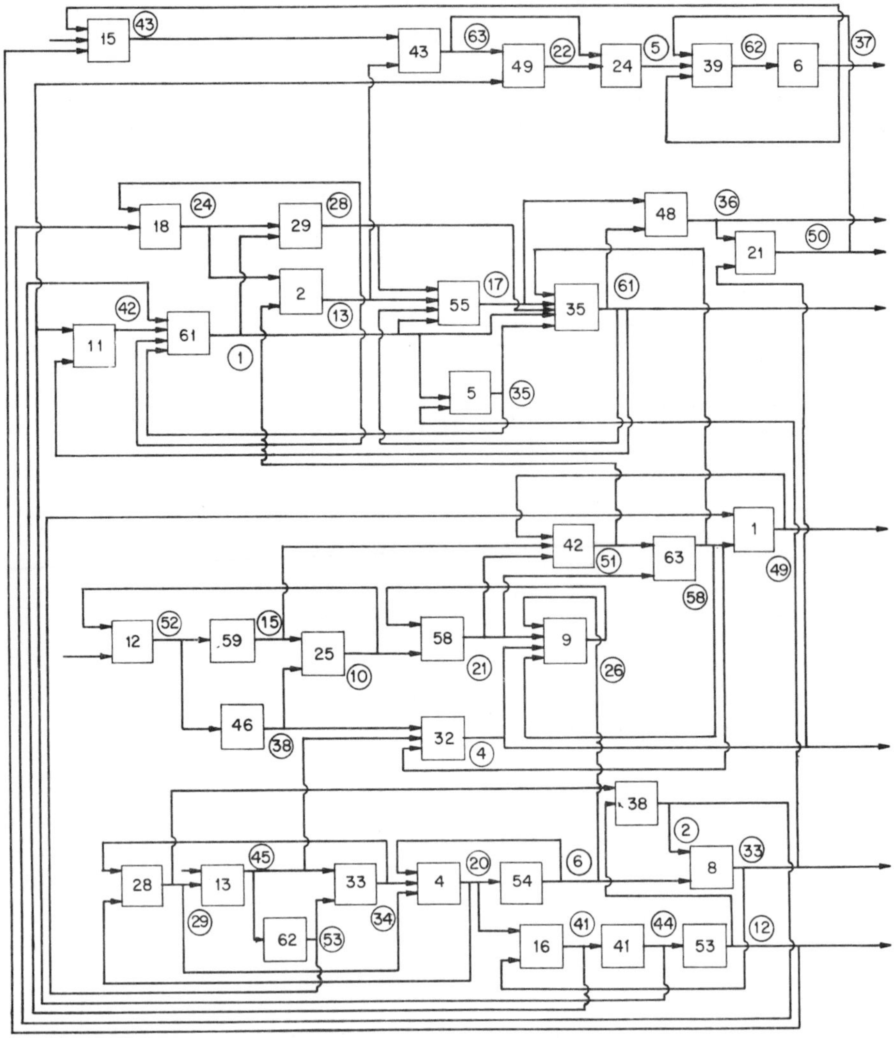

Fig. 11 - Part of a large scale process flow sheet

Fig. 12a illustrates the hierarchical structure associated with the eight irreducible subsystems comprising the first disjoint system of total two. Here, 1.01 and 1.02 identify the irreducible subsystems belonging to the first hierarchical level whereas 6.01 identifies the irreducible subsystem belonging to the sixth and last hierarchical level shown in Fig. 15. In this manner the study of the large scale system in Fig. 11 has been reduced to a separate study of the eight irreducible subsystems, in the first disjoint system, and the two subsystems in the second disjoint system.

```
H
 L        1 1 2 2 3 4 5 6
  .       . . . . . . . .
   I      0 0 0 0 0 0 0 0
    S     1 2 1 2 1 1 1 1
         -----
1.01     |θ |
1.02     | θ|
         ---------
2.01     |*  |θ |
2.02     |* *| θ|
         -----------
3.01     |   |* *|θ|
         -------------
4.01     |   |   |*|θ|
         ---------------
5.01     |   |  *| |*|θ|
         -----------------
6.01     |   |*  |*| |*|θ|
         -----------------
```

FIGURE 12a

DISJOINT SYSTEM 1 OF 2

```
H
 L        1 2
  .       . .
   I      0 0
    S     1 1
         ---
1.01     |θ|
         -----
2.01     |*|θ|
         -----
```

FIGURE 12b

DISJOINT SYSTEM 2 OF 2

*** THE IRREDUCIBLE SUBSYSTEMS ARRANGED IN A HIERARCHICAL STRUCTURE ***

*** The horizontal chained lines separate the levels in the hierarchical structure.

*** The HL in HL.IS stands for hierarchical level and IS stands for irreducible subsystem. Thus the 3 in 3.01 indicates the third hierarchical level and 01 indicates the first subsystem in that level.

*** The symbol θ, on the main diagonal, represents the irreducible subsystem corresponding to the particular row and column.

*** The symbol *, on the i-th row and j-th column represents the connection -from- the irreducible subsystem associated with the j-th column -to- that associated with the i-th row.

The computer output in Figs. 13a, ..., 13h, 14a and 14b furnishes more detailed information regarding each irreducible subsystem. For example, Fig. 13a shows the individual process units and the associated streams which comprise the irreducible subsystem 1.01 in the first disjoint system. The individual units, here, are arranged such that the minimum number of streams which are common to all the recycles in the subsystem are also identified. The merits of such an occurrence matrix will be more evidently shown by KEVORKIAN (1972).

DISJθINT SYSTEM 1 θF 2

```
           FUNCTIθN θR INDIVIDUAL UNIT
      VAR | 0 0 0 0 0 0
       θR | 5 2 1 6 3 0
HL.IS STRM| 4 8 3 2 3 4
--------------------------
1.01   20 | * *       I |
       34 |   *     I * |
       53 |       θ *   |
       45 |     θ * *   |
       29 |   θ *     * |
        6 | θ         * |
--------------------------
```

THE IRREDUCIBLE SUBSYSTEM 1.01

FIGURE 13a

```
           FUNCTIθN θR INDIVIDUAL UNIT
      VAR | 0 0 0 0
       θR | 1 4 5 2
HL.IS STRM| 2 6 9 5
----------------------
1.02   10 | *     I |
       15 |     θ * |
       38 |   θ   * |
       52 | θ * *   |
----------------------
```

THE IRREDUCIBLE SUBSYSTEM 1.02

FIGURE 13b

```
           FUNCTIθN θR INDIVIDUAL UNIT
      VAR | 0 0 0 0 0
       θR | 4 5 3 0 1
HL.IS STRM| 1 3 8 8 6
------------------------
2.01   41 | *       I |
       33 |       θ * |
        2 |     θ *   |
       12 |   θ *     |
       44 | θ *       |
------------------------
1.01   20 |         * |
       29 |     *     |
        6 |       *   |
------------------------
```

THE IRREDUCIBLE SUBSYSTEM 2.01

FIGURE 13c

```
           FUNCTIθN θR INDIVIDUAL UNIT
      VAR | 0 0 0 0 0 0
       θR | 5 0 4 6 3 0
HL.IS STRM| 8 1 2 3 2 9
--------------------------
2.02   26 | *         I |
        4 |         θ * |
       58 |   *   I * * |
       51 |     θ *     |
       49 |   θ *       |
       21 | θ   * *   * |
--------------------------
1.01   53 |   *         |
       45 |         *   |
        6 |           * |
--------------------------
1.02   10 | *           |
       15 |     *       |
       38 |         *   |
--------------------------
```

THE IRREDUCIBLE SUBSYSTEM 2.02

FIGURE 13d

```
           FUNCTIθN θR INDIVIDUAL UNIT
      VAR | 0 0 0 0 0 0 0 0
       θR | 1 0 0 3 5 1 6 2
HL.IS STRM| 8 2 5 5 5 1 1 9
------------------------------
3.01   28 | *     * *   * I |
        1 |     * * *   I * |
       42 |           θ *   |
       17 |       * I       |
       61 |       θ * *     |
       35 |     θ *     *   |
       13 |   θ     *       |
       24 | θ *           * |
------------------------------
2.01   41 |             *   |
       33 |     *           |
        2 | *               |
       44 |           *     |
------------------------------
2.02   58 |       *         |
       51 |   *             |
------------------------------
```

THE IRREDUCIBLE SUBSYSTEM 3.01

FIGURE 13e

*** VAR OR STRM reads: variable or information stream.

*** The symbols θ and I on the main diagonal, indicate the output variables or streams of the corresponding functions or individual units respectively.
Moreover the symbol I indicates an iterate (function formulation) or a recycle (individual unit formulation).

*** The symbol * on the i-th row and j-th column represents the variable or stream associated with the i-th row as the input to the function or individual unit associated with the j-th column.

DISJØINT SYSTEM 1 ØF 2

FUNCTIØN ØR INDIVIDUAL UNIT

HL.IS	STRM	048
4.01	36	Ø
3.01	17	*
	61	*

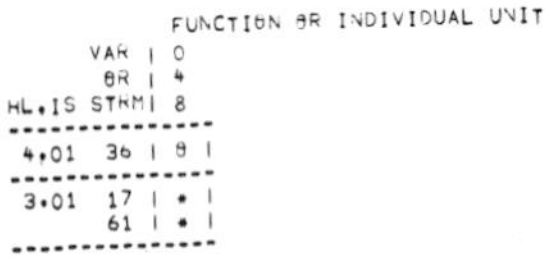

THE IRREDUCIBLE SUBSYSTEM 4.01

FIGURE 13f

FUNCTIØN ØR INDIVIDUAL UNIT

HL.IS	STRM	021
5.01	50	Ø
2.02	4	*
4.01	36	*

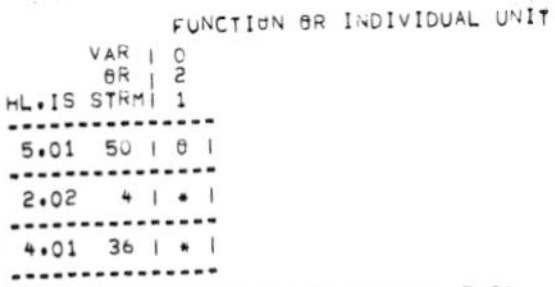

THE IRREDUCIBLE SUBSYSTEM 5.01

FIGURE 13g

FUNCTIØN ØR INDIVIDUAL UNIT

HL.IS	STRM	006	015	043	049	024	039
6.01	62	*					I
	5					Ø	*
	22				Ø	*	
	63			Ø	*	*	
	43		Ø	*			
	37	Ø	*				*
2.01	12		*				
	44				*		
3.01	13			*			
5.01	50						*

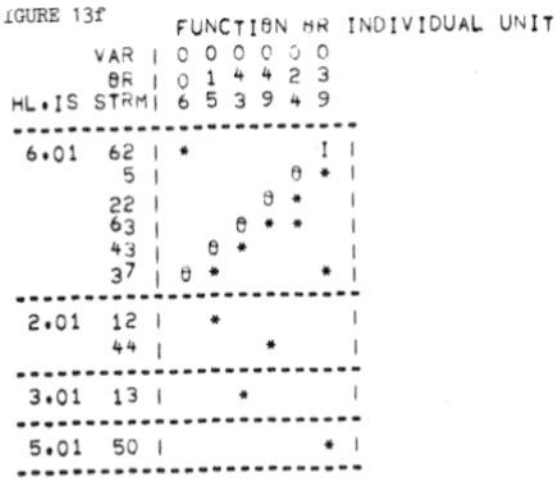

THE IRREDUCIBLE SUBSYSTEM 6.01

FIGURE 13h

DISJØINT SYSTEM 2 ØF 2

FUNCTIØN ØR INDIVIDUAL UNIT

HL.IS	STRM	014	027
1.01	31	*	I
	9	Ø	*

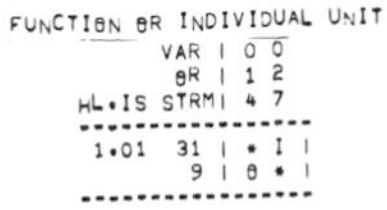

THE IRREDUCIBLE SUBSYSTEM 1.01

FIGURE 14a

FUNCTIØN ØR INDIVIDUAL UNIT

HL.IS	STRM	040	060	047	044	026	007	019	022	031	030	003	017	050	052	037	056	020	045	010	051	057	034	023	036
2.01	54						*			*															I
	25																							Ø	*
	16																						Ø		*
	30																					Ø	*		
	19	*						*					*								I				
	39		*	*						*				*						I					
	46																		Ø	*			*		
	60																	Ø	*						
	3																Ø	*							
	47															Ø	*				*				
	11														Ø	*									
	18													Ø	*										
	23												Ø	*							*			*	
	55											Ø	*		*										
	59										Ø			*										*	
	14									Ø		*				*	*								
	56								Ø		*											*			
	57							Ø	*	*															
	40						Ø	*																	
	48					Ø	*																		
	27				Ø	*	*																		
	32			Ø							*														
	8		Ø		*																				
	7	Ø			*													*		*					
1.01	31								*														*		
	9											*			*										

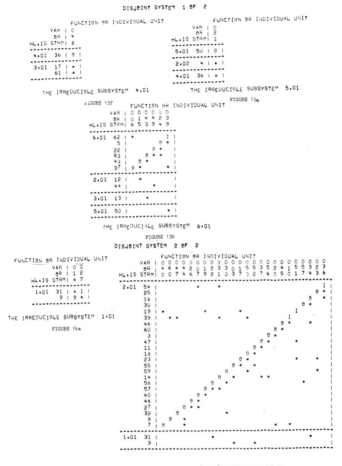

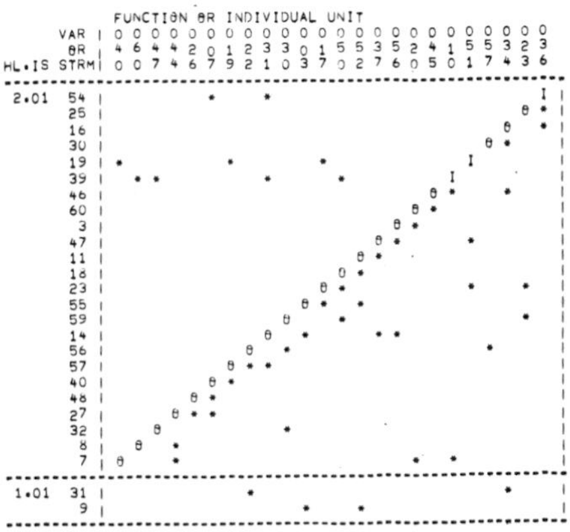

THE IRREDUCIBLE SUBSYSTEM 2.01

FIGURE 14b

Table II illustrates the storage requirements and computational efficiency of the program DECOMP.

TABLE II

Size of problem, n	Execution time (s) on RXDS $\Sigma 7$ digital computer	
	Partitioning	Disjointing and constructing hierarchical systems
31	0.18	0.06
63	0.84	0.06
112	1.74	0.18
139*	13.14	0.
Storage requirement (including tearing)		$n^2 + 5n + 4970$

* one irreducible subsystem

It should be noted that the given execution times depend on the inherent structure of the particular problem. It is for this reason that no attempt was made here to compare the execution times given in Table II with those given by LEDET (1970).

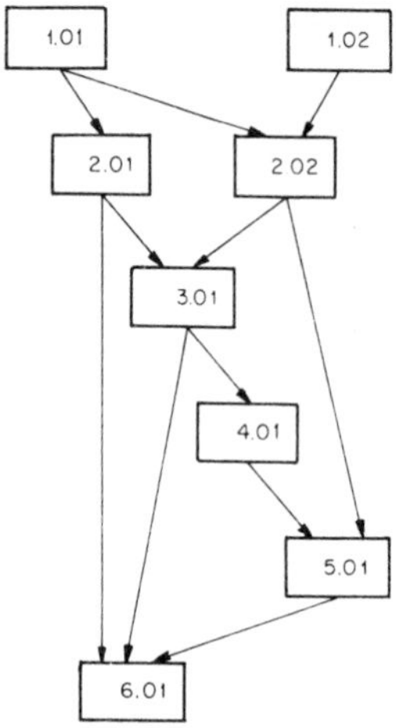

Fig. 15 - The hierarchical structure associated with the system in Fig. 10

12. CONCLUSION

A unified concept in decomposition theory is presented whereby a large scale system S can be systematically partitioned, disjointed, and constructed in a hierarchical form. The new procedure uses the information gained in the partitioning phase to identify the separate hierarchical structures which are associated with the possible disjoint systems in S.

The user-oriented digital program DECOMP, which implements the concept presented, has been successfully applied to problems exceeding 100 non-linear simultaneous equations. The efficiency of the program DECOMP is evident from its low computer storage requirements and fast execution times.

NOMENCLATURE

σ	set of individual units or blocks constituting a process flow sheet
x_i	an unknown variable
x	set of n unknown variables
u	set of m known variables
f	set of n non-linear functions containing x and u
I_n	the set of integers {1, 2, ..., n}
C	occurrence matrix
C_i	the ith row of the occurrence matrix C
$\overline{C}$	the characteristic occurrence matrix
$\mathcal{C}$	quasi-triangular form of C
$\mathcal{C}_{ij}$	cells of the quasi-triangular matrix $\mathcal{C}$
R^n	n-dimensional Euclidean space
g_i and z_i	subsets of the sets f and x, respectively
$\overline{z}_j$	solution of $g_j = 0$
A	adjacency matrix
A_i	the ith row of the adjacency matrix A
g_j^i and z_j^i	subsets of equations and variables belonging to the sets h_i and y_i, respectively
$h_i = \{g_1^i, g_2^i, \ldots, g_{\rho_i}^i\}$	
$y_i = \{z_1^i, z_2^i, \ldots, z_{\rho_i}^i\}$	
D	quasi-diagonal form of C
D_{ii}	diagonal cells of D
D_{qj}^i	cells of the matrix D_{ii}
$d_{jj}^{i,r}$	sub-cells of the diagonal cell D_{rr}^i
$(\mathbf{d}_{r\ r-1}^i)_q$	denotes the qth row of the sub-diagonal cell $D_{r\ r-1}^i$
S	the system associated with f and x
S_j^i	an irreducible subsystem which is associated with g_j^i and z_j^i
V_i	a disjoint system, in S, containing the irreducible subsystems $S_1^i, S_2^i, \ldots, S_{\rho_i}^i$
$S_j^{i,r}$	an irreducible subsystem which is associated with the rth level of the hierarchical structure corresponding to V_i
$\mathcal{L}$	maximal loop
$W^{i,r} = \{S_1^{i,r}, S_2^{i,r}, \ldots, S_{\tau_r}^{i,r}\}$	
Λ_j	set containing variables which directly influence the variable x_j
$\hat{\Lambda}_j$	set containing x_j and all variables which influence the variable x_j directly or indirectly. In other words: $\Lambda_j \subset \hat{\Lambda}_j$
$\hat{\Lambda}_{1,i}, \ldots, \hat{\Lambda}_{\theta_i,i}$	denotes the characteristic sets associated with the θ_i functions and variables describing g_i

NOMENCLATURE

(contd.)

(x_j, f_k)	an output pair which reads: equation f_k is solved for variable x_j
(x,f)	output pairs set
$\emptyset$	empty or null set, that is, a set which contains no elements
Ω_j	set containing the subsets of variables which directly influence z_j
$\hat{\Omega}_j$	set containing the subsets of variables which influence z_j directly or indirectly. In other words: $\Omega_j \subset \hat{\Omega}_j$
ρ_i	number of irreducible subsystems belonging to the ith disjoint system
k	number of irreducible subsystems belonging to S
p	number of disjoint systems which constitute the system S
θ_i	number of equations involved in the ith irreducible subset g_i
τ_r	number of irreducible subsystems belonging to the rth level of the hierarchical structure
l	number of levels in a hierarchical structure

SYMBOLS

$\{x_1, x_2, \ldots, x_n\}$	represents a set whose elements are separated by commas and enclosed in braces { }
$x_j \varepsilon x$	reads: the variable x_j is an element of the set x. (Slash across this symbol denotes its negation.)
$h_i \supset g_j^i$	reads: the set h_i contains the subset g_j^i
$x_j \mathsf{E} f_i$	reads: x_j is contained in the function f_i
$z_i \sqsubset g_i$	reads: the subset of variables z_i is contained in the subset of functions g_i. Slash across any of the last two symbols denotes their negation.
$x = \bigcup_{i=1}^{n} x_i$	denotes the union of the n elements, $x_1, x_2, .., x_n$
$\cap$	denotes the intersection of two sets or subsets
$A \setminus B$	set of elements which belong to the set A but which do not belong to the set B
$\sigma_j \to \sigma_i$	denotes a path of information flow from unit σ_j to unit σ_i
$\dot{+}$	boolean union

REFERENCES

A.L. DULMAGE and N.S. MENDELSOHN (1962) "On the inversion of sparse matrices", Math. Comput., 16, p. 494.

F. HARARY (1960) "A graph theoretic method for the complete reduction of a matrix with a view toward finding its eigenvalues", J. Math. and Phys., 38, p. 104.

F. HARARY (1962) "A graph theoretic approach to matrix inversion by partitioning", Numerische Math., 4, p. 128.

D.M. HIMMELBLAU (1967) "Decomposition of large-scale systems. II. Systems containing non-linear elements", Chem. Eng. Sci. 22, p. 883.

A.K. KEVORKIAN and J. SNOEK (1972) "Decomposition in large scale systems: Theory and applications in solving of large sets of non-linear simultaneous equations", NATO ADVANCED STUDY INSTITUTE ON DECOMPOSITION AS A TOOL FOR SOLVING LARGE SCALE PROBLEMS, Cambridge University.

W.P. LEDET and D.M. HIMMELBLAU (1970) "Decomposition procedures for the solving of large scale systems", Advan. Chem. Eng. 8, p. 185.

D.V. STEWARD (1962) "On an approach to techniques for the analysis of the structure of large systems of equations", SIAM Review, 4, p. 321.

D.V. STEWARD (1965) "Partitioning and tearing systems of equations", J. SIAM Numer. Anal. Ser. B, 2, p. 345.

DECOMPOSITION IN LARGE SCALE SYSTEMS
THEORY AND APPLICATIONS IN SOLVING LARGE SETS OF NON-LINEAR SIMULTANEOUS EQUATIONS

by

A.K. KEVORKIAN and J. SNOEK

KONINKLIJKE/SHELL-LABORATORIUM, AMSTERDAM

(Shell Research N.V.)

Abstract - The steady-state design problem of large scale systems can be formulated as the solving of a set of non-linear simultaneous equations:

$$f_i(x_1, x_2, \ldots, x_n, u_1, u_2, \ldots, u_m) = 0 \qquad i \epsilon I_n$$

in which x is a set of n unknown variables and u is a set of m known variables. Here, most of the algorithms existing in the literature make use of an nth order convergence matrix f_x. The computation times involved in evaluating f_x and its inverse tend to be excessive when the order n becomes large.

In this article a newly developed and tested iterative algorithm is introduced. The concept is based on a new Tearing procedure, which is coupled with an elementary sensitivity analysis. The procedure, given an initial condition of the unknown vector x, improves the rate of convergence while using a minimum order convergence matrix.

The complexity of the developed algorithm is reviewed in comparison with existing techniques known from the literature. The practical applicability of the digital program that implements the presented algorithm is tested by solving for the steady-state of an existing non-linear process.

1. INTRODUCTION

The steady-state design and dynamic simulation of large scale non-linear physical processes requires the solving of a large set of non-linear simultaneous equations of the form:

$$f_i(x_1, x_2, \ldots, x_n, u_1, u_2, \ldots, u_m) = 0 \quad i \epsilon I_n \tag{1.1}$$

in which x is a set of n unknown variables and u is a set of m known variables. This formulation is in fact identical to the finding of the extremum of a real differentiable function $F(x_1, x_2, \ldots, x_n)$ where the local extrema are stationary points satisfying the following n equations in n unknowns:

$$\partial F/\partial x_i = 0 \quad i \epsilon I_n \tag{1.2}$$

The algorithms developed until now, to solve (1.1) or (1.2), generally assume an initial vector of all the variables in x and use different approaches to iterate x so as to achieve a solution within a given relative accuracy. The most well-known of these algorithms is Newton's iteration method, TRAUB (1964), which, similar to the method of steepest descent, ARROW (1958), requires the evaluation of an nth order

Jacobian matrix

$$f_x = \begin{bmatrix} \partial f_1/\partial x_1 & \dots & \partial f_1/\partial x_n \\ \cdot & \cdot & \cdot \quad \cdot \\ \partial f_n/\partial x_1 & \dots & \partial f_n/\partial x_n \end{bmatrix} \tag{1.3}$$

and the computation of its inverse f_x^{-1}.

The excessive computational time involved in evaluating large order matrices f_x and f_x^{-1} has necessitated the development of computationally more efficient algorithms. An example is the Newton-Raphson method coupled with an optimally ordered triangular factorization of the system equations, PESCHON (1968), as applied to power system problems. Recently, attention has been given to a class of methods referred to as the "Quasi-Newton" algorithms such as the Secant method, SARGENT (1971), where f_x and f_x^{-1} are built approximately rather than computed. However, almost all the available iterative algorithms have the major drawback of iterating all the n unknown variables in x and thus necessitating the availability of an nth order convergence matrix f_x. When n becomes large, all these methods may run into difficulties because of the large digital computer storage requirements and considerable computation time needed for solving the entire system of equations simultaneously.

The application of decomposition techniques, partitioning in particular, breaks down a large set of simultaneous equations to a number, k, of irreducible subsets of equations and indicates the sequence in which each of these k subsets can be solved independently of the remaining ones. In this way the nth order Jacobian matrix, f_x, required for solving equation (1.1) is replaced by k smaller Jacobian matrices $f_x{}^1$, $f_x{}^2$, ..., $f_x{}^k$,

where
$$\bigcup_{i=1}^{k} x^i = x \quad \text{and} \quad x^i \bigcap_{i \neq j} x^j = \emptyset \tag{1.4}$$

However, when no effective partitioning is possible, that is one or more of the irreducible subsets of equations are still large, the difficulty associated with computer storage requirement and computation time remains prohibitive.

The concept of tearing, as formulated in Section 2, corresponds to the optimal ordering of the equations, f_i where $i \varepsilon I_\theta$, constituting an irreducible subset* of θ equations such that the unknown variables x_1, x_2, ..., x_θ are solved serially, one at a time, by iterating on the minimum number, β, of variables. In other words, tearing is a decomposition procedure where a θth order Jacobian matrix, required for the solving of an irreducible set f_i $i\varepsilon I_\theta$, is reduced to its minimum order β. Clearly, tearing should be considered superfluous when each of the equations f_i, $i\varepsilon I_\theta$, contains all the unknown variables x_i, $i\varepsilon I_\theta$.

The developed tearing algorithm is presented, in Section 3, in the form of several observations. The difference here between the presented concept and that used in the algorithms of LEDET (1970), STEWARD (1962) and others is emphasized.

It is important to note that the significance of sensitivity analysis in tearing procedures has already been emphasized by RINARD (1964), HIMMELBLAU (1967), LEDET (1970), WESTERBERG (1971-I, II), KEVORKIAN (1972-II) and many others. However, the recommendations in the first three references mentioned above, are not sufficient for solving large sets of non-linear simultaneous equations since much intuitive assumptions are taken into consideration. WESTERBERG and KEVORKIAN illustrate the importance of the sensitivity analysis in choosing the proper output set in order to improve the convergence properties in solving irreducible sets of equations. The objectives WESTERBERG uses in choosing an output set are justified only when the Successive Substitution method is used as the iteration procedure. His assumption, however, that these objectives will also improve the convergence characteristics of other, more sophisticated, iteration procedures seems to be doubtful.

---

*henceforth denoted, for the sake of convenience, as a set of θ equations

In this article the objective in choosing an output set is adapted to the iteration algorithm as derived in Section 4. This algorithm makes use of the minimum order Jacobian-type matrix, denoted by g_z. It is shown that the correction required in the assumed, βth order, iterative vector $\hat{z}$ is given by

$$\Delta\hat{z} \;\alpha\; (I-g_z)^{-1} \tag{1.5}$$

In Section 5 it is shown that the use of the least sensitive output pairs set, in which each element such as (x_l, f_k) satisfies

$$|\hat{x}_l(\partial f_k/\partial x_l)_{\hat{x}}|>|\hat{x}_j(\partial f_k/\partial x_j)_{\hat{x}}| \;, \quad j=1, \ldots, l-1, l+1, \ldots, \theta \tag{1.6}$$

results in a convergence matrix in which the elements of the Jacobian-type matrix g_z tend to zero and thus improve the rate of convergence of the iterative procedure given in Section 4.

The main features of the digital program DECOMP which implements the tearing and sensitivity concepts developed here, together with the structural analysis techniques given by KEVORKIAN (1972-I), is described in Section 6. The iterative algorithm, in Section 4, is incorporated in a digital program SITER (Sequential ITERation).

The application of the user-oriented digital programs DECOMP and SITER is illustrated in Section 6 by solving the steady-state of an existing non-linear process.

2. SIMPLIFICATION IN THE SOLVING OF AN IRREDUCIBLE SET OF EQUATIONS - TEARING

Let the occurrence matrix associated with the irreducible set of non-linear (linear) simultaneous equations

$$f_i(x_1, x_2, \ldots, x_\theta) = 0 \quad i\varepsilon I_\theta \tag{2.1}$$

with an assigned output pairs set [KEVORKIAN (1972-I)]

$$(x, f) = \{(x_1, f_1), (x_2, f_2), \ldots, (x_\theta, f_\theta)\} \tag{2.2}$$

be arranged such that a minimum number of columns, β, have nonzero elements above the main diagonal while subject to the restriction that (x, f) corresponds to the main diagonal. A schematic diagram of the resulting matrix is shown in Fig. 1 where only the shaded area may contain nonzero elements.

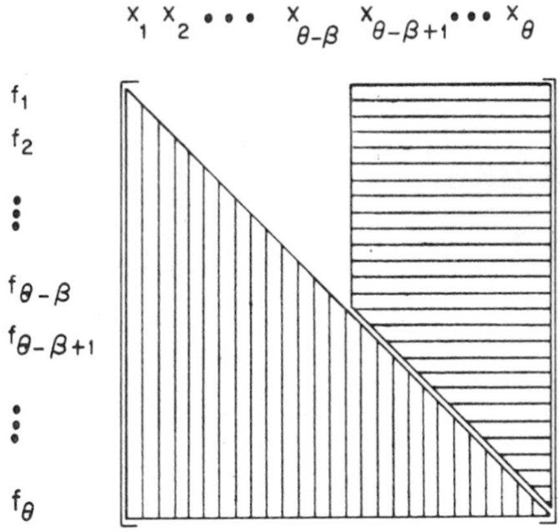

FIGURE 1
A SCHEMATIC DIAGRAM OF THE REARRANGED OCCURRENCE MATRIX

The equations associated with the matrix shown in Fig. 1 can be written in the form:

$$f_i(x_1, x_2, \ldots, x_i, x_{\theta-\beta+1}, \ldots, x_\theta) = 0 \quad i=1, 2, \ldots, \theta-\beta \tag{2.3}$$

$$f_j(x_1, x_2, \ldots, x_j, \ldots, x_\theta) = 0 \quad j=\theta-\beta+1, \ldots, \theta \tag{2.4}$$

in which β will always satisfy the following inequality:

$$1 \leqslant \beta \leqslant (\theta-1) \quad \theta = 2, 3, \ldots \tag{2.5}$$

It should be noted that β may never be zero since this contradicts the definition of an irreducible set of equations [KEVORKIAN (1972-I)].

Let us now define the set

$$v = \{x_{\theta-\beta+1}, x_{\theta-\beta+2}, \ldots, x_\theta\} \tag{2.6}$$

such that $$v \supset v_i \quad \text{where} \quad v_i [f_i \tag{2.7}$$

The re-arranging of equations (2.3) and (2.4) into the following

$$x_i = f_i' (x_1, x_2, \ldots, x_{i-1}, \hat{v}_i) \quad i=1, 2, \ldots, \theta-\beta \tag{2.8}$$

$$x_j = f_j' (x_1, x_2, \ldots, x_{\theta-\beta}, \hat{w}_j) \quad j=\theta-\beta+1, \ldots, \theta \tag{2.9}$$

where $$w_j = v_j \setminus \{x_j\} \tag{2.10}$$

corresponds to the breaking of all the loops, or closed paths of information flow, which exist in the flow graph associated with equation (2.1) in which the output pairs set is given by equation (2.2). As a result, the formulation given by equations (2.8) and (2.9) corresponds to the iterative scheme shown in Fig. 2.

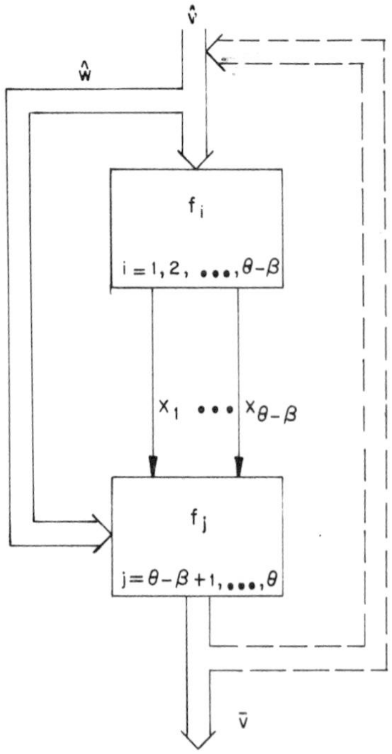

FIGURE 2
THE ITERATIVE SCHEME ASSOCIATED WITH EQUATIONS (2.8) AND (2.9)

Here, the unknown variables x_1, x_2 ..., x_θ are solved serially, one at a time, by iterating on the β variables belonging to the set v which is henceforth denoted as the "Iterative set".

The minimum number of iterates β, defined above, depends on* the output set assigned to an irreducible set of equations [KEVORKIAN (1972-II As a result, when

$$(x, f)_1, (x, f)_2, \ldots, (x, f)_q$$

represent the possible output pairs sets, which could be chosen from f, then there exists an absolute minimu number of iterates $\check{\beta}$ which is given by:

$$\check{\beta} = \min\{\beta_1, \beta_2, \ldots, \beta_q\} \tag{2.11}$$

in which β_j, where $j \varepsilon I_q$, corresponds to the output pairs set $(x, f)_j$. The determination of $\check{\beta}$, as suggested by LEDET (1970), is equivalent to the arranging of an occurrence matrix into the form shown in Fig. 1 while subject to the restriction that the resulting diagonal elements are nonzero. In this way the variables and

---

* except when $\theta = 2, 3$

functions associated with the diagonal elements correspond to the desired output pairs set $(x, f)_k$ which governs the information flow in a graph in which all the loops are torn by breaking the paths, leaving the β_k vertices where $\beta_k = \check{\beta}$.

The procedure discussed in this section has been defined, STEWARD (1965), LEDET (1970), as Tearing. Also the elements of the iterative set v, and the equations for which they are solved, have been defined as the torn variables and torn equations, respectively.

3. THE NEW TEARING CONCEPT

An algorithm may be considered a suitable tool for the tearing of a large scale system (over 100 variables) only when the main concept in the algorithm is, as much as possible, independent of the size of the problem. In this way the computer execution time required for the tearing of a system approaching 1000 variables may not be prohibitive.

In the literature there are several algorithms and formulations for identifying the minimum number of paths which are common to all the loops in an irreducible set f, STEWARD (1962), or for obtaining the minimum number of iterates in f, LEDET (1970). The main aim, in both of the algorithms, is to reduce the size of the convergence matrix required to solve f.

STEWARD's algorithm, which is executed on the system adjacency matrix, requires the tracing of all the loops associated with f and also the storage of the equations belonging to each traced loop. As a result, the efficiency of the procedure heavily depends on the size of the system. Therefore it could be expected that as f becomes large the computer execution time tends to be prohibitive.

LEDET's algorithm, which is executed on the system occurrence matrix, is carried out in two phases. The first phase performs an initial tearing while the second phase improves the results. The three criteria, which are needed for performing the first phase, require an exhaustive search throughout the occurrence matrix before a variable could be considered an iterate. The second phase is considered to be quite complex. In fact the implementation of both phases in the algorithm will, according to LEDET, result in prohibitive computation times. He points out, however, that in most cases the first phase of the algorithm will yield the minimum number of iterates, or nearly the minimum. Nevertheless, the procedure, here, is again strongly dependent on the size of the problem. The computational requirement of LEDET's procedure will be more evident in Section 6.

The newly developed tearing procedure, which is executed on the system occurrence matrix, identifies an iterative variable by inspecting a minimum number of rows and columns in the occurrence matrix. The algorithm is presented below in the form of several observations. The clarification and proof of these observations is facilitated by referring to the flow graph representation, $\mathcal{G}$, of the irreducible set.

Here it is considered, for convenience, that the outgoing information flow at vertex f_i, $i \varepsilon I_\theta$, in $\mathcal{G}$ is given by the variable x_i where

$$x_i = f_i'(x_1, \ldots, x_{i-1}, x_{i+1}, \ldots, x_\theta)$$

Theorem I: The existence of the loop $f_p \Longleftrightarrow f_r$ where

$$f_r \not\longrightarrow f_j \text{ (or } f_j \not\longrightarrow f_r) \quad j=1, \ldots, p-1, p+1, \ldots, \theta \tag{3.1}$$

implies that

$$x_p \varepsilon v \tag{3.2}$$

Proof: The loop $f_p \Longleftrightarrow f_r$, as shown in Fig. 3, can be broken only when $x_p \varepsilon v$ or $x_r \varepsilon v$. From equation (3.1), however, any loop which contains f_r will also contain f_p, whereas any loop which contains f_p need not contain f_r. As a result $x_p \varepsilon v$.

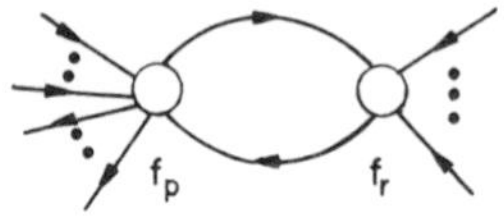

FIGURE 3

As discussed in Section 2, the relationship $x_p \varepsilon v$ implies that all the paths $f_p \longrightarrow f_j$, $j \varepsilon I_\theta$, in $\mathcal{G}$ should be broken. As a result, the paths which are connected to f_p, namely $f_i \longrightarrow f_p$ where $i \varepsilon I_\theta$, can never belong to a loop anymore and, hence are broken too.

Theorem I may be extended by relating the loop shown in Fig. 3, denoted as the "Loose-end" loop, to flow graphs which exhibit complicated configurations. This gives rise to the following observations.

Corollary I: The existence of the loops $f_p \Longleftrightarrow f_s$ and $f_p, f_s \Longleftrightarrow f_r$ where

$$f_r \not\longrightarrow f_j \text{ (or } f_j \not\longrightarrow f_r\text{)} \quad j \neq p \neq s \quad \text{and} \quad j \varepsilon I_\theta \tag{3.3}$$

implies that

$$x_p, x_s \varepsilon v \tag{3.4}$$

Proof: The loops in Fig. 4 can be broken by tearing the outputs of at least two vertices. Moreover, the loops in the flow graph $\mathcal{G}$ which contain f_r will also contain f_p and/or f_s whereas any loop which contains f_p and/or f_s need not contain f_r. As a result, x_p and x_s, which are the output variables assigned to equations (vertices) f_p and f_s respectively, belong to the iterative set v.

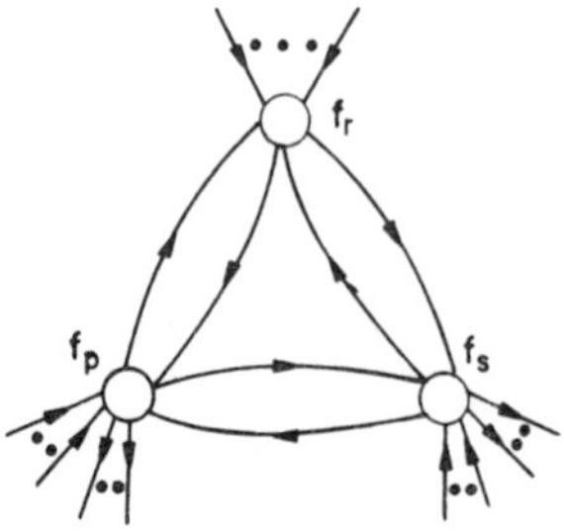

FIGURE 4

Corollary II: The existence of the loops $f_p, f_q \Longleftrightarrow f_r, f_s$ where

$$f_p \not\Longleftrightarrow f_q \tag{3.5}$$

and

$$\left.\begin{array}{l} f_p \not\longrightarrow f_j \text{ (or } f_j \not\longrightarrow f_p\text{)} \\ f_q \not\longrightarrow f_j \text{ (or } f_j \not\longrightarrow f_q\text{)} \end{array}\right\} \quad j \neq r \neq s \quad \text{and} \quad j \varepsilon I_\theta \tag{3.6}$$

implies that

$$x_r, x_s \varepsilon v \tag{3.7}$$

The loops $f_p, f_q \Longleftrightarrow f_r, f_s$ subject to the conditions (3.5) and (3.6) are illustrated in Fig. 5. The proof of (3.7) is analogous to the proof of (3.4).

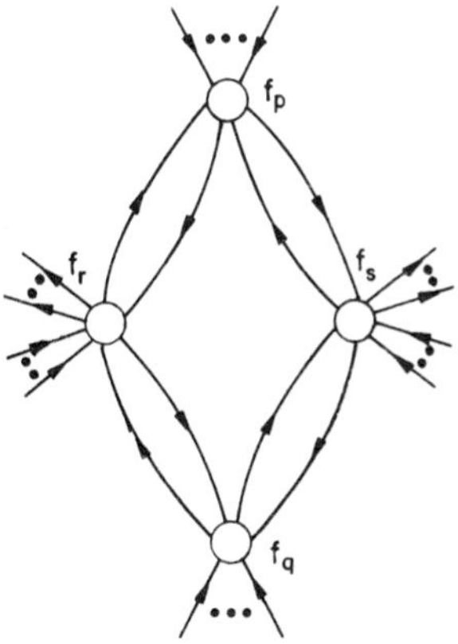

FIGURE 5

Corollary III: The existence of the loops f_p, $f_q \Longleftrightarrow f_r$, f_s and $f_r \Longleftrightarrow f_s$ where

$$f_s \not\longrightarrow f_j \ (\text{or } f_j \not\longrightarrow f_s) \ j \neq p \neq q \neq r \quad \text{and} \quad j \varepsilon I_\theta \tag{3.8}$$

$$\text{implies that} \quad x_r \varepsilon v \tag{3.9}$$

Proof: Consider the case when $x_p \notin v$. In such a situation the loop $f_r \Longleftrightarrow f_p$ can only be broken when $x_r \varepsilon v$. Let us now consider the situation where $x_p \varepsilon v$. In such a case Fig. 6 reduces to the configuration shown in Fig. 4 and thus from corollary I we have $x_r \varepsilon v$. As a result, the relationship (3.9) will always hold.

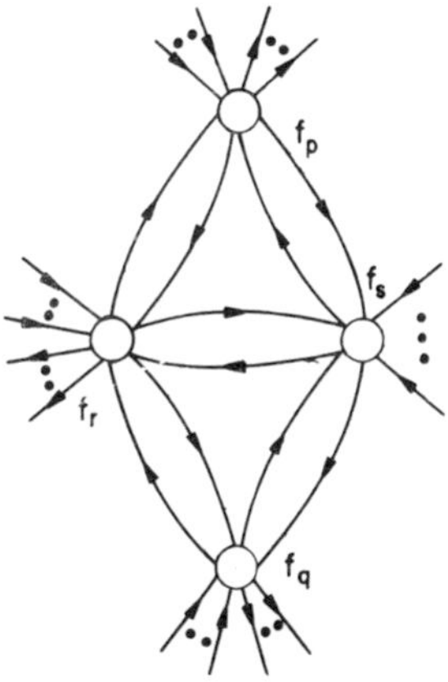

FIGURE 6

Theorem I and the associated corollaries identify the variables which must belong to the iterative set v. Here, the tearing of an output variable, belonging to the iterative set, further reduces the size of the flow graph $\mathcal{G}$. However, $\mathcal{G}$ may also be simplified by elimination of those vertices and paths which may never influence the iterative set v. The identification of these redundant vertices and paths can be achieved by making use of the following observations:

Lemma I: The existence of the path $f_r \longrightarrow f_s$ (or $f_s \longrightarrow f_r$)

$$\left.\begin{array}{ll} \text{where} & f_r \Longleftrightarrow f_{p_1}, f_{p_2}, \ldots, f_{p_j} \\ \text{and} & f_i \not\longrightarrow f_r \text{ (or } f_r \not\longrightarrow f_i) \quad i \neq p_1 \neq \ldots \neq p_j \text{ and } i \varepsilon I_\theta \end{array}\right\} \quad (3.10)$$

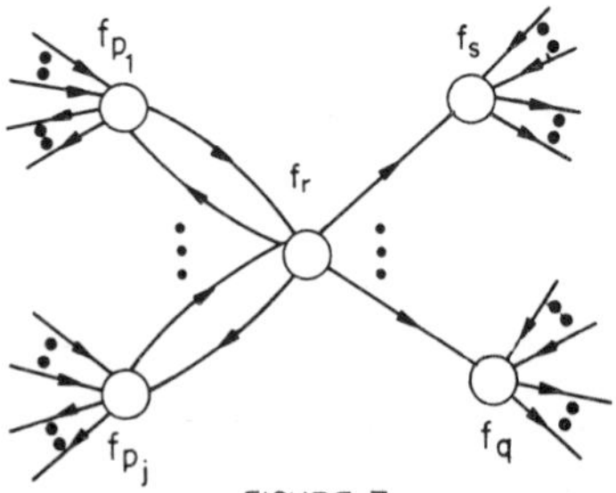

FIGURE 7

implies that the removing of the path $f_r \longrightarrow f_s$ (or $f_s \longrightarrow f_r$) from the flow graph $\mathcal{G}$ will not eliminate a variable which might have belonged to the iterative set v.

Proof: Condition (3.10) indicates that any loop which contains the path $f_r \longrightarrow f_s$ (or $f_s \longrightarrow f_r$), must contain at least one of the vertices $f_{p_1}, \ldots, f_{p_j}$. On the other hand, the loops $f_r \Longleftrightarrow f_{p_1}, \ldots, f_r \Longleftrightarrow f_{p_j}$ can be broken only by tearing the outputs assigned to f_r or $f_{p_1}, \ldots, f_{p_j}$, respectively. These two possibilities in tearing, however, imply that the path $f_r \longrightarrow f_s$ (or $f_s \longrightarrow f_r$) can never form anymore a loop in Fig. 7. As a result, the path $f_r \longrightarrow f_s$ can be removed from the flow graph $\mathcal{G}$ without eliminating an iterative variable belonging to the set v.

Lemma II-a: The existence of the paths $f_r \longrightarrow f_s$ and $f_{p_1}, \ldots, f_{p_j} \longrightarrow f_r$

$$\text{where} \qquad f_r \not\longrightarrow f_i \quad i \neq s,\ i \varepsilon I_\theta \ \text{ and } \ f_r \not\Longleftrightarrow f_s \qquad (3.11)$$

implies that the vertex f_r can be removed from $\mathcal{G}$ by replacing $f_{p_1}, \ldots, f_{p_j} \longrightarrow f_r \longrightarrow f_s$ by $f_{p_1}, \ldots, f_{p_j} \longrightarrow f_s$.

Proof: Any loop which contains the vertex f_r, in Fig. 8a, must also contain f_s. Moreover, since $f_s \not\longrightarrow f_r$, the vertex f_r can be merged into f_s by replacing the two edged paths

$f_{p_1}, \ldots, f_{p_j} \longrightarrow f_r \longrightarrow f_s$ by

$f_{p_1}, \ldots, f_{p_j} \longrightarrow f_s$.

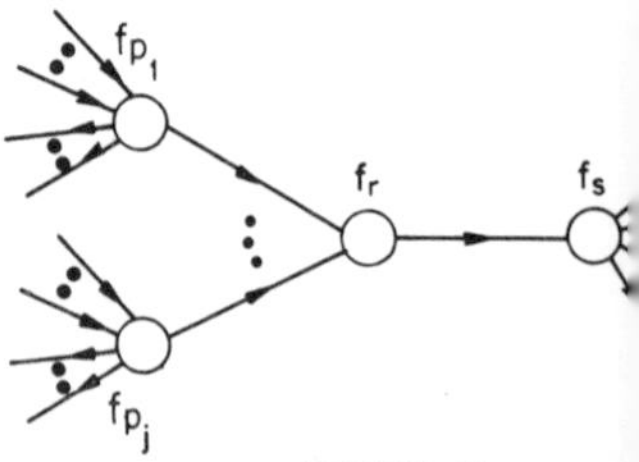

FIGURE 8a

Lemma II-b: The existence of the paths $f_s \longrightarrow f_r$ and $f_r \longrightarrow f_{p_1}, \ldots, f_{p_j}$

$$\text{where} \qquad f_i \not\longrightarrow f_r \quad i \neq s,\ i \varepsilon I_\theta \ \text{ and } \ f_r \not\Longleftrightarrow f_s \qquad (3.12)$$

implies that the vertex f_r can be removed from $\mathcal{G}$ by replacing $f_s \longrightarrow f_r \longrightarrow f_{p_1}$, $\ldots, f_{p_j}$ by $f_s \longrightarrow f_{p_1}, \ldots, f_{p_j}$.

The proof, here, is analogous to that of Lemma II-a.

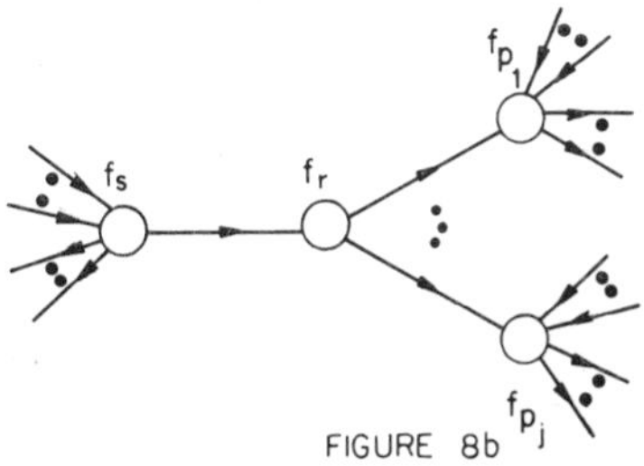

FIGURE 8b

Lemma III-a: The existence of the paths $f_{p_1}, \ldots, f_{p_j} \longrightarrow f_r$ and $f_r \longrightarrow f_s$

where $f_r \not\Longleftrightarrow f_{p_1}, \ldots, f_{p_j}$ (3.13)

and $f_{p_1}, \ldots, f_{p_j} \longrightarrow f_s$ (3.14)

implies the removing of the path $f_r \longrightarrow f_s$ will not eliminate a variable which might have belonged to the iterative set v.

Proof: For each of the two-edged paths $f_{p_1}, \ldots, f_{p_j} \longrightarrow f_r \longrightarrow f_s$ there exists a by-pass $f_{p_1}, \ldots, f_{p_j} \longrightarrow f_s$. As a result, the path $f_r \longrightarrow f_s$ is superfluous as far as tearing is concerned.

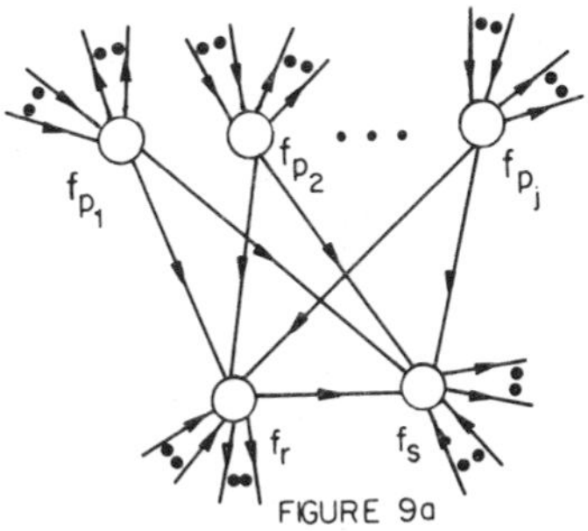

FIGURE 9a

Lemma III-b: The existence of the paths $f_r \longrightarrow f_s$ and $f_s \longrightarrow f_{p_1}, \ldots, f_{p_j}$

where $f_s \not\Longleftrightarrow f_{p_1}, \ldots, f_{p_j}$ (3.15)

and $f_r \longrightarrow f_{p_1}, \ldots, f_{p_j}$ (3.16)

implies that the removing of the path $f_r \longrightarrow f_s$ will not eliminate a variable which might have belonged to the iterative set v.

The proof, here, is similar to that given in Lemma III-a.

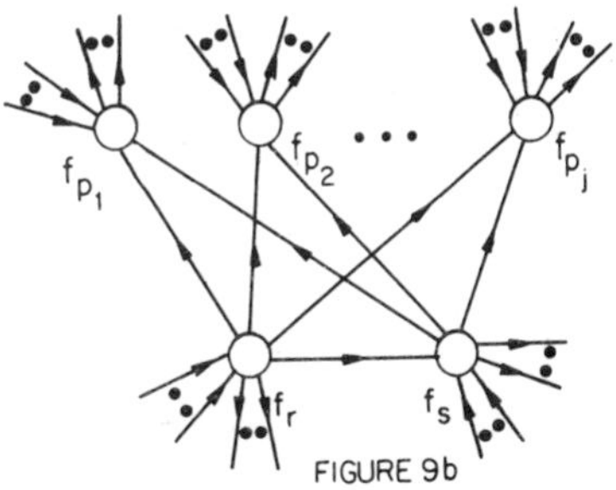

FIGURE 9b

Theorem I and Lemmas I and II are usually sufficient for the identification of the β variables belonging to an iterative set v. If these observations are not sufficient for the complete identification of v, corollaries I, ..., III and Lemma III are used for the breaking of the remaining loops in $\mathcal{G}$. However, it may happen, although in extreme* cases, that the presented observations are not sufficient for breaking all the loops in $\mathcal{G}$. In such a situation, tearing is continued by the application of the following observation:

Lemma IV: The existence of the loop $f_p \Longleftrightarrow f_q$

* Here, several equations in f may contain almost all the variables in x in which case no effective tearing may take place.

where $$x_q \notin v \tag{3.17}$$

implies that $$x_p \in v \tag{3.18}$$

Proof: A second-order loop such as $f_p \Longleftrightarrow f_q$ can be broken only when $x_q \in v$ or $x_p \in v$. This completes the proof.

Lemma IV is applied such that the possibilities $x_q \in v$ and $x_p \in v$ are both considered.

4. THE ITERATION METHOD USED FOR SOLVING x

In the scheme shown in Fig. 2 the calculated iterative variables $\bar{x}_j$, belonging to the set $\bar{v}$, may be related to the assumed iterative variables $\hat{x}_j$, belonging to the set $\hat{v}$, through the following equality:

$$\bar{z} = g\ (\hat{z}) \tag{4.1}$$

where the elements of the β order vector z are defined by:

$$z_i = x_{\theta-\beta+i} \ / \ (\hat{x}_{\theta-\beta+i})^\circ \quad i\varepsilon I_\beta \tag{4.2}$$

in which the constants $(\hat{x}_{\theta-\beta+i})^\circ$, $i\varepsilon I_\beta$, are the initially assumed solutions of the iterative variables. Here a solution of the corresponding irreducible set of equations, within a specified relative accuracy, is obtained when:

$$\bar{z} = \hat{z} \tag{4.3}$$

Making use of the above equation in a Taylor series expansion of equation (4.1), about the assumed solution $(\hat{x}_{\theta-\beta+i})^\circ$, $i\varepsilon I_\beta$, results in

$$(\hat{z})^\circ + \Delta\hat{z} = g((\hat{z})^\circ) + g_z\Delta\hat{z} \tag{4.4}$$

where $(\hat{z})^\circ$ is a unit vector, as evident from equation (4.2), and g_z is a Jacobian-type matrix which is evaluated numerically from the following expression:

$$g_z = \frac{\delta\bar{z}_i}{\delta\hat{z}_j} = \frac{\delta\bar{x}_{\theta-\beta+i}/(\hat{x}_{\theta-\beta+i})^\circ}{\delta\hat{x}_{\theta-\beta+j}/(\hat{x}_{\theta-\beta+j})^\circ} \quad i,j\varepsilon I_\beta \tag{4.5}$$

where $\delta\hat{x}_{\theta-\beta+j}$ is a small specified perturbation in the assumed value of the iterative variable $\hat{x}_{\theta-\beta+j}$ and $\delta\bar{x}_{\theta-\beta+i}$ is the corresponding deviation in the calculated value of the iterative variable $\bar{x}_{\theta-\beta+i}$. The correction in the assumed iterative vector $\hat{z}$, as calculated from equation (4.4), becomes

$$\Delta\hat{z} = (I-g_z)^{-1}\ (g((\hat{z})^\circ) - (\hat{z})^\circ) \tag{4.6}$$

in which I is an identity matrix. In this way the new assumed solution takes the form:

$$(\hat{z})^1 = (\hat{z})^\circ + \Delta\hat{z} \tag{4.7}$$

In the following Section it is shown that the rate of convergence of the presented iterative algorithm is improved when the elements of the g_z matrix tend to zero, namely

$$(\delta\bar{z}_i/\delta\hat{z}_j) \approx 0 \quad i,j\varepsilon I_\beta \tag{4.8}$$

Moreover, the requirement given above automatically guarantees the non-singularity of the convergence matrix $(I-g_z)$ which appears in equation (4.6) .

5. THE SENSITIVITY ANALYSIS

Consider a Taylor series expansion of the irreducible set of equations given by (2.1)

$$f_k(\hat{x}) + \left(\frac{\partial f_k}{\partial x_1}\right)_{\hat{x}} \Delta x_1 + \ldots + \left(\frac{\partial f_k}{\partial x_l}\right)_{\hat{x}} \Delta x_l + \ldots + \left(\frac{\partial f_k}{\partial x_\theta}\right)_{\hat{x}} \Delta x_\theta = 0 \quad k\varepsilon I_\theta \tag{5.1}$$

in which second and higher order terms are neglected. The re-arranging of the above equation in the form given below, where $c_\circ$ represents a constant term,

$$\left(\frac{\Delta x_l}{\hat{x}_l}\right) = - \sum_{\substack{j=1 \\ j\neq l}}^{\theta} \frac{\hat{x}_j(\partial f_k/\partial x_j)_{\hat{x}}}{\hat{x}_l(\partial f_k/\partial x_l)_{\hat{x}}} \left(\frac{\Delta x_j}{\hat{x}_j}\right) - c_\circ \tag{5.2}$$

while subject to the following condition

$$|\hat{x}_l(\partial f_k/\partial x_l)_{\hat{x}}| > |\hat{x}_j(\partial f_k/\partial x_j)_{\hat{x}}| \quad j=1, \ldots, l-1, l+1, \ldots, \theta \tag{5.3}$$

implies that a small relative change in the variable x_j, where x_j is contained in equation f_k, will yield a relative change in the variable x_l such that

$$\left|\frac{\Delta x_l}{\hat{x}_l}\right| < \left|\frac{\Delta x_j}{\hat{x}_j}\right| \quad j=1, \ldots, l-1, l+1, \ldots, \theta \tag{5.4}$$

An output pairs set in which each element such as (x_l, f_k) where

$$x_l = f_k'(x_1, \ldots, x_{l-1}, x_{l+1}, \ldots, x_\theta) , \tag{5.5}$$

satisfies condition (5.4), and subject to the constraint that each output variable, is assigned to only one equation, is defined as the "Least sensitive output pairs set".

Let us now define the relative gain K_{pq} associated with a path linking a vertex f_p, with an assigned output variable x_p, directly to another vertex f_q by the following relationship:

$$K_{pq} = \frac{|\hat{x}_p(\partial f_q/\partial x_p)_{\hat{x}}|}{|\hat{x}_q(\partial f_q/\partial x_q)_{\hat{x}}|} \quad x_p \in f_q \tag{5.6}$$

in which x_q is the output variable assigned to equation f_q. As a result, a flow graph in which the information flow is governed by the least sensitive output pairs set represents a network in which the relative gain associated with each path directly linking two vertices is less than unity. In other words, the overall relative gain associated with a path $f_{p_1} \longrightarrow f_{p_2} \longrightarrow \ldots \longrightarrow f_{p_r}$ which indirectly links the two vertices f_{p_1} and f_{p_r} will be much less than unity. Thus in an iterative scheme such as shown in Fig. 2, in which (x, f) corresponds to the least sensitive output pairs set, the ultimate relative gain between the equations in which the assumed iterative variables are substituted and those from which the calculated iterative variables are obtained will in general be much less than unity, namely

$$\left|\frac{\Delta\bar{x}_{\theta-\beta+i}/(\hat{x}_{\theta-\beta+i})^\circ}{\Delta\hat{x}_{\theta-\beta+j}/(\hat{x}_{\theta-\beta+j})^\circ}\right| \ll 1 \quad i,j\varepsilon I_\beta \tag{5.7}$$

and thus satisfy the condition imposed by equation (4.8).

The condition given above indicates that the elements in the resulting Jacobian-type matrix g_z will usually have very small numerical values, in other words, the identity matrix in $(I-g_z)$ will dominate over the Jacobian g_z. Thus it could be concluded that the constraint given by (5.7) reduces the sensitivity of the $(I-g_z)$ matrix to variation in $\hat{z}$ and thus improves the rate of convergence of the iterative procedure presented in Section 4.

However, the generation of an output pairs set in which each element satisfies (5.3) may not always be possible. This can be illustrated by considering the case where a variable such as $x_l \epsilon x$ is the least sensitive variable in more than one equation in f, namely $f_1, \ldots, f_r$ where $1 < r \leqslant \theta$. This implies, immediately, that (r-1) equations will be constrainted to generate output pairs (x_i, f_j), where $i \epsilon I_\theta$, j=1, ..., k-1, k+1, ..., r, in which x_i is the second least sensitive variable in f_j. However the situation might arise again such that x_i is the second least sensitive variable in more than one of the remaining equations $f_1, \ldots, f_{k-1}, f_{k+1}, \ldots, f_r$. These assumptions may proceed until one or more of the system equations are eventually constrained to generate very sensitive output variables.

It is important to note that in the process of assigning least sensitive output variables it can happen that a variable such as x_s, which corresponds to a sensitiv variable in f_p, may not occur in any of the remaining system equations. In such a case (x_s, f_p) is forced to be an element of the desired output pairs set. Hence, it could be concluded that the least sensitive output pairs set may contain one or mor elements which do not satisfy the relationship given by (5.3). Thus in the extreme case where the generated output pairs set does not satisfy (5.7) this is an indication that the corresponding irreducible system is inherently sensitive in the neigh bourhood of the initially assumed solution.

6. THE DIGITAL PROGRAMS "DECOMP" AND "SITER"

The tearing and sensitivity concepts developed here and the structural analysis techniques presented by KEVORKIAN (1972-I) have been included in the Fortran Progra DECOMP. The program is initiated by inserting into one of its subroutines the following set of equations:

$$f(x, u) = 0 \tag{6.1}$$

together with an assumed value of the vector x. On the basis of this information DECOMP:

(a) generates the system occurrence matrix;

(b) separates the given set of equations into irreducible subsets (partitioning);

(c) groups the irreducible subsets into disjoint sets (disjointing);

(d) arranges the irreducible subsets belonging to each disjoint set into a minimall levelled hierarchical structure;

(e) generates the least sensitive output pairs set;

(f) identifies the iteration variables (tearing) and thus constructs the re-arrange occurrence matrix (Fig. 1).

A typical output of DECOMP is illustrated in Figs. 10-a, b and c, where the decomposed structure represents the illustrative example discussed in the following Section. Fig. 10-a shows the hierarchical structure associated with the seven irreducible subsystems (subsets) constituting the system considered. Figs. 10-b and c furnish more detailed information regarding each irreducible subsystem. Here, each row is associated with a variable, as given on the extreme left, while each column is associated with an equation, as given on the top.

The least sensitive output set is indicated by the symbols **θ** and I which appear on a diagonal of each illustration in Figs. 10-b and c. **θ** identifies the equations given by (2.8) while I identifies those given by (2.9). The connections between the various irreducible subsystems as indicated in each illustration of Fig. 10-c has been discussed by KEVORKIAN (1972-I).

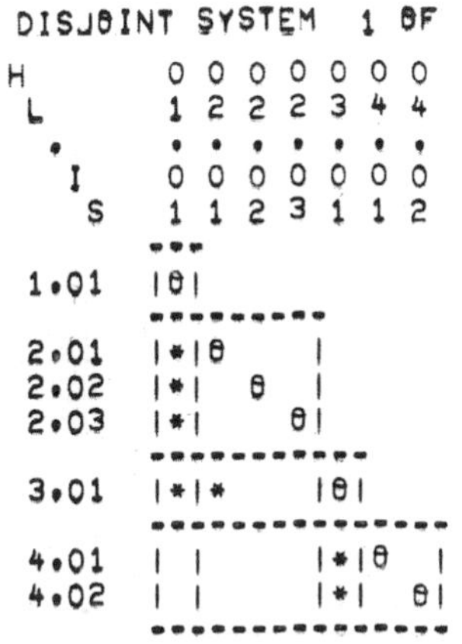

```
DISJOINT SYSTEM  1 OF  1
H        0 0 0 0 0 0 0
L        1 2 2 2 3 4 4
  .      . . . . . . .
  I      0 0 0 0 0 0 0
   S     1 1 2 3 1 1 2
         - - -
1.01    |θ|
        - - - - - - - - - -
2.01    |*|θ    |
2.02    |*|  θ  |
2.03    |*|    θ|
        - - - - - - - - - - - -
3.01    |*|*    |θ|
        - - - - - - - - - - - - - -
4.01    | |    |*|θ  |
4.02    | |    |*|  θ|
        - - - - - - - - - - - - - - - - -
```

Figure 10a

THE IRREDUCIBLE SUBSYSTEMS ARRANGED IN A HIERARCHICAL STRUCTURE

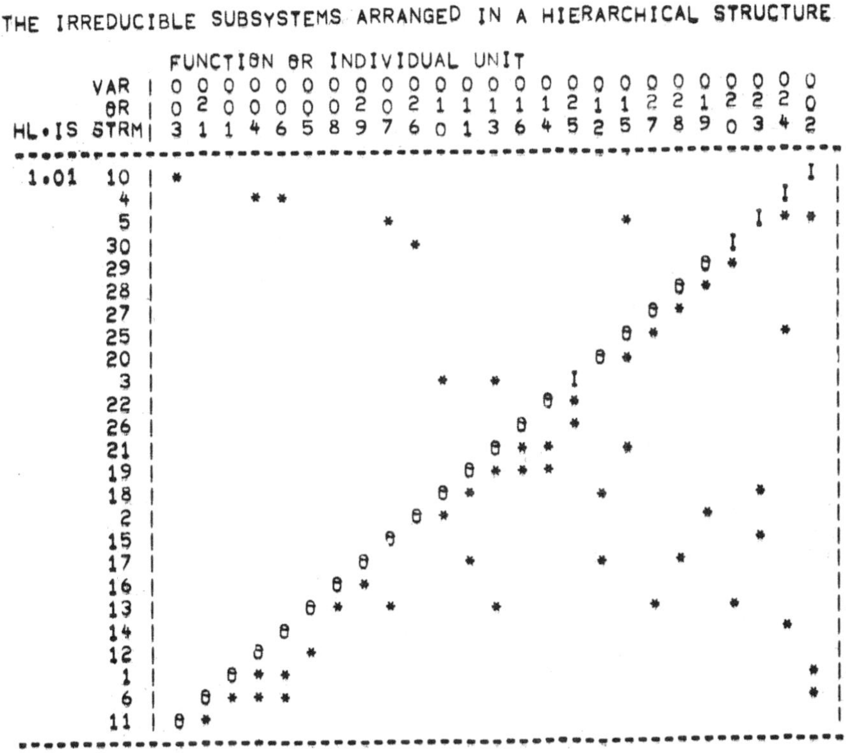

Figure 10b – The irreducible subsystem 1.01,
belonging to the first hierarchical level, in detail

```
          FUNCTIØN ØR INDIVIDUAL UNIT
     VAR | 0
      ØR | 1
HL.IS STRM| 7
--------------
 2.01    7 | Ø |
--------------
 1.01   13 | * |
--------------
```

THE IRREDUCIBLE SUBSYSTEM 2.01

```
          FUNCTIØN ØR INDIVIDUAL UNIT
     VAR | 0
      ØR | 1
HL.IS STRM| 8
--------------
 3.01    8 | Ø |
--------------
 1.01   28 | * |
--------------
 2.01    7 | * |
--------------
```

THE IRREDUCIBLE SUBSYSTEM 3.01

```
          FUNCTIØN ØR INDIVIDUAL UNIT
     VAR | 0
      ØR | 2
HL.IS STRM| 2
--------------
 2.02   24 | Ø |
--------------
 1.01   28 | * |
         2 | * |
--------------
```

THE IRREDUCIBLE SUBSYSTEM 2.02

```
          FUNCTIØN ØR INDIVIDUAL UNIT
     VAR | 0
      ØR | 0
HL.IS STRM| 9
--------------
 4.01    9 | Ø |
--------------
 3.01    8 | * |
--------------
```

THE IRREDUCIBLE SUBSYSTEM 4.01

```
          FUNCTIØN ØR INDIVIDUAL UNIT
     VAR | 0
      ØR | 3
HL.IS STRM| 1
--------------
 2.03   31 | Ø |
--------------
 1.01   10 | * |
        12 | * |
         1 | * |
         6 | * |
--------------
```

THE IRREDUCIBLE SUBSYSTEM 2.03

```
          FUNCTIØN ØR INDIVIDUAL UNIT
     VAR | 0
      ØR | 3
HL.IS STRM| 0
--------------
 4.02   23 | Ø |
--------------
 3.01    8 | * |
--------------
```

THE IRREDUCIBLE SUBSYSTEM 4.02

Figure 10c - The irreducible subsystems
belonging to the second and lower hierarchical levels in detail

Table I gives the storage requirements and the computational efficiency of DECOMP. It is noted that the execution time is influenced by both the size of the problem and its inherent structure. However, the significance of the presented tearing concept is clearly shown.

For comparison a problem (Hanford N-reactor) comprising 112 equations, obtained from LEDET (1970), has been inserted in the program DECOMP. As shown in Table I the total execution time required for partitioning, disjointing, constructing a hierarchical structure and tearing, using an RXDS-Σ7 digital computer, is 4.08 seconds. The total execution time required by LEDET's procedure, using a CDC-6600 digital computer, is cited as 52.0 seconds [LEDET (1970)].

The iterative algorithm, presented in Section 4, has been included into the Fortran Program SITER (Sequential ITERation). Here the irreducible subsystems, obtained from DECOMP, are solved, one at a time, starting with those belonging to the first hierarchical level.

The computational efficiency of this program will be discussed with reference to the illustrative example given in the following Section.

7. AN ILLUSTRATIVE EXAMPLE

The example considers an existing marine steam boiler whose steady-state is described by a set of 31 non-linear simultaneous equations of the form given by eq. (1.1). The insertion of these equations, with an assumed solution, $\hat{x}$, of the 31 unknown variables, in the digital program DECOMP results in the modified occurrence matrix shown in Fig. 11. Here a non-zero element in the row and column corresponding to equation f_k and variable x_i, respectively, is given by the "Sensitivity ratio" S_{ki}, where

$$S_{ki} = \frac{|\hat{x}_l(\partial f_k/\partial x_l)_{\hat{x}}|}{|\hat{x}_i(\partial f_k/\partial x_i)_{\hat{x}}|} \quad i\varepsilon I_\theta \tag{7.1}$$

in which $$|\hat{x}_l(\partial f_k/\partial x_l)_{\hat{x}}| \geqslant |\hat{x}_j(\partial f_k/\partial x_j)_{\hat{x}}| \quad j=1, \ldots, l-1, l+1, \ldots, \theta \tag{7.2}$$

As indicated in Fig. 11, the sensitivity ratios are rounded off in the program to the nearest integer number. Hence, the least sensitive variable x_l in equation f_k, $k\varepsilon I_\theta$, will have a sensitivity ratio given by

$$S_{kl} = 1 \quad k,l\varepsilon I_\theta \tag{7.3}$$

Thus the generation of the least sensitive output pairs set, subject to the constraint that each output variable is assigned only to one equation, corresponds to the row and column permutation of the modified occurrence matrix shown in Fig. 11 such that the elements with the smallest, non-zero sensitivity ratios occur on the main diagonal. The resulting least sensitive output pairs set is illustrated in the first two columns of Table III while the sensitivity ratios associated with each output pair is given in the last column of the same table.

The decomposed structure of this example, as discussed in Section 6, is given by Figs. 10-a, b and c. The irreducible subsystem 1.01 which comprises 25 simultaneous equations, can be solved by iterating on the five variables x_3, x_{30}, x_5, x_4 and x_{10}.

The information obtained from Fig. 10-b permits the rewriting of the system equations in the form and sequence as given in Table II, which are then inserted in SITER. Figs. 12-a and b illustrate the iterative solutions, within a specified accuracy of 10^{-5}, for initial guesses of twice and one-tenth of the final solution, respectively. It should be noted that in general one run is sufficient for DECOMP to calculate the steady-state solution for various input vectors u.

The significance of solving non-linear simultaneous equations by the presented tearing and least sensitivity output pairs set concepts may be demonstrated by considering the three methods as listed in Table IV. Method A here represents the

VARIABLES

EQUATIONS	1	2	3	4	5	6	7	8	9	10	11	12	13	14	15	16	17	18	19	20	21	22	23	24	25	26	27	28	29	30	31
1	1	0	0	0	0	1	0	0	0	0	0	0	0	0	0	0	0	0	0	0	0	0	0	0	0	0	0	0	0	0	0
2	2	0	0	0	3	5	0	0	0	1	0	0	0	0	0	0	0	0	0	0	0	0	0	0	0	0	0	0	0	0	0
3	0	0	0	0	0	0	0	0	0	2	1	0	0	0	0	0	0	0	0	0	0	0	0	0	0	0	0	0	0	0	0
4	1	0	0	9	0	1	0	0	0	0	0	1	0	0	0	0	0	0	0	0	0	0	0	0	0	0	0	0	0	0	0
5	0	0	0	0	0	0	0	0	0	0	0	1	1	0	0	0	0	0	0	0	0	0	0	0	0	0	0	0	0	0	0
6	1	0	0	9	0	1	0	0	0	0	0	0	0	1	0	0	0	0	0	0	0	0	0	0	0	0	0	0	0	0	0
7	0	0	0	0	2	0	0	0	0	0	0	0	1	0	2	0	0	0	0	0	0	0	0	0	0	0	0	0	0	0	0
8	0	0	0	0	0	0	0	0	0	0	0	0	1	0	0	1	0	0	0	0	0	0	0	0	0	0	0	0	0	0	0
9	0	0	0	0	0	0	0	1	1	0	0	0	0	0	0	0	0	0	0	0	0	0	0	0	0	0	0	0	0	0	0
10	0	1	4	0	0	0	0	0	0	0	0	0	0	0	0	0	0	1	0	0	0	0	0	0	0	0	0	0	0	0	0
11	0	0	0	0	0	0	0	0	0	0	0	0	0	0	0	0	1	1	3	0	0	0	0	0	0	0	0	0	0	0	0
12	0	0	0	0	0	0	0	0	0	0	0	0	0	0	0	0	1	6	0	1	0	0	0	0	0	0	0	0	0	0	0
13	0	0	4	0	0	0	0	0	0	0	0	0	1	0	0	0	0	0	200	0	2	0	0	0	0	0	0	0	0	0	0
14	0	0	0	0	0	0	0	0	0	0	0	0	0	0	0	0	0	0	101	0	1	1	0	0	0	0	0	0	0	0	0
15	0	0	0	0	1	0	0	0	0	0	0	0	0	0	0	0	0	0	0	5	1	0	0	0	1	0	0	0	0	0	0
16	0	0	0	0	0	0	0	0	0	0	0	0	0	0	0	0	0	0	101	0	1	0	0	0	0	2	0	0	0	0	0
17	0	0	0	0	0	0	1	0	0	0	0	0	1	0	0	0	0	0	0	0	0	0	0	0	0	0	0	0	0	0	0
18	0	0	0	0	0	0	1	2	0	0	0	0	0	0	0	0	0	0	0	0	0	0	0	0	0	0	0	22	0	0	0
19	0	1	0	0	0	0	0	0	0	0	0	0	0	0	0	0	0	0	0	0	0	0	0	0	0	0	0	3	1	0	0
20	0	0	0	0	0	0	0	0	0	0	0	0	1	0	0	0	0	0	0	0	0	0	0	0	0	0	0	0	48	2	0
21	0	0	0	0	0	1	0	0	0	0	1	0	0	0	0	0	0	0	0	0	0	0	0	0	0	0	0	0	0	0	0
22	0	1	0	0	0	0	0	0	0	0	0	0	0	0	0	0	0	0	0	0	0	0	0	6	0	0	0	1	0	0	0
23	0	0	0	0	1	0	0	0	0	0	0	0	0	0	32	0	0	14	0	0	0	0	0	0	0	0	0	0	0	0	0
24	0	0	0	1	1	0	0	0	0	0	0	0	0	1	0	0	0	0	0	0	0	0	0	0	2	0	0	0	0	0	0
25	0	0	1	0	0	0	0	0	0	0	0	0	0	0	0	0	0	0	0	0	0	1	0	0	0	5	0	0	0	0	0
26	0	1	0	0	0	0	0	0	0	0	0	0	0	0	0	0	0	0	0	0	0	0	0	0	0	0	0	0	0	2	0
27	0	0	0	0	0	0	0	0	0	0	0	0	1	0	0	0	0	0	0	0	0	0	0	0	1	0	1	0	0	0	0
28	0	0	0	0	0	0	0	0	0	0	0	0	0	0	0	0	2	0	0	0	0	0	0	0	0	0	1	6	0	0	0
29	0	0	0	0	0	0	0	0	0	0	0	0	0	0	0	1	6	0	0	0	0	0	0	0	0	0	0	0	0	0	0
30	0	0	0	0	0	0	0	1	0	0	0	0	0	0	0	0	0	0	0	0	0	0	1	0	0	0	0	0	0	0	0
31	1	0	0	0	0	1	0	0	0	1	0	1	0	0	0	0	0	0	0	0	0	0	0	0	0	0	0	0	0	0	2

Figure 11 - A modified occurrence matrix in which the non-zero elements represent the defined "Sensitivity ratios"

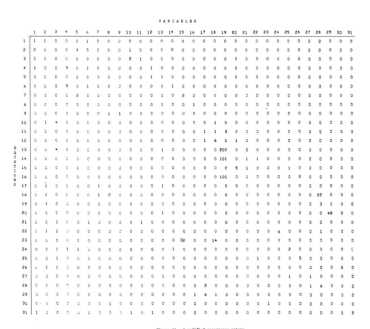

OUTPUT OF THE ITERATIVE PROGRAM SITER

```
                                ITERATIØN    INITIAL AND CALCULATED VALUES
                                  0 ***
                                                 1             2              3             4             5
(Initially Assumed Solution)           9.6000E 02   6.6000E 01  -5.8000E 00   2.0000E 00   7.2000E 00
(Corresponding Calculated Values)      3.9689E 02   3.4992E 01  -8.9187E 00   1.2712E-01   3.3086E 00
                                  1 ***
                                                 1             2              3             4             5
                                       4.6240E 02   3.2726E 01  -3.6254E 00   6.1469E-01   3.5045E 00
                                       4.8110E 02   3.3016E 01  -2.1323E 00   1.4207E 00   3.4396E 00
                                  2 ***
                                                 1             2              3             4             5
                                       4.7848E 02   3.3270E 01  -2.8907E 00   9.7984E-01   3.5608E 00
                                       4.7848E 02   3.3270E 01  -2.8214E 00   1.0662E 00   3.5575E 00
                                  3 ***
                                                 1             2              3             4             5
                                       4.7844E 02   3.3294E 01  -2.8761E 00   1.0229E 00   3.5631E 00
                                       4.7844E 02   3.3294E 01  -2.8757E 00   1.0236E 00   3.5631E 00
                                  4 ***
                                                 1             2              3             4             5
                                       4.7844E 02   3.3294E 01  -2.8761E 00   1.0232E 00   3.5631E 00
(Final Solution)                       4.7844E 02   3.3294E 01  -2.8761E 00   1.0232E 00   3.5631E 00
                                *STØP* 0
```

Figure 12a - The initially assumed solution
is approximately twice the final solution

ITERATION INITIAL AND CALCULATED VALUES

	ITERATION	1	2	3	4	5
(Initially Assumed Solution)	0 ***	4.8000E 01	3.3000E 00	-2.9000E-01	1.0000E-01	3.6000E-01
(Corresponding Calculated Values)		7.0733E 02	4.2557E 01	5.9040E 00	3.7645E 00	3.6660E 00
	1 ***	2.9079E 02	1.2378E 01	1.1482E 00	6.3753E-01	3.9703E 00
		5.5800E 02	3.5947E 01	2.2761E 00	2.2305E 00	3.8847E 00
	2 ***	4.5801E 02	2.9796E 01	-2.7925E 00	9.1875E-01	3.5462E 00
		4.8738E 02	3.3358E 01	-2.0287E 00	1.1474E 00	3.5503E 00
	3 ***	4.7817E 02	3.3279E 01	-2.8811E 00	1.0200E 00	3.5628E 00
		4.7849E 02	3.3292E 01	-2.8687E 00	1.0258E 00	3.5622E 00
	4 ***	4.7844E 02	3.3294E 01	-2.8761E 00	1.0233E 00	3.5631E 00
		4.7844E 02	3.3294E 01	-2.8761E 00	1.0232E 00	3.5631E 00
	5 ***	4.7844E 02	3.3294E 01	-2.8761E 00	1.0232E 00	3.5631E 00
(Final Solution)		4.7844E 02	3.3294E 01	-2.8761E 00	1.0232E 00	3.5631E 00

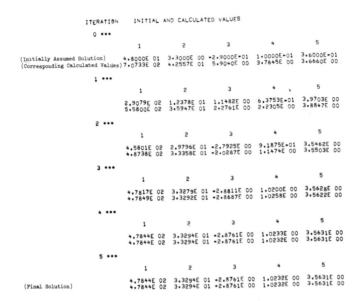

Figure 12b - The initially assumed solution is approximately one tenth the final solution

iterative algorithm in which the tearing procedure is coupled with the least sensitive output pairs set. Method B is associated with an iterative procedure in which which an output set is chosen which corresponds to the absolute minimum of iteration variables $\check{\beta} = 3$, neglecting the least sensitive output pairs set. Finally, method C corresponds to the conventional Newton-Raphson method. The computer execution times, for the case when the initial guess $\hat{x}$ is twice the final solution, are given in the second column of this Table.

8. CONCLUSION

A newly developed and tested iterative algorithm, which solves large sets of non-linear simultaneous equation, is introduced. The concept, here, is based on a newly formulated tearing procedure which is coupled with the sensitivity analysis of the system equations in the neighbourhood of an assumed solution.

The tearing procedure, which has been included in the user-oriented digital program DECOMP, tore problems exceeding 100 variables, in a fraction of the execution times cited in the literature.

The user-oriented digital program SITER, which implements the iterative algorithm, has been successfully applied for the steady-state solving of several existing non-linear processes. The efficiency of the presented iterative algorithm is evident from its low computer storage requirements and fast execution times.

TABLE I

Size of problem, n	Execution time on RXDS-Σ7 digital computer, s					
	Total	Occurrence matrix	Partitioning	Disjointing and constructing hierarchical structure	Least sensitive output pairs set	Tearing
31*	0.90	0.18	0.18	0.06	0.30	0.24
63	5.70	0.66	0.84	0.06	3.66	0.48
112**	4.08	-	1.74	0.18	-	2.16
139***	82.68	8.70	13.14	0.0	55.80	5.04
Storage requirement			$n^2 + 5n + 4970$			

* The illustrative example
** The occurrence matrix and output set were specified
*** One irreducible system

TABLE IV

	Iteration algorithm	Execution time on RXDS-Σ7 digital computer, s	Size of convergence matrix used in the iteration algorithm	Number of iterations	Type of solution
A	Developed procedure (Decomposition coupled with the sensitivity concept)	0.36*	5 x 5	4	Feasible
B	Decomposition without sensitivity concept	2.28*	3 x 3	28	Infeasible**
C	Newton-Raphson	16.98	31 x 31	5	Infeasible**

* DECOMP's execution time, for the given example, is 0.9 seconds
** Physically not possible

TABLE II

Irreducible subsystems	Iteration variables	Equations
		$x_{11} = f'_3 \; (x_{10})$
		$x_6 = f'_{21} \; (x_{11})$
		$x_1 = f'_1 \; (x_6)$
		$x_{12} = f'_4 \; (x_1, x_4, x_6)$
		$x_{14} = f'_6 \; (x_1, x_4, x_6)$
		$x_{13} = f'_5 \; (x_{12})$
		$x_{16} = f'_8 \; (x_{13})$
		$x_{17} = f'_{29} \; (x_{16})$
		$x_{15} = f'_7 \; (x_5, x_{13})$
		$x_2 = f'_{26} \; (x_{30})$
1.01		$x_{18} = f'_{10} \; (x_2, x_3)$
		$x_{19} = f'_{11} \; (x_{17}, x_{18})$
		$x_{21} = f'_{13} \; (x_3, x_{13}, x_{19})$
		$x_{26} = f'_{16} \; (x_{19}, x_{21})$
		$x_{22} = f'_{14} \; (x_{19}, x_{21})$
		$x_{20} = f'_{12} \; (x_{17}, x_{18})$
		$x_{25} = f'_{15} \; (x_5, x_{20}, x_{21})$
		$x_{27} = f'_{27} \; (x_{13}, x_{25})$
		$x_{28} = f'_{28} \; (x_{17}, x_{27})$
		$x_{29} = f'_{19} \; (x_2, x_{28})$
	z_1	$x_3 = f'_{25} \; (x_{22}, x_{26})$
	z_2	$x_{30} = f'_{20} \; (x_{13}, x_{29})$
	z_3	$x_5 = f'_{23} \; (x_{15}, x_8)$
	z_4	$x_4 = f'_{24} \; (x_5, x_{14}, x_{25})$
	z_5	$x_{10} = f'_2 \; (x_1, x_5, x_6)$
2.01		$x_7 = f'_{17} \; (x_{13})$
2.02		$x_{24} = f'_{22} \; (x_2, x_{28})$
2.03		$x_{31} = f'_{31} \; (x_1, x_6, x_{10}, x_{12})$
3.01		$x_8 = f'_{18} \; (x_7, x_{28})$
4.01		$x_9 = f'_9 \; (x_8)$
4.02		$x_{23} = f'_{30} \; (x_8)$

TABLE III

FUNCTION	OUTP.VAR.	SENSITIVITY RATIO
1	1	1
2	10	1
3	11	1
4	12	1
5	13	1
6	14	1
7	15	2
8	16	1
9	9	1
10	18	1
11	19	3
12	20	1
13	21	2
14	22	1
15	25	1
16	26	2
17	7	1
18	8	2
19	29	1
20	30	2
21	6	1
22	24	6
23	5	1
24	4	1
25	3	1
26	2	1
27	27	1
28	28	6
29	17	6
30	23	1
31	31	2

NOMENCLATURE

x_i	an unknown variable
x	set of unknown variables
$x^1, \ldots, x^k$	subsets of the set x
u	set of known variables
f	set of non-linear (linear) equations containing x and u
F	real differentiable function
f_x and f_x^{-1}	Jacobian matrix and its inverse
θ	number of unknown variables associated with an irreducible set of equations
I_θ	the set of integers $\{1, 2, \ldots, \theta\}$
(x, f)	output pairs set
(x_i, f_j)	output pair; x_i is the assigned output variable in equation f_j
β	minimum number of iterates
$\check{\beta}$	absolute minimum number of iterates
v	iterative set
v_j	subset of iterative variables occurring in equation f_j
$w_j = v_j \setminus \{x_j\}$ where x_j is the output variable in f_j	
$\hat{v}$ and $\bar{v}$	sets containing the assumed and calculated values of the iterative variables respectively
$\mathcal{G}$	flow graph (directed graph)
z	β-order vector whose elements constitute the iterative variables
$\hat{z}$ and $\bar{z}$	assumed and calculated values of the vector z
$(\hat{z})^\circ$	unit vector
$(\hat{x}_{\theta-\beta+i})^\circ$	initially assumed solution of an iterative variable
$\delta\bar{z}_i = \delta\bar{x}_{\theta-\beta+i}/(\hat{x}_{\theta-\beta+i})^\circ$	
$\delta\hat{z}_i = \delta\hat{x}_{\theta-\beta+i}/(\hat{x}_{\theta-\beta+i})^\circ$	
$\delta\hat{x}_{\theta-\beta+i}$	a small specified perturbation in $\hat{x}_{\theta-\beta+i}$
$\delta\bar{x}_{\theta-\beta+i}$	deviation in $\bar{x}_{\theta-\beta+i}$ due to the small change of $\delta\hat{x}_{\theta-\beta+i}$ in $\hat{x}_{\theta-\beta+i}$
Δz	correction term
$(\hat{z})^1 = (\hat{z})^\circ + \Delta z$	
I	identity matrix
g_z	β-order Jacobian-type matrix
$(\partial f_k/\partial x_l)_{\hat{x}}$	partial derivative of f_k, with respect to x_l, evaluated at $\hat{x}$
$\hat{x}$	an assumed solution of the elements in x
$\Delta x_l = x_l - \hat{x}_l$	
K_{pq}	the gain associated with a path linking at vertex f_p, in a flow graph, directly to another vertex f_q

NOMENCLATURE

ϕ	empty or null set
c_o	a constant
S_{ki}	a sensitivity ratio which is associated with the variable x_i occurring in equation f_k

SYMBOLS

$\{x_1, x_2, \ldots, x_n\}$	represents a set whose elements are separated by commas and enclosed in braces { }
$x_j \varepsilon x$	reads: the variable x_j is an element of the set x. (Slash across this symbol denotes its negation.)
$x \supset x^i$	reads: the set x contains the subset x^i
$x_j \mathbf{E} f_i$	reads: the variable x_j is contained in equation f_i
$v_i [f_i$	reads: the subset of iterative variables v_i is contained in equation f_i
$A \setminus B$	set of elements which belong to the set A but which do not belong to the set B
$f_i \longrightarrow f_j$	denotes the path of information flow from vertex f_i to vertex f_j in $\mathcal{G}$
$f_i \Longleftrightarrow f_j$	denotes the closed path of information flow linking vertices f_i and f_j in $\mathcal{G}$. (Slash across any of the last two symbols denote their negation.)
α	proportional
$x = \bigcup_{i=1}^{k} x^i$	denotes the union of the k subsets $x^1, x^2, \ldots, x^k$
$\bigcap$	denotes the intersection of two sets or subsets

REFERENCES

K.J. ARROW, L. HURWICK and H. UZAWA (1958), "Studies in linear and non-linear programming", Stanford University Press, Stanford.

D.M. HIMMELBLAU (1967), "Decomposition of large-scale systems. II. Systems containing non-linear elements", Chem. Eng. Sci., 22, p. 883.

A.K. KEVORKIAN and J. SNOEK (1972-I), "Decomposition in large scale systems: Theory and applications of structural analysis in partitioning, disjointing and constructing hierarchical systems", NATO ADVANCED STUDY INSTITUTE ON DECOMPOSITION AS A TOOL FOR SOLVING LARGE SCALE PROBLEMS", Cambridge University.

*)A.K. KEVORKIAN and J. SNOEK (1972-II), "Discussion for decomposition study and applications to large-scale non-linear processes", Contribution to "Optimization in Technological Design" edited by M. Avriel, M.J. Rijckaert and D.J. Wilde, Prentice-Hall, Inc.

W.P. LEDET and D.M. HIMMELBLAU (1970), "Decomposition procedures for the solving of large-scale systems", Advan. Chem. Eng., 8, p. 185.

J. PESCHON (1968), "Case study of optimum control of electric power systems", Case studies in system control, University of Michigan, p. 5.

I.H. RINARD and D.L. RIPPS (1964), "The steady-state simulation of continuous chemical processes", Chem. Eng. Progr. Symposium, Series No. 55, 61, p. 34.

R.W.H. SARGENT (1971), "Minimization without constraints", NATO International Summer School on the "Impact of Optimization Theory on Technological Design", Louvain, Belgium.

D.V. STEWARD (1962), "On an approach to techniques for the analysis of the structure of large systems of equations", SIAM REVIEW, 4, No. 4, p. 321.

D.V. STEWARD (1965), "Partitioning and tearing systems of equations", J. SIAM Numer. Anal., Ser. B, 2, No. 2, p. 345.

J.F. TRAUB (1964), "Iterative methods for the solution of equations", Prentice-Hall, Englewood Cliffs, New Jersey.

A.W. WESTERBERG and F.C. EDIE (1971-I), "Computer-aided design, Part 1. Enhancing convergence properties by the choice of output variable assignments in the solution of sparse equation sets", Chem. Eng. J., 2, p. 9.

A.W. WESTERBERG and F.C. EDIE (1971-II), "Computer-aided design, Part 2. An approach to convergence and tearing in the solution of sparse equation sets", Chem. Eng. J., 2, p. 17.

*) The title "Optimation in Technological Design" should read: "Perspectives on Optimal Engineering Design".

ON THE APPLICATION OF DECOMPOSITION TECHNIQUES IN GEOMETRIC PROGRAMMING

M.J. RIJCKAERT and L.J. HELLINCKX

Instituut voor Chemie-Ingenieurstechniek
Katholieke Universiteit Leuven,Belgium

Abstract : First a survey of existing decomposition techniques for geometric programming is presented together with some computational experience with these techniques. Then a numerical procedure is described for computing dual equilibrium solutions, which uses decomposition principles. A numerical example is given.

Geometric programming is a non-linear programming technique, developped by DUFFIN, PETERSON and ZENER (1967). In his most generalized form, it is actually able to determine the positive N-dimensional vector x, which minimizes the real valued function

$$g_0(x) \qquad (1)$$

while satisfying the M constraints

$$\sigma_m \, g_m(x)^{\sigma_m} \leqslant 1 \qquad m = 1,\ldots,M \qquad (2)$$

The functions g_m should have a form, which can be written as

$$g_m(x) = \sum_{t=1}^{T_m} \sigma_{mt} \, c_{mt} \prod_{n=1}^{N} x_n^{a_{mtn}} \qquad m = 0,1,\ldots,M$$

Since the sign of each term of g_m is expressed explicitly in the signum function σ_{mt}, the constants c_{mt} can be restricted to positive values.

(Note : the first index of $c_{mt}, \sigma_{mt}, a_{mtn}$ and of the later to be defined u_{mt} and ω_{mt} refers to the constraint, the second one to the term in a constraint. If a third index is present, as in a_{mtn}, it is related to a component of x.)

In the original geometric programming theory, the signum functions σ_{mt} and σ_m were fixed at +1. The functions g_m are then called posynomials and are the sum of posynomial terms

$$u_{mt} = c_{mt} \prod_{n=1}^{N} x_n^{a_{mtn}} \qquad (3)$$

From a theoretical point of view, posynomial programs show very interesting properties. They can be transformed by simple logarithmic transformation into convex programs so that the Kuhn-Tucker optimality conditions are necessary and sufficient as well. Unfortunately, practical applications for posynomial programs are quite less numerous than for signomial programs, in which some σ_{mt} and σ_m are equal to -1. Programs with some negative σ_m are termed reverse geometric programs by DUFFIN (1970a) since

$$-1 \, (- g_m(x))^{-1} \leqslant 1$$

corresponds to

$$g_m(x) \geqslant 1$$

Due to the fact that the exponents a_{mtn} can be any real number, geometric programming has found one of his major field of applications in engineering problems. A survey of such applications can be found in RIJCKAERT (1971). Also ZENER (1971) presents some engineering design problems solved efficiently by geometric programming.

For posynomial programs an extremally useful duality theory was developped by DUFFIN, PETERSON and ZENER (1967). For signomial programs and algebraic programs, which can be reduced to it (DUFFIN, 1970c), the presence of negative σ_m or σ_{mt} throws them out of the class of convex programs. Hence in this case, only a quasidual can be constructed which lacks many of the powerful primal-dual properties of convex programs. But still then, geometric programming remains one of the most specific examples in mathematical programming for which the dual problem tends to be more easily solved than the primal one. The geometric dual - or quasidual for signomial problems - has a set of constraints consisting of the following equations, which are all linear in the non-negative dual variables ω_{mt} : (as indicated by the indices, each ω_{mt} corresponds to one primal term u_{mt})

- a single normality condition

$$\sum_{t=1}^{T_0} \sigma_{0t}\,\omega_{0t} = \sigma_0 \tag{4}$$

in which σ_0 equals the sign of min $g_0(x)$. This sign needs to be known in advance. If not, an educated guess has to be made and if the results don't confirm this guess, the alternative sign should be tried out too.

- the orthogonality conditions. (Total number N i.e. one for each primal variable x_n.)

$$\sum_{m=0}^{M} \sum_{t=1}^{T_m} a_{mtn}\,\sigma_{mt}\,\omega_{mt} = 0 \qquad n = 1,\dots,N \tag{5}$$

The dual objective function is

$$\sigma_0 \left[\prod_{m=0}^{M} \prod_{t=1}^{T_m} \left(\frac{c_{mt}\,\omega_{m0}}{\omega_{mt}}\right)^{\sigma_{mt}\,\omega_{mt}} \right]^{\sigma_0} \tag{6}$$

For a posynomial program this function needs to be maximized to get the dual solution corresponding to the global minimum of the primal objective function. (A local minimum is also a global minimum because of convexity conditions). For a signomial program, a local minimum of the primal program corresponds to a stationnary point of the dual objective function, since the Kuhn-Tucker optimality conditions are no longer sufficient. Such a solution that satisfies the necessary Kuhn-Tucker conditions is called a pseudominimum by PASSY (1967) or a quasiminimum by AVRIEL (1970) or an equilibrium solution.
The so far undefined variables ω_{m0} are proportional to the Lagrange multipliers of the primal constraints and can be expressed as linear functions of the dual variables ω_{mt}

$$\omega_{m0} = \sigma_m \sum_{t=1}^{T_m} \sigma_{mt}\,\omega_{mt} \qquad m = 0,1,\dots,M \tag{7}$$

In contrast to the variables ω_{mt} $t = 1,\dots, T_m$; $m = 0,1,\dots,M$, which are called independent dual variables, the variables ω_{m0} $m = 0,1,\dots,M$ are termed dependent dual variables. It also follows immediately from (5) and (7) that $\omega_{00} = 1$.

For brevity, the equations (7) will be referenced to as the LM-equations. Furthermore, it will be computationally advantageous to assume that in the optimum all constraints are tight i.e. satisfied as equalities instead of inequalities. This can always be guaranteed by adding "slack terms" to the constraints and to the objective function (1), as indicated by DUFFIN (1970b).

Primal decomposition techniques.

A few successful attempts have been made in the past to apply decomposition techniques to geometric programming in posynomial form.
MINE (1970) proposed a decomposition scheme for programs, which can be written in block diagonal form, which means that the vector variable x can be subdivided into P distinct vector variables $X_1,\dots,X_p$ such that

$$X_1 = (x_{m_1}, \dots, x_{n_1})$$
$$\dots$$
$$X_p = (x_{m_p}, \dots, x_{n_p})$$
$$\dots$$
$$X_P = (x_{m_P}, \dots, x_{n_P}$$

with $n_0 = 0$, $m_p = n_{p-1} + 1$, $p = 1,\dots,P$ and $n = N$.
The vectors X_p $p = 1,\dots,P$ must be chosen such that

$$g_0(x) = \sum_{p=1}^{P} g_{0p}(X_p)$$

is minimized while satisfying

$$g_m(x) = \sum_{p=1}^{P} g_{mp}(X_p) \leqslant 1 \qquad m = 1,\dots,M_0$$

$$g_m(x) = g_{mp}(X_p) \leqslant 1 \qquad m = N_p,\dots,M_p,\ p = 1,\dots,P$$

with $N_p = M_{p-1} + 1$, $p = 1,\dots,P$ and $M_P = M$.
The functions g_{mp} $m = 0,1,\dots,M$; $p = 1,\dots,P$ are posynomial terms in the vector variable X_p.
This block diagonal problem is then decomposed into smaller ones which are tied together by a dynamic programming type of recurrence relation. Each individual subproblem is a parametric linear programming problem. It is clear that practical problems having the above diagional form will be seldomly encountered, if at all. Besides for the prototype problem mentionned in their paper

$$\text{Min } 2.0\, x_1^{0.9} x_2^{-1.5} x_3^{-3.0} + 5.0\, x_4^{-0.3} x_5^{2.6} + 4.7\, x_6^{-1.8} x_7^{-0.5} x_8^{1.0}$$

$$7.2\, x_1^{-3.8} x_2^{2.2} x_3^{4.3} + 0.5\, x_4^{-0.7} x_5^{-1.6} + 0.2\, x_6^{4.3} x_7^{-1.9} x_8^{8.5} \leqslant 1$$

$$10.0\, x_1^{2.3} x_2^{1.7} x_3^{4.5} \leqslant 1 \qquad 0.2\, x_4^{-2.1} x_5^{0.4} \leqslant 1 \qquad 6.2\, x_6^{4.2} x_7^{-2.7} x_8^{-0.6} \leqslant 1$$

$$3.1\, x_1^{1.6} x_2^{0.4} x_3^{3.8} \leqslant 1 \qquad 3.7\, x_4^{5.4} x_5^{1.3} \leqslant 1 \qquad 0.3\, x_6^{-1.1} x_7^{7.3} x_8^{-5.6} \leqslant 1$$

$$x_n \geqslant 0 \quad ,\ n = 1,\dots,8$$

our computational experience didn't show any advantage for this decomposition schema to other methods. (See RIJCKAERT, 1971) The importance of the above decomposition method therefore seems to be mainly theoretical.
HEYMANN (1969) proposed a decomposition scheme for systems, which can be considered as consisting of interrelated sybsystems. With each subsystem $p = 1,\ldots,P$ a positive vector X_p is associated together with a positive coupling vector X_0 common to all subsystems. The objective function to be minimized has components originating from the various subsystems and from the coupling subsystem :

$$\text{Min } g_0(x) = g_{00}(X_0) + \sum_{p=1}^{P} g_{0p}(X_0,X_p)$$

The primal constraints for the coupling system are

$$g_{m0}(X_0) \leqslant 1 \qquad m = 1,\ldots,M_0$$

and for the p-th, $p = 1,\ldots,P$ subsystem

$$g_{mp}(X_0,X_p) \leqslant 1 \qquad m = N_p,\ldots, M_p;\ p = 1,\ldots,P$$

with $N_p = M_{p-1} + 1$, $p = 1,\ldots,P$ and $M_P = M$.
The functions g_{mp} $m = 0,1,\ldots,M$; $p = 1,\ldots,P$ are in this case required to be sums of posynomial terms. The above problem is subsequently decomposed into P subproblems

$$\begin{aligned} &\text{Min } g_{00}(X_0) + g_{0p}(X_0,X_p) \\ &\text{s.t.} \quad g_{m0}(X_0) \leqslant 1 \qquad m = 1,\ldots,M_0 \\ &\qquad\ \ g_{mp}(X_0,X_p) \leqslant 1 \qquad m = N_p,\ldots,M_p \\ &\qquad\qquad X_0 \geqslant 0 \quad \text{and} \quad X_p \geqslant 0 \end{aligned}$$

This decomposition is valid when each of the above P subproblems has zero degrees of difficulty. The number of degrees of difficulty is a very important characteristic of geometric programs and is given by

$$D = \sum_{m=0}^{M} T_m - (N+1) \tag{8}$$

It is the difference between the number of dual variables and the number of linear normality and orthogonality conditions. When $D = 0$, the dual constraint set consists of a square system of linear equations and the dual solution can be found without any formal optimization by solving a system of linear equations.
Since the overall solution can be obtained by straightforward algebraic manipulations out of the optimal solutions of the P subproblems, it is obvious that, when the above condition is fullfilled the method will work excellently.

Decomposition in dual geometric programming.

For problems with a positive number of degrees of difficulty, additional conditions can be developped for the dual variables. These new equations are non-linear and form with the linear orthogonality, normality and LM-equations a system with exactly just as many equations as unknowns. The solutions of these equations - multiple solutions are possible, since part of the system consists of non-linear equations - satisfy the Kuhn-Tucker conditions.

These nonlinear equations (originally called maximizing equations in the case of posynomial programs) are called equilibrium conditions and are derived by setting to zero those directional derivatives of the dual objective function for which the positivity, normality, and orthogonality conditions are not violated.

The normality and orthogonality conditions can be represented as

$$A\omega = b \tag{9}$$

where A is a matrix of dimension (N+1) by T $(= \sum_{m=0}^{M} T_m)$

ω is the T-dimensional vector composed by the independent dual variables.

b is a N+1-dimensional vector.

To state the equilibrium equations, the vector b is replaced by a null-vector of the same dimension, so that one gets

$$A\omega = 0 \tag{10}$$

which is a system of N+1 homogeneous linear equations in T (> N+1) unknowns, for which at least D (=T-N-1) linearly independent solutions exist. The way in which these equations are derived is essential for the further development of the paper and will therefore be discussed in detail.

By eliminating D columns from the matrix A, a nonsingular (N+1)-dimensional matrix B can be constructed. For notational simplicity, we suppose that B is formed from the first N+1 columns of A. If not, one can interchange the order of the columns of A. The matrix formed by the D discarded columns is called C.

If the N+2-nd column of A is denoted A_{N+2} and if Ω is the (N+1)-dimensional solution vector of

$$B\Omega = - A_{N+2} \tag{11}$$

then the T-dimensional vector $\nu = (\Omega,+1,0,\ldots,0)$ will be a solution of

$$A\omega = 0 \tag{10}$$

Such a vector ν is called a nullity vector. (DUFFIN, 1967)

D linearly independent solutions of equations (10) can be obtained by solving once the following multiple right handside system of linear equations

$$B\Omega^d = - A_{N+1+d} \qquad d = 1,\ldots,D \tag{12}$$

This gives the following D linearly independent nullity vectors

$$\nu^d = (\Omega^d, \varepsilon^d) \qquad d = 1,\ldots,D$$

with ε^d a D-dimensional vector, whose d-th component is 1. All other components being zero.

The linear independence of the vectors $\nu^1,\ldots, \nu^D$ can be proved by looking at the last D components.

The equilibrium conditions are (PASSY, 1969)

$$\prod_{m=0}^{M} \prod_{t=1}^{T_m} \left(\frac{\omega_{mt}}{\omega_{m0}}\right)^{\nu^d_{mt}\sigma_{mt}} = \prod_{m=0}^{M} \prod_{t=1}^{T_m} c_{mt}^{\nu^d_{mt}\sigma_{mt}} \qquad d = 1,\ldots,D \tag{13}$$

Taking logarithms

$$\sum_{m=0}^{M}\sum_{t=1}^{T_m} \sigma_{mt}\,\nu_{mt}^{d}\,\log\frac{\omega_{mt}}{\omega_{m0}} = \sum_{m=0}^{M}\sum_{t=1}^{T_m} \sigma_{mt}\,\nu_{mt}^{d}\,\log c_{mt} \qquad d = 1,\dots,D \tag{14}$$

Hence the equilibrium conditions are linear in the logarithms of the dual variables $\omega(\omega_{mt},\ t = 1,\dots,T_m\ ;\ m = 0,1,\dots,M)$ and $\omega^\circ(\omega_{m0},\ m = 1,\dots,M)$.
Noting that the variables ω_{m0} appear each only once in the M linear LM-equations (7), the complete system to determine the equilibrium solutions can be schematically represented as

$$\begin{array}{c} \\ N+1 \\ M \end{array}\begin{array}{c} \begin{array}{ccc} N+1 & D & M \end{array} \\ \left[\begin{array}{cc|c|c} B & & C & 0 \\ \hline & F & & I \end{array}\right] \end{array} \begin{bmatrix} \omega \\ \omega^\circ \end{bmatrix} = \begin{bmatrix} b \\ 0 \end{bmatrix} \qquad \begin{array}{c} (15a) \\ (15b) \end{array}$$

$$D \quad \left[\begin{array}{c|c|c} G & I & G^\circ \end{array}\right] \begin{bmatrix} \log\omega \\ \log\omega^\circ \end{bmatrix} = \begin{bmatrix} c \end{bmatrix} \tag{15c}$$

where I as a unity matrix of the required dimensions. G, G° are matrices composed by the coefficients of $\log\omega$ and $\log\omega^\circ$ in (14). F is the matrix formed by the coefficients of ω in (7).

The system (15) can be solved by a generalized block Gauss-Seidel procedure Assuming values for D independent dual variables, (15a) can be solved for the remaining (N+1) independent variables. Subsequently ω° can be obtained from (15b). Note that the coefficients of ω° form a unity matrix. Finally from (15c) one can calculate the logarithms of the D independent variables for which values were guessed in (15a). If the results don't agree with the assumed ones, one restart the procedure with the present values.

Hence during the iterative procedure the system of equations (15) in decomposed into 3 separate subsystems of which 2 are solved automatically.
Indeed if the last D components of ω are selected as non-basic in (15a), then the coefficients of the variables to be computed from (15b) and (15c) form a unity matrix.

The efficiency of the above procedure will be improved by making an educated guess for the last D components of ω. In contrast to the primal program, such an educated guess is rather easy to be made in dual geometric programming because of the linearity of the constraints.

For posynomials for instance, a good starting point can usually be located by a (crude) line search for the dual objective between two extreme points of the convex set of points satisfying the orthogonality and normality conditions. Such an extreme point is obtainable by linear programming after introducing a linear objective function. For instance : minimizing and maximizing the linear part of the logarithm of (6)

$$\sum_{m=0}^{M}\sum_{t=1}^{T_m} \omega_{mt}\,\log c_{mt}$$

It is also easy to find an upper and lower bound for each dual variable by considering only the normality and orthogonality conditions. Such a procedure may seem very elaborate for large problems, for which the above solution scheme is essentially designed. However in practice, the number of optimizations to be carried out will be quite reduced since the constraint set remains the same in each problem. Furthermore after solving one problem, one can check all remaining objective functions and eliminate these from future consideration, which have their optimum in the same extreme point.

Appendix

To illustrate the above method for readers unfamiliar with geometric programming, the above equations will be derived for a small example. It computes the optimal pressure at which a gas should be stored in order to minimize storage costs. The problem is described in full detail by JEN (1968) and HELLINCKX (1972).

$$\text{Min } c_{01}x_1^{0.85} + c_{02}x_1 + c_{03}x_2^{-0.75}$$

$$\text{s.t. } c_{11}x_2^{0.05} - c_{12}x_1 \leqslant 1$$

$$x_1 \, , \, x_2 > 0$$

with $c_{01} = 3.6996$ $\quad$ $c_{11} = 0.76727$

$c_{02} = 1.9576$ $\quad$ $c_{03} = 700.30$ $\quad$ $c_{12} = 0.05$

The normality condition (4) is :

$$\omega_{01} + \omega_{02} + \omega_{03} = 1$$

The orthogonality conditions (5) are :

$$0.85\,\omega_{01} + \omega_{02} - \omega_{12} = 0$$

$$-\,0.75\,\omega_{03} + 0.05\,\omega_{11} = 0$$

The LM-equation (7) is :

$$-\,\omega_{11} + \omega_{12} + \omega_{10} = 0$$

Two nullity vectors are :

$$(1, -1, 0\ , 0, -0.15)$$

$$(0, \tfrac{-1}{15}, \tfrac{1}{15}, 1, \tfrac{-1}{15})$$

Hence the equilibrium conditions are (14) :

$$\log\omega_{01} - \log\omega_{02} \qquad 0.15\log\omega_{12} - 0.15\log\omega_{10} = 0.187$$

$$-\frac{1}{15}\log\omega_{02} + \frac{1}{15}\log\omega_{03} + \log\omega_{11} + \frac{1}{15}\log\omega_{12} - \frac{16}{15}\log\omega_{10} = -\,0.0727$$

One gets the following equilibrium solution :

$\omega_{01} = 0.2614$ $\quad$ $\omega_{11} = 9.0648$

$\omega_{02} = 0.1342$ $\quad$ $\omega_{12} = 0.3565$

$\omega_{03} = 0.6043$ $\quad$ $\omega_{13} = 8.7083$

As described in WILDE (1967), the primal variables can be calculated from the value of the dual objective and the dual variables. This is done by the following formulae

$$u_{mt} = (\text{optimal value dual objective}) \cdot \omega_{mt} \qquad m = 0$$

$$u_{mt} = \frac{\omega_{mt}}{\omega_{m0}} \qquad m = 1,\ldots,M$$

This gives

$x_1 = 0.348$ $\qquad x_2 = 446$

Both variables can be easily calculated from the terms of the primal objective, which are all functions of only one primal variable.

References

AVRIEL M., WILLIAMS A.C. (1970), Complementary geometric programming, SIAM J.Appl.Math., *19*, 125.

DUFFIN R.J., PETERSON E.L., ZENER C.M. (1967), Geometric programming - Theory and applications, J. Wiley, N.Y..

DUFFIN R.J. (1970a), Linearizing geometric programs, SIAM Review, *12*, 221.

DUFFIN R.J., PETERSON E.L. (1970b), Geometric programs treated with slack variables, Carnegie-Mellon Univ., Report 70-45.

DUFFIN R.J., PETERSON E.L. (1970c), Geometric programming with signomials, Carnegie-Mellon Univ., Report 70-58.

HELLINCKX L.J., RIJCKAERT M.J. (1972), Optimal capacities of production facilities. An application of geometric programming, Can.J.Chem.Eng., *50*, 148.

HEYMANN M., AVRIEL M. (1969), On a decomposition for a special class of geometric programming problems, J.O.T.A., *6*, 392.

JEN F.C., PEGELS G.C., DUPUIS T.M. (1968), Optimal capacities of production facilities, Man.Sci., *14*, B573.

MINE H., OHNO K. (1970), Decomposition of mathematical programming problems by dynamic programming and its application to block-diagonal geometric programs, J.Math.Anal.Appl., *32*, 370.

PASSY U., WILDE D.J. (1967), Generalized polynomial optimization, SIAM J.Appl.Math., *15*, 1344.

PASSY U., WILDE D.J. (1969), Mass action and polynomial optimization, J.Eng.Math., *3*, 325.

RIJCKAERT M.J. (1971), Engineering applications of geometric programming, in Perspectives of optimal engineering design, M.Avriel, M.J. Rijckaert, D.J.Wilde (Eds.), Prentice Hall, H.J., to appear.

WILDE D.J., BEIGHTLER C.S. (1967), Foundations of optimization, Prentice Hall, N.J..

ZENER C.M. (1971), Optimal engineering design by geometric programming, J. Wiley, N.Y..

THE DECOMPOSITION OF LARGE (GENERALIZED) GEOMETRIC PROGRAMMING PROBLEMS BY TEARING

ELMOR L. PETERSON
Department of Industrial Engineering/Management Sciences,
and Department of Mathematics
Northwestern University, Evanston, Illinois, 60201, U.S.A.

Abstract: Decomposition principles are developed for those (generalized) geometric programming problems whose matrices have a certain type of block diagonal structure and whose functions are (at least partially) separable (in a certain compatible way).

1. INTRODUCTION

Geometric programming in its most general form provides a mechanism for reformulating many important inseparable optimization problems as separable (generalized) geometric programming problems.

The key to such reformulations is the exploitation of the linearities that are present in a given optimization problem. Such linearities frequently appear as linear equations or linear inequalities, but they can also appear in much more subtle guises as matrices associated with nonlinearities.

The geometric programming problems that result from such reformulations are (at least partially) separable, and many that arise from the modeling of large systems have sparse matrices. Moreover, such separability and sparsity can frequently be exploited by using the decomposition principles to be developed here.

Initial work on geometric programming decomposition was done by Zener (1964), but only for a special type of unconstrained "posynomial" problem that occurs in the design of heat exchangers. However, subsequent extensions and generalizations to various types of constrained posynomial problems have been made by Heymann and Avriel (1969), Ecker (1972), and Rijckaert (1973).

This paper presents two geometric programming decomposition principles that transcend the special class of constrained posynomial problems. The first principle is the most basic and forms an integral part of the second principle. The second principle can be viewed as primarily another manifestation of the general method of "diakoptics" as introduced by Kron (1953). (Diakoptics is a term that stems from the Greek word "kopto" meaning to break or tear apart and the Greek prefix "dia" which reinforces the word that it precedes). The second principle can further be viewed as a nonlinear generalization of the "dualplex method" of Gass (1966), which (according to Dantzig (1970)) is related to the "partition programming" of Rosen (1964) in dual form.

The author suspects that the second principle also generalizes and/or unifies many nonlinear programming decomposition principles that appear in diverse disciplines and fields of application. In fact, the author hopes that this paper will stimulate and facilitate a unification and further development of such principles.

With that goal in mind this paper has been made as self-contained as possible, with a minimum of detailed technical developments. In particular, the geometric programming approach is introduced in section 2 in the simplest possible setting - programming without explicit constraints. Two decomposable problem types are then described in section 3 and are subsequently decomposed in section 4. The first type is decomposed simply by inspection, and the second type is reduced to the first type by the method of tearing (i.e. diakoptics), guided in some cases by geometric programming duality. The resulting decomposition principles are then specialized in section 5 to the seemingly more general setting of programming with explicit constraints.

2. THE GEOMETRIC PROGRAMMING APPROACH

Classical optimization theory and ordinary mathematical programming are concerned with the minimization (or maximization) of an arbitrary real-valued function g over some given subset $\mathcal{S}$ of the functions nonempty domain $\mathcal{C}$. For pedagogical simplicity we shall restrict our attention to the finite-dimensional case in which $\mathcal{C}$ is itself a subset of n-dimensional Euclidean space E_n.

In (generalized) geometric programming, as defined by Peterson (1970), the subset $\mathcal{S}$ is required to be the intersect of $\mathcal{C}$ with an arbitrary cone $\mathcal{X} \subseteq E_n$. However, for both practical and theoretical reasons, this problem of minimizing g over $\mathcal{X} \cap \mathcal{C}$ is not studied in isolation. It is first embedded in the family $\mathcal{Q}$ of closely related minimization problems $\mathcal{Q}(u)$ that are generated by simply translating (the domain $\mathcal{C}$ of) g through all possible displacements $-u \in E_n$ while keeping $\mathcal{X}$ fixed. (For gaining insight we recommend making a sketch of a typical case in which $\mathcal{X}$ is a one dimensional subspace of E_2. In most known cases of practical significance the cone $\mathcal{X}$ is in fact a subspace of E_n.) The problem of minimizing g over $\mathcal{X} \cap \mathcal{C}$ appears in the family $\mathcal{Q}$ as problem $\mathcal{Q}(0)$ and is studied in relation to all other problems $\mathcal{Q}(u)$, with special attention given to those problems $\mathcal{Q}(u)$ that are close to $\mathcal{Q}(0)$ in the sense that (the norm of) u is small.

Each problem $\mathcal{Q}(u)$ is said to be a <u>geometric programming problem</u>, and the family $\mathcal{Q}$ of all such problems (for fixed $g:\mathcal{C}$ and $\mathcal{X}$) is termed a <u>geometric programming family</u>. For purposes of easy reference and mathematical precision, problem $\mathcal{Q}(u)$ is now given the following formal definition in terms of classical terminology and notation.

PROBLEM $\mathcal{Q}(u)$. <u>Using the "feasible solution" set</u>

$$\mathcal{S}(u) \triangleq \mathcal{X} \cap (\mathcal{C} - u),$$

<u>calculate both the "problem infimum"</u>

$$\varphi(u) \triangleq \inf_{x \in \mathcal{S}(u)} g(x+u)$$

<u>and the "optimal solution" set</u>

$$\mathcal{S}^*(u) \triangleq \{x \in \mathcal{S}(u) \mid g(x+u) = \varphi(u)\}.$$

For a given u, problem $\mathcal{Q}(u)$ is either "consistent" or "inconsistent", depending on whether the feasible solution set $\mathcal{S}(u)$ is nonempty or empty. It is of course obvious that the family $\mathcal{Q}$ contains infinitely many consistent problems $\mathcal{Q}(u)$. The domain of the infimum function φ is taken to be the corresponding nonempty set

$$\mathcal{U} \triangleq \{u \in E_n \mid \text{problem } \mathcal{Q}(u) \text{ is consistent}\}.$$

Thus, the range of φ may contain the point $-\infty$; but if $\varphi(u) = -\infty$, then the optimal solution set $\mathcal{S}^*(u)$ is clearly empty.

Each optimization problem can generally be put into the form of the geometric programming problem $\mathcal{Q}(0)$ in more than one way by suitably choosing the function g and the cone $\mathcal{X}$. For example, one can always let g be the "objective function" for the given optimization problem simply by choosing $\mathcal{X}$ to be E_n, but we shall soon see that such a choice is generally not the best possible choice for most important optimization problems. The reason is that most such problems involve a certain amount of linearity (due to the presence of linear equations, linear inequalities, matrices, etc.), which can be conveniently handled through the introduction of an appropriate <u>nontrivial</u> subcone $\mathcal{X} \subset E_n$. The presence of such a subcone $\mathcal{X}$ is one of the distinguishing features of the geometric programming point of view.

Due to the pre-eminence of problem $\mathcal{A}(0)$ we shall find it useful to interpret problem $\mathcal{A}(u)$ as a perturbed version of $\mathcal{A}(0)$, so we term the set $\mathcal{U}$ the feasible perturbation set for problem $\mathcal{A}(0)$. Actually, a rather elementary argument (given essentially by Peterson (1970)) shows that

$$\mathcal{U} = \mathcal{C} - \mathcal{X}.$$

This kind of perturbation analysis is intimately related to the geometric programming duality previously studied by Peterson (1970), and it is the only ingredient that must be combined with our first decomposition principle to obtain our second decomposition principle. In fact, for a given problem $\mathcal{A}(0)$ that is of a certain decomposable type 2 the infimum function $\varphi^1:\mathcal{U}^1$ that results from replacing $\mathcal{X}$ with an appropriate "separable subcone" $\mathcal{X}^1 \subset \mathcal{X}$ turns out to be the objective function for a corresponding "master problem" when the domain $\mathcal{U}^1$ is restricted to its intersect with an appropriate "coupling subcone" $\mathcal{X}_0 \subset \mathcal{X}$ whose sum with $\mathcal{X}^1$ is $\mathcal{X}$.

This kind of perturbation analysis is also of direct practical interest. In particular, the following examples of important problem classes indicate that the functions $\varphi:\mathcal{U}$ and $g^*:\mathcal{U}$ often show the dependence of optimality on actual external influences, such as available resources, design requirements, materiel costs, and the data being optimally approximated in linear regression analysis.

Example 1. Perhaps the most striking example of the utility of the geometric programming approach comes from using it to study the minimization of "signomials". This was first done by Zener (1961, 1962) and Duffin (1962), and served as the initial development (as well as the main stimulus for subsequent developments) of geometric programming.

A signomial (sometimes termed a "generalized polynomial") is any function with the form

$$P(t) = \sum_{i=1}^{n} c_i t_1^{a_{i1}} t_2^{a_{i2}} \cdots t_m^{a_{im}},$$

where the coefficients c_i and the exponents a_{ij} are arbitrary real constants but the independent variables t_j are restricted to be positive. After much experience in the physical sciences and engineering Zener clearly recognized that many optimization problems of practical importance can be accurately modeled with such functions. In many cases they come directly from the laws of nature and/or the laws of economics. In other cases this functional form gives a good fit to empirical data over a wide range of the variables t_j.

The presence of the "exponent matrix" (a_{ij}) (which is of course associated with algebraic nonlinearities) is the key to applying geometric programming to signomial optimization. To effectively place the problem of minimizing $P(t)$ in the format of problem $\mathcal{A}(0)$, simply make the change of variables

$$x_i = \sum_{j=1}^{m} a_{ij} \log t_j, \quad i = 1,2,\ldots,n;$$

and then use the laws of exponents to infer that minimizing $P(t)$ is equivalent to solving problem $\mathcal{A}(0)$ when

$$g(x) \triangleq \sum_{i=1}^{n} c_i e^{x_i}$$

and

$$\mathcal{X} \triangleq \text{column space of } (a_{ij}).$$

The advantages of studying this problem $\mathcal{A}(0)$ rather than its signomial predecessor come mainly from the fact that, unlike the signomial P, the exponential function g is completely separable (in that it is the sum of terms, each of which depends on only a single independent variable x_i). Notice also that the parameter u_i is a logarithmic perturbation of the $|c_i|$, generally a very useful type of perturbation to consider because the coefficients c_i are typically costs per unit quantity of materiel and hence tend to vary somewhat. The exponents a_{ij} are usually geometrical constants or are fixed by the laws of nature and/or economics; so they do not tend to vary, and hence there is little lost in not studying their perturbations.

Example 2. Our second example comes from the minimization of quadratic functions

$$Q(z) = (1/2) < z, Hz > + < h, z >,$$

where H is an arbitrary constant matrix and h is an arbitrary constant vector.

A factorization of the coefficient matrix H (which is of course associated with quadratic nonlinearities) is the key to effectively applying geometric programming to quadratic programming. More specifically, linear algebra is used to compute matrices D and $\mathcal{D}$ such that

$$H = D^t D - \mathcal{D}^t \mathcal{D},$$

where ^{t} indicates the transpose operation. In terms of D and $\mathcal{D}$ the quadratic function

$$Q(z) = (1/2)(< Dz, Dz > - < \mathcal{D}z, \mathcal{D}z >) + < h, z >.$$

Of course, the expression $< Dz, Dz >$ is not present when Q(z) is negative semidefinite; and the expression $-< \mathcal{D}z, \mathcal{D}z >$ is not present when Q(z) is positive semidefinite (i.e. a convex function).

From elementary linear algebra we now infer that minimizing Q(z) is equivalent to solving problem $\mathcal{A}(0)$ when

$$g(x) \triangleq (1/2)\left(\sum_{i=1}^{m} x_i^2 - \sum_{i=m+1}^{2m} x_i^2\right) + x_{2m+1}$$

and

$$\mathcal{X} \triangleq \text{column space of} \begin{bmatrix} D \\ \mathcal{D} \\ h \end{bmatrix}.$$

Notice that, unlike the quadratic function Q, the quadratic function g is completely separable, a fact that can be exploited both theoretically and computationally.

It is useful to introduce some additional parameters into the preceding function g so that a much broader class of optimization problems can be studied. In particular, we redefine g so that

$$g(x) \triangleq \sum_{i=1}^{m} p_i^{-1} |x_i - b_i|^{p_i} - \sum_{i=m+1}^{2m} p_i^{-1} |x_i - b_i|^{p_i} + x_{2m+1} - b_{2m+1}$$

where b_i and p_i are arbitrary constants. Notice that the function g is still completely separable and can be specialized to the quadratic case by choosing $b_i = 0$ and $p_i = 2$ for each i.

Another interesting specialization is obtained by choosing $p_i = p$ for each i, while choosing $\mathcal{D} = 0$ and $h = 0$. The resulting problem consists essentially of finding the "best ℓ_p-norm approximation" to the fixed vector $(b_1, \ldots, b_m)$ by vectors in the column space of the matrix D, a fundamental problem in linear regression analysis. Finally, it is probably worth noting that the parameter u is then a perturbation of the vector $(b_1, \ldots, b_m)$ of data being optimally approximated, a rather useful type of perturbation to consider.

Other important examples that can be effectively treated by geometric programming (but have usually been studied without explicitly mentioning the subject) are: generalized "Weber problems" that are of central importance in location theory, discrete optimal control problems with linear dynamics (sometimes called dynamic programming problems with linear transition functions), and various nonlinear network flow problems that arise in the analysis of electrical networks and the analysis of certain (multicommodity) transportation networks.

For those examples the functions $\varphi:\mathcal{U}$ and $S^*:\mathcal{U}$ show the dependence of optimality on the following external influences: existing facility locations in location theory, the control and state sets as well as the initial state and final target sets in control theory, the input currents and the input potential differences in electrical network theory, and the travel demands in transportation network theory.

For a more thorough discussion of such examples, see the recent expository paper by Peterson (1973), and the references cited therein.

In concluding our present discussion of important examples it is probably worth mentioning that the modeling of certain transportation networks that contain at least a single one-way automobile artery gives rise to cones $\mathcal{X}$ that are generally not subspaces. Consequently, the added generality built into this presentation of geometric programming is not without practical significance.

3. TWO DECOMPOSABLE PROBLEM TYPES

In all geometric programming problems $\mathcal{A}(0)$ known to the author to be of practical significance the cone $\mathcal{X}$ is polyhedral and hence finitely generated. We therefore suppose without any known loss of practical significance that there is at least one $n{\times}m$ matrix $\mathcal{M}$ with a corresponding index set $\mathcal{P} \subseteq \{1,2,\ldots,m\}$ for which

$$\mathcal{X} = \{\, x \in E_n \mid x = \mathcal{M}z \text{ for at least one } z \in E_m \text{ for which } z_j \geq 0,\ j \in \mathcal{P} \,\}.$$

The index set $\mathcal{P}$ can of course be taken to be the empty set when $\mathcal{X}$ is in fact a subspace of E_n.

We also suppose that it has been possible to choose the matrix representation $\mathcal{M}$ and its corresponding index set $\mathcal{P}$ so that $\mathcal{M}$ is sparse with a block diagonal structure that is one of the two types illustrated in Figure 1.

Type 1

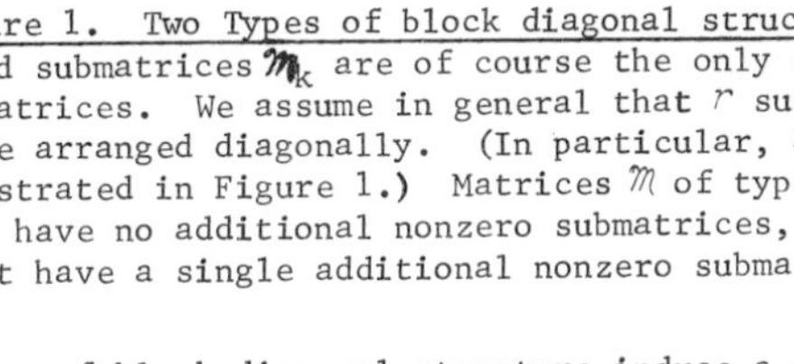

Type 2

Figure 1. Two Types of block diagonal structure for $\mathcal{M}$.

The enumerated submatrices $\mathcal{M}_k$ are of course the only submatrices of $\mathcal{M}$ that need not be zero matrices. We assume in general that r such submatrices $\mathcal{M}_1, \mathcal{M}_2, \ldots, \mathcal{M}_r$ with $r \geq 2$ are arranged diagonally. (In particular, r is 4 for both of the examples illustrated in Figure 1.) Matrices $\mathcal{M}$ of type 1 can then be characterized as those that have no additional nonzero submatrices, and matrices $\mathcal{M}$ of type 2 are those that have a single additional nonzero submatrix $\mathcal{M}_0$ consisting of entire columns of $\mathcal{M}$.

Both types of block diagonal structure induce a partitioning of the rows of $\mathcal{M}$ and hence the components of x in such a way that

$$x = (x^1, x^2, \ldots, x^r),$$

where the components of the vector variable x^k are enumerated exactly the same as those rows of $\mathcal{M}$ that contain rows of the submatrix $\mathcal{M}_k$.

Similarly, each of the two types of block diagonal structure induces a partitioning of the columns of $\mathcal{M}$ and hence the components of z in such a way that

$$z = (z^1, z^2, \ldots, z^r) \text{ for problems of type 1}$$

while

$$z = (z^0, z^1, z^2, \ldots, z^r) \text{ for problems of type 2,}$$

where the components of the vector variable z^k are enumerated exactly the same as those columns of $\mathcal{M}$ that contain columns of submatrix $\mathcal{M}_k$. The components of the additional vector variable z^0 in a given problem of type 2 are of course the "coupling variables" that must be contended with in reducing such a problem to an equivalent problem of type 1.

To render both problem types amenable to decomposition, we further suppose that the function $g:\mathcal{C}$ is at least partially separable and that its partial separability is compatible with the preceding partitioning of the components of x. In particular then, we assume that there are functions $g_k:\mathcal{C}_k$, $k = 1,2,\ldots,r$, such that

$$\mathcal{C} = \mathop{\times}_{k=1}^{r} \mathcal{C}_k \qquad \text{and} \qquad g(x) = \sum_{k=1}^{r} g_k(x^k).$$

This assumption is of course automatically satisfied when $g:\mathcal{C}$ is completely separable, a condition that holds for examples 1 and 2 in secton 2.

It is important to note that the preceding assumptions about problem $\mathcal{a}(0)$ are inherited by all problems $\mathcal{a}(u)$ in the geometric programming family $\mathcal{a}$. In particular, the cone $\mathfrak{X}$ and hence the block diagonal structure of its matrix representation $\mathcal{M}$ remain invariant of u; and the separability of the function $g:\mathcal{C}$ is clearly inherited by all functions $g(\,\cdot + u) : (\mathcal{C} - u)$. Consequently, the decomposition principles to be developed in the context of treating problem $\mathcal{a}(0)$ are just as applicable to all other problems $\mathcal{a}(u)$ in the family $\mathcal{a}$.

In treating such problems $\mathcal{a}(u)$ we shall need to partition the components of u in the same way that the components of x have been partitioned, namely,

$$u = (u^1, u^2, \ldots, u^r),$$

where the components of the vector variable u^k are enumerated exactly the same as those rows of $\mathcal{M}$ that contain rows of the submatrix $\mathcal{M}_k$.

4. DECOMPOSITION BY TEARING

The decomposition principles to be developed here utilize the cones

$$\mathfrak{X}_k \triangleq \{x^k \in E_{n_k} \mid x^k = \mathcal{M}_k z^k \text{ for at least one } z^k \in E_{m_k} \text{ for which } z_j^k \geq 0,\ j \in \mathcal{P}\}.$$

There are of course r such cones $\mathfrak{X}_k$, $k = 1,2,\ldots,r$ for a problem of type 1, and $r + 1$ such cones $\mathfrak{X}_k$, $k = 0,1,2,\ldots,r$ for a problem of type 2. Only the extra (coupling) cone $\mathfrak{X}_0$ for a problem of type 2 is a subcone of $\mathfrak{X}$.

Problems of type 1. We begin by observing that the cone $\mathfrak{X}$ is separable, in that

$$x \in \mathfrak{X} \text{ if and only if } x^k \in \mathfrak{X}_k,\ k = 1,2,\ldots,r.$$

This separation of the cone $\mathfrak{X}$ into the direct sum of the cones $\mathfrak{X}_k$, $k = 1,2,\ldots,r$, and the separation of the function $g:\mathcal{C}$ into a sum of the functions $g_k:\mathcal{C}_k$, $k = 1,2,\ldots,r$, immediately imply that problem $\mathcal{a}(0)$, now designated problem $\mathcal{a}^1(0)$, can be solved by solving the smaller geometric programming problems $\mathcal{a}_k(0)$ that are constructed from the respective functions $g_k:\mathcal{C}_k$ and the respective cones $\mathfrak{X}_k$, $k = 1,2,\ldots r$. In particular, the (desired) infimum $\varphi^1(0)$ for problem $\mathcal{A}^1(0)$ can clearly be determined from the infima $\varphi_k(0)$ for the respective problems $\mathcal{A}_k(0)$, $k = 1,2,\ldots,r$ by the formula

$$\varphi^1(0) = \sum_{k=1}^{r} \varphi_k(0).$$

Moreover, the (desired) optimal solution set $\mathcal{S}^{1*}(0)$ for problem $\mathcal{A}^1(0)$ can obviously be determined from the optimal solution sets $\mathcal{S}_k^*(0)$ for the respective problems $\mathcal{A}_k(0)$, $k = 1,2,\ldots, r$ by the formula

$$\mathcal{S}^{1*}(0) = \mathop{\times}_{k=1}^{r} \mathcal{S}_k^*(0).$$

This direct decomposition of problem $\mathcal{A}^1(0)$ into r smaller problems $\mathcal{A}_k(0)$, $k = 1,2,\ldots,r$ generally increases computational efficiency and can in fact be a necessity when $\mathcal{A}^1(0)$ is itself too large for computer storage.

Using the previously noted fact that the same decomposition principle can be applied to each problem $\mathcal{A}(u)$ in a given family $\mathcal{A}$, we infer that the (desired) functions $\varphi^1: \mathcal{U}^1$ and $\mathcal{S}^{1*}:\mathcal{U}^1$ are determined by the corresponding functions $\varphi_k: \mathcal{U}_k$ and $\mathcal{S}_k^*: \mathcal{U}_k$ that are associated with the geometric programming families $\mathcal{A}_k$ (that are of course constructed from the respective functions $g_k:\mathcal{C}_k$ and the respective cones $\mathcal{X}_k$, $k = 1,2,\ldots, r$). This direct decomposition of the family $\mathcal{A}^1$ into r smaller families $\mathcal{A}_k$, $k = 1,2,\ldots,r$ can be concisely described by the formulas

$$\mathcal{U}^1 = \mathop{\times}_{k=1}^{r} \mathcal{U}_k,$$

$$\varphi^1(u) = \sum_{k=1}^{r} \varphi_k(u^k),$$

and

$$\mathcal{S}^{1*}(u) = \mathop{\times}_{k=1}^{r} \mathcal{S}_k^*(u^k),$$

where

$$\mathcal{U}_k = \mathcal{C}_k - \mathcal{X}_k, \qquad k = 1,2,\ldots, r.$$

Consequently, the functions $\varphi^1:\mathcal{U}^1$ and $\mathcal{S}^{1*}:\mathcal{U}^1$ can be studied by studying the functions $\varphi_k:\mathcal{U}_k$ and $\mathcal{S}_k^*:\mathcal{U}_k$, $k = 1,2,\ldots, r$.

In particular, given that the (one-sided) directional derivative $D_{d^k}\varphi_k(u^k)$ of the function $\varphi_k:\mathcal{U}_k$ (at a point $u^k \in \mathcal{U}_k$ in the direction $d^k \in E_{n_k}$) exists for $k = 1,2,\ldots, r$, and given that both $u = (u^1,u^2,\ldots,u^r)$ and $d = (d^1,d^2,\ldots,d^r)$, the (one-sided) directional derivative $D_d\varphi^1(u)$ of the function $\varphi^1:\mathcal{U}^1$ (at the point $u \in \mathcal{U}^1$ in the direction $d \in E_n$) clearly exists and is given by the formula

$$D_d\, \varphi^1(u) = \sum_{k=1}^{r} D_{d^k}\, \varphi_k(u^k).$$

It is only on rare occasions that the functions $\varphi_k:\mathcal{U}_k$, $k = 1,2,\ldots,r$ can be obtained in terms of elementary formulas. Consequently, the directional derivatives $D_{d^k}\varphi_k(u^k)$, $k = 1,2,\ldots,r$ usually have to be determined by numerical differentiation or other numerical methods.

If a given function $\varphi_k:\mathcal{U}_k$ is convex, then the "convex analysis" of Fenchel (1949, 1951) and Rockafellar (1970) shows that for almost every $u^k \in \mathcal{U}_k$ the directional derivative $D_{d^k}\varphi_k(u^k)$ exists for all the "feasible directions" $d^k \in E_{n_k}$ (i.e. all the directions $d^k \in E_{n_k}$ for which $u^k + sd^k \in \mathcal{U}_k$ for sufficiently small $s > 0$). On the other hand, the paper by Peterson (1970) shows that $\varphi_k:\mathcal{U}_k$ is in fact convex if the function $g_k:\mathcal{C}_k$ is assumed to be convex, an assumption that seems much less restrictive when viewed in the light of some recent work of Duffin and Peterson (1973, 1972). In any event, the paper by Peterson (1970) also shows that this assumption and other relatively mild assumptions actually guarantee that

$$D_{d^k}\varphi_k(u^k) = \max_{y^k \in \mathcal{T}_k^*(0,u^k)} \langle d^k, y^k\rangle,$$

where $\mathcal{T}_k^*(0;u^k)$ is the optimal solution set for the "geometric dual" $\mathcal{B}_k(0;u^k)$ of problem $\mathcal{A}_k(u^k)$. Since the paper by Peterson (1970) also provides "extremality conditions" that frequently lead to a relatively simple computation of $\mathcal{T}_k^*(0;u^k)$ from the knowledge of only a single optimal solution $x^{k*} \in \mathcal{S}_k^*(u^k)$, and since $\mathcal{T}_k^*(0;u^k)$ is frequently polyhedral (and sometimes a singleton), the directional derivatives $D_d\varphi^1(u)$ can often be determined by the preceding two displayed formulas by little more than linear programming (and sometimes much less). This means of course that first-order methods can often be used in conjunction with the preceding decomposition principle to minimize $\varphi^1(u)$ over a given subset of $\mathcal{U}^1$ - a fundamental technique to be employed in developing a decomposition principle for problems of type 2.

<u>Problems of type 2.</u> We begin by observing that a given problem $\mathcal{A}(0)$ of type 2, now designated problem $\mathcal{A}^2(0)$, reduces to a problem $\mathcal{A}^1(u)$ of type 1 when the coupling vector variable z^0 is (temporarily) fixed and u is chosen to be $\mathcal{M}_0 z^0$. Using very elementary arguments, we then infer that the (desired) infimum $\varphi^2(0)$ for problem $\mathcal{A}^2(0)$ can in fact be determined by the formula

$$\varphi^2(0) = \inf_{u \in \mathcal{X}_0 \cap \mathcal{U}^1} \varphi^1(u).$$

The minimization problem that appears in this formula is obviously a geometric programming problem and is termed the <u>master problem</u>. Its objective function $\varphi^1 : \mathcal{U}^1$ has of course already been (partially) separated into a sum of infima functions $\varphi_k : \mathcal{U}_k$, $k = 1,2,\ldots,r$ by the decomposition principle previously developed for problems of type 1. That decomposition principle is of course to be used here not only to calculate the functional values $\varphi^1(u)$ but also to calculate any directional derivatives $D_d\varphi^1(u)$ that are needed to implement an appropriate algorithm for solving the master problem. Once the master problem's optimal solution set

$$\mathcal{U}^* \triangleq \{u \in \mathcal{X}_0 \cap \mathcal{U}^1 \mid \varphi^1(u) = \varphi^2(0)\}$$

has been obtained, the (desired) optimal solution set $\mathcal{S}^{2*}(0)$ for problem $\mathcal{A}^2(0)$ can clearly be determined by the formula

$$\mathcal{S}^{2*}(0) = \bigcup_{u^* \in \mathcal{U}^*} \{u^* + \mathop{\times}_{k=1}^{r} \mathcal{S}_k^*(u^{*k})\}.$$

In the process of determining a $u^* \in \mathcal{U}^*$ with the aid of this tearing procedure, one can of course expect to determine an $x^{k*} \in \mathcal{S}_k^*(u^{*k})$, $k = 1,2,\ldots,r$; in which event $u^* + (x^{1*}, x^{2*}, \ldots, x^{r*})$ is one of the desired optimal solutions to problem $\mathcal{A}^2(0)$. This assumes of course that such optimal solutions exist.

In summary, problem $\mathcal{A}^2(0)$ has been torn into r smaller problems $\mathcal{A}_k(u^k)$, $k = 1,2,\ldots,r$ that are judiciously selected by the master problem, with the possible help of geometric programming duality in the convex case. This tearing <u>may</u> increase computational efficiency but can in fact be a necessity when $\mathcal{A}^2(0)$ is itself too large for computer storage.

As previously noted, the same decomposition principle can be applied to each problem $\mathcal{A}(u)$ in a given family $\mathcal{A}$. The obvious result of such an application here is notationally somewhat cumbersome to describe and is left to the imagination of the interested reader.

Although developed in the context of problems without explicit constraints, the preceding decomposition principles apply equally well to problems with explicit constraints - by virtue of the relations to be given in the next section.

5. PROBLEMS WITH EXPLICIT CONSTRAINTS

To keep track of the explicit constraints, we introduce two nonintersecting (possibly empty) positive-integer index sets I and J with finite cardinality o(I) and o(J) respectively. We also introduce the following notation and hypotheses:

(1) For each $k \in \{0\} \cup I \cup J$ suppose that g_k is a function with a nonempty

domain $C_k \subseteq E_{n_k}$, and for each $j \in J$ suppose that D_j is a nonempty subset of E_{n_j}.

(2) For each $k \in \{0\} \cup I \cup J$ let u^k be an independent vector parameter in E_{n_k}, and let μ be an independent vector parameter with components μ_i for each $i \in I$.

(3) Denote the cartesian product of the vector parameters u^i, $i \in I$, by the symbol u^I, and denote the cartesian product of the vector parameters u^j, $j \in J$, by the symbol u^J. Then the cartesian product $u \triangleq (u^0, u^I, u^J)$ of the vector parameters, u^0, u^I, and u^J is an independent vector parameter in E_n, where

$$n \triangleq n_0 + \sum_I n_i + \sum_J n_j \; .$$

(4) For each $k \in \{0\} \cup I \cup J$ let x^k be an independent vector variable in E_{n_k}, and let $\varkappa$ be an independent vector variable with components $\varkappa_j$ for each $j \in J$.

(5) Denote the cartesian product of the vector variables x^i, $i \in I$, by the symbol x^I, and denote the cartesian product of the vector variables x^j, $j \in J$, by the symbol x^J. Then the cartesian product $x \triangleq (x^0, x^I, x^J)$ of the vector variables x^0, x^I, and x^J is an independent vector variable in E_n.

(6) Suppose that X is a cone in E_n.

Now, consider the following <u>geometric programming family</u> A of <u>geometric programming problems</u> $A(u,\mu)$ that were first investigated by Peterson (1973, 1974).

PROBLEM $A(u,\mu)$. <u>Consider the objective function</u> $G(\cdot + u, x)$ <u>whose domain</u>

$$C(u) \triangleq \{(x,\varkappa) \mid x^k + u^k \in C_k,\ k \in \{0\} \cup I,\ \underline{\text{and}}\ (x^j + u^j, \varkappa_j) \in C_j^+,\ j \in J\},$$

<u>and whose functional value</u>

$$G(x + u, \varkappa) \triangleq g_0(x^0 + u^0) + \sum_J g_j^+(x^j + u^j, \varkappa_j),$$

<u>where</u>

$$C_j^+ \triangleq \{(z^j, \varkappa_j) \mid \underline{\text{either}}\ \varkappa_j = 0\ \underline{\text{and}} \sup_{d^j \in D_j} \langle z^j, d^j \rangle < +\infty,\ \underline{\text{or}}\ \varkappa_j > 0\ \underline{\text{and}}\ z^j \in \varkappa_j C_j\}$$

<u>and</u>

$$g_j^+(z^j, \varkappa^j) \triangleq \begin{cases} \sup\limits_{d^j \in D_j} \langle z^j, d^j \rangle & \underline{\text{if}}\ \varkappa_j = 0\ \underline{\text{and}} \sup\limits_{d^j \in D_j} \langle z^j, d^j \rangle < +\infty \\ \varkappa_j g_j(z^j/\varkappa_j) & \underline{\text{if}}\ \varkappa_j > 0\ \underline{\text{and}}\ z^j \in \varkappa_j C_j . \end{cases}$$

<u>Using the feasible solution set</u>

$$S(u,\mu) \triangleq \{(x,\varkappa) \in C(u) \mid x \in X,\ \underline{\text{and}}\ g_i(x^i + u^i) + \mu_i \le 0,\ i \in I\},$$

<u>calculate both the problem infimum</u>

$$\varphi(u,\mu) \triangleq \inf_{(x,\varkappa) \in S(u,\mu)} G(x + u, \varkappa)$$

<u>and the optimal solution set</u>

$$S^*(u,\mu) \triangleq \{(x,\varkappa) \in S(u,\mu) \mid G(x + u, \varkappa) = \varphi(u,\mu)\} .$$

In defining the feasible solution set $S(u,\mu)$, it is important to make a sharp distinction between the cone condition $x \in X$ and the constraints $g_i(x^i + u^i) + \mu_i \le 0$, $i \in I$, both of which restrict the vector variable $(x,\varkappa)$. The cone X is frequently finitely generated, and hence the cone condition can frequently be eliminated by a linear transformation of the vector variable x; but the (generally nonlinear) constraints usually can not be eliminated by even a nonlinear transformation. Nevertheless, we never explicitly eliminate the cone condition because such a linear transformation would introduce a common vector variable into the arguments of g_0, g_i, and g_j^+. Such a common vector variable would of course only tend to camouflage the separability that is built into problem $A(u,\mu)$.

Analogous to the unconstrained case introduced in section 2, we shall find it useful to interpret problem $A(u,\mu)$ as a perturbed version of $A(0,0)$, so we term the set

$$U \triangleq \{(u,\mu) \mid \text{problem } A(u,\mu) \text{ is consistent}\}$$

the feasible perturbation set for problem $A(0,0)$. It is important to note that there are no perturbations associated with the variables $\varkappa_j$; the reason is that such perturbations would clearly have no effect. Note also that the special perturbations μ_i appear only with the constraints and hence have no counterpart in the unconstrained case. The unconstrained case can of course be obtained from the present constrained case by simply letting both index sets I and J be the empty set.

The extreme flexibility of the present geometric programming formulation is clearly illustrated by the following examples of important problem classes.

Example 0. Linear programming can be viewed as a special case of geometric programming in various ways. Probably the most direct way is to let J be the empty set and choose the functions $g_k : C_k$, $k \in \{0\} \cup I$ by letting

$$C_0 \triangleq E_1 \text{ with } g_0(x^0) \triangleq x^0,$$

and

$$C_i \triangleq E_1 \text{ with } g_i(x^i) \triangleq x^i - b_i,\ i \in I,$$

where the numbers b_i are components of a given vector b. In addition, choose the cone X by letting

$$X \triangleq \{(x^0, x^I) \in E_n \mid x^0 = \langle a,z\rangle \text{ and } x^I = Mz \text{ for at least one } z \in E_m \text{ for which } z_j \geq 0,\ j \in P\},$$

where a is a given vector, M is a given matrix, and P is a given subset of $\{1,2,\ldots,m\}$.

Problem $A(0,0)$ then clearly consists of minimizing the linear function $\langle a,z\rangle$ subject to both the linear constraints $Mz \leq b$ and the nonegativity conditions $z_j \geq 0$, $j \in P$. This is of course the most general linear programming problem, and it is worth noting that while the effect of the parameter u^0 is obvious the parameters u^i and μ_i both perturb the constraint upper bound b_i.

Other ways of viewing linear programming as a special case of geometric programming are given in the expository paper by Peterson (1973).

Example 0+. To view ordinary mathematical programming as a special case of geometric programming, let I be the set $\{1,2,\ldots,p\}$ and let J be the empty set. Then, choose the cone X to be the column space of a specially structured matrix, namely,

$$X \triangleq \text{column space of } \begin{bmatrix} \mathcal{I} \\ \mathcal{I} \\ \vdots \\ \mathcal{I} \end{bmatrix}$$

where the $m \times m$ identity matrix $\mathcal{I}$ appears in a total of $1 + p$ positions.

Problem $A(0,0)$ then clearly consists of minimizing $g_0(z)$ subject to both the constraints $g_i(z) \leq 0$, $i = 1,2,\ldots,p$ and the domain condition $z \in \bigcap_{k=0}^{p} C_k$. It is worth noting that the vector parameter u^k perturbs the domain C_k by translation while the parameter μ_k perturbs the constraint upper bound 0. The subfamily of A that results from choosing the vector parameters $u^k = 0$ has been termed the "ordinary mathematical programming" family by Rockafellar (1970), because its problem $A(0,0)$ and its vector parameter μ have been the focus of most of the past work in mathematical programming.

Actually, there may be advantages in reformulating the ordinary mathematical programming problem as the following geometric programming problem:

$$\text{Minimize } g_0(x^0) \text{ subject to}$$

$$g_1(x^1) \le 0, \quad g_2(x^2) \le 0, \quad \ldots, \quad g_p(x^p) \le 0$$

$$x^0 - x^1 = 0$$

$$x^1 - x^2 = 0$$

$$\cdots \quad x^{p-1} - x^p = 0$$

$$x^0 \in C_0, \quad x^1 \in C_1, \quad \ldots, \quad x^p \in C_p.$$

In this geometric programming formulation the ordinary programming problem is at least partially separable.

In any event, most important optimization problems have much more structure than that which is present in the ordinary programming problem. Moreover, that extra structure can frequently be exploited by choosing the cone X differently from the way it has been chosen in this example. This fact is clearly illustrated by the following extensions of examples 1 and 2 from section 2.

Example 1. To treat the minimization of signomials subject to signomial inequality constraints, let I be the set $\{1,2,\ldots,p\}$ and let J be the empty set. Then, choose the functions $g_k{:}C_k$, $k = 0,1,2,\ldots,p$ by letting

$$C_k \triangleq E_{n_k} \text{ with } g_k(x^k) \triangleq \sum_{[k]} c_i e^{x_i} - b_k,$$

where the numbers c_i and b_k are given constants and the index set

$$[k] \triangleq \{m_k, m_k + 1, \ldots, n_k\},$$

with the understanding that

$$1 \triangleq m_0 \le n_0, \quad n_0 + 1 \triangleq m_1 \le n_1, \quad \ldots, \quad n_{p-1} + 1 \triangleq m_p \le n_p = n.$$

Finally, choose the cone X by letting

$$X \triangleq \text{column space of } (a_{ij}),$$

where (a_{ij}) is an arbitrary $n \times m$ real matrix.

Making the change of variables

$$x_i = \sum_{j=1}^{m} a_{ij} \log t_j \text{ where } t_j > 0,\ j = 1,2,\ldots,m \text{ and } i = 1,2,\ldots,n,$$

we easily infer that problem A(0,0) consists essentially of minimizing the signomial $\sum_{[0]} c_i \prod_{j=1}^{m} t_j^{a_{ij}}$ subject to the signomial inequality constraints $\sum_{[k]} c_i \prod_{j=1}^{m} t_j^{a_{ij}} \le b_k$, $k = 1,2,\ldots,p$ (and of course the conditions $t_j > 0$, $j = 1,2,\ldots,m$). It is worth noting that the parameter u_i perturbs the coefficient c_i (actually, the logarithm of the $|c_i|$) while the parameter μ_k perturbs the constraint upper bound b_k.

Of course, the main reason for studying the preceding geometric programming formulation of signomial optimization is the complete separability that is introduced. For a more thorough discussion of this topic see the recent papers by Duffin and Peterson (1973, 1972), and the references cited therein. Parts of those papers show that all "algebraic programming" problems can actually be reduced to the "posynomial" case in which all coefficients c_i are positive and each posynomial has at most two terms. To avoid possible confusion, it should also be mentioned that in those papers, as well as the predecessors by Duffin and Peterson (1966) and Duffin, Peterson, and Zener (1967), it has been advantageous to replace the separable functions $\sum_{[k]} c_i e^{x_i}$ with the (technically) inseparable functions $\log(\sum_{[k]} c_i e^{x_i})$ for the study of geometric programming duality.

Example 2. To treat both quadratic programming with quadratic constraints and ℓ_p - regression problems with ℓ_p - norm constraints, let I be the set $\{1,2,\dots,r\}$ and let J be the empty set. Then, choose the functions $g_k : C_k$, $k = 0,1,2,\dots r$ by letting

$$C_k \triangleq E_{n_k} \text{ with } g_k(x^k) \triangleq \sum_{[k]^+} \frac{1}{p_i} | x_i - b_i |^{p_i} - \sum_{[k]^-} \frac{1}{p_i} | x_i - b_i |^{p_i} + x_{]k[} - b_{]k[},$$

where the numbers b_i, $b_{]k[}$, and p_i are given constants and the index sets are

$$[k] \triangleq \{m_k^+, m_k^+ + 1,\dots,n_k^+\} ; [k]^- \triangleq \{m_k^-, m_k^- + 1,\dots,n_k^-\} ; \text{ and }]k[\triangleq n_k,$$

with the understanding that

$$1 \triangleq m_0^+ \le n_0^+, \quad n_0^+ + 1 \triangleq m_0^- \le n_0^-, \quad n_0^- + 1 \triangleq n_0, \quad n_0 + 1 \triangleq m_1^+ \le n_1^+,\dots,n_r^- + 1 \triangleq n.$$

Finally, choose the cone X by letting

$$X \triangleq \text{column space of } \begin{bmatrix} D_0 \\ \mathcal{D}_0 \\ h_0 \\ \cdot \\ \cdot \\ \cdot \\ D_r \\ \mathcal{D}_r \\ h_r \end{bmatrix}$$

where D_k is an $(n_k^+ - n_{k-1}) \times m$ real matrix, $\mathcal{D}_k$ is an $(n_k^- - n_k^+) \times m$ real matrix, and h_k is a $1 \times m$ real matrix.

Making the usual change of variables

$$x = Mz$$

where M denotes the preceding partitioned matrix, we obtain quadratic programming and ℓ_p - regression analysis with the following two specializations.

To obtain all quadratic programming problems, let $b_i = 0$ and $p_i = 2$ for $i \in [k]^+ \cup [k]^-$. Then, problem A(0,0) consists essentially of minimizing the quadratic function $(1/2) < z, H_0 z > + < h_0, z > - b_{]0[}$ subject to the quadratic constraints $(1/2) < z, H_k z > + < h_k, z > \le b_{]k[}$, $k = 1,2,\dots,r$, where $H_k \triangleq D_k^t D_k - \mathcal{D}_k^t \mathcal{D}_k$. Although the effect of the parameter u_i is rather uninteresting and complicated, the parameter μ_k perturbs the constraint upper bound $b_{]k[}$.

To obtain all ℓ_p - regression problems, take $p_i = p^k$ for $i \in [k]^+ \cup [k]^-$; let $b_{]0[} = 0$ and $b_i = 0$ for $i \in [k]^-$; and choose $\mathcal{D}_k = 0$ and $h_k = 0$. Then, problem A(0,0) consists essentially of minimizing the $\ell_p 0$ - norm function $(1/p^0)(\| D_0 z - b^k \|_{p^0})^{p^0}$ subject to the $\ell_p k$ - norm constraints $(1/p^k)(\| D_k z - b^k \|_{p^k})^{p^k} \le b_{]k[}$, $k = 1,2,\dots,r$, where $\| \cdot \|_{p^k}$ denotes the $\ell_p k$ - norm and $b^k \triangleq (b_{m_k^+},\dots,b_{n_k^+})$. While the vector parameter $u^k \triangleq (u_{m_k^+},\dots,u_{n_k^+})$ perturbs the vector b^k, both the parameters $u_{]k[}$ and μ_k perturb the constraint upper bound $b_{]k[}$.

Of course, the main reason for studying the preceding geometric programming formulation of both quadratic programming and ℓ_p - regression analysis is the complete separability that is introduced. For a more thorough discussion of this topic in the convex case (in which $\mathcal{D}_k = 0$ for $k = 0,1,2,\dots,r$) see the papers by Peterson and Ecker (1968, 1970, 1969, 1970).

Other important examples that can be effectively treated by geometric programming (but have usually been studied without explicitly mentioning the subject) are discussed in the recent expository paper by Peterson (1973), and the references cited therein. In one of those examples, the "generalized chemical equilibrium problem", the index set J is not the empty set. That example and an alternative way of viewing linear programming as a special case of geometric programming are at this time the only practical justifications for including the index set J (and hence its resulting complications) in the present formulation of geometric programming. However, "geometric programming duality" as developed by Peterson (1973, 1974) provides an aesthetic justification for doing so.

We now show how geometric programming with explicit constraints can be viewed as a special case of geometric programming without explicit constraints. In doing so, we see how to apply the decomposition principles developed in section 4 to the problems discussed in this section.

Introducing an additional independent vector variable α with components α_i for each $i \in I$, we let the functional domain

$$\mathcal{C} \triangleq \{(x^0, x^I, \alpha, x^J, \kappa) \in E_\eta \mid x^0 \in C_0;\ x^i \in C_i, \alpha_i \in E_1, \text{ and } g_i(x^i) + \alpha_i \leq 0,\ i \in I;\ (x^j, \kappa_j) \in C_j^+,\ j \in J\};$$

and we let the functional value

$$\mathcal{G}(x^0, x^I, \alpha, x^J, \kappa) \triangleq g_0(x^0) + \sum_J g_j^+(x^j, \kappa_j) \triangleq G(x, \kappa).$$

We also let the cone

$$\mathfrak{X} \triangleq \{(x^0, x^I, \alpha, x^J, \kappa) \in E_\eta \mid (x^0, x^I, x^J) \in X;\ \alpha = 0;\ \kappa \in E_{o(J)}\}.$$

Then, problem A(0,0) is clearly identical to problem $\mathcal{A}(0)$. In fact, it is easy to see that problem $A(u,\mu)$ is identical to problem $\mathcal{A}(u)$, with the parameter μ_i being identified with the parameter u_i that corresponds to α_i. Of course, the parameter u_j that corresponds to κ_j has no effect and can be set equal to zero.

It is important to realize that the components of $x = (x^0, x^I, \alpha, x^J, \kappa)$ can be placed in any order to achieve a block diagonal structure for some matrix representation $\mathcal{M}$ of $\mathfrak{X}$. Moreover, it is clear from the definition of $\mathfrak{X}$ that the possibility of achieving such a block diagonal structure for some $\mathcal{M}$ depends entirely on the possibility of ordering the components of $x = (x^0, x^I, x^J)$ in such a way that a block diagonal structure is achieved for some matrix representation M of X.

It is equally important to realize that the function $\mathcal{G}:\mathcal{C}$ inherits any separability that is present in the functions $g_0:C_0$ and $g_j^+:C_j^+$. Unfortunately, $\mathcal{G}:\mathcal{C}$ clearly does not generally inherit any of the separability that is present in the functions $g_i:C_i$; unless there are corresponding Kuhn-Tucker (Lagrange) multipliers λ_i, in which case the constraints $g_i(x^i) + \alpha_i \leq 0$, $i \in I$ in the defining equation for $\mathcal{C}$ can be deleted, provided that the expression $\sum_I \lambda_i [g_i(x^i) + \alpha_i]$ is added to the defining equation for $\mathcal{G}(x^0, x^I, \alpha, x^J, \kappa)$. If desired, this manipulation can of course be limited to only some of the constraints for which there are corresponding Kuhn-Tucker multipliers.

Other ramifications of the preceding relation between constrained and unconstrained geometric programming are given in the recent work of Peterson (1974).

6. CONCLUDING REMARKS

Decomposition principles can of course be developed for geometric programming problems $\mathcal{A}(0)$ that possess matrix representations $\mathcal{M}$ of $\mathfrak{X}$ with block structures that are not one of the two types treated here. In fact, a future paper will be devoted to developing decomposition principles for those problems $\mathcal{A}(0)$ for which $\mathcal{M}$ has one of the two additional types of block diagonal structure illustrated in Figure 2.

Type 3

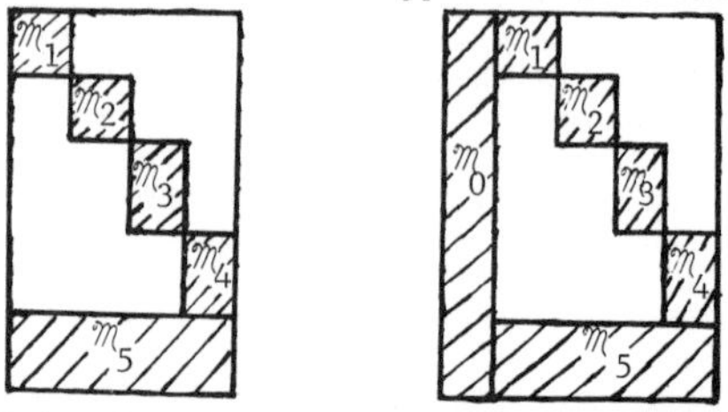

Type 4

Figure 2. Two additional types of block diagonal structure for $\mathcal{M}$.

Geometric programming duality plays a much more fundamental role in the treatment of problems $\mathcal{A}(0)$ of type 3 than it has played here in the treatment of problems $\mathcal{A}(0)$ of types 1 and 2. Problems $\mathcal{A}(0)$ of type 4 are of course a combination of problems $\mathcal{A}(0)$ of types 2 and 3. In fact, problems $\mathcal{A}(0)$ of type 4 can be treated by combining the decomposition principles that have been developed here for problems $\mathcal{A}(0)$ of type 2 with those to be developed elsewhere for problems $\mathcal{A}(0)$ of type 3. The order in which the principles are to be combined is indicated by the partitioning of the matrix $\mathcal{M}$ between the submatrices $\mathcal{M}_0$ and $\mathcal{M}_5$ in illustration **2** of Figure 2.

REFERENCES

Dantzig, G.B., (1970), Large scale systems and the computer revolution, in Proc. Princeton Symp. Math. Prog., (1967) ed. H.W. Kuhn, Princeton University Press, p. 51.

Duffin, R.J., (1962), Dual programs and minimum cost, SIAM Jour. Appl. Math., 10, p. 119.

Duffin, R.J., (1962), Cost minimization problems treated by geometric means, ORSA Jour., 10, p. 668.

Duffin, R.J., and E.L. Peterson, (1966), Duality theory for geometric programming, SIAM Jour. Appl. Math., 14, p. 1307.

Duffin, R.J., and E.L. Peterson, (1973), Geometric programming with signomials, Jour. Opt. Th. Appls., 11, in press.

Duffin, R.J., and E.L. Peterson, (1972), Reversed geometric programs treated by harmonic means, Indiana University Math. Jour., to appear.

Duffin, R.J., and E.L. Peterson, (1972), The proximity of (algebraic) geometric programming to linear programming, Math. Prog., to appear.

Duffin, R.J., E.L. Peterson, and C. Zener, (1967), Geometric Programming, Wiley, New York.

Ecker, J.G., (1972), Decomposition in separable geometric programming, Jour. Opt. Th. Appls., 9, p. 176.

Fenchel, W., (1949), On conjugate convex functions, Canadian Jour. Math., 1, p. 73.

Fenchel, W., (1951), Convex Cones, Sets, and Functions, Math. Dept. mineographed lecture notes, Princeton University.

Gass, S.I., (1966), The dualplex method for large-scale linear programs, Operations Research Center, 1966-15, University of California, Berkeley.

Heymann, M., and M. Avriel, (1969), On a decomposition for a special class of geometric programming problems, Jour. Opt. Th. Appls., 3, p.

Kron, G., (1953), A set of principles to interconnect the solutions of physical systems, Jour. Appl. Phys., 24, p. 965.

Kron, G., (1963), Diakoptics: the Piecewise Solution of Large-Scale Systems, MacDonald, London.

Peterson, E.L., (1970), Symmetric duality for generalized unconstrained geometric programming, SIAM Jour. Appl. Math., 19, p. 487.

Peterson, E.L., (1973), Geometric programming and some of its extensions, in Perspectives on Optimization, ed. M. Avriel, M. Rijckaert, and D. Wilde, Prentice-Hall, to appear.

Peterson, E.L., (1974), Generalization and symmetrization of duality in geometric programming, SIAM Jour. Appl. Math., to be submitted.
Peterson, E.L., and J.G. Ecker, (1968), Geometric programming: a unified duality theory for quadratically constrained quadratic programs and ℓ_p - constrained ℓ_p - approximation problems, Bull. Am. Math. Soc., 74, p. 316.
Peterson, E.L., and J.G. Ecker, (1970), Geometric programming: duality in quadratic programming and ℓ_p - approximation I, in Proc. Princeton Symp. Math. Prog. (1967), ed. H.W. Kuhn, Princeton University Press, p. 445.
Peterson, E.L., and J.G. Ecker, (1969), Geometric programming: duality in quadratic programming and ℓ_p - approximation II (canonical programs), SIAM Jour. Appl. Math., 17, p. 317.
Peterson, E.L., and J.G. Ecker, (1970), Geometric programming: duality in quadratic programming and ℓ_p - approximation III (degenerate programs), Jour. Math. Anal. Appls., 29, p. 365.
Rijckaert, M., (1973), Decomposition applied to dual geometric programming, in Proc. NATO Inst. on Decomposition (1972), ed. D.M. Himmelblau, North Holland Publ. Com., p.
Rockafellar, R.T., (1970), Convex Analysis, Princeton University Press.
Rosen, J.B., (1964), Primal partition programming for block diagonal matrices, Numerische Math., 6, p. 250.
Zener, C., (1961), A mathematical aid in optimizing engineering designs, Proc. Nat. Acad. Sci., 47, p. 537.
Zener, C., (1962), A further mathematical aid in optimizing engineering designs, Proc. Nat. Acad. Sci., 48, p. 518.
Zener, C., (1962), Minimization of system costs in terms of subsystem costs, Proc. Nat. Acad. Sci., 51, p. 162.
Zener, C., (1971), Engineering Design by Geometric Programming, Wiley, New York.

A GENERAL SYMMETRIC NONLINEAR DECOMPOSITION THEORY

by

T.O.M. Kronsjö *

University of Birmingham

Great Britain

0. Introduction

There exists to-day a great number of decomposition schemes. Considerable interest is attached to the possibility of viewing them as special cases of a general decomposition theory.

This general decomposition theory may be derived in the following way:

A pair of symmetric primal and dual programmes is chosen as the starting point.

The primal problem is treated as a two-stage problem, which may be replaced by a dual master problem and a dual subproblem.

The dual problem is treated as a two-stage problem,which may be replaced by a primal master problem and a primal subproblem.

A general primal and dual decomposition theory is obtained by letting the dual master serve as a primal subproblem to the primal master, and the primal master serves as a dual subproblem to the dual master. The result is an entirely symmetric treatment of master and subproblems, yielding a decomposition theory based on master type problems.

The elements of a proof of convergence are indicated.

Repeated application of this theory to nonlinear programming problems of triangular structure leads to the occurrence of three types of master problems: a primal master, primal-and-dual master(s) and a dual master.

A number of specializing assumptions may then be made with respect to
1) functional form of the original problem;
2) special structure of the original problem;
3) number of generated variables and generated constraints used by primal, dual and primal-and-dual master problem(s);
4) approximation method of a nonlinear function;
5) algorithmic method of solving the master problem and generating subproblem solutions.

An economic interpretation of each major aspect of the theory is briefly indicated at the appropriate stage.

---

* The author wishes to express his gratitude to C.L. Sandblom for critical scrutiny and suggestions for possible improvement.

1. The Problem Formulation

G.B. Dantzig, E. Eisenberg and R.W. Cottle (1965) and A. Whinston (1967) have considered the saddlefunction $K(x,y)$, where K is convex in x for each y, concave in y for each x. An exposition is given in K.P. Wong (1970).

This function may be economically interpreted to be a total cost function consisting of external costs, e.g. for labour, and internal costs, e.g. for resources. The exact mathematical expressions reflecting these constituent cost elements will be outlined in the following.

The above mentioned authors have defined a pair of associated primal and dual problems, viz.:

The Primal

(1) Find $(x,y) \geq 0$ and Min F such that

$$F = K(x,y) - y^T K_y(x,y),\ K_y(x,y) \leq 0$$

The Dual

(2) Find $(x,y) \geq 0$ and Max G such that

$$G = K(x,y) - x^T K_x(x,y),\ K_x(x,y) \geq 0$$

where T denotes transposition, $K_y(x,y)$ and $K_x(x,y)$ denote vectors of partial derivatives.

The primal and dual problems may be given an economic interpretation as follows:

x represents a vector of activity levels, y a vector of prices or marginal costs of a vector of resource constraints $K_y(x,y) \leq 0$, $K(x,y) - y^T K_y(x,y)$ a labour cost function.

It follows that total cost, i.e. labour cost and resource cost, is $K(x,y) - y^T K_y(x,y) + y^T K_y(x,y) = K(x,y)$. The vector of marginal labour and resource cost of a vector of activities x is therefore $K_x(x,y)$. The vector of constraints $K_x(x,y) \geq 0$ requires that activities are run at no or positive marginal cost, $K(x,y) - x^T K_x(x,y)$ represents the total income of labour and resource holders less a deduction for inefficiency amounting to the marginal loss of each activity times the activity level.

The primal represents an economy in which society or entrepreneurs own all material resources and minimize labour costs by adjusting activity levels and prices.

The dual represents an economy in which society or entrepreneurs operate a nonprofit economy and maximize the reward of labour and the value of resources less a deduction for inefficiency.

Dantzig, Eisenberg and Cottle prove that for any extremal or nonextremal solution x,y of the primal and u,v of the dual the relation $F(x,y) \geq G(u,v)$ holds and that there exists a common extremal solution x^o,y^o to both the primal and dual systems with F = G when an extremal solution x^o,y^o to the primal exists, and that K, twice differentiable, has the property at x^o,y^o that its matrix of second partials with respect to y is negative definite.

Whinston made an important extension of the symmetric duality theory by eliminating the requirement of Dantzig, Eisenberg and Cottle of negative definiteness of the matrix of second partial derivatives and replacing it by the weaker assumption of convexity in x and concavity in y.

Notational conventions

subscript	indicates	the indices of a vector of variables, odd for primal, even for dual variables;
superscript	"	the particular solution number of the vector of variables;
z	"	a row vector of variables, to be specified by one or more subscripts, e.g. $z_{14} \equiv (z_1, z_4)$ where both z_1 and z_4 are row vectors;
K	"	a saddle function of subscripted variables e.g. $K_{1324} \equiv K_{1324}(z_1,z_3,z_2,z_4)$, $K_{12} \equiv K_{12}(z_1,z_2)$;
Ki	"	the partial derivative of the saddle value function K with respect to the vector of variables z_i, Ki is a column vector. e.g. $K4^{jj}_{14} \equiv K_{z_4}(z^j_1, z^j_4)$.

The traditional notational system was discarded for use in this paper, as it leads to excessively unwieldy formulas using up to three times as many symbols.

In the continuation we shall assume K(x,y) to be of the separable structure

$$(3) \quad K(x,y) \equiv K(x_1,x_2,y_1,y_2) \equiv K(z_1,z_3,z_2,z_4) \equiv K_{1324} \equiv K_{12} + K_{32} + K_{14}$$

The linkages between the variables z_1, z_3, z_2, z_4 in the saddle value function K is most easily seen from the following figure:

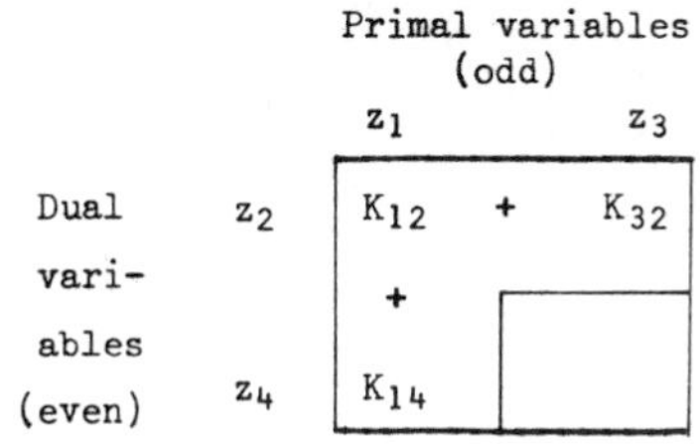

Fig. 1. The assumed structure of the saddle function K_{1324}.

The corresponding partial derivatives of K are

(4)

Partial derivative using conventional notation	Exact form using the above notation	Abbreviated form
$K_{y_1} \equiv K_{z_2}$	$K2_{12} + K2_{32}$	$K2_{132}$
$K_{y_2} \equiv K_{z_4}$	$K4_{14}$	$K4_{14}$
$K_{x_1} \equiv K_{z_1}$	$K1_{12} + K1_{14}$	$K1_{124}$
$K_{x_2} \equiv K_{z_3}$	$K3_{32}$	$K3_{32}$

Inserting (3) and (4) into (1) and (2), defining the primal to be equal to h^o and the dual equal to f^o we obtain:

The Structured Primal Problem

$$(5) \quad h^o = \underset{z_{1324}}{\mathrm{Min}}\{K_{1324} - z_2K2_{132} - z_4K4_{14} \mid K2_{132} \leq 0,\ K4_{14} \leq 0,\ z_{1324} \geq 0\}$$

The Structured Dual Problem

$$(6) \quad f^o = \underset{z_{1324}}{\mathrm{Max}}\{K_{1324} - z_1K1_{124} - z_3K3_{32} \mid K1_{124} \geq 0,\ K3_{32} \geq 0,\ z_{1324} \geq 0\}$$

where z_{1324}^{oooo} denotes an extremal solution to (5) or (6) and $h^o = f^o$ given the validity of the assumptions required by G.B. Dantzig, E. Eisenberg and R.W. Cottle or A. Whinston.

2. The Two-Stage Formulation of the Structured Primal and Dual Problems

Considering the Structured Primal Problem, we may formulate it as a Two-Stage Problem using the following reasoning. For any given z_{14} we choose such z_{32} that minimize the corresponding cost functions. The problem may now be considered to be entirely a function h of z_{14}, viz.:

$$h_{14} = -z_4K4_{14} + \underset{z_{32}}{\mathrm{Min}}\{K_{1324} - z_2K2_{132} \mid K2_{132} \leq 0,\ z_{32} \geq 0\} \mid$$
$$K4_{14} \leq 0,\ z_{14} \geq 0$$

This may economically be interpreted as a division of responsibilities between one organization which decides about activity levels z_1 and prices z_4, and another organization which for the given z_1 and z_4 has to minimize the resulting cost function and to meet the resulting requirements for goods by deciding about the activity levels z_3 and prices z_2. The overall minimum is obtained by the first organization systematically varying its choice of z_1 and z_4 until the total cost is minimized.

We select such z_{14} that minimize the function h_{14}. C.-L. Sandblom (1972) demonstrates that the resulting two-stage problem and the original problem have the same optimal solutions in common.

$$h^{o} = \underset{z_{14}}{\mathrm{Min}}\{-z_4 K4_{14} + \underbrace{\underset{z_{32}}{\mathrm{Min}}\{K_{1324} - z_2 K2_{132} \mid K2_{132} \leq 0,\ z_{32} \geq 0\}} \mid K4_{14} \leq 0,\ z_{14} \geq 0\}$$

The inner primal problem may be replaced by its dual

$$\underset{z_{32}}{\mathrm{Max}}\{K_{1324} - z_3 K3_{32} \mid K3_{32} \geq 0,\ z_{32} \geq 0\}$$

If we know feasible solutions z_{32}^{jj} $j=1,\ldots,l$ of the inner constraints, then we can enumerate the corresponding values of the objective function, viz.:

$$K_{1324}^{jj} - z_3^{j} K3_{32}^{jj} \qquad j=1,\ldots,l$$

A variable q may be required to be larger than or equal to each one of the above values

$$q \geq K_{1324}^{jj} - z_3^{j} K3_{32}^{jj} \quad j=1,\ldots,l$$

By minimizing q we may then obtain the maximum value of the above objective function values. We can then approximate the inner maximization by

$$\underset{q}{\mathrm{Min}}\{q \mid q \geq K_{1324}^{jj} - z_3^{j} K3_{32}^{jj} \quad j=1,\ldots,l\}$$

and obtain the <u>Dual Master (Primal Formulation)</u>

(1)	$h = \underset{z_{14},q}{\mathrm{Min}}\{-z_4 K4_{14} + q \mid$
(2)	$K_{1324}^{jj} - z_3^{j} K3_{32}^{jj} - q \leq 0 \quad j=1,\ldots,l$
(3)-(4)	$K4_{14} \leq 0,\ z_{14} \geq 0\}$
	with solution h^{l+1}, $z_1^{l+1}{}_4^{l+1}$, q^{l+1}; letting $l+1 \to l$
	we may denote the solution h^{l}, z_{14}^{ll}, q^{l}.

In general we shall not know all feasible solutions, which usually will not be enumerable. Therefore we require that the feasible set for the variables of a subproblem is closed. Further we require that the objective function has finite steepness over the relevant set. Given these assumptions we can enumerate feasible solutions arbitrary near to each other within the closed sets. As we increase the closeness of the feasible solutions the maximum of the corresponding objective function values will tend to a limiting value. Finally this should be reflected by replacing max by sup, which is here omitted in order to keep the exposition brief.

The Dual Master is named the Full Dual Master if it contains a finite or an infinite sequence of constraints corresponding to solution points which come arbitrarily close to any arbitrary point, and a Restricted Dual Master if it only contains a subset.

If we do not know the aforementioned finite or infinite sequence z_{32}^{jj} we may determine the one limiting q by solving the <u>Dual Subproblem</u>

(5)-(7)

$$dh = \underset{z_{32}}{\mathrm{Max}}\{K^{1}_{1\,32\,4}{}^{1} - z_3 K3_{32} - q^{1} | K3_{32} \geq 0,\ z_{32} \geq 0\}$$

with solution dh^{11}, z^{11}_{32}.

The Dual Master provides an under- or correct estimation of the true cost function of the two-stage formulation of the Structured Primal Problem.

The Structured Dual Problem may in the same way be seen as a Two-Stage Problem in z_{32} and z_{14}.

This may be economically interpreted as a division of responsibilities between one organization which decides about prices z_2 and activity levels z_3 such that all activities are undertaken with marginal losses or marginally break even, and another organization which for given z_2 and z_3 decides about prices z_4 and activity levels z_1 so as to maximize the total cost of labour and resources less a penalty for inefficiency, while ensuring that all its own activities are undertaken with marginal losses or marginally break even. The overall maximum is obtained by the first organization systematically varying its choice of z_2 and z_3 until the total payment to labour and resources is maximized less the penalty on any inefficiency of the other organization less a similar penalty on any inefficiency of its own decisions.

$$f^{o} = \underset{z_{32}}{\mathrm{Max}}\{-z_3 K3_{32} + \underbrace{\underset{z_{14}}{\mathrm{Max}}\{K_{1\,32\,4} - z_1 K1_{124} | K1_{124} \geq 0,\ z_{14} \geq 0\}}\ \ K3_{32} \geq 0,\ z_{32} \geq 0\}$$

the inner dual problem may be replaced by its primal

$$\underset{z_{14}}{\mathrm{Min}}\{K_{1\,32\,4} - z_4 K4_{14} | K4_{14} \leq 0,\ z_{14} \geq 0\}$$

If we know feasible solutions z^{ii}_{14} $i=1,\ldots,k$ of the inner constraints, then we can enumerate the corresponding values of the objective function, viz.:

$$K^{i}_{1\,32\,4}{}^{i} - z^{i}_{4} K4^{ii}_{14} \qquad i=1,\ \ldots,k$$

A variable p may be required to be less than or equal to each one of the above values

$$p \leq K^{i}_{1\,32\,4}{}^{i} - z^{i}_{4} K4^{ii}_{14} \qquad i=1,\ldots,k$$

By maximizing p we may then obtain the minimum value of the above objective function values. We can then approximate the inner minimization by

$$\underset{p}{\mathrm{Max}}\{p | p \leq K^{i}_{1\,32\,4}{}^{i} - z^{i}_{4} K4^{ii}_{14} \qquad i=1,\ldots,k\}$$

and obtain the <u>Primal Master (Dual Formulation)</u>

(8) $f = \underset{z_{32},p}{\text{Max}}\{-z_3K3_{32} + p|$

(9) $K_{1}{}^{i}{}_{32}{}^{i}{}_{4} - z_4^iK4_{14}^{ii} - p \geq 0 \qquad i=1,\ldots,k$

(10)-(11) $K3_{32} \geq 0,\ z_{32} \geq 0\}$

with solution f^{k+1}, $z_3^{k+1}{}_2^{k+1}$, p^{k+1}; letting $k+1 \rightarrow k$

we may denote the solution f^k, z_{32}^{kk}, p^k.

Again in general we shall not know the finite or infinite sequence of solutions and have to proceed as in the ease of the Dual Master.

The Primal Master is named the Full Primal Master if it contains a finite or an infinite sequence of constraints corresponding to solution points which come arbitrary close to any arbitrary point, and a Restricted Primal Master if it only contains some subset.

If we do not know the constraints corresponding to the finite or infinite sequence z_{14}^{ii}, we may determine the one limiting p by solving the Primal Subproblem

(12)-(14) $df = \underset{z_{14}}{\text{Min}}\{K_{1}{}^{kk}_{324} - z_4K4_{14} - p^k|K4_{14} \leq 0,\ z_{14} \geq 0\}$

with solution df^{kk}, z_{14}^{kk}.

3. The Primal and Dual Masters Serving as Each Other's Subproblem

We obtain a symmetric formulation of Master and Subproblem by letting the Dual Master serve as the Primal Subproblem of the Primal Master, and the Primal Master serve as a Dual Subproblem to the Dual Master.

We note that in the special case that the Dual (Primal) Master only contains one constraint based on the last Primal (Dual) Master solution, its optimal solution will be identical with that of the corresponding Primal (Dual) Subproblem.

In the continuation we shall show that such a scheme of iterations converges to an optimal solution of the original problem.

4. The Main Steps of a Proof of Convergence

The proof of convergence is based upon the following steps:

i) Deriving that the objective function of the primal master f^k is greater than or equal to that of the optimal solution of the dual problem f^o which is equal to the optimal solution of the primal problem h^o which is greater than or equal to the objective function of the dual master problem h^1, viz.:

$$f^k \geq f^o = h^o \geq h^1$$

ii) Establishing that f^k for increasing k is a non-increasing sequence, i.e.

$$f^k \geq f^{k+1}$$

iii) Establishing that h^l for increasing l is a non-decreasing sequence, i.e.

$$h^{l+1} \geq h^l$$

iv) Deriving that provided certain conditions are satisfied then for a subsequence $k' \to \infty$ and $l' \to \infty$

$$\lim_{k' \to \infty, l' \to \infty} f^{k'} - h^{l'} = 0$$

v) Concluding from i)-iv) that the limit of the entire sequence must be

$$\lim_{k \to \infty} f^k = f^o \quad \text{and} \quad \lim_{l \to \infty} h^l = h^o \quad \text{where} \quad f^o = h^o$$

5. <u>The Detailed Arguments of a Proof of Convergence</u>

i) <u>The derivation of</u>

(1)
$$\underbrace{f^k \geq \overbrace{f^o = h^o}^{(2)}}_{(3)} \; \underbrace{\geq h^1}_{(4)}$$

The equality (2) of the Structured Primal and Dual Problems follows from the Dantzig-Eisenberg-Cottle Theorem 2 or the Whinston Theorem 4.2.

The left inequality (3) follows from that any optimal solution of a Restricted Primal Master (f^k), which does not contain the above-mentioned finite or infinite sequence of constraints, will have a maximum which is greater or equal to that of the Full Primal Master, which contains the above mentioned finite or infinite sequence, and which has the same optimal solution as the original Structured Dual Problems (f^o).

The right inequality (4) follows from that any optimal solution of a Restricted Dual Master (h^1), which does not contain the above mentioned finite or infinite sequence of constraints, will have a minimum which is smaller or equal to that of the Full Dual Master, which contains the above mentioned finite or infinite sequence, and which has the same optimal solution as the original Structured Primal Problems (h^o).

ii) <u>Establishing</u>

(5)
$$f^k \geq f^{k+1}$$

If all active constraints are always retained in the Restricted Primal Master, then any additional constraint can only serve to decrease or leave unchanged its maximum solution.

iii) <u>Similarly</u>

(6)
$$h^{l+1} \geq h^l$$

If all active constraints are always retained in the Restricted Dual Master, then any additional constraint can only serve to increase or leave unchanged its minimum solution.

iv) Demonstrating that for a subsequence $k'\to\infty$ and $l'\to\infty$

$$\lim_{k'\to\infty, l'\to\infty} f^{k'} - h^{l'} = 0$$

The demonstration requires the following steps:

a) Noting that the current

(7) Primal Master solution is f^k, z_{32}^{kk}, p^k

(8) Dual Master solution is h^l, z_{14}^{ll}, q^l

b) Establishing that $f^k - h^l$ equals minus the value of the objective function of the Primal Subproblem (df^{kl}) plus the value of the objective function of the Dual Subproblem (dh^{kl}), viz.:

$$f^k - h^l = -df^{kl} + dh^{kl}$$

The value of the Primal Subproblem objective function 2-(12) for the feasible Primal Subproblem solution provided by the current Dual Master solution (8) is

$$(9) \quad df^{kl} = \underbrace{K_{1324}^{lkkl}}_{K^{kl}} - \underbrace{z_4^l K 4_{14}^{ll}}_{0} - \underbrace{p^k}_{f^k}$$

Abbreviated notation cf. (11) below	Kuhn-Tucker conditions for Dual Master	Kuhn-Tucker conditions for Primal Master	for proof see C.-L. Sandblom (1972)

$$= K^{kl} - f^k$$

The value of the Dual Subproblem objective function 2-(5) for the feasible Dual Subproblem solution provided by the current Primal Master solution (7) is

$$(10) \quad dh^{kl} = \underbrace{K_{1324}^{lkkl}}_{K^{kl}} - \underbrace{z_3^k K 3_{32}^{kk}}_{0} - \underbrace{q^l}_{h^l}$$

Abbreviated notation of (11) below	Kuhn-Tucker conditions for Primal Master	Kuhn-Tucker conditions for Dual Master	for proof see C.-L. Sandblom (1972)

$$= K^{kl} - h^l$$

where

$$(11) \quad K^{kl} = K_{1324}^{lkkl}$$

but

(12) $f^k - h^l = f^k - h^l$	Identity
$= -(K^{kl} - f^k) + (K^{kl} - h^l$	Subtracting and adding K^{kl}
$= -df^{kl} + dh^{kl}$	By (9) and (10)

c) Noting that there are two different modes of iteration, viz.:

The Simultaneous Mode:

The current solution of the Dual Master (l) is transferred to the Primal Master, simultaneously, the current solution of the Primal Master (k) is transferred to the Dual Master. Thereafter an improved solution of the

Primal Master (k+*1*) is obtained, simultaneously, an improved solution of the Dual Master (l+*1*) is obtained.

The Sequential Mode:

The current solution of the Dual Master (l) is transferred to the Primal Master, an improved solution of the Primal Master (k+*1*) is obtained, this Primal Master solution is transferred to the Dual Master, an improved solution of the Dual Master (l+*1*) is obtained, etc.

d) Establishing for the Simultaneous Mode that as long as the objective functions of the Primal and Dual Masters differ in value the following relationships hold: If the Dual (Primal) Master solution may not be used to improve the Primal (Dual) Master, i.e. $df^{kl} \geq 0$ ($dh^{kl} \leq 0$) then the Primal (Dual) Master solution may be used to improve the Dual (Primal) Master, i.e. $dh^{kl} > 0$ ($df^{kl} < 0$); and that if the value of the objective function of the Dual (Primal) Subproblem, i.e. dh^{kl} (df^{kl}) is greater (less) than the difference between the objective functions of the Primal (Dual) and Dual (Primal) Master, i.e. $f^k - h^l$ ($h^l - f^k$), then the Dual (Primal) Master solution may not improve the Primal (Dual) Master, i.e. $df^{kl} \geq 0$ ($dh^{kl} \leq 0$), viz.:

$$(13) \qquad \underbrace{f^k > h^l}_{(14)} \quad \text{and} \quad \underbrace{df^{kl} \geq 0}_{(15)} \iff \underbrace{dh^{kl} \geq f^k}_{(16)}\!\!\underbrace{{} - h^l > 0}_{(14)}$$

$$(17) \qquad \underbrace{f^k > h^l}_{(18)} \quad \text{and} \quad \underbrace{dh^{kl} \leq 0}_{(19)} \iff \underbrace{df^{kl} \leq h^l}_{(20)}\!\!\underbrace{{} - f^k < 0}_{(18)}$$

Relationship (16) may be obtained by using (12) to express dh^{kl}

$$dh^{kl} = f^k - h^l + \underbrace{df^{kl}}_{\geq 0 \text{ by } (15)} \geq f^k - h^l$$

Vice versa, if (16) holds it follows from (12) that

$$df^{kl} = -(f^k - h^l) + \underbrace{dh^{kl}}_{\geq f^k - h^l} \geq -(f^k - h^l) + f^k - h^l = 0$$

The other relationship (17) is proven in an analogous way.

e) Establishing for the Sequential Mode:

If the Dual Master solution can improve the Primal Master solution (i.e. $df^{kl} < 0$) then after the improvement of the Primal Master solution then the same Dual Master solution can no longer improve the Primal Master (i.e. $df^{k+1,l} = 0$) and the Primal Master solution can improve the Dual Master solution (i.e. $dh^{k+1,l} = f^{k+1} - h^l > 0$) unless both masters have attained the optimum solution to the original problem (i.e. $dh^{k+1,l} = f^{k+1} - h^l = 0$), viz.:

$$(21)\qquad \underbrace{df^{kl} < 0}_{(22)} \Rightarrow \underbrace{df^{k+1,l} \geq 0}_{(23)},\ \underbrace{dh^{k+1,l} \geq f^{k+1}}_{(24)} \underbrace{- h^{l} \geq 0}_{(25)}$$

Given that $df^{kl} < 0$, we form an additional Primal Master constraint 2-(9)

$$K^{1\ \ 1}_{1324} - z^{1}_{4}K4^{11}_{14} - p \geq 0$$

We note that for the current Primal Master solution z^{kk}_{32}, p^{k} the left hand side becomes

$$\underbrace{K^{1kkl}_{1324} - z^{1}_{4}K4^{11}_{14} - p^{k}}_{df^{kl}\ <\ 0 \text{ by 2-(12) and (22)}}$$

and the additional constraint is not satisfied by the current Primal Master solution. The new optimal solution of the Primal Master $z_3^{k+1}\ {}_2^{k+1}$, p^{k+1} will satisfy the new constraint which will assume the value

$$\underbrace{K^{1\ k+1\ k+1\ 1}_{1\ 3\ \ \ 2\ \ \ 4} - z^{1}_{4}K4^{11}_{14} - p^{k+1} \geq 0}_{df^{k+1,l} \text{ by 2-(12)}}$$

We note that the left side of the constraint is identical with the value $df^{k+1,l}$ of the Primal Subproblem objective function for the current Dual Master solution.

Hence $df^{k+1,l} \geq 0$ and it follows from (13) that $dh^{k+1,l} \geq f^{k+1} - h^{l} > 0$.

Similarly, if the Primal Master solution can improve the Dual Master solution (i.e. $dh^{kl} > 0$) then after the improvement of the Dual Master solution, then the same Dual Master solution can no longer improve the Dual Master (i.e. $dh^{k,l+1} \leq 0$) and the Dual Master solution can improve the Primal Master solution (i.e. $df^{k,l+1} = h^{l+1} - f^{k} \leq 0$) unless both Masters have attained the overall optimum to the original problem, viz.:

$$(26)\qquad \underbrace{dh^{kl} > 0}_{(27)} \Rightarrow \underbrace{dh^{k,l+1} \leq 0}_{(28)},\ \underbrace{df^{k,l+1} \leq h^{l+1}}_{(29)} \underbrace{- f^{k} \leq 0}_{(30)}$$

f) The obtained results ensure that it will be possible to continue to improve the masters unless the optimal solution to the original problem has been attained by both masters.

g) It still remains to be shown that

$$\lim_{k'\to\infty,\,l'\to\infty} (f^{k'} - h^{l'}) = \lim_{k'\to\infty,\,l'\to\infty} - df^{k'l'} + \lim_{l'\to\infty,\,l'\to\infty} dh^{k'l'} = 0$$

For this the following arguments are used:

1) The process either leads to that df (dh) both become equal to zero in a finite number of iterations or always remain less (greater) than zero leading to an infinite number of iterations.

2) The function K is required to be such that all variables z_{1324}, p, q are bounded. C.-L. Sandblom (1972) has formulated the detailed properties of the saddlefunction K. Special linear and nonlinear cases of the method have been elaborated in earlier papers by the author, Kronsjö (1968, 1969, 1972).

3) Using a lemma by Bolzano and Weierstrass stated in Fichtenholz (1963) it is possible to extract from an infinite sequence S of points within a bounded set, a subsequence $S^{\sim}$ which is such that

(31) $$(z_{1324}^{lkkl}, p^{l}, q^{l}) \rightarrow (z_{1324}^{\sim\sim\sim\sim}, p^{\sim}, q^{\sim}) \quad \text{when } k,l \rightarrow \infty,\ k,l \in S^{\sim}.$$

4) It follows from the assumption of an infinite sequence of iterations and the continuity and boundedness assumptions that the limiting value of the objective function of the Primal Subproblem 2-(12) is

$$\lim_{k,l \rightarrow \infty} \{df^{kl} | k,l \in S^{\sim}\} = df^{\sim\sim} \leq 0$$

A related function occurs, however, at a later iteration k' > k where k', k $\in S^{\sim}$ as a satisfied Primal Master constraint 2-(9), hence

$$\lim_{k'>k;\ k,l \rightarrow \infty} \{df^{kl} | k', k, l \in S^{\sim}\} = df^{\sim\sim} \geq 0$$

It follows that

(32) $$\lim_{k,l \rightarrow \infty} \{df^{kl} | k,l \in S^{\sim}\} = df^{\sim\sim} = 0$$

Similarly, the limiting value of the objective function of the Dual Subproblem 2-(5) is

$$\lim_{k,l \rightarrow \infty} \{dh^{kl} | k,l \in S^{\sim}\} = dh^{\sim\sim} \geq 0$$

A related function occurs, however, at a later iteration l' > l where l', l $\in S^{\sim}$ as a satisfied Dual Master constraint 2-(2), hence

$$\lim_{l'>l;\ k,l \rightarrow \infty} \{dh^{kl} | l', l, k \in S^{\sim}\} = dh^{\sim\sim} \leq 0$$

It follows that

(33) $$\lim_{k,l \rightarrow \infty} \{dh^{kl} | k,l \in S^{\sim}\} = dh^{\sim\sim} = 0$$

5) The convergence of f^k and h^l can then be established by considering

$$\underbrace{\lim_{k,l\to\infty} \{f^k - h^l | k,l \in S^{\sim}\}}_{f^{\sim} - h^{\sim}} = \underbrace{\lim_{k,l\to\infty} \{-df^{kl} + dh^{kl} | k,l \in S^{\sim}\}}_{\underbrace{\underbrace{-df^{\sim\sim}}_{0 \text{ by } (32)} \underbrace{+dh^{\sim\sim}}_{0 \text{ by } (33)}}_{0} \text{ due to } (31)}$$

Hence $f^{\sim} = h^{\sim}$ which together with the inequality (1) gives

$$f^{\sim} = f^o = h^o = h^{\sim}$$

Thus the solution of the Primal and Dual Masters converges towards the optimal solution of the original problem at least for a subsequence of solutions.

As, however, the f^k is a nonincreasing sequence, and h^l is a nondecreasing sequence for any k and l, it follows that the entire series of values of f^k and h^l must converge towards $f^o = h^o$.

6. The Primal Master in Primal Formulation

In order to see more clearly the relationship of the earlier obtained Primal Master formulation to the usual formulation of primal master problems, e.g. that of Dantzig-Wolfe, we proceed to express the dual of the primal master problem 2-(8)-(11). As p is unconstrained in sign, and our symmetric duality theory only is formulated for nonnegative variables, we split the variable p into a positive component p^+ and a negative component $-p^-$ where p^+ and p^- are non-negative, i.e.

(1) $$p = p^+ - p^-, \quad p^+ \geq 0, \quad p^- \geq 0$$

We use this formula to eliminate p in 2-(8)-(11). After adding dual variables $t_I = (t_1,\ldots,t_k)$, the problem becomes

$$f = \underset{z_{32},t_I,p^+,p^-}{\text{Max}} \{-z_3K3_{32} + p^+ - p^- |$$
$$K^i_{1\,32\,4}{}^{i} - z^i_4K4^{ii}_{14} - p^+ + p^- \geq 0 \quad i=1,\ldots,k$$
$$K3_{32} \geq 0,$$
$$z_{32} \geq 0,\ p^+ \geq 0,\ p^- \geq 0,\ t_i \geq 0 \quad i=1,\ldots,k\}$$

We consider that the problem originates from a saddle function H, and that the Primal Master (Dual Formulation) is the detailed formulation of the dual problem

$$f = \underset{z_{32},t_I,p^+,p^-}{\text{Max}} \{H(z_{32},p^+,p^-,t_I) - t_IH_{t_I} - z_3H_{z_3} |$$
$$H_{t_I} \geq 0,\ H_{z_3} \geq 0,\ z_{32} \geq 0,\ p^+ \geq 0,\ p^- \geq 0,\ t_I \geq 0\}$$

where

$$t_I \equiv (t_1, \ldots, t_k) \quad \text{and} \quad H_{t_I} = \begin{bmatrix} H_{t_1} \\ \vdots \\ H_{t_k} \end{bmatrix}$$

It follows that

$$H_{t_i} = K^i_{1\,32}{}^i_4 - z^i_4 K4^{ii}_{14} - p^+ + p^- \geq 0 \qquad i = 1, \ldots, k$$

$$H_{z_3} = \sum_i t_i K3_{32} \geq 0$$

$$= \quad K3_{32} \geq 0$$

as all $t_i \geq 0$ we have that either $\sum_i t_i > 0$ or $\sum_i t_i = 0$. It follows that in the first case $H_{z_3} = K3_{32}$ as required, and that in the second case $H_{z_3} = 0$.

$$H = -z_3 K3_{32} + p^+ - p^- + \sum_i t_i (K^i_{1\,32}{}^i_4 - z^i_4 K4^{ii}_{14} - p^+ + p^-) + z_3 K3_{32}$$

$$= \sum_i t_i (K^i_{1\,32}{}^i_4 - z^i_4 K4^{ii}_{14} - p^+ + p^-) + p^+ - p^-$$

The corresponding primal problem is

$$f = \underset{z_{32}, t_I, p^+, p^-}{\text{Min}} \{ H(z_{32}, p^+, p^-, t_I) - z_2 H_{z_2} - p^+ H_{p^+} - p^- H_{p^-} \mid$$
$$H_{z_2} \leq 0,\ H_{p^+} \leq 0,\ H_{p^-} \leq 0,\ z_{32} \geq 0,\ p^+ \geq 0,\ p^- \geq 0,\ t_I \geq 0 \}$$

where $H_{z_2} = \sum_i t_i K2^i_{1\,32} \leq 0$

$$\left.\begin{aligned} H_{p^+} &= -\sum_i t_i + 1 \leq 0 \\ H_{p^-} &= +\sum_i t_i - 1 \leq 0 \end{aligned}\right\} \left.\begin{aligned} \sum_i t_i - 1 &\geq 0 \\ \sum_i t_i - 1 &\geq 0 \end{aligned}\right\} H_p = \sum_i t_i - 1 = 0$$

and we obtain

<u>the Primal Master (Primal Formulation)</u>

(2) $$f = \underset{z_{32}, t_I, p}{\text{Min}} \{ \sum_i t_i (K^i_{1\,32}{}^i_4 - z_2 K2^i_{1\,32} - z^i_4 K4^{ii}_{14}) \mid$$

(3) $$\sum_i t_i K2^i_{1\,32} \leq 0$$

(4) $$\sum_i t_i = 1$$

(5)-(6) $$t_i \geq 0 \quad i = 1, \ldots, k, \quad z_{32} \geq 0 \ \}$$

It is easy to recognize that this formulation is closely related to the classical formulation of Dantzig-Wolfe.

7. An Outline of the Master Problems Obtained by Repeated Application of the Decomposition Procedure on a Block Triangular Structure

In order to be able to apply the decomposition procedure again, we assume that the saddle function K is a function K_{135246} as indicated below.

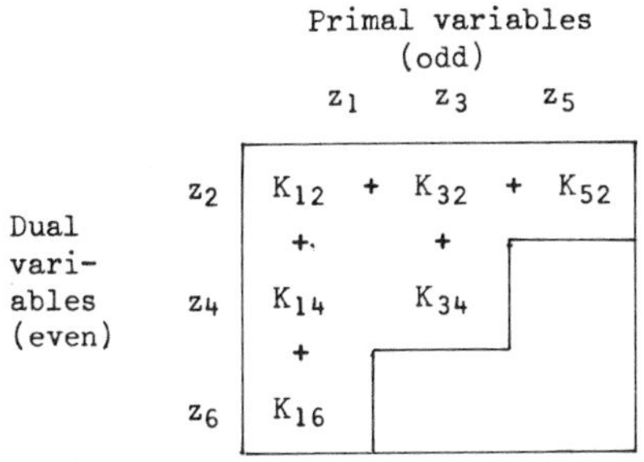

Fig. 2. The assumed structure of the saddle function K_{135246}.

It follows that the corresponding partial derivatives now become

(1)

Partial Derivative using conventional notation	Exact Form using the above notation	Abbreviated Form
$K_{y_1} \equiv K_{z_2}$	$K2_{12} + K2_{32} + K2_{52}$	$K2_{1352}$
$K_{y_2} \equiv K_{z_4}$	$K4_{14} + K4_{34}$	$K4_{134}$
$K_{y_3} \equiv K_{z_6}$	$K6_{16}$	$K6_{16}$
$K_{x_1} \equiv K_{z_1}$	$K1_{12} + K1_{14} + K1_{16}$	$K1_{1246}$
$K_{x_2} \equiv K_{z_3}$	$K3_{32} + K3_{34}$	$K3_{324}$
$K_{x_3} \equiv K_{z_5}$	$K5_{52}$	$K5_{52}$

There are several possible ways in which we can derive the equations of the Master Problems resulting from repeated application of the above outlined decomposition procedure.

The following way seems to give the simplest equations:

The problem is first decomposed into a Dual Master and a Dual Subproblem as indicated in the figure below.

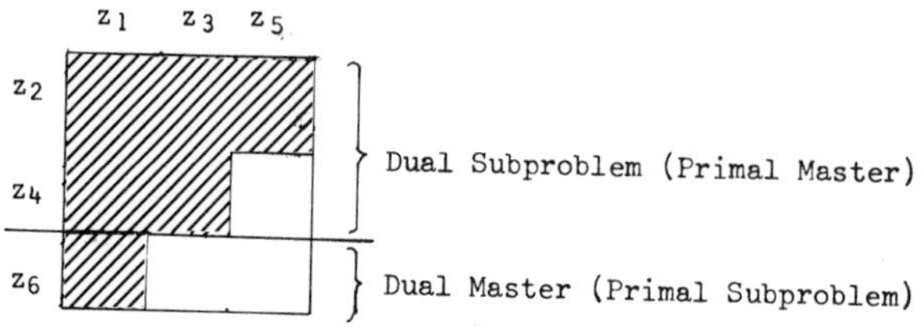

Fig. 3. The decomposition into a Dual (Primal) Master and a Dual (Primal Subproblem.

The appropriate Dual Master equations in Primal Formulation are obtained from 2-(1)-(4) replacing z_3 by z_3 and z_5; z_4 by z_6; z_2 by z_2 and z_4; and appropriate modifications of the functions of partial derivatives.

<u>The Dual Master (Primal Formulation)</u>

(2) $$h = \underset{z_{16},q}{\text{Min}} \{-z_6 K6_{16} + q \mid$$

(3) $$K_{135246}^{jjjj} - z_3^j K3_{324}^{jjj} - z_5^j K5_{52}^{jj} - q \leq 0 \quad j=1,\dots,l$$

(4)-(5) $$K6_{16} \leq 0, \; z_{16} \leq 0 \; \}$$

with solution h^{l+1}, $z_1^{l+1}{}_6^{l+1}$, q^{l+1}; letting $l+1 \rightarrow l$ we may denote the solution h^l, z_{16}^{ll}, q^l.

The original problem is then decomposed using 2-(8)-(11) to obtain a Primal Master (Dual Formulation)

$$f = \underset{\substack{z_{3524},\\ p,t_I}}{\text{Max}} \{-z_3 K3_{324} - z_5 K5_{52} + p \mid$$
$$K_{135246}^{i \quad i} - z_6^i K6_{16}^{ii} - p \geq 0 \qquad i=1,\dots,k$$
$$K3_{324} \geq 0, \; K5_{52} \geq 0, \; z_{3524} \geq 0, \; t_i \geq 0 \quad i=1,\dots,k\}$$

from which we obtain a Primal Master (Primal Formulation)

$$f = \underset{\substack{z_{3524},\\ t_I,p}}{\text{Min}} \{\sum_i t_i (K_{135246}^{i \quad i} - z_2 K2_{1352}^i - z_4 K4_{134}^i - z_6^i K6_{16}^{ii}) \mid$$
$$\sum_i t_i K2_{1352}^i \leq 0$$
$$\sum_i t_i K4_{134}^i \leq 0$$
$$\sum_i t_i = 1$$
$$t_i \geq 0 \quad i=1,\dots,k, \; z_{3524} \geq 0 \; \}$$

The last problem is decomposed into a Dual Master and a Dual Subproblem as sketched in the following figure.

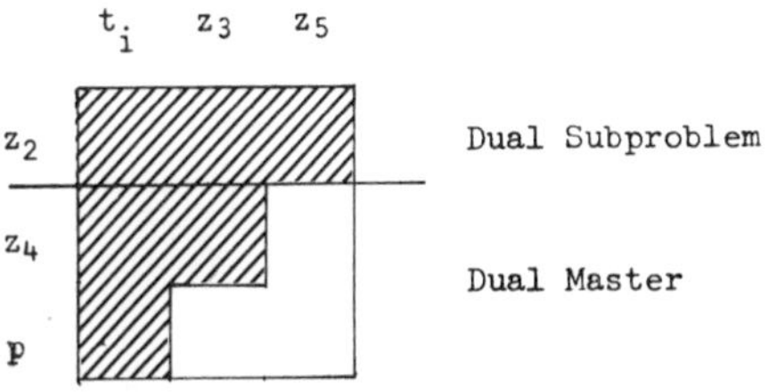

Fig. 4. The decomposition of the Primal Master into a Dual Master and Dual Subproblem.

Using the last formulation we express the two-stage problem

$$f = \min_{t_I, z_{34}, p} \Big\{ \sum_i t_i (z_4 K4^i_{134} - z_6^i K6^{ii}_{16}) + \min_{z_{52}} \Big\{ \sum_i t_i (K^i_{135246} - z_2 K2^i_{1352}) \,\Big|\, \sum_i t_i K2^i_{1352} \le 0,\ z_{52} \ge 0 \Big\} \,\Big|\, \sum_i t_i K4^i_{134} \le 0, \sum_i t_i = 1,\ t_i \ge 0 \quad i=1,\dots,k,\ z_{34} \ge 0 \Big\}$$

In order to formulate the dual of the inner minimization problem, we assume that it comes from a saddle function H, with primal and dual

$$\min_{z_{52}} \{ H - z_2 H_{z_2} \mid H_{z_2} \le 0, z_{52} \ge 0 \} = \max_{z_{52}} \{ H - z_5 H_{z_5} \mid H_{z_5} \ge 0,\ z_{52} \ge 0 \}$$

It follows that $H = \sum_i t_i K^i_{135246}$ and that the required dual problem is

$$\max_{z_{52}} \Big\{ \sum_i t_i (K^i_{135246} - z_5 K5_{52}) \,\Big|\, (\sum_i t_i K5_{52} =) K5_{52} \ge 0,\ z_{52} \ge 0 \Big\}$$

Again assuming the solution sequence z^{jj}_{52} $j=1,\dots,l$, we may approximate the maximal objective function value by solving

$$\min_q \Big\{ q \,\Big|\, q \ge \sum_i t_i (K^{i\,jj\,i}_{135246} - z_5^j K5^{jj}_{52}) \qquad j=1,\dots,l \Big\}$$

which gives <u>the Primal-and-Dual Master (Primal Formulation)</u>

(6) $$f = \min_{t_I, z_{34}, q, v_J, p} \Big\{ \sum_i t_i (z_4 K4^i_{134} - z_6^i K6^{ii}_{16}) + q \,\Big|$$

(7) $$\sum_i t_i (K^{i\,jj\,i}_{135246} - z_5^j K5^{jj}_{52}) - q \le 0 \quad j=1,\dots,l$$

(8) $$\sum_i t_i\, K4^i_{134} \le 0$$

(9) $$\sum_i t_i = 1$$

(10)-(11) $$t_i \ge 0 \quad i=1,\dots,k;\ z_{34} \ge 0\ v_j \ge 0\ j=1,\dots,l \Big\}$$

with optimal solution t_I^m, v_J^m, z_{34}^{mm}, q^m, p^m.

The Primal-and-Dual Master does not give any values z_{16}^{mm} and z_{52}^{mm} as required by the Primal Master and the Dual Master. We use, however, the information given by the t_i^m and v_j^m variables to construct the appropriate values, using the formulas

(12)-(13) $$z_{16}^{mm} = \sum_i t_i^m z_{16}^{ii} \quad \text{and} \quad z_{52}^{mm} = \sum_j v_j^m z_{52}^{jj}$$

The obtained z_{16}^{mm} values will satisfy the subproblem constraints $K6_{16} \leq 0$, $z_{16} \geq 0$ provided these are convex in all variables. Similarly, the obtained z_{52}^{mm} values will satisfy the subproblem constraints $K5_{52} \geq 0$, $z_{52} \geq 0$ provided these are concave in all variables.

This can easily be demonstrated by setting

(14)-(16) $$z_{16} = \sum_i t_i z_{16}^{ii} \mid \sum_i t_i = 1, \; t_i \geq 0 \qquad i=1,\dots,k$$

where z_{16}^{ii} satisfies

(17)-(18) $$K6_{16}^{ii} \leq 0, \quad z_{16}^{ii} \geq 0 \qquad i=1,\dots,k$$

Multiplying each vector in (17)-(18) by t_i and summing the resultant vectors we obtain

(19)-(20) $$\sum_i t_i K6_{16}^{ii} \leq (\sum_i t_i 0 =)0 \qquad \sum_i t_i z_{16}^{ii} \geq (\sum_i t_i 0 =)0$$

If $K6_{16}$ is a convex function in all dimensions we have

(21) $$K6(\sum_i t_i z_{16}^{ii}) \leq \sum_i t_i K6_{16}^{ii}$$

From (21), (14), (19) and (20), (14) we obtain

(22) $$K6_{16} \leq 0, \qquad z_{16} \geq 0$$

which proves that the constructed solution satisfies the subproblem constraints.

A similar demonstration can be made for z_{52} only exchanging the requirement of convexity for that of concavity.

The original problem is finally decomposed into a Primal Master and Primal Subproblem as indicated in the figure below.

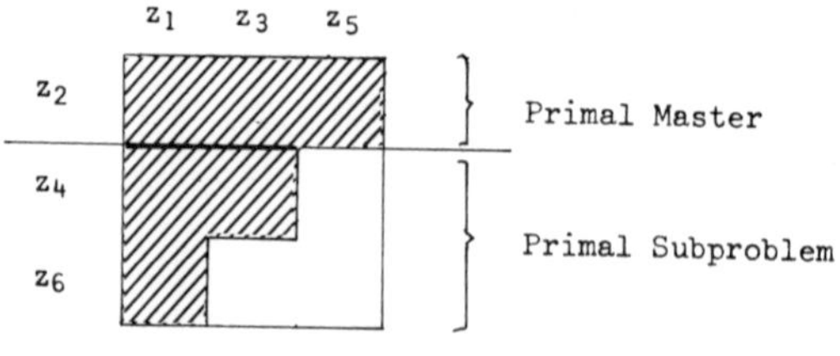

Fig.5 . Decomposition into Primal Master and Primal Subproblem.

The appropriate Primal Master equations in Primal Form are obtained from 2-(8)-(11) by replacing z_3 by z_5; z_1 by z_1 and z_3; z_4 by z_4 and z_6 and appropriate modifications of the functions of partial derivatives.

The Primal Master (in Primal Form)

$$f = \underset{z_{52},t_I,p}{\mathrm{Min}} \ \{\sum_i t_i (K^{ii\ ii}_{135246} - z_2 K2^{ii}_{1352} - z_4^i K4^{iii}_{134} - z_6^i K6^{ii}_{16}) \mid \tag{23}$$

$$\sum_i t_i K2^{ii}_{1352} \leq 0 \tag{24}$$

$$\sum t_i = 1 \tag{25}$$

$$t_i \geq 0 \quad i=1,\dots,k; \quad z_{52} \geq 0, \quad p \gtreqless 0 \ \} \tag{26-28}$$

with optimal solution $z_5^{k+1}\,{}_2^{k+1}$, t_I^{k+1}, p^{k+1}; letting $k+1 \to k$ we may denote the solution z_{52}^{kk}, t_I^k, p^k.

8. The Possible Introduction of Specializing and Supplementary Assumptions

After the general framework has been drafted it is near at hand to investigate the possibility of introducing specializing and supplementary assumptions with respect to

1) functional form of the original problem, e.g.:

a)	linear	$K = cx + yb - yAx$
b)	quadratic	$K = \frac{1}{2}(x^T Dx - yEy^T) + cx + yb - yAx$
c)	fractional	$K = cx/dx + yb - yAx$
d)	nonlinear objective function	$K = f(x) + yb - yAx$
e)	nonlinear-linear	$K = f(x_1) + cx_2 - yg(x_1) - yAx_2$
f)	nonlinear	$K = f(x) + yg(x)$

2) special structure of the original problem, e.g.:

a) 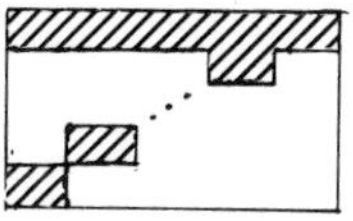

b) 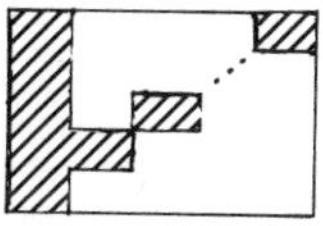

c) 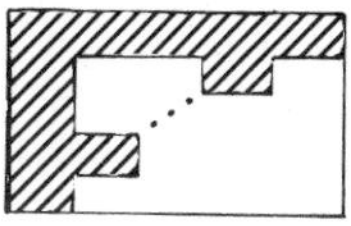

d) 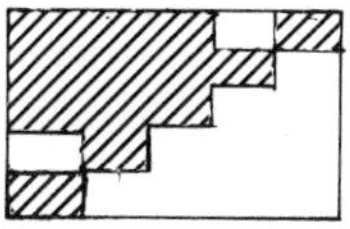

e) 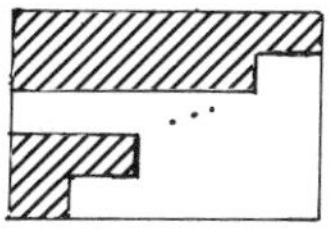

3) <u>number of generated variables or constraints, e.g.:</u>

Case		Dual Master 1	2	many
Primal Master	1	a)	d)	g)
	2	b)	e)	h)
	many	c)	f)	i)

j) Primal Master 1, Primal-and-Dual Master Many, Dual Master 1

k) Primal Master Many, Primal-and-Dual Master 1, Dual Master Many

l) Primal Master Many, Primal-and-Dual Master Many, Dual Master Many

4) <u>approximation method of a nonlinear function, e.g.:</u>

a) linearization

5) <u>algorithmic method of solving a master problem and generating subproblem solutions, e.g.:</u>

a) Primal Simplex b) Dual Simplex c) Self-Dual Parametric Simplex

d) Primal-Dual Simplex e) Dual-Primal Simplex (i.e. the dual of the Primal-Dual Simplex Method)

f) Lagrange Function g) Gradient of Lagrange Function

Any attempt to construct an ultimate classification is likely to be made incomplete by the invention of new possibilities. The compilation of a classification, though temporary, of existing decompositional methods should be useful in surveying explored variants and indicating possibly promising future variants.

References

1) G.B. Dantzig, E. Eisenberg and R.W. Cottle, (1965), Symmetric Dual Nonlinear Programs, Pacific Journal of Mathematics, Vol. 15, No. 3, pp.809-812.

2) G.M.Fichtenholz, (1963), Kurs differentsial'nogo i integral'nogo ischisleniia, t.I, Fizmatgiz, Moskva.

3) T.O.M. Kronsjö, (1968), Centralization and Decentralization of Decision Making. The Decomposition of Any Linear Programme in Primal and Dual Directions - To Obtain a Primal and a Dual Master Solved in Parallel and One or More Common Subproblems -, in Revue Francaise Informatique et de Recherche Operationelle, Paris, 1968, 2e année, No. 10, pp. 73-114.

4) T.O.M. Kronsjö, (1969), Optimal Co-ordination of a Large Convex Economic System (Decomposition of a Nonlinear Convex Separable Economic System in Primal and Dual Directions to Obtain a Common Subproblem), in Jahrbücher für National-ökonomie und Statistik, Band 183, Heft 3, Gustav Fischer Verlag, Stuttgart, pp. 378-400.

5) T.O.M. Kronsjö, (1972), A Decentralized Economic Planning System Based on Decomposition into Only Master Problems, in Jahrbuch der Wirtschaft Osteuropas, Band 3, Günter Olzog Verlag München-Wien, pp. 137-151.

6) C.-L. Sandblom, (1972), Theorems Related to Symmetric Nonlinear Decomposition, University of Birmingham, National Economic Planning Research Reports, RC/A 69.

7) A. Whinston, (1967), Some Applications of the Conjugate Function Theory to Duality, in J. Abadie (ed.), Nonlinear Programming, North-Holland Publishing Company, Amsterdam, pp. 75-96.

8) K.P. Wong, (1970), Decentralised Planning by Vertical Decomposition of an Economic System: A Nonlinear Programming Approach, National Economic Planning Unit, University of Birmingham, Great Britain, Chapter on Duality in Nonlinear Programming.

AUTHOR INDEX

SUBJECT INDEX